Basic Analysis IV

Basic Analysis IV: Measure Theory and Integration

The giant squids are interested in advanced mathematical training and Jim is happy to help!

James K. Peterson
Department of Mathematical Sciences
Clemson University

CRC Press is an imprint of the
Taylor & Francis Group, an **informa** business

A CHAPMAN & HALL BOOK

First edition published 2021
by CRC Press
6000 Broken Sound Parkway NW, Suite 300, Boca Raton, FL 33487-2742

and by CRC Press
2 Park Square, Milton Park, Abingdon, Oxon, OX14 4RN

© 2020 Taylor & Francis Group, LLC

CRC Press is an imprint of Taylor & Francis Group

Reasonable efforts have been made to publish reliable data and information, but the author and publisher cannot assume responsibility for the validity of all materials or the consequences of their use. The authors and publishers have attempted to trace the copyright holders of all material reproduced in this publication and apologize to copyright holders if permission to publish in this form has not been obtained. If any copyright material has not been acknowledged please write and let us know so we may rectify in any future reprint.

Except as permitted under U.S. Copyright Law, no part of this book may be reprinted, reproduced, transmitted, or utilized in any form by any electronic, mechanical, or other means, now known or hereafter invented, including photocopying, microfilming, and recording, or in any information storage or retrieval system, without written permission from the publishers.

For permission to photocopy or use material electronically from this work, access www.copyright.com or contact the Copyright Clearance Center, Inc. (CCC), 222 Rosewood Drive, Danvers, MA 01923, 978-750-8400. For works that are not available on CCC please contact mpkbookspermissions@tandf.co.uk

Trademark notice: Product or corporate names may be trademarks or registered trademarks, and are used only for identification and explanation without intent to infringe.

ISBN: 978-1-138-05511-7 (hbk)
ISBN: 978-1-315-16618-6 (ebk)
LCCN: 2019059882

Dedication We dedicate this work to all of our students who have been learning these ideas of analysis through our courses. We have learned as much from them as we hope they have from us. We are a firm believer that all our students are capable of excellence and that the only path to excellence is through discipline and study. We have always been proud of our students for doing so well on this journey. We hope these notes in turn make you proud of our efforts.

Abstract This book introduces graduate students in mathematics concepts from measure theory and also, continues their training in the *abstract* way of looking at the world. We feel that is a most important skill to have when your life's work will involve quantitative modeling to gain insight into the real world.

Acknowledgments I want to acknowledge the great debt I have to my wife, Pauli, for her patience in dealing with the long hours spent in typing and thinking. You are the love of my life.

The cover for this book is an original painting by me done in July 2017. It shows the moment when the giant squids reached out to me to learn advanced mathematics.

Table of Contents

I Introductory Matter   1

1 Introduction   3
- 1.1 The Analysis Courses 3
  - 1.1.1 Senior Level Analysis 4
  - 1.1.2 The Graduate Analysis Courses 4
  - 1.1.3 More Advanced Courses 7
- 1.2 Teaching the Measure and Integration Course 8
- 1.3 Table of Contents 9
- 1.4 Acknowledgments 11

II Classical Riemann Integration   13

2 An Overview of Riemann Integration   15
- 2.1 Integration 15
  - 2.1.1 The Riemann Integral as a Limit 17
  - 2.1.2 The Fundamental Theorem of Calculus 19
  - 2.1.3 The Cauchy Fundamental Theorem of Calculus 22
- 2.2 Handling Jumps 24
  - 2.2.1 Removable Discontinuity 24
  - 2.2.2 Jump Discontinuity 25

3 Bounded Variation   29
- 3.1 Partitions 30
- 3.2 Monotone 31
  - 3.2.1 The Saltus Function 36
  - 3.2.2 The Continuous Part of a Monotone Function 38
- 3.3 Bounded Variation 44
- 3.4 The Total Variation Function 48
- 3.5 Continuous Functions of Bounded Variation 51

4 Riemann Integration   55
- 4.1 Definition 55
- 4.2 Existence 58
- 4.3 Properties 66
- 4.4 Riemann Integrable? 71
- 4.5 More Properties 73
- 4.6 Fundamental Theorem 77
- 4.7 Substitution 85
- 4.8 Same Integral? 88

vii

viii   *TABLE OF CONTENTS*

5 Further Riemann Results   **93**
   5.1   Limit Interchange 93
   5.2   Riemann Integrable? 99
   5.3   Content Zero 100

III   Riemann - Stieltjes Integration   **109**

6 The Riemann - Stieltjes Integral   **111**
   6.1   Properties 112
   6.2   Step Integrators 115
   6.3   Monotone Integrators 121
   6.4   Equivalence Theorem 123
   6.5   Further Properties 124
   6.6   Bounded Variation Integrators 127

7 Further Riemann - Stieltjes Results   **133**
   7.1   Fundamental Theorem 133
   7.2   Existence 136
   7.3   Computations 139

IV   Abstract Measure Theory One   **147**

8 Measurability   **149**
   8.1   Borel Sigma-Algebra 152
   8.2   The Extended Borel Sigma-Algebra 153
   8.3   Measurable Functions 155
       8.3.1   Examples 157
   8.4   Properties 158
   8.5   Extended Real-Valued 161
   8.6   Extended Properties 163
   8.7   Continuous Compositions 167
       8.7.1   The Composition with Finite Measurable Functions 168
       8.7.2   The Approximation of Non-Negative Measurable Functions 168
       8.7.3   Continuous Functions of Extended Real-Valued Measurable Functions . . 170

9 Abstract Integration   **173**
   9.1   Measures 174
   9.2   Properties 175
   9.3   Sequences of Sets 178
   9.4   Integration 182
   9.5   Integration Properties 186
   9.6   Equality a.e. Problems 190
   9.7   Complete Measures 191
   9.8   Convergence Theorems 194
       9.8.1   Monotone Convergence Theorems 194
       9.8.2   Fatou's Lemma 198
   9.9   The Absolute Continuity of a Measure 200
   9.10   Summable Functions 202
   9.11   Extended Integrands 202
   9.12   Levi's Theorem 205

TABLE OF CONTENTS

9.13	Constructing Charges	207
9.14	Properties of Summable Functions	209
9.15	The Dominated Convergence Theorem	211
9.16	Alternative Abstract Integration Schemes	214
9.16.1	Properties of the Darboux Integral	218

10 The $\mathcal{L}_p$ Spaces — 223

10.1	The General L_p Spaces	227
10.2	The World of Counting Measure	238
10.3	Essentially Bounded Functions	240
10.4	The Hilbert Space L_2	247

V   Constructing Measures — 249

11 Building Measures — 251

11.1	Measures from Outer Measure	251
11.2	The Properties of the Outer Measure	254
11.3	Measures Induced by Outer Measures	257
11.4	Measures from Metric Outer Measures	258
11.5	Constructing Outer Measure	264

12 Lebesgue Measure — 273

12.1	Outer Measure	273
12.2	Lebesgue Outer Measure is a Metric Outer Measure	283
12.3	Lebesgue Measure is Regular	287
12.4	Approximation Results	288
12.4.1	Approximating Measurable Sets	288
12.4.2	Approximating Measurable Functions	292
12.5	The Summable Functions are Separable	294
12.6	The Existence of Non-Lebesgue Measurable Sets	295
12.7	Metric Spaces	297

13 Cantor Sets — 301

13.1	Generalized	301
13.2	Representation	304
13.3	The Cantor Functions	305
13.4	Consequences	307

14 Lebesgue - Stieltjes Measure — 309

14.1	Lebesgue - Stieltjes Outer Measure and Measure	311
14.2	Approximation Results	317
14.3	Properties	318

VI   Abstract Measure Theory Two — 323

15 Convergence Modes — 325

15.1	Extracting Subsequences	328
15.2	Egoroff's Theorem	336
15.3	Vitali's Theorem	339
15.4	Summary	344

TABLE OF CONTENTS

16 Decomposing Measures   **349**
- 16.1   Jordan Decomposition 349
- 16.2   Hahn Decomposition 354
- 16.3   Variation 356
- 16.4   Absolute Continuity 359
- 16.5   Radon-Nikodym 362
- 16.6   Lebesgue Decomposition 370

17 Connections to Riemann Integration   **375**

18 Fubini Type Results   **379**
- 18.1   The Riemann Setting 379
  - 18.1.1   Fubini on a Rectangle 380
- 18.2   The Lebesgue Setting 387

19 Differentiation   **397**
- 19.1   Absolutely Continuous Functions 397
- 19.2   LS and AC 398
- 19.3   Bounded Variation Derivatives 401
- 19.4   Measure Estimates 404
- 19.5   Extending the Fundamental Theorem of Calculus 408
- 19.6   Charges Induced by Absolutely Continuous Functions 414

VII   Summing It All Up   417

20 Summing It All Up   **419**

VIII   References   423

IX   Detailed Index   427

X   Appendix: Undergraduate Analysis Background Check   443

A Undergraduate Analysis Part One   **445**
- A-1   Sample Exams 448
  - A-1.1   Exam 1 448
  - A-1.2   Exam 2 449
  - A-1.3   Exam 3 450
  - A-1.4   Final 451

B Undergraduate Analysis Part Two   **453**
- B-1   Sample Exams 457
  - B-1.1   Exam 1 457
  - B-1.2   Exam 2 458
  - B-1.3   Exam 3 459
  - B-1.4   Final 460

XI Appendix: Linear Analysis Background Check 463

C Linear Analysis **465**

C-1 Sample Exams 467

C-1.1 Exam 1 467

C-1.2 Exam 2 468

C-1.3 Exam 3 469

C-1.4 Final 470

XII Appendix: Preliminary Examination Check 475

D The Preliminary Examination in Analysis **477**

D-1 Sample Exams 477

D-1.1 Exam 1 477

D-1.2 Exam 2 479

D-1.3 Exam 3 481

D-1.4 Exam 4 482

D-1.5 Exam 5 483

D-1.6 Exam 6 484

D-1.7 Exam 7 485

D-1.8 Exam 8 487

Part I

Introductory Matter

Chapter 1

Introduction

We believe that all students who are seriously interested in mathematics at the master's and doctoral level should have a passion for analysis even if it is not the primary focus of their own research interests. So you should all understand that my own passion for the subject will shine though in the notes that follow! And, it goes without saying that we assume that you are all mature mathematically and eager and interested in the material! Now, the present text focuses on the topics of *Measure* and *Integration* from a very abstract point of view, but it is very helpful to place this course into its proper context. Also, for those of you who are preparing to take the qualifying examination in analysis, the overview below will help you see why all this material fits together into a very interesting web of ideas.

1.1 The Analysis Courses

In outline form, the classical material on analysis would cover the material using textbooks equivalent to the ones listed below:

(A): Undergraduate Analysis, text **Advanced Calculus: An Introduction to Analysis**, by Watson Fulks. Our take on this material can be seen in (Peterson (8) 2020). Additional material, not covered in courses now but very useful is discussed in (Peterson (10) 2020).

(B): Introduction to Abstract Spaces, text **Introduction to Functional Analysis and Applications**, by Ervin Kreyszig. Our take on this material is seen in (Peterson (9) 2020).

(C): Measure Theory and Abstract Integration, texts **General Theory of Functions and Integration**, by Angus Taylor and **Real Analysis**, by Royden are classics but actually very hard to read and absorb. Our take on this material is the text you are currently reading.

(D:) Topology and Functional Analysis. A classical text is **Topology and Analysis** by Simmons. Our take on this material, and other topics, is seen in (Peterson (6) 2020).

In addition, a nice book that organizes the many interesting examples and counterexamples in this area is good to have on your shelf. We recommend the text **Counterexamples in Analysis** by Gelbaum and Olmstead. There are thus essentially five courses required to teach you enough of the concepts of mathematical analysis to enable you to read technical literature (such as engineering, control, physics, mathematics, statistics and so forth) at the beginning research level. Here are some more details about these courses.

1.1.1 Senior Level Analysis

Typically, this is a full two semester sequence that discusses thoroughly what we would call the analysis of functions of a real variable. This two semester sequence covers the following:

Advanced Calculus I: This course studies sequences and functions whose domain is simply the real line. There are, of course, many complicated ideas, but everything we do here involves things that act on real numbers to produce real numbers. If we call these things that act on other things, **OPERATORS**, we see that this course is really about real-valued operators on real numbers. This course invests a lot of time in learning how to be precise with the notion of convergence of sequences of objects, that happen to be real numbers, to other numbers.

1. Basic Logic, Inequalities for Real Numbers, Functions
2. Sequences of Real Numbers, Convergence of Sequences
3. Subsequences and the Bolzano - Weierstrass Theorem
4. Cauchy sequences
5. Continuity of Functions
6. Consequences of Continuity
7. Uniform Continuity
8. Differentiability of Functions
9. Consequences of Differentiability
10. Taylor Series Approximations

Advanced Calculus II: In this course, we rapidly become more abstract. First, we develop carefully the concept of the Riemann Integral. We show that although differentiation is intellectually quite a different type of limit process, it is intimately connected with the Riemann integral. Also, for the first time, we begin to explore the idea that we could have sequences of objects other than real numbers. We study carefully their convergence properties. We learn about two fundamental concepts: pointwise and uniform convergence of sequences of objects called functions. We are beginning to see the need to think about sets of objects, such as functions, and how to define the notions of convergence and so forth in this setting.

1. The Riemann Integral
2. Sequences of Functions
3. Uniform Convergence of Sequence of Functions
4. Series of Functions

Our version of this course, (Peterson (8) 2020) adds two variable calculus, convex analysis in $\Re$ and a nice discussion of Fourier Series. It turns out many of our mathematics majors do not take an advanced engineering mathematics class, so they are not necessarily exposed to Fourier Series tools. And there is a lot of nice mathematics in that!

1.1.2 The Graduate Analysis Courses

There are three basic courses at most universities. First, linear analysis, then measure and integration and finally, functional analysis. Linear analysis is a core analysis course and all master's students usually take it. Also, linear analysis and measure theory and integration form the two courses which we test prospective Ph.D. students on as part of the analysis preliminary examination. The content of these courses, also must fit within a web of other responsibilities. Many students are typically weak

INTRODUCTION

in abstraction coming in, so if we teach the material too fast, we lose them. Now if 20 students take linear analysis, usually 15 or 75% are already committed to an master's program which emphasizes Operations Research, Statistics, Algebra/ Combinatorics or Computation in addition to applied Analysis. Hence, currently, there are only about 5 students in linear analysis who might be interested in an master's specialization in analysis. The other students typically either don't like analysis at all and are only there because they have to be or they like analysis but it is part of their studies in number theory, partial differential equations for the computation area, and so forth. Either way, the students will not continue to study analysis for a degree specialization. However, we think it is important for all students to both know and appreciate this material. Traditionally, there are several ways to go.

The Cynical Approach: Nothing you can do will make students who don't like analysis change their mind. So teach the material hard and fast and target the 2 - 3 students who can benefit. The rest will come along for the ride and leave the course convinced that analysis is just like they thought – too hard and too complicated. If you do this approach, you can pick about any book you like. Most books for our students are too abstract and so are very hard for them to read. But the 2 -3 students who can benefit from material at this level, will be happy with the book. *We admit this is not our style although some think it is a good way to find the really bright analysis students.* We prefer the alternate Enthusiastic *"maybe I can get them interested anyway"* Approach: The instructor scours the available literature in order to make up notes to lead the students *"gently"* into the required abstract way of thinking. We haven't had much luck finding a published book for this, so since this is our preferred plan of action: we have worked on notes such as the ones you have in your hand for a long time. These notes start out handwritten and slowly mature into the typed versions. Many students have suffered with the earlier versions and we thank them for that! We believe it is important to actively try to get all the students interested but, of course, this is never completely successful. However, we still think there is great value in this approach and it is the one we have been trying for many years.

Introductory Linear Analysis: Our constraints here are to choose content so the students are adequately exposed to a more abstract way of thinking about the world. We generally cover

- Metric Spaces.
- Vector Spaces with a Norm.
- Vector Spaces with an Inner Product.
- Linear Operators.
- Basic Linear Functional Analysis such as Hahn - Banach Theorems and so forth.
- The Baire Category, the Open Mapping and Closed Graph Theorems.

It doesn't sound like much but there is a lot of material in here the students haven't seen. For example, we typically focus a lot on how we are really talking about sets of objects with some additional structure. A set plus a way to measure distance between objects gives a metric space; if we can add and scale objects, we get a vector space; if we have a vector space and add the structure that allows us to project a vector to a subspace, we get an inner product space. We also mention we could have a set of objects and define one operation to obtain a group or if we define a special collection of sets we call open, we get a topological space and so forth. If we work hard, we can help open their minds to the fact that each of the many sub-disciplines in the Mathematical Sciences focuses on special structure we add to a set to help us solve problems in that arena.

There are lots of ways to cover the important material in these topic areas and even many ways to decide on exactly what is important from metric, normed and inner product spaces. So there is that kind of freedom, but not so much freedom that you can decide to drop say, inner product spaces. For example, we could use Stürm - Liouville systems as an example when the discussion turns to eigenvalues of operators. It is nice to use projection theorems in an inner product setting as a big finishing application, but remember the students are weak in background, e.g. their knowledge of ordinary differential equations and Calculus in $\Re^n$ is normally weak. So we are limited in our coverage of the completeness of an orthonormal sequence in an inner product space in many respects. In fact, since the students don't know measure theory (which is in this text) the discussion of Hilbert spaces of functions is inherently weak. However, we decided to address this by constructing the reals from scratch and then showing how to build a set of functions $\mathbb{L}_2([a, b])$ which is a Hilbert space without having to resort to measure theory. Of course, this is all equivalent to what you would get with a measure theory approach, but it has the advantage of allowing the students to see how all the details work without brushing stuff under the rug, so to speak! However, we can now develop this material in another way using the ideas from measure theory which is the content of the present text. Also, if you look carefully at that material, you need to cover some elementary versions of the Hahn - Banach theorem to do it right which is why we have placed it the linear analysis course. In general, in the class situation, we run out of time to cover such advanced topics. The trade-off seems to be between thorough coverage of a small number of topics or rapid coverage of many topics superficially. However, we can do it all in a book treatment and we have designed all of the basic analysis books for self-study too.

We believe this course is a very critical one in the evolution of the student's ability to think abstractly and so teaching there is great value in teaching very, very carefully the basics of this material. This course takes a huge amount of time for lecture preparation and student interaction in the office, so when we teach this material, we slow down in our research output!

In more detail, in this text, we begin to rephrase all of our knowledge about convergence of sequence of objects in a much more general setting.

1. Metric Spaces: A set of objects and a way of measuring distance between objects which satisfies certain special properties. This function is called a **metric** and its properties were chosen to mimic the properties that the absolute value function has on the real line. We learn to understand convergence of objects in a general metric space. It is really important to note that there is NO additional structure imposed on this set of objects; no linear structure (i.e. vector space structure), no notion of a special set of elements called a basis which we can use to represent arbitrary elements of the set. The **metric** in a sense generalizes the notion of distance between numbers. We can't really measure the size of an object by itself, so we do not yet have a way of generalizing the idea of size or length.

   A fundamentally important concept now emerges: the notion of completeness and how it is related to our choice of metric on a set of objects. We learn a clever way of constructing an abstract representation of the completion of any metric space, but at this time, we have no practical way of seeing this representation.

2. Normed Spaces: We add linear structure to the set of objects and a way of measuring the magnitude of an object; that is, there is now an operation we think of as addition and another operation which allows us to scale objects and a special function called a **norm** whose value for a given object can be thought of as the object's magnitude. We then develop what we mean by convergence in this setting. Since we have a vector space structure, we can now begin to talk about a special subset of objects called a **basis** which can be used to find a useful way of representing an arbitrary object in the space.

   Another most important concept now emerges: the cardinality of this basis may be finite or infinite. We begin to explore the consequences of a space being finite versus infinite dimensional.

INTRODUCTION

3. Inner Product Spaces: To a set of objects with vector space structure, we add a function called an **inner product** which generalizes the notion of dot product of vectors. This has the extremely important consequence of allowing the inner product of two objects to be zero even though the objects are not the same. Hence, we can develop an abstract notion of the orthogonality of two objects. This leads to the idea of a basis for the set of objects in which all the elements are mutually orthogonal. We then finally can learn how to build representations of arbitrary objects efficiently.

4. Completions: We learn how to complete an arbitrary metric, normed or inner product space in an abstract way, but we know very little about the practical representations of such completions.

5. Linear Operators: We study a little about functions whose domain is one set of objects and whose range is another. These functions are typically called operators. We learn a little about them here.

6. Linear Functionals: We begin to learn the special role that real-valued functions acting on objects play in analysis. These types of functions are called linear functionals and learning how to characterize them is the first step in learning how to use them. We just barely begin to learn about this here.

We have implemented this approach in (Peterson (9) 2020).

Measure Theory: This course generalizes the notion of integration to a very abstract setting. The set of notes you are reading is a textbook for this material. Roughly speaking, we first realize that the Riemann integral is a linear mapping from the space of bounded real-valued functions on a compact interval into the reals which has a number of interesting properties. We then study how we can generalize such mappings so that they can be applied to arbitrary sets X, a special collection of subsets of X called a sigma-algebra and a new type of mapping called a measure which on $\Re$ generalizes our usual notion of the length of an interval. In this class, we discuss the following:

1. The Riemann Integral.

2. Measures on a sigma-algebra S in the set X and integration with respect to the measure.

3. Measures specialized to sigma-algebras on the set $\Re^n$ and integrations with respect to these measures. The canonical example of this is Lebesgue measure on $\Re^n$.

4. Differentiation and integration in these abstract settings and their connections.

This is what we cover in the notes you are reading now and some other things!

1.1.3 More Advanced Courses

It is also recommended that students consider taking a course in what is called Functional Analysis. Our version is (Peterson (6) 2020) which tries hard to put topology, analysis and algebra ideas together into one pot. It is carefully designed for self-study as this course is just not being offered as much as it should be. While not part of the qualifying examination, in this course, we can finally develop in a careful way the necessary tools to work with linear operators, weak convergence and so forth. This is a huge area of mathematics, so there are many possible ways to design an introductory course. A typical such course would cover:

1. An Introduction to General Operator Theory.

2. Topological Vector Spaces and Distributions.

3. An Introduction to the Spectral Theory of Linear Operators; this is the study of the eigenvalues and eigenobjects for a given linear operator; there are lots of applications here!

4. Differential Geometry ideas.

5. Degree Theory ideas.

6. Some advanced topics using these ideas: possibilities include

    (a) Existence Theory of Boundary Value Problems.

    (b) Existence Theory for Integral Equations.

    (c) Existence Theory in Control.

1.2   Teaching the Measure and Integration Course

So now that you have seen how the analysis courses all fit together, it is time for the main course. So roll up your sleeves and prepare to work! Let's start with a few more details on what this course on *Measure* and *Integration* will cover.

In this course, we assume mathematical maturity and we tend to follow *The Enthusiastic "maybe I can get them interested anyway" Approach* in lecturing (so, be warned)! It is difficult to decide where to start in this course. There is usually a reasonable fraction of you who have never seen an adequate treatment of Riemann Integration. For example, not everyone may have seen the equivalent of the second part of undergraduate analysis where Riemann integration is carefully discussed. We therefore have taught several versions of this course and we have divide the material into *blocks* as follows: We believe there are a lot of advantages in treating integration abstractly. So, if we covered the Lebesgue integral on $\Re$ right away, we can take advantage of a lot of the special structure $\Re$ has which we don't have in general. It is better for long-term intellectual development to see measure and integration approached without using such special structure. Also, all of the standard theorems we want to do are just as easy to prove in the abstract setting, so why specialize to $\Re$? So we tend to do abstract measure stuff first. The core material for **Block 1** is as follows:

1. Abstract measure ν on a sigma-algebra S of subsets of a universe X.

2. Measurable functions with respect to a measure ν; these are also called *random variables* when ν is a probability measure.

3. Integration $\int f d\nu$.

4. Convergence results: monotone convergence theorem, dominated convergence theorem, etc.

Then we develop the Lebesgue Integral in $\Re^n$ via outer measures as the great example of a nontrivial measure. So **Block 2** of material is thus:

1. Outer measures in $\Re^n$.

2. Caratheodory conditions for measurable sets.

3. Construction of the Lebesgue sigma-algebra.

4. Connections to Borel sets.

To fill out the course, we pick topics from the following:

INTRODUCTION

1. Riemann and Riemann - Stieltjes integration. This would go before **Block 1** if we do it. Call it block **Riemann**.

2. Decomposition of measures – I love this material so this is after **Block 2**. Call it block **Decomposition**.

3. Connection to Riemann integration via absolute continuity of functions. This is actually hard stuff and takes about 3 weeks to cover nicely. Call it **Block Riemann and Lebesgue**. If this is done without **Block Riemann**, you have to do a quick review of Riemann stuff so they can follow the proofs.

4. Fubini type theorems. This would go after **Block 2**. Call this Block **Fubini**.

5. Differentiation via the Vitali approach. This is pretty hard too. Call this **Differentiation**.

6. Treatment of the usual L_p spaces. Call this **Block L_p**.

7. More convergence stuff like convergence in measure, L_p convergence implies convergence of a subsequence pointwise, etc. These are hard theorems and to do them right requires a lot of time. Call this **More Convergence**.

We have taught this in at least the following ways. And always, lots of homework and projects, as we believe only hands-on work really makes this stuff sink in.

Way 1: Block Riemann, Block 1, Block 2 and Block Decomposition.

Way 2: Block 1, Block 2, Block Decomposition and Block Riemann and Lebesgue.

Way 3: Block 1, Block 2, Block Decomposition and Differentiation.

Way 4: Block 1, Block 2, Block L_p, Block More Convergence and Block Decomposition.

Way 5: Block 1, Block 2, Block Fubini, Block More Convergence and Block Decomposition.

In book form, we can cover all of these topics in some detail; not as much as we like, but enough to get you started on your lifelong journey into these areas.

We use Octave (Eaton et al. (3) 2020), which is an open source GPL licensed (Free Software Foundation (4) 2020) clone of MATLAB®, as a computational engine and we are not afraid to use it as an adjunct to all the theory we go over. Of course, you can use MATLAB® (MATLAB (5) 2018 - 2020) also if your university or workplace has a site license or if you have a personal license. Get used to it: theory, computation and science go hand in hand! Well, buckle up and let's get started!

1.3 Table of Contents

This text is based on quite a few years of teaching graduate courses in analysis and began from handwritten notes starting roughly in the late 1990's but with material being added all the time. Along the way, these notes have been helped by the students we have taught and also by our own research interests as research informs teaching and teaching informs research in return.

In this text, we go over the following blocks of material.

Part One: These are our beginning remarks you are reading now which are in Chapter 1.

Part Two: Classical Riemann Integration Here we are concerned with classic Riemann integration and the new class of functions called functions of bounded variation.

- In Chapter 2, we provide a quick overview of Riemann Integration just to set the stage.
- Chapter 3 discusses a new topic: functions of bounded variation.
- In Chapter 4, we carefully go over the basics of the theory of Riemann integration. We want you to see how these proofs are done, so that is easy to see how they are modified for the case of Riemann - Stieltjes integration.
- Chapter 5 goes over further important details of classical Riemann integration theory.

Part Three: Riemann - Stieltjes Integration We extend Riemann integration to Riemann - Stieltjes integration.

- Chapter 6 provides the basics of Riemann - Stieltjes integration theory.
- In Chapter 7, we add further details about Riemann - Stieltjes integrals.

Part Four: Abstract Measure Theory One We define what a measure is and then develop a corresponding theory of integration based on this measure.

- Chapter 8 defines measures and their associated spaces. We do not know how to construct measures yet, but we can work out many of the consequences of a measure if we assume it is there to use.
- Chapter 9 develops the theory of integration based on a measure.
- In Chapter 10, we generalize the discussion of the ℓ^p sequence spaces to the more general setting of equivalence classes of functions whose p^{th} power is integrable in this more general sense.

Part Five: Constructing Measures We learn how to construct measures in various ways.

- Chapter 11 discusses how to build or construct measures in very general ways.
- In Chapter 12, we carefully construct what is called Lebesgue measure.
- Chapter 13 shows you how to build specialized subsets of the real line called Cantor sets which are the source of many examples and ideas. This chapter is a series of exercises which as you finish them, you see how these sets are constructed via limiting processes and what their properties are.
- In Chapter 14, we extend the construction of Lebesgue measures to Lebesgue - Stieltjes measures.

Part Six: Abstract Measure Theory Two We want to finish with general comments on a variety of topics: the different kinds of convergence of sequences of functions we now have, the way measures can be decomposed into fundamental modes, how the Lebesgue integral connects to the Riemann Integral, generalized Fubini results and finally some thoughts on differentiation in this context. This last part allows us to state and prove an extension of the Fundamental Theorem of Calculus.

- In Chapter 15, we discuss carefully the many types of convergence of sequences of functions we have in this context.
- Chapter 16 explains how we can write measures as sums of simpler parts giving us a variety of decomposition theorems.
- In Chapter 17, we develop connections between Lebesgue and Riemann Integration.
- Chapter 18 looks at the idea of multidimensional abstract integration in terms of iterated integrals: i.e. extensions of the classical Fubini Theorems.
- Chapter 19 develops the idea of absolutely continuous functions and measures which we use to extend the Fundamental Theorem of Calculus.

Part Seven: Summing It All Up In Chapter 20, we talk about the things you have learned here and where you should go next to continue learning more analysis.

There is much more we could have done, but these topics are a nice introduction into the further use of abstraction in your way of thinking about models and should prepare you well for the next step, which is functional analysis and topology.

1.4 Acknowledgments

I have been steadily shrinking since 1996 and currently I can only be seen when standing on a large box. This has forced me to alter my lecture style somewhat so that my students can see me and I am considering the use of platform shoes. However, if I can ever remember the code to my orbiting spaceship, I will be able to access the shape changing facilities on board and solve the problem permanently. However, at the moment, I do what I can with this handicap.

There were many problems with the handwritten version of these notes, but with the help of my many students in this class over the years, they have gotten better. However, being written in my handwriting was not a good thing. In fact, one student some years ago was so unhappy with the handwritten notes, that he sent me the names of four engineering professors who actually typed their class notes nicely. He told me I could learn from them how to be a better teacher. Well, it has taken me awhile, but at least the notes are now typed. The teaching part, though, is another matter.

I will leave you with an easy proposition to prove; consider it your first mathematical test!

Proposition 1.4.1 My Proposition

My proposition is this: chocolate makes one happier!

Proof 1.4.1
Eat a piece of chocolate and the proof is there! ∎

And, of course, there is a corollary involving donuts which I will leave to you!

Jim Peterson
School of Mathematical and Statistical Sciences
Clemson University

Part II

Classical Riemann Integration

Chapter 2

An Overview of Riemann Integration

In this chapter, we will give you a quick overview of Riemann integration. There are few real proofs but it is useful to have a quick tour before we get on with the job of extending this material to a more abstract setting.

2.1 Integration

There are two intellectually separate ideas here:

1. The idea of a **Primitive** or **Antiderivative** of a function f. This is any function F which is differentiable and satisfies $F'(t) = f(t)$ at all points in the domain of f. Normally, the domain of f is a finite interval of the form $[a, b]$, although it could also be an infinite interval like all of $\Re$ or $[1, \infty)$ and so on. Note that an *antiderivative* does not require any understanding of the process of Riemann integration at all – only what differentiation is!

2. The idea of the Riemann integral of a function. You should have been exposed to this in your first Calculus course and perhaps a bit more rigorously in your undergraduate second semester analysis course.

Let's review what Riemann Integration involves. First, we start with a bounded function f on a finite interval $[a, b]$. This kind of function f need not be continuous! Then select a finite number of points from the interval $[a, b]$, $\{x_0, x_1, , \ldots, x_{n-1}, x_n\}$. We don't know how many points there are, so a different selection from the interval would possibly give us more or less points. But for convenience, we will just call the last point x_n and the first point x_0. These points are not arbitrary – x_0 is always a, x_n is always b and they are ordered like this:

$$x_0 = a < x_1 < x_2 < \ldots < x_{n-1} < x_n = b$$

The collection of points from the interval $[a, b]$ is called a Partition of $[a, b]$ and is denoted by some letter – here we will use the letter π. So if we say π is a partition of $[a, b]$, we know it will have $n + 1$ points in it, they will be labeled from x_0 to x_n, and they will be ordered left to right with strict inequalities. But, we will not know what value the positive integer n actually is. The simplest Partition π is the two point partition $\{a, b\}$. Note these things also:

1. Each partition of $n + 1$ points determines n subintervals of $[a, b]$.

2. The lengths of these subintervals always add up to the length of $[a, b]$ itself, $b - a$.

3. These subintervals can be represented as

$$\{[x_0, x_1], [x_1, x_2], \ldots, [x_{n-1}, x_n]\}$$

BASIC ANALYSIS IV: MEASURE THEORY AND INTEGRATION

or more abstractly as $[x_i, x_{i+1}]$ where the index i ranges from 0 to $n - 1$.

4. The length of each subinterval is $x_{i+1} - x_i$ for the indices i in the range 0 to $n - 1$.

Now from each subinterval $[x_i, x_{i+1}]$ determined by the Partition π, select any point you want and call it s_i. This will give us the points s_0 from $[x_0, x_1]$, s_1 from $[x_1, x_2]$ and so on up to the last point, s_{n-1} from $[x_{n-1}, x_n]$. At each of these points, we can evaluate the function f to get the value $f(s_j)$. Call these points an **Evaluation Set** for the partition π. Let's denote such an evaluation set by the letter σ. Note there are many such evaluation sets that can be chosen from a given partition π. We will leave it up to you to remember that when we use the symbol σ, you must remember it is associated with some partition.

If the function f was nice enough to be positive always and continuous, then the product $f(s_i) \times (x_{i+1} - x_i)$ can be interpreted as the area of a rectangle. Then, if we add up all these rectangle areas we get a sum which is useful enough to be given a special name: the Riemann sum for the function f associated with the Partition π and our choice of evaluation set $\sigma = \{s_0, \ldots, s_{n-1}\}$. This sum is represented by the symbol $S(f, \pi, \sigma)$ where the things inside the parentheses are there to remind us that this sum depends on our choice of the function f, the partition π and the evaluations set σ.

Definition 2.1.1 Riemann Sum

> The Riemann sum for the bounded function f, the partition π and the evaluation set $\sigma = \{s_0, \ldots, s_{n-1}\}$ from $\pi\{x_0, x_1, \ldots, x_{n-1}, x_n\}$ is defined by
>
> $$S(f, \pi, \sigma) = \sum_{i=0}^{n-1} f(s_i)(x_{i+1} - x_i)$$
>
> It is pretty misleading to write the Riemann sum this way as it can make us think that the n is always the same when in fact it can change value each time we select a different partition π. So many of us write the definition this way instead
>
> $$S(f, \pi, \sigma) = \sum_{i \in \pi} f(s_i)(x_{i+1} - x_i) = \sum_{\pi} f(s_i)(x_{i+1} - x_i)$$
>
> and we just remember that the choice of π will determine the size of n.

Homework

Exercise 2.1.1 Let $f(t) = 2t^2 + 5$ and let $P = \{0.0, 1.0, 1.5, 2.0\}$ be a partition of the interval $[0.0, 2.0]$. Let $E = \{0.5, 1.3, 1.8\}$ be the evaluation set. Calculate $S(f, P, E)$ and draw the picture that represents this Riemann sum.

Exercise 2.1.2 Let $f(t) = |\cos(2t)|$ and let $P = \{-1.0, 2.0, 2.5, 5.0\}$ be a partition of the interval $[-1.0, 5.0]$. Let $E = \{1.5, 2.3, 3.9\}$ be the evaluation set. Calculate $S(f, P, E)$ and draw the picture that represents this Riemann sum.

Exercise 2.1.3 Let $f(t) = 4t^2 + 2$ on the interval $[1, 3]$ with $P = \{1, 1.5, 2.0, 2.5, 3.0\}$ and $E = \{1.2, 1.8, 2.3, 2.8\}$. Do this one in MATLAB.

Exercise 2.1.4 Let $f(t) = t^2 + 13t + 2$ on the interval $[1, 3]$ with $P = \{1, 1.6, 2.3, 2.8, 3.0\}$ and $E = \{1.2, 1.9, 2.5, 2.85\}$. Do this one in MATLAB.

Exercise 2.1.5 Let $f(t) = 3t^2 + 2t^4$ on the interval $[1, 2]$ with $P = \{1, 1.2, 1.5, 1.8, 2.0\}$ and $E = \{1.1, 1.3, 1.7, 1.9\}$. Do this one in MATLAB.

2.1.1 The Riemann Integral as a Limit

We can construct many different Riemann sums for a given function f. To define the **Riemann Integral** of f, we only need a few more things:

1. Each partition π has a maximum subinterval length – let's use the symbol $\| \pi \|$ to denote this length. We read the symbol $\| \pi \|$ as the **norm** or **gauge** of π.

2. Each partition π and evaluation set σ determines the number $S(f, \pi, \sigma)$ by a simple calculation.

3. So if we took a collection of partitions π_1, π_2 and so on with associated evaluation sets σ_1, σ_2, etc., we would construct a sequence of real numbers $\{S(f, \pi_1, \sigma_1), \ S(f, \pi_2, \sigma_2), \ldots, , \ S(f, \pi_n, \sigma_n), \ldots, \}$. Let's assume the norm of the partition π_n gets smaller all the time; i.e. $\lim_{n \to \infty} \| \pi_n \| = 0$. We could then ask if this sequence of numbers converges to something.

What if the sequence of Riemann sums we construct above converged to the same number I no matter what sequence of partitions whose norm goes to zero and associated evaluation sets we chose?

Then, we would have that the value of this limit is *independent* of the choices above. This is indeed what we mean by the **Riemann Integral** of f on the interval $[a, b]$.

Definition 2.1.2 Riemann Integrability of a Bounded Function

> *Let f be a bounded function on the finite interval $[a, b]$. if there is a number I so that*
>
> $$\lim_{n \to \infty} S(f, \pi_n, \sigma_n) = I$$
>
> *no matter what sequence of partitions $\{\pi_n\}$ with associated sequence of evaluation sets $\{\sigma_n\}$ we choose as long as $\lim_{n \to \infty} \| \pi_n \| = 0$, we will say that the Riemann Integral of f on $[a, b]$ exists and equals the value I.*

The value I is dependent on the choice of f and interval $[a, b]$. So we often denote this value by $I(f, [a, b])$ or more simply as $I(f, a, b)$. Historically, the idea of the Riemann integral was developed using area approximation as an application, so the summing nature of the Riemann sum was denoted by the 16th century *letter S* which resembled an elongated or stretched letter S which looked like what we call the integral sign $\int$. Hence, the common notation for the Riemann Integral of f on $[a, b]$, when this value exists, is $\int_a^b f$. We usually want to remember what the independent variable of f is also and we want to remind ourselves that this value is obtained as we let the norm of the partitions go to zero. The symbol dx for the independent variable x is used as a reminder that $x_{i+1} - x_i$ is going to zero as the norm of the partitions goes to zero. So it has been very convenient to add to the symbol $\int_a^b f$ this information and use the augmented symbol $\int_a^b f(x)\, dx$ instead. Hence, if the independent variable was t instead of x, we would use $\int_a^b f(t)\, dt$. Since for a function f, the name we give to the independent variable is a matter of personal choice, we see that the choice of variable name we use in the symbol $\int_a^b f(t)\, dt$ is very arbitrary. Hence, it is common to refer to the independent variable we use in the symbol $\int_a^b f(t)\, dt$ as the dummy variable of integration.

We need a few more facts. We shall prove later the following things are true about the Riemann Integral of a bounded function. First, we know when a bounded function actually has a Riemann integral.

Theorem Existence of the Riemann Integral

> Let f be a bounded function on the finite interval $[a, b]$. Then the Riemann integral of f on $[a, b]$, $\int_a^b f(t)dt$ exists if
>
> 1. f is continuous on $[a, b]$
>
> 2. f is continuous except at a finite number of points on $[a, b]$.
>
> Further, if f and g are both Riemann integrable on $[a, b]$ and they match at all but a finite number of points, then their Riemann integrals match; i.e. $\int_a^b f(t)dt$ equals $\int_a^b g(t)dt$.

Most of the functions we want to work with do have a lot of *smoothness*, i.e. continuity and even differentiability on the intervals we are interested in. Hence, the existence theorem will apply. Here are some examples:

1. If $f(t)$ is t^2 on the interval $[-2, 4]$, then $\int_{-2}^4 f(t)dt$ does exist as f is continuous on this interval.

2. If g was defined by

$$g(t) = \begin{cases} t^2 & -2 \leq t < 1 \text{ and } 1 < t \leq 4 \\ 5 & t = 1 \end{cases}$$

we see g is not continuous at only one point and so it is Riemann integrable on $[-2, 4]$. Moreover, since f and g are both integrable and match at all but one point, their Riemann integrals are equal.

However, with that said, in this course, we want to relax the smoothness requirements on the functions f we work with and define a more general type of integral for this less restricted class of functions.

In Riemann integration, we subdivide the domain of f which is assumed to be a finite interval into pieces which are disjoint except at the endpoints where they touch, pick a function value in each of these pieces and compute the Riemann sum. We could do this a different way. We could slice the range of f into disjoint values c_i and look at the inverse image sets $f^{-1}(c_i)$. If we had a way of measuring the size of the sets $f^{-1}(c_i)$, a different type of approximation would be the sum $\sum c_i \, \mu(f^{-1}(c_i))$ where μ is a function which assigns a positive number to the set $f^{-1}(c_i)$. If the original subdivision gave us the sets $E_i = [t_i, t_{i+1}]$, then the corresponding Riemann sum would be $\sum f(s_i)\mu([t_i, t_{i+1}])$ where $\mu([t_i, t_{i+1}]) = t_{i+1} - t_i$ is our usual way of defining the lengths of a finite interval. However the inverse image $f^{-1}(c)$ for a real number c does have to be a simple finite interval. So this new way of slicing up the range immediately hits a snag. We need to figure out a way to assign a non-negative number to arbitrary subsets of real numbers. This is not so easy to do, and this entire text is about how to go about doing that.

Homework

Exercise 2.1.6 *If f is RI on $[-1, 4]$ and g is RI on $[-1, 4]$, prove linear combinations of f and g are also RI on $[-1, 4]$.*

Exercise 2.1.7 *Prove the constant function $f(x)$ is RI for any finite interval $[a, b]$ and compute its value.*

2.1.2 The Fundamental Theorem of Calculus

There is a big connection between the idea of the *antiderivative* of a function f and its Riemann integral. For a positive function f on the finite interval $[a, b]$, we can construct the area under the curve function $F(x) = \int_a^x f(t)\,dt$ where for convenience we choose an x in the open interval (a, b). We show $F(x)$ and $F(x + h)$ for a small positive h in Figure 2.1. Let's look at the difference in these areas:

$$
\begin{aligned}
F(x + h) - F(x) &= \int_a^{x+h} f(t)\,dt - \int_a^x f(t)\,dt \\
&= \int_a^x f(t)\,dt + \int_x^{x+h} f(t)\,dt - \int_a^x f(t)\,dt \\
&= \int_x^{x+h} f(t)\,dt
\end{aligned}
$$

where we have used standard properties of the Riemann integral to write the first integral as two pieces and then do a subtraction. Now divide this difference by the change in x which is h. We find

$$
\frac{F(x + h) - F(x)}{h} = \frac{1}{h} \int_x^{x+h} f(t)\,dt \tag{2.1}
$$

The difference in area, $\int_x^{x+h} f(t)\,dt$, is the second shaded area in Figure 2.1. Clearly, we have

$$
F(x + h) - F(x) = \int_x^{x+h} f(t)\,dt \tag{2.2}
$$

We know that f is bounded on $[a, b]$; hence, there is a number B so that $f(t) \leq B$ for all t in $[a, b]$. Thus, using Equation 2.2, we see

$$
0 \leq F(x + h) - F(x) \leq \int_x^{x+h} B\,dt = B\,h \tag{2.3}
$$

From this we have

$$
\begin{aligned}
0 \leq \lim_{h \to 0} \left(F(x + h) - F(x) \right) &\leq \lim_{h \to 0} B\,h \\
&= 0
\end{aligned}
$$

We conclude that F is continuous at each x in $[a, b]$ as

$$
\lim_{h \to 0} \left(F(x + h) - F(x) \right) = 0
$$

It seems that the new function F we construct by integrating the function f in this manner, always builds a new function that is continuous. Is F differentiable at x? If f is continuous at x, then given a positive ϵ, there is a positive δ so that

$$
f(x) - \epsilon < f(t) < f(x) + \epsilon \text{ if } x - \delta < t < x + \delta
$$

and t is in $[a, b]$. So, if h is less than δ, we have

$$
\frac{1}{h} \int_x^{x+h} (f(x) - \epsilon) < \frac{F(x + h) - F(x)}{h} = \frac{1}{h} \int_x^{x+h} f(t)\,dt < \frac{1}{h} \int_x^{x+h} (f(x) + \epsilon)
$$

This is easily evaluated to give

$$f(x) - \epsilon < \frac{F(x+h) - F(x)}{h} = \int_x^{x+h} f(t)\, dt < f(x) + \epsilon$$

if h is less than δ. This shows that

$$\lim_{h \to 0^+} \frac{F(x+h) - F(x)}{h} = f(x)$$

You should be able to believe that a similar argument would work for negative values of h: i.e.,

$$\lim_{h \to 0^-} \frac{F(x+h) - F(x)}{h} = f(x)$$

This tells us that $F'(x)$ exists and equals $f(x)$ as long as f is continuous at x as

$$F'(x^+) = \lim_{h \to 0^+} \frac{F(x+h) - F(x)}{h} = f(x)$$

$$F'(x^-) = \lim_{h \to 0^-} \frac{F(x+h) - F(x)}{h} = f(x)$$

This relationship is called the **Fundamental Theorem of Calculus**. The same sort of argument works for x equals a or b, but we only need to look at the derivative from one side. We will prove this sort of theorem using fairly relaxed assumptions on f for the interval $[a, b]$ in the later chapters. Even if we just consider the world of Riemann Integration, we only need to assume that f is Riemann Integrable on $[a, b]$, which allows for jumps in the function.

Theorem Fundamental Theorem of Calculus

> *Let f be Riemann Integrable on $[a, b]$. Then the function F defined on $[a, b]$ by $F(x) = \int_a^x f(t)\, dt$ satisfies*
>
> *1. F is continuous on all of $[a, b]$*
>
> *2. F is differentiable at each point x in $[a, b]$ where f is continuous and $F'(x) = f(x)$.*

Using the same f as before, suppose G was defined on $[a, b]$ as follows

$$G(x) = \int_x^b f(t)\, dt$$

Note that

$$F(x) + G(x) = \int_a^x f(t)\, dt + \int_x^b f(t)\, dt$$

$$= \int_a^b f(t)\, dt$$

Since the Fundamental Theorem of Calculus tells us F is differentiable, we see $G(x) = \int_a^b f(t)\, dt - F(x)$ must also be differentiable. It follows that

$$G'(x) = -F'(x) = -f(x)$$

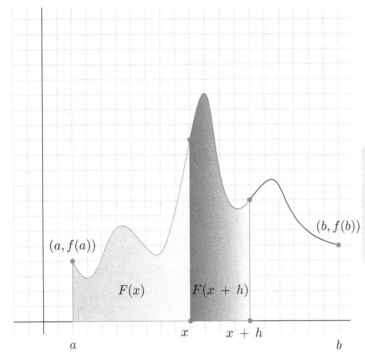

Figure 2.1: The function $F(x)$.

Let's state this as a variant of the Fundamental Theorem of Calculus, the *Reversed Fundamental Theorem of Calculus* so to speak.

Theorem Fundamental Theorem of Calculus Reversed

Let f be Riemann Integrable on $[a,b]$. Then the function F defined on $[a,b]$ by $F(x) = \int_x^b f(t)\,dt$ satisfies

1. *F is continuous on all of $[a,b]$*
2. *F is differentiable at each point x in $[a,b]$ where f is continuous and $F'(x) = -f(x)$.*

Homework

Exercise 2.1.8 *Prove f is in $RI([0,1])$ where f is defined by*

$$f(x) = \begin{cases} x\,\sin(1/x), & x \in (0,1] \\ 0, & x = 0 \end{cases}$$

Exercise 2.1.9 *Let $f : \Re^2 \to \Re$ be continuous on $[0,2] \times [-10,10]$. Define F by $F(t) = \int_{-10}^t f(s, x(s))ds$ for any $x \in C([-10,10])$. Is F continuous on $[-10,10]$?*

Exercise 2.1.10 *Consider the differential equation $x' = 3x$ with $x(0) = 2$. If x is a differentiable function on $[0,T]$ which satisfies this differential equation, it is clear x' must be continuous and*

22   BASIC ANALYSIS IV: MEASURE THEORY AND INTEGRATION

$x(t) = \int_0^t 3x(s)ds$.

Exercise 2.1.11 *Let $f : \Re^2 \to \Re$ be continuous on $[0, 2] \times [-10, 10]$. Consider a differentiable function x on $[-10, 10]$ defined by $x'(t) = f(t, x(t))$. Note this forces x to have a continuous derivative. Let $T : C([-10, 10]) \to C^1([-10, 10])$ be defined by $(T(x))(t) = x(-10) + \int_0^t f(s, x(s))ds$. Show the existence of a solution to this differential equation to showing there is a fixed point to the operator equation $x = T(x)$. Make sure you understand proving such a fixed point exists is not easy. This is one of things we do in a standard course on the theory of differential equations.*

2.1.3   The Cauchy Fundamental Theorem of Calculus

We can use the Fundamental Theorem of Calculus to learn how to evaluate many Riemann integrals. Let G be an antiderivative of the function f on $[a, b]$. Then, by definition, $G'(x) = f(x)$ and so we know G is continuous at each x. But we still don't know that f itself is continuous. However, if we assume f *is* continuous, then if we define F on $[a, b]$ by

$$F(x) \;=\; f(a) + \int_a^x f(t)\, dt$$

the Fundamental Theorem of Calculus is applicable. Thus, $F'(x) = f(x)$ at each point. But that means $F' = G' = f$ at each point. Functions whose derivatives are the same must differ by a constant. Call this constant C. We thus have $F(x) = G(x) + C$. So, we have

$$F(b) \;=\; f(a) + \int_a^b f(t)dt = G(b) + C$$
$$F(a) \;=\; f(a) + \int_a^a f(t)dt = G(a) + C$$

But $\int_a^a f(t)\, dt$ is zero, so we conclude after some rewriting

$$G(b) \;=\; f(a) + \int_a^b f(t)dt + C$$
$$G(a) \;=\; f(a) + C$$

And after subtracting, we find the important result

$$G(b) - G(a) \;=\; \int_a^b f(t)dt$$

This is what tells us how to integrate many functions. For example, if $f(t) = t^3$, we can guess the antiderivatives have the form $t^4/4 + C$ for an arbitrary constant C. Thus, since $f(t) = t^3$ is continuous, the result above applies. We can therefore calculate Riemann integrals of polynomials, combinations of trigonometric functions and so forth. Let's formalize this as a theorem called the **Cauchy Fundamental Theorem of Calculus**. All we really need to prove this result is that f is Riemann integrable on $[a, b]$, which is true if our function f is continuous.

Theorem Cauchy Fundamental Theorem of Calculus

Let G be any antiderivative of the Riemann integrable function f on the interval $[a, b]$. Then $G(b) - G(a) = \int_a^b f(t)\, dt$.

The common symbol for the antiderivative of f has evolved to be $\int f$ because of the close connection between the antiderivative of f and the Riemann integral of f which is given in the Cauchy Fundamental Theorem of Calculus. The usual Riemann integral, $\int_a^b f(t)\, dt$ of f on $[a, b]$ computes a *definite* value – hence, the symbol $\int_a^b f(t)\, dt$ is usually referred to as the *definite integral of f on $[a, b]$* to contrast it with the *family* of functions represented by the antiderivative $\int f$. Since the antiderivatives are arbitrary up to a constant, most of us refer to the antiderivative as the *indefinite integral of f*. Also, we hardly ever say "let's find the antiderivative of f" – instead, we just say, "let's integrate f". We will begin using this shorthand now!

Homework

Exercise 2.1.12 *The idea of an antiderivative is intellectually distinct from the Riemann integral of a bounded function f. Let*

$$f(x) \;=\; \begin{cases} x^2 \sin(1/x^2), & x \neq 0,\ x \in [-1, 1] \\ 0, & x = 0 \end{cases}$$

- *Prove this function has a removable discontinuity at 0.*

- *Prove f is even differentiable on $[-1, 1]$ with derivative*

$$f'(x) \;=\; \begin{cases} 2x \sin(1/x^2) - (2/x)\cos(1/x^2), & x \neq 0,\ x \in [-1, 1] \\ 0, & x = 0 \end{cases}$$

- *Prove f' is* not *bounded on $[-1, 1]$ and hence it cannot be Riemann Integrable.*

- *Now connect this to the idea of antiderivatives, by letting g be defined by*

$$g(x) \;=\; \begin{cases} 2x \sin(1/x^2) - (2/x)\cos(1/x^2), & x \neq 0,\ x \in [-1, 1] \\ 0, & x = 0 \end{cases}$$

Note g is f'.

Define G by

$$G(x) \;=\; \begin{cases} x^2 \sin(1/x^2), & x \neq 0,\ x \in [-1, 1] \\ 0, & x = 0 \end{cases}$$

Prove G is the antiderivative of g even though g itself does not have a Riemann integral.

Again, the point is that the idea of the antiderivative of a function is intellectually distinct from that of being Riemann integrable.

Exercise 2.1.13 *Repeat the arguments of the previous exercise for the bounded function f. Let*

$$f(x) \;=\; \begin{cases} 3x^2 \sin(5/x^2), & x \neq 0,\ x \in [-1, 1] \\ 0, & x = 0 \end{cases}$$

Exercise 2.1.14 *The fundamental theorem of calculus is often the only antiderivative we have. Illustrate this point by looking at the antiderivatives of the functions h_{ab} infinitely differentiable bump functions we defined in (Peterson (8) 2020).*

2.2 The Riemann Integral of Functions with Jumps

Now let's look at the Riemann integral of functions which have points of discontinuity.

2.2.1 Removable Discontinuity

Consider the function f defined on $[-2, 5]$ by

$$f(t) \;=\; \begin{cases} 2t & -2 \le t < 0 \\ 1 & t = 0 \\ (1/5)t^2 & 0 < t \le 5 \end{cases}$$

Let's calculate $F(t) = \int_{-2}^{t} f(s)\, ds$. This will have to be done in several parts because of the way f is defined.

1. On the interval $[-2, 0]$, note that f is continuous except at one point, $t = 0$. Hence, f is Riemann integrable by the existence theorem. Also, the function $2t$ is continuous on this interval and so is also Riemann integrable. Then since f on $[-2, 0]$ and $2t$ match at all but one point on $[-2, 0]$, their Riemann integrals must match. Hence, if t is in $[-2, 0]$, we compute F as follows:

$$\begin{aligned} F(t) &= \int_{-2}^{t} f(s)\, ds \\ &= \int_{-2}^{t} 2s\, ds \\ &= s^2 \Big|_{-2}^{t} \\ &= t^2 - (-2)^2 = t^2 - 4 \end{aligned}$$

2. On the interval $[0, 5]$, note that f is continuous except at one point, $t = 0$. Hence, f is Riemann integrable. Also, the function $(1/5)t^2$ is continuous on this interval and is therefore also Riemann integrable. Then since f on $[0, 5]$ and $(1/5)t^2$ match at all but one point on $[0, 5]$, their Riemann integrals must match. Hence, if t is in $[0, 5]$, we compute F as follows:

$$\begin{aligned} F(t) &= \int_{-2}^{t} f(s)\, ds \\ &= \int_{-2}^{0} f(s)\, ds + \int_{0}^{t} f(s)\, ds \\ &= \int_{-2}^{0} 2s\, ds + \int_{0}^{t} (1/5)s^2\, ds \\ &= s^2 \Big|_{-2}^{0} + (1/15)s^3 \Big|_{0}^{t} \\ &= -4 + t^3/15 \end{aligned}$$

Thus, we have found that

$$F(t) \;=\; \begin{cases} t^2 - 4 & -2 \le t < 0 \\ t^3/15 - 4 & 0 < t \le 5 \end{cases}$$

Note, we didn't define F at $t = 0$ yet. Since f is Riemann Integrable on $[-2, 5]$, we know from the Fundamental Theorem of Calculus that F must be continuous. Let's check. F is clearly continuous on either side of 0 and we note that $\lim_{t \to 0^-} F(t)$ which is $F(0^-)$ is -4 which is exactly the value of $F(0^+)$. Hence, F is indeed continuous at 0 and we can write

$$F(t) = \begin{cases} t^2 - 4 & -2 \le t \le 0 \\ t^3/15 - 4 & 0 \le t \le 5 \end{cases}$$

What about the differentiability of F? The Fundamental Theorem of Calculus guarantees that F has a derivative at each point where f is continuous and at those points $F'(t) = f(t)$. Hence, we know this is true at all t except 0. Note at those t, we find

$$F'(t) = \begin{cases} 2t & -2 \le t < 0 \\ (1/5)t^2 & 0 < t \le 5 \end{cases}$$

which is exactly what we expect. Also, note $F'(0^-) = 0$ and $F'(0^+) = 0$ as well. Hence, since the right- and left-hand derivatives match, we see $F'(0)$ does exist and has the value 0. But this is not the same as $f(0) = 1$. Note, F is **not** the antiderivative of f on $[-2, 5]$ because of this mismatch.

2.2.2 Jump Discontinuity

Now consider the function f defined on $[-2, 5]$ by

$$f(t) = \begin{cases} 2t & -2 \le t < 0 \\ 1 & t = 0 \\ 2 + (1/5)t^2 & 0 < t \le 5 \end{cases}$$

Let's calculate $F(t) = \int_{-2}^{t} f(s)\, ds$. Again, this will have to be done in several parts because of the way f is defined.

1. On the interval $[-2, 0]$, note that f is continuous except at one point, $t = 0$. Hence, f is Riemann integrable. Also, the function $2t$ is continuous on this interval and hence is also Riemann integrable. Then since f on $[-2, 0]$ and $2t$ match at all but one point on $[-2, 0]$, their Riemann integrals must match. Hence, if t is in $[-2, 0]$, we compute F as follows:

$$\begin{aligned} F(t) &= \int_{-2}^{t} f(s)\, ds \\ &= \int_{-2}^{t} 2s\, ds \\ &= s^2 \Big|_{-2}^{t} \\ &= t^2 - (-2)^2 = t^2 - 4 \end{aligned}$$

2. On the interval $[0, 5]$, note that f is continuous except at one point, $t = 0$. Hence, f is Riemann integrable. Also, the function $2 + (1/5)t^2$ is continuous on this interval and so is also Riemann integrable. Then since f on $[0, 5]$ and $2 + (1/5)t^2$ match at all but one point on $[0, 5]$, their Riemann integrals must match. Hence, if t is in $[0, 5]$, we compute F as follows:

$$F(t) = \int_{-2}^{t} f(s)\, ds$$

$$
\begin{aligned}
&= \int_{-2}^{0} f(s)\, ds + \int_{0}^{t} f(s)\, ds \\
&= \int_{-2}^{0} 2s\, ds + \int_{0}^{t} (2 + (1/5)s^2)\, ds \\
&= s^2 \Big|_{-2}^{0} + (2s + (1/15)s^3)\Big|_{0}^{t} \\
&= -4 + 2t + t^3/15
\end{aligned}
$$

Thus, we have found that

$$
F(t) = \begin{cases} t^2 - 4 & -2 \le t < 0 \\ -4 + 2t + t^3/15 & 0 < t \le 5 \end{cases}
$$

As before, we didn't define F at $t = 0$ yet. Since f is Riemann Integrable on $[-2, 5]$, we know from the Fundamental Theorem of Calculus that F must be continuous. F is clearly continuous on either side of 0 and we note that $\lim_{t \to 0^-} F(t)$ which is $F(0^-)$ is -4, which is exactly the value of $F(0^+)$. Hence, F is indeed continuous at 0 and we can write

$$
F(t) = \begin{cases} t^2 - 4 & -2 \le t \le 0 \\ -4 + 2t + t^3/15 & 0 \le t \le 5 \end{cases}
$$

What about the differentiability of F? The Fundamental Theorem of Calculus guarantees that F has a derivative at each point where f is continuous and at those points $F'(t) = f(t)$. Hence, we know this is true at all t except 0. Note at those t, we find

$$
F'(t) = \begin{cases} 2t & -2 \le t < 0 \\ 2 + (1/5)t^2 & 0 < t \le 5 \end{cases}
$$

which is exactly what we expect. However, when we look at the one-sided derivatives, we find $F'(0^-) = 0$ and $F'(0^+) = 2$. Hence, since the right- and left-hand derivatives do not match, we see $F'(0)$ does not exist. Finally, note F is **not** the antiderivative of f on $[-2, 5]$ because of this mismatch.

Homework

Exercise 2.2.1 *Compute $\int_{-3}^{t} f(s)\, ds$ for*

$$
f(t) = \begin{cases} 3t & -3 \le t < 0 \\ 6 & t = 0 \\ (1/6)t^2 & 0 < t \le 6 \end{cases}
$$

1. *Graph f and F carefully labeling all interesting points.*

2. *Verify that F is continuous and differentiable at all points but $F'(0)$ does not match $f(0)$ and so F is not the antiderivative of f on $[-3, 6]$.*

Exercise 2.2.2 *Compute $\int_{-3}^{t} f(s)\, ds$ for*

$$
f(t) = \begin{cases} 3t & -3 \le t < 0 \\ 6 & t = 0 \\ (1/6)t^2 + 2 & 0 < t \le 6 \end{cases}
$$

1. *Graph f and F carefully labeling all interesting points.*

2. *Verify that F is continuous and differentiable at all points except 0 and so F is not the antiderivative of f on $[-3, 6]$.*

Chapter 3

Functions of Bounded Variation

Now that we have seen a quick overview of what Riemann Integration entails, let's go back and look at it very carefully. This will enable us to extend it to a more general form of integration called **Riemann - Stieljtes**. From what we already know about Riemann integrals, the Riemann integral is a mapping ϕ which is linear and whose domain is some subspace of the vector space of all bounded functions. Let $B([a, b])$ denote this vector space which is a normed linear space using the usual infinity norm. The set of all Riemann Integrable functions can be denoted by the symbol $RI([a, b])$ and we know it is a subspace of $B([a, b])$. We also know that the subspace $C([a, b])$ of all continuous functions on $[a, b]$ is contained in $RI([a, b])$. In fact, if $PC([a, b])$ is the set of all functions on $[a, b]$ that are piecewise continuous, then $PC([a, b])$ is also a vector subspace contained in $RI([a, b])$. Hence, we know $\phi : RI([a, b]) \subseteq B([a, b]) \to \Re$ is a linear functional on the subspace $RI([a, b])$. Also, if f is not zero, then

$$\frac{\mid \int_a^b f(t) \, dt \mid}{\| f \|_\infty} \leq \frac{\int_a^b \mid f(t) \mid \, dt}{\| f \|_\infty}$$
$$\leq \frac{\int_a^b \| f \|_\infty \, dt}{\| f \|_\infty} = b - a$$

Thus, we see that $\| \phi \|_{op}$ is finite and ϕ is a bounded linear functional on a subspace of $B[a, b]$ if we use the infinity norm on $RI([a, b])$. But of course, we can choose other norms. There are clearly many functions in $B([a, b])$ that do not fit nicely into the development process for the Riemann Integral. So let $NI([a, b])$ denote a new subspace of functions which contains $RI([a, b])$. We know that the Riemann integral satisfies an important idea in analysis called **limit interchange**. That is, if a sequence of functions $\{f_n\}$ from $RI([a, b])$ converges in infinity norm to f that the following facts hold:

1. f is also in $RI([a, b])$.

2. The classic limit interchange holds:

$$\lim_{n \to \infty} \int_a^b f_n(t) \, dt = \int_a^b \left(\lim_{n \to \infty} f_n(t) \right) \, dt$$

We can say this more abstractly as this: if $f_n \to f$ in $\| \cdot \|_\infty$ in $RI([a, b])$, then f remains in $RI([a, b])$ and

$$\lim_{n \to \infty} \phi(f_n) = \phi \left(\lim_{n \to \infty} f_n \right)$$

But if we wanted to extend ϕ to the larger subspace $NI([a,b])$ in such a way that it remained a bounded linear functional, we would also want to know what kind of sequence convergence we should use in order for the interchange ideas to work. There are lots of questions:

1. Do we need to impose a norm on our larger subspace $NI([a,b])$?

2. Can we characterize the subspace $NI([a,b])$ in some fashion?

3. If the extension is called $\hat{\phi}$, we want to make sure that $\hat{\phi}$ is exactly ϕ when we restrict our attention to functions in $RI([a,b])$?

Also, do we have to develop integration only on finite intervals $[a,b]$ of $\Re$? How do we extend the ideas of integration to other kinds of subsets? How do we even extend traditional Riemann integration to unbounded intervals of $\Re$? All of these questions will be answered in the upcoming chapters, but first we will see how far we can go with the traditional Riemann approach. We will also see where the Riemann integral approach breaks down and makes us start to think of more general tools so that we can get our work done.

3.1 Partitions

Definition 3.1.1 Partition

> A **partition** of the finite interval $[a,b]$ is a finite collection of points, $\{x_0, \ldots, x_n\}$, ordered so that $a = x_0 < x_1 < \cdots < x_n = b$. We denote the partition by π and call each point x_i a partition point.

For each $j = 1, \ldots, n-1$, we let $\Delta x_j = x_{j+1} - x_j$. The collection of all finite partitions of $[a,b]$ is denoted $\Pi[a,b]$.

Definition 3.1.2 Partition Refinements

> The partition $\pi_1 = \{y_0, \ldots, y_m\}$ is said to be a **refinement** of the partition $\pi_2 = \{x_0, \ldots, x_n\}$ if every partition point $x_j \in \pi_2$ is also in π_1. If this is the case, then we write $\pi_2 \preceq \pi_1$, and we say that π_1 is finer than π_2 or π_2 is coarser than π_1.

Definition 3.1.3 Common Refinement

> Given $\pi_1, \pi_2 \in \Pi[a,b]$, there is a partition $\pi_3 \in \Pi[a,b]$ which is formed by taking the union of π_1 and π_2 and using common points only once. We call this partition the common refinement of π_1 and π_2 and denote it by $\pi_3 = \pi_1 \vee \pi_2$.

Comment 3.1.1 *The relation $\preceq$ is a partial ordering of $\Pi[a,b]$. It is not a total ordering, since not all partitions are comparable. There is a coarsest partition, also called the trivial partition. It is given by $\pi_0 = \{a,b\}$. We may also consider uniform partitions of order k. Let $h = (b-a)/k$. Then $\pi = \{x_0 = a, x_0 + h, x_0 + 2h, \ldots, x_{k-1} = x_0 + (k-1)h, x_k = b\}$.*

Proposition 3.1.1 Refinements and Common Refinements

> If $\pi_1, \pi_2 \in \Pi[a,b]$, then $\pi_1 \preceq \pi_2$ if and only if $\pi_1 \vee \pi_2 = \pi_2$.

Proof 3.1.1
If $\pi_1 \preceq \pi_2$, then $\pi_1 = \{x_0, \ldots, x_p\} \subset \{y_0, \ldots, y_q\} = \pi_2$. Thus, $\pi_1 \cup \pi_2 = \pi_2$, and we have $\pi_1 \vee \pi_2 = \pi_2$. Conversely, suppose $\pi_1 \vee \pi_2 = \pi_2$. By definition, every point of π_1 is also a point of $\pi_1 \vee \pi_2 = \pi_2$. So, $\pi_1 \preceq \pi_2$. ∎

Definition 3.1.4 The Gauge or Norm of a Partition

For $\pi \in \Pi[a, b]$, we define the **gauge** of π, denoted $\|\pi\|$, by $\|\pi\| = \max\{\Delta x_j : 1 \leq j \leq p\}$.

Homework

Exercise 3.1.1 *Prove that the relation $\preceq$ is a partial ordering of $\Pi[a, b]$.*

Exercise 3.1.2 *Fix $\pi_1 \in \Pi[a, b]$. The set $C(\pi_1) = \{\pi \in \Pi[a, b] : \pi_1 \preceq \pi\}$ is called the **core** determined by π_1. It is the set of all partitions of $[a, b]$ that contain (or are finer than) π_1.*

1. *Prove that if $\pi_1 \preceq \pi_2$, then $C(\pi_2) \subset C(\pi_1)$.*

2. *Prove that if $\|\pi_1\| < \epsilon$, then $\|\pi\| < \epsilon$ for all $\pi \in C(\pi_1)$.*

3. *Prove that if $\|\pi_1\| < \epsilon$ and $\pi_2 \in \Pi[a, b]$, then $\|\pi_1 \vee \pi_2\| < \epsilon$.*

3.2 Monotone Functions

In our investigations of how monotone functions behave, we will need two fundamental facts about *infimum* and *supremum* of a set of numbers which are given in Lemma 3.2.1 and Lemma 3.2.2. First, recall the Infimum Tolerance Lemma:

Lemma 3.2.1 The Infimum Tolerance Lemma

Let S be a nonempty set of numbers that is bounded below. Then given any tolerance ϵ, there is at least one element s in S so that

$$\inf(S) \leq s < \inf(S) + \epsilon$$

Proof 3.2.1
This is an easy proof by contradiction. Assume there is some ϵ so that no matter what s from S we choose, we have

$$s \geq \inf(S) + \epsilon$$

This says that $\inf(S) + \epsilon$ is a lower bound for S and so by definition, $\inf(S)$ must be bigger than or equal to this lower bound. But this is clearly not possible. So the assumption that such a tolerance ϵ exists is wrong and the conclusion follows. ∎

Then, recall the Supremum Tolerance Lemma:

Lemma 3.2.2 The Supremum Tolerance Lemma

Let T be a nonempty set of numbers that is bounded above. Then given any tolerance ϵ, there is at least one element t in T so that

$$\sup(T) - \epsilon < t \leq \sup(T)$$

Proof 3.2.2
This again is an easy proof by contradiction and we include it for completeness. Assume there is

32   BASIC ANALYSIS IV: MEASURE THEORY AND INTEGRATION

some ϵ so that no matter what t from T we choose, we have

$$t \leq \sup(T) - \epsilon$$

This says that $\sup(T) - \epsilon$ is an upper bound for T and so by definition, $\sup(T)$ must be less than or equal to this upper bound. But this is clearly not possible. So the assumption that such a tolerance ϵ exists is wrong and the conclusion must follow. ∎

We are now in a position to discuss carefully monotone functions and other functions built from them. We follow discussions in (Douglas (2) 1996) at various places.

Definition 3.2.1 Monotone Functions

A real-valued function $f : [a, b] \to \Re$ is said to be increasing (respectively, strictly increasing) if $x_1, x_2 \in [a, b]$, $x_1 < x_2 \Rightarrow f(x_1) \leq f(x_2)$ (respectively, $f(x_1) < f(x_2)$). Similar definitions hold for decreasing and strictly decreasing functions.

Theorem 3.2.3 A Monotone Function Estimate

Let f be increasing on $[a, b]$, and let $\pi = \{x_0, \ldots, x_p\}$ be in $\Pi[a, b]$. For any $c \in [a, b]$, define

$$f(c^+) = \lim_{x \to c^+} f(x) \quad and \quad f(c^-) = \lim_{x \to c^-} f(x)$$

where we define $f(a^-) = f(a)$ and $f(b^+) = f(b)$. Then

$$\sum_{j=0}^{p} [f(x_j^+) - f(x_j^-)] \leq f(b) - f(a)$$

Proof 3.2.3

First, we note that $f(x^+)$ and $f(x^-)$ always exist. The proof of this is straightforward. For $x \in (a, b]$, let $T_x = \{f(y) : a \leq y < x\}$. Then T_x is bounded above by $f(x)$, since f is monotone increasing. Hence, T_x has a well-defined supremum. Let $\epsilon > 0$ be given. Then, using the Supremum Tolerance Lemma, Lemma 3.2.2, there is a $y^* \in [a, x)$ such that $\sup T_x - \epsilon < f(y^*) \leq \sup T_x$. For any $y \in (y^*, x)$, we have $f(y^*) \leq f(y)$ since f is increasing. Thus, $0 \leq (\sup T_x - f(y)) \leq (\sup T_x - f(y^*)) < \epsilon$ for $y \in (y^*, x)$. Let $\delta = (x - y^*)/2$. Then, if $0 < x - y < \delta$, $\sup T_x - f(y) < \epsilon$. Since ϵ was arbitrary, this shows that $\lim_{y \to x^-} f(y) = \sup T_x$. The proof for $f(x^+)$ is similar, using the Infimum Tolerance Lemma, Lemma 3.2.1. You should be able to see that $f(x^-)$ is less than or equal to $f(x^+)$ for all x. We will define $f(a^-) = f(a)$ and $f(b^+) = f(b)$ since f is not defined prior to a or after b.

To prove the stated result holds, first choose an arbitrary $y_j \in (x_j, x_{j+1})$ for each $j = 0, \ldots, p - 1$. Then, since f is increasing, for each $j = 1, \ldots, p$, we have $f(y_{j-1}) \leq f(x_j^-) \leq f(x_j^+) \leq f(y_j)$. Thus,

$$f(x_j^+) - f(x_j^-) \leq f(y_j) - f(y_{j-1}) \tag{3.1}$$

We also have $f(a) \leq f(a^+) \leq f(y_0)$ and $f(y_{p-1}) \leq f(b^-) \leq f(b)$. Thus, it follows that

$$
\begin{aligned}
\sum_{j=0}^{p} \left(f(x_j^+) - f(x_j^-) \right) &= f(x_0^+) - f(x_0^-) + \sum_{j=1}^{p-1} [f(x_j^+) - f(x_j^-)] + f(x_p^+) - f(x_p^-) \\
&\leq f(a^+) - f(a^-) + \sum_{j=1}^{p-1} [f(y_j) - f(y_{j-1})] + f(b^+) - f(b^-)
\end{aligned}
$$

Using Equation 3.1 and replacing x_0 by a and x_p with b, we then note the sum on the right-hand side collapses to $f(y_{p-1}) - f(y_0)$. Finally, since $f(a^-) = f(a)$ and $f(b^+) = f(b)$, we obtain

$$
\begin{aligned}
\sum_{j=0}^{p} \left(f(x_j^+) - f(x_j^-) \right) &\leq f(a^+) - f(a) + f(y_{p-1}) - f(y_0) + f(b) - f(b^-) \\
&\leq f(y_0) - f(a) + f(y_{p-1}) - f(b^-) + f(b) - f(y_0) \\
&\leq f(b) - f(a) + f(y_{p-1}) - f(b^-)
\end{aligned}
$$

But $f(y_{p-1}) - f(b^-) \leq 0$, so

$$
\sum_{j=0}^{p} \left(f(x_j^+) - f(x_j^-) \right) \leq f(b) - f(a)
$$

∎

Theorem 3.2.4 A Monotone Function Has a Countable Number of Discontinuities

If f is monotone on $[a, b]$, the set of discontinuities of f is countable.

Proof 3.2.4
For concreteness, we assume f is monotone increasing. The decreasing case is shown similarly. Since f is monotone increasing, the only types of discontinuities it can have are jump discontinuities. If $x \in [a, b]$ is a point of discontinuity, then the size of the jump is given by $f(x^+) - f(x^-)$. Define $D_k = \{x \in (a, b) : f(x^+) - f(x^-) > 1/k\}$, for each integer $k \geq 1$. We want to show that D_k is finite.

Select any finite subset S of D_k and label the points in S by $x_1, \ldots, x_p$ with $x_1 < x_2 < \cdots < x_p$. If we add the point $x_0 = a$ and $x_{p+1} = b$, these points determine a partition π. Hence, by Theorem 3.2.3, we know that

$$
\sum_{j=1}^{p} [f(x_j^+) - f(x_j^-)] \leq \sum_{\pi} [f(x_j^+) - f(x_j^-)] \leq f(b) - f(a)
$$

But each jump satisfies $f(x_j^+) - f(x_j^-) > 1/k$ and there are a total of p such points in S. Thus, we must have

$$
p/k < \sum_{j=1}^{p} [f(x_j^+) - f(x_j^-)] \leq f(b) - f(a)
$$

Hence, $p/k < f(b) - f(a)$, implying that $p < k[f(b) - f(a)]$. Thus, the cardinality of S is bounded *above by the fixed constant $k[f(b) - f(a)]$. Let $\hat{N}$ be the first positive integer bigger than or equal to $k[f(b) - f(a)]$. If the cardinality of D_k were infinite, then there would be a subset T of D_k with cardinality $\hat{N} + 1$. The argument above would then tell us that $\hat{N} + 1 \le k[f(b) - f(a)] \le \hat{N}$ giving a contradiction. Thus, D_k must be a finite set. This means that $D = \cup_{k=1}^{\infty} D_k$ is countable also.*

Finally, if x is a point where f is not continuous, then $f(x^+) - f(x^-) > 0$. Hence, there is a positive integer k_0 so that $f(x^+) - f(x^-) > 1/k_0$. This means x is in D_{k_0} and so is in D. ■

Definition 3.2.2 The Discontinuity Set of a Monotone Function

Let f be monotone increasing on $[a, b]$. We will let S denote the set of discontinuities of f on $[a, b]$. We know this set is countable by Theorem 3.2.4 so we can label it as $S = \{x_j\}$. Define functions u and v on $[a, b]$ by

$$u(x) = \begin{cases} 0, & x = a \\ f(x) - f(x^-), & x \in (a, b] \end{cases}$$

$$v(x) = \begin{cases} f(x^+) - f(x), & x \in [a, b) \\ 0, & x = b \end{cases}$$

In Figure 3.1, we show a monotone increasing function with several jumps. You should be able to compute u and v easily at these jumps.

There are several very important points to make about these functions u and v which are listed below.

Comment 3.2.1

1. *Note that $u(x)$ is the left-hand jump of f at $x \in (a, b]$ and $v(x)$ is the right-hand jump of f at $x \in [a, b)$.*

2. *Both u and v are non-negative functions and $u(x) + v(x) = f(x^+) - f(x^-)$ is the total jump in f at x, for $x \in (a, b)$.*

3. *Moreover, f is continuous at x from the left if and only if $u(x) = 0$, and f is continuous from the right at x if and only if $v(x) = 0$.*

4. *Finally, f is continuous on $[a, b]$ if and only if $u(x) = v(x) = 0$ on $[a, b]$.*

Homework

Exercise 3.2.1 *Let f be defined by*

$$f(x) = \begin{cases} 2x, & 0 \le x < 1 \\ 3, & 1 = x \\ 5x^2 + 1, & 1 < x \le 2 \end{cases}$$

Find the set of discontinuities of f and compute u and v for f.

Exercise 3.2.2 *Let f be defined by*

$$f(x) = \begin{cases} 2, & x = -1 \\ x+4, & -1 < x < 1 \\ 6, & 1 = x \\ x^2 + 7, & 1 < x < 2 \\ 12, & x = 2 \\ 2x^3, & 2 < x \leq 3 \end{cases}$$

Find the set of discontinuities of f and compute u and v for f.

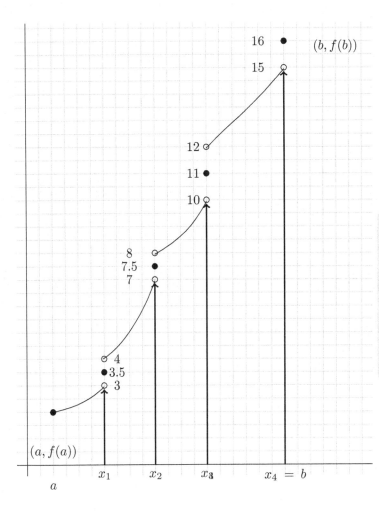

Figure 3.1: The function $F(x)$.

Exercise 3.2.3 *Let f_n, for $n \geq 2$, be defined by*

$$f_n(x) = \begin{cases} 1, & x = 1 \\ 1/(k+1), & 1/(k+1) \leq x < 1/k,\ 1 \leq k < n \\ 0, & 0 \leq x < 1/n \end{cases}$$

Find the set of discontinuities of f_n and compute u and v for f_n.

BASIC ANALYSIS IV: MEASURE THEORY AND INTEGRATION

Exercise 3.2.4 *Let f be the pointwise limit of the sequence of functions (f_n) for $n \geq 2$ defined by*

$$f_n(x) \;=\; \begin{cases} 1, & x = 1 \\ 1/(k+1), & 1/(k+1) \leq x < 1/k, \; 1 \leq k < n \\ 0, & 0 \leq x < 1/n \end{cases}$$

- *Prove f is monotonic.*

- *Find the set of discontinuities of f and compute u and v for f.*

3.2.1 The Saltus Function

Now, let S_0 be any finite subset of S. From Theorem 3.2.3, we have

$$\sum_{x \in S_0} f(x^+) - f(x^-) \;\leq\; f(b) - f(a)$$

This implies

$$\sum_{x \in S_0} u(x) + v(x) \;\leq\; f(b) - f(a)$$

$$\sum_{x \in S_0} u(x) \;+\; \sum_{x \in S_0} v(x) \;\leq\; f(b) - f(a)$$

The above tells us that the set of numbers we get by evaluating this sum over finite subsets of S is bounded above by the number $f(b) - f(a)$. Hence, $\sum_{j=1}^{n} u(x_j)$ and $\sum_{j=1}^{n} v(x_j)$ are bounded above by $f(b) - f(a)$ for all n. Thus, these sets of numbers have a finite supremum. But u and v are non-negative functions, so these sums form monotonically increasing sequences. Hence, these sequences converge to their supremum which we label as $\sum_{j=1}^{\infty} u(x_j)$ and $\sum_{j=1}^{\infty} v(x_j)$.

Now, consider a nonempty subset, T, of $[a, b]$, and suppose $F \subset S \cap T$ is finite. Then, by the arguments already presented, we know that

$$\sum_{x_j \in F} u(x_j) + \sum_{x_j \in F} v(x_j) \leq f(b) - f(a) \tag{3.2}$$

This implies

$$\sum_{x_j \in F} u(x_j) \;\leq\; f(b) - f(a) \qquad \text{and} \qquad \sum_{x_j \in F} v(x_j) \;\leq\; f(b) - f(a)$$

From this, it follows that

$$\sum_{x_j \in S \cap T} u(x_j) \;=\; \sup\{ \sum_{x_j \in F} u(x_j) \;:\; F \subset S \cap T, \; F \text{ finite}\}$$

Likewise, we also have

$$\sum_{x_j \in S \cap T} v(x_j) \;=\; \sup\{ \sum_{x_j \in F} v(x_j) \;:\; F \subset S \cap T, F \text{ finite}\}$$

Definition 3.2.3 The Saltus Function Associated with a Monotone Function

For $x, y \in [a, b]$ with $x < y$, define

$$\begin{aligned}
S[x, y] &= S \cap [x, y], \; S[x, y) = S \cap [x, y) \\
S(x, y] &= S \cap (x, y] \; S(x, y) = S \cap (x, y)
\end{aligned}$$

Then, define the function $S_f : [a, b] \to \Re$ by

$$S_f(x) = \begin{cases} f(a), & x = a \\ f(a) + \sum_{x_j \in S(a, x]} u(x_j) + \sum_{x_j \in S[a, x)} v(x_j), & a < x \le b \end{cases}$$

We call S_f the **Saltus Function** associated with the monotone increasing function f.

Intuitively, $S_f(x)$ is the sum of all of the jumps (i.e. discontinuities) up to and including the left-hand jump at x. In essence, it is a generalization of the idea of a *step function*.

Theorem 3.2.5 Properties of the Saltus Function

Let $f : [a, b] \to \Re$ be monotone increasing. Then

1. S_f is monotone increasing on $[a, b]$;

2. if $x < y$, with $x, y \in [a, b]$, then $0 \le S_f(y) - S_f(x) \le f(y) - f(x)$;

3. S_f is continuous on $S^c \cap [a, b]$ where S^c is the complement of the set S.

Proof 3.2.5
Suppose $x < y$. Then

$$\begin{aligned}
S_f(y) - S_f(x) &= \sum_{x_j \in S(a, y]} u(x_j) - \sum_{x_j \in S(a, x]} u(x_j) + \sum_{x_j \in S[a, y)} v(x_j) - \sum_{x_j \in S[a, x)} v(x_j) \\
&= \sum_{x_j \in S(x, y]} u(x_j) + \sum_{x_j \in S[x, y)} v(x_j) \\
&\ge 0
\end{aligned}$$

This proves the first statement. Now, suppose $x, y \in [a, b]$ with $x < y$. Let F be a subset of $[a, b]$ that consists of a finite number of points of the form $F = \{x_0 = x, x_1, \ldots, x_p = y\}$, such that $x = x_0 < x_1 < \cdots < x_p = y$. In other words, F is a partition of $[x, y]$. Then, by Equation 3.2 we know

$$\sum_{x_j \in F \cap S(x, y]} u(x_j) + \sum_{x_j \in F \cap S[x, y)} v(x_j) \le f(y) - f(x)$$

Taking the supremum of the left-hand side over all such sets, F, we obtain

$$\sum_{x_j \in S(x, y]} u(x_j) + \sum_{x_j \in S[x, y)} v(x_j) \le f(y) - f(x)$$

But by the remarks made in the first part of this proof, this sum is exactly $S_f(y) - S_f(x)$. We conclude that $S_f(y) - S_f(x) \le f(y) - f(x)$ as desired.

Finally, let x be a point in $S^c \cap [a, b]$. Then f is continuous at x, so, given $\epsilon > 0$, there is a $\delta > 0$ such that $y \in [a, b]$ and $|x - y| < \delta \Rightarrow |f(x) - f(y)| < \epsilon$. But by the second part of this proof, we *have $|S_f(x) - S_f(y)| \leq |f(y) - f(x)| < \epsilon$. Thus, S_f is continuous at x.* ∎

Homework

Exercise 3.2.5 *Let f be defined by*

$$f(x) = \begin{cases} 2x, & 0 \leq x < 1 \\ 3, & 1 = x \\ 5x^2 + 1, & 1 < x \leq 2 \end{cases}$$

Compute the Saltus function for f.

Exercise 3.2.6 *Let f be defined by*

$$f(x) = \begin{cases} 2, & x = -1 \\ x + 4, & -1 < x < 1 \\ 6, & 1 = x \\ x^2 + 7, & 1 < x < 2 \\ 12, & x = 2 \\ 2x^3, & 2 < x \leq 3 \end{cases}$$

Compute the Saltus function for f.

Exercise 3.2.7 *Let f_n, for $n \geq 2$, be defined by*

$$f_n(x) = \begin{cases} 1, & x = 1 \\ 1/(k+1), & 1/(k+1) \leq x < 1/k, \ 1 \leq k < n \\ 0, & 0 \leq x < 1/n \end{cases}$$

Compute the Saltus function for f_n.

Exercise 3.2.8 *Let f be the pointwise limit of the sequence of functions (f_n) for $n \geq 2$ defined by*

$$f_n(x) = \begin{cases} 1, & x = 1 \\ 1/(k+1), & 1/(k+1) \leq x < 1/k, \ 1 \leq k < n \\ 0, & 0 \leq x < 1/n \end{cases}$$

Compute the Saltus function for f.

So, why do we care about S_f? The function S_f measures, in a sense, the degree to which f fails to be continuous. If we subtract S_f from f, we would be subtracting its discontinuities, resulting in a continuous function that behaves similarly to f.

3.2.2 The Continuous Part of a Monotone Function

Definition 3.2.4 The Continuous Part of a Monotone Function

Define $f_c : [a, b] \to \Re$ by $f_c(x) = f(x) - S_f(x)$.

Theorem 3.2.6 Properties of f_c

1. f_c *is monotone on $[a, b]$.*

2. f_c *is continuous also.*

BOUNDED VARIATION

Proof 3.2.6

The proof that f_c is monotone is left to you as an exercise with this generous hint:

Hint 3.2.1 *Note if $x < y$ in $[a, b]$, then*

$$f_c(y) - f_c(x) = (f(y) - f(x)) - (S_f(y) - S_f(x))$$

The right-hand side is non-negative by Theorem 3.2.5. □

To prove f_c is continuous is a bit tricky. We will do most of the proof but leave a few parts for you to fill in.

Pick any x in $[a, b)$ and any positive ϵ. Since the $f(x^+)$ exists, there is a positive δ so that $0 \leq f(y) - f(x^+) < \epsilon$ if $x < y < x + \delta$. Thus, for such y,

$$
\begin{aligned}
f_c(y) - f_c(x) &= [f(y) - S_f(y)] - [f(x) - S_f(x)] \\
&= f(y) - \left\{ \sum_{x_j \in S(a,y]} u(x_j) + \sum_{x_j \in S[a,y)} v(x_j) \right\} \\
&\quad - f(x) + \left\{ \sum_{x_j \in S(a,x]} u(x_j) + \sum_{x_j \in S[a,x)} v(x_j) \right\}
\end{aligned}
$$

Recall, $S(a, y] = S(a, x] \cup S(x, y]$ and $S[a, y) = S[a, x) \cup S[x, y)$. So,

$$
f_c(y) - f_c(x) = f(y) - \left\{ \sum_{x_j \in S(x,y]} u(x_j) + \sum_{x_j \in S[x,y)} v(x_j) \right\} - f(x)
$$

Now, the argument reduces to two cases:

1. *If y and x are points of discontinuity, we get*

$$
\begin{aligned}
f_c(y) - f_c(x) &= f(y) - u(y) - \left\{ \sum_{x_j \in S(x,y)} u(x_j) + \sum_{x_j \in S(x,y)} v(x_j) \right\} - f(x) - v(x) \\
&= f(y) - (f(y) - f(y^-)) - \left\{ \sum_{x_j \in S(x,y)} u(x_j) + \sum_{x_j \in S(x,y)} v(x_j) \right\} \\
&\quad - f(x) - (f(x^+) - f(x)) \\
&\leq f(y^-) - f(x^+) \\
&\leq f(y) - f(x^+) < \epsilon
\end{aligned}
$$

2. *If either x and/ or y are not a point of discontinuity, a similar argument holds*

Thus, we see f_c is continuous from the right at this x. Now use a similar argument to show continuity from the left at x. Together, these arguments show f_c is continuous at x. ∎

Homework

40 BASIC ANALYSIS IV: MEASURE THEORY AND INTEGRATION

Exercise 3.2.9 *Let f be defined by*

$$f(x) \;=\; \begin{cases} 2x, & 0 \le x < 1 \\ 3, & 1 = x \\ 5x^2 + 1, & 1 < x \le 2 \end{cases}$$

Compute f_c.

Exercise 3.2.10 *Let f be defined by*

$$f(x) \;=\; \begin{cases} 2, & x = -1 \\ x + 4, & -1 < x < 1 \\ 6, & 1 = x \\ x^2 + 7, & 1 < x < 2 \\ 12, & x = 2 \\ 2x^3, & 2 < x \le 3 \end{cases}$$

Compute f_c.

Exercise 3.2.11 *Let f_n, for $n \ge 2$, be defined by*

$$f_n(x) \;=\; \begin{cases} 1, & x = 1 \\ 1/(k+1), & 1/(k+1) \le x < 1/k,\ 1 \le k < n \\ 0, & 0 \le x < 1/n \end{cases}$$

Compute $(f_n)_c$.

Exercise 3.2.12 *Let f be the pointwise limit of the sequence of functions (f_n) for $n \ge 2$ defined by*

$$f_n(x) \;=\; \begin{cases} 1, & x = 1 \\ 1/(k+1), & 1/(k+1) \le x < 1/k,\ 1 \le k < n \\ 0, & 0 \le x < 1/n \end{cases}$$

Compute the Saltus function for f_c.

Example 3.2.1 *Let's define f on $[0, 2]$ by*

$$f(x) \;=\; \begin{cases} -2 & x = 0 \\ x^3 & 0 < x < 1 \\ 9/8 & x = 1 \\ x^4/4 + 1 & 1 < x < 2 \\ 7 & x = 2 \end{cases}$$

1. *Find u and v.*

2. *Find S_f.*

3. *Find f_c.*

First, note $f(0^-) = -2$, $f(0) = -2$ and $f(0^+) = 0$ and so 0 is a point of discontinuity. Further, $f(1^-) = 1$, $f(1) = 9/8$ and $f(1^+) = 5/4$ giving another point of discontinuity at 1. Finally, since $f(2^-) = 5$, $f(2) = 7$ and $f(2^+) = 7$, there is a third point of discontinuity at 2. So, the set of discontinuities of f is $S = \{0, 1, 2\}$. Thus,

$$S(0,x] = \begin{cases} \emptyset & 0 < x < 1 \\ \{1\} & 1 \leq x < 2 \\ \{1,2\} & 2 = x \end{cases} \quad \text{and } S[0,x) = \begin{cases} \{0\} & 0 < x \leq 1 \\ \{0,1\} & 1 < x \leq 2 \end{cases}$$

Also,

$$u(x) = \begin{cases} 0 & x = 0 \\ 0 & 0 < x < 1 \\ 9/8 - 1 = 1/8 & x = 1 \\ 0 & 1 < x < 2 \\ 7 - 5 = 2 & 2 = x \end{cases} \quad \text{and } v(x) = \begin{cases} 0 - (-2) = 2 & x = 0 \\ 0 & 0 < x < 1 \\ 5/4 - 9/8 = 1/8 & x = 1 \\ 0 & 1 < x < 2 \\ 0 & 2 = x \end{cases}$$

Now, here

$$S_f(x) = \begin{cases} f(0) = -2, & x = 0 \\ f(0) + \sum_{x_j \in S(0,x]} u(x_j) + \sum_{x_j \in S[0,x)} v(x_j) & 0 < x \leq 2 \end{cases}$$

Thus,

$$S_f(x) = \begin{cases} -2, & x = 0 \\ -2 + v(0) = -2 + 2 = 0 & 0 < x < 1 \\ -2 + u(1) + v(0) = -2 + 1/8 + 2 = 1/8 & x = 1 \\ -2 + u(1) + v(0) + v(1) = -2 + 1/8 + 2 + 1/8 = 1/4 & 1 < x < 2 \\ -2 + u(1) + u(2) + v(0) + v(1) = -2 + 1/8 + 2 + 2 + 1/8 = 9/4 & x = 2 \end{cases}$$

So, S_f is the nice step function and $f_c = f - S_f$ gives

$$S_f(x) = \begin{cases} -2, & x = 0 \\ 0 & 0 < x < 1 \\ 1/8 & x = 1 \\ 1/4 & 1 < x < 2 \\ 9/4 & x = 2 \end{cases} \quad \text{and } f_c(x) = \begin{cases} -2 - (-2) = 0 & x = 0 \\ x^3 - 0 = x/3 & 0 < x < 1 \\ 9/8 - 1/8 = 1 & x = 1 \\ x^4/4 + 1 - 1/4 = x^4/4 + 3/4 & 1 < x < 2 \\ 7 - 9/4 = 19/4 & x = 2 \end{cases}$$

We see f_c is continuous on $[0,2]$.

Example 3.2.2 *Now, we can compute the Riemann integral of the function from the previous example $[0,2]$; i.e. let's calculate $F(t) = \int_0^t f(x)\,dx$. This will have to be done in several parts because of the way f is defined.*

1. *On the interval $[0,1]$, note that f is continuous except at two points, $x = 0$ and $x = 1$. Hence, f is Riemann integrable. Also, the function x^3 is continuous on this interval and so is also Riemann integrable. Then since f on $[0,1]$ and x^3 match at all but two points on $[0,2]$, their Riemann integrals must match. Hence, if t is in $[-2,0]$, we compute F as follows:*

$$F(t) = \int_0^t f(x)\,dx = \int_0^t x^3\,dx = x^4/4 \Big|_0^t = t^4/4$$

2. *On the interval $[1,2]$, note that f is continuous except at the two points, $x = 1$ and $x = 2$. Hence, f is Riemann integrable. Also, the function $1 + x^4/4$ is continuous on this interval and so is also Riemann integrable. Then since f on $[1,2]$ and $1 + x^4/4$ match at all but two points on $[1,2]$, their Riemann integrals must match. Hence, if t is in $[1,2]$, we compute F as*

follows:

$$F(t) = \int_0^t f(x)\, dx = \int_0^1 f(x)\, dx + \int_1^t f(s)\, ds$$

$$= \int_0^1 x^3\, dx + \int_1^t (1 + x^4/4)\, dx = x^4/4 \, |_0^1 + (x + x^5/5)\, |_1^t$$

$$= 1/4 + (t + t^5/5) - (1 + 1/5) = t^5/5 + t - 19/20$$

Thus, we have found that

$$F(t) = \begin{cases} t^4/4 & 0 \le t \le 1 \\ t^5/5 + t - 19/20 & 1 \le t \le 2 \end{cases}$$

Note, we know from the Fundamental Theorem of Calculus that F must be continuous. To check this at an interesting point such as $t = 1$, note F is clearly continuous on either side of 1 and we note that $\lim_{t \to 1^-} F(t)$ which is $F(1^-)$ is 1/4, which is exactly the value of $F(1^+)$. Hence, F is indeed continuous at 1!

Example 3.2.3 *For the F found above, what about its differentiability? The Fundamental Theorem of Calculus guarantees that F has a derivative at each point where f is continuous and at those points $F'(t) = f(t)$. Hence, we know this is true at all t except 0, 1 and 2 because these are points of discontinuity of f. F' is nicely defined at 0 and 1 as a one-sided derivative and at all other t save 1 by*

$$F'(t) = \begin{cases} t^3 & 0 \le t < 1 \\ t^4 + 1 & 0 < t \le 2 \end{cases}$$

*However, when we look at the one-sided derivatives, we find $F'(0^+) = 0 \ne f(0) = -2$, $F'(2^-) = 17 \ne f(2) = 7$ and $F'(1^-) = 1$ and $F'(1^+) = 2$ giving $F'(1)$ does not even exist. Thus, note F is **not** the antiderivative of f on $[0, 2]$ because of this mismatch.*

Homework

Exercise 3.2.13 *Prove f_c is monotone.*

Exercise 3.2.14 *Let's define f by*

$$f(x) = \begin{cases} -1 & x = 0 \\ x^2 & 0 < x < 1 \\ 7/4 & x = 1 \\ \sqrt{x + 3} & 1 < x < 2 \\ 3 & x = 2 \end{cases}$$

1. *Find u and v.*

2. *Find S_f.*

3. *Find f_c.*

4. *Do a nice graph of u, v, f, f_c and S_f.*

5. *Following the discussion in Section 2.2, explain how to compute the Riemann Integral of f and find its value (yes, this is in the careful rigorous section, and so this problem is a bit out of place, but we will be dotting all of our i's and crossing all of our t's soon enough!).*

BOUNDED VARIATION

Exercise 3.2.15 *Let's define f by*

$$f(x) = \begin{cases} -1 & x = -5 \\ 2x^2 & -5 < x < 5 \\ 60 & x = 5 \\ 20x + 3 & 5 < x < 10 \\ 300 & x = 10 \end{cases}$$

1. *Find u and v.*

2. *Find S_f.*

3. *Find f_c.*

4. *Do a nice graph of u, v, f, f_c and S_f.*

5. *Following the discussion in Section 2.2, explain how to compute the Riemann Integral of f and find its value.*

Exercise 3.2.16 *Let f be defined by*

$$f(x) = \begin{cases} 2x, & 0 \le x < 1 \\ 3, & 1 = x \\ 5x^2 + 1, & 1 < x \le 2 \end{cases}$$

Find the set of discontinuities of f and compute u and v for f.

Exercise 3.2.17 *Let f be defined by*

$$f(x) = \begin{cases} 2, & x = -1 \\ x + 4, & -1 < x < 1 \\ 6, & 1 = x \\ x^2 + 7, & 1 < x < 2 \\ 12, & x = 2 \\ 2x^3, & 2 < x \le 3 \end{cases}$$

Find the set of discontinuities of f and compute u and v for f.

Exercise 3.2.18 *Let f_n, for $n \ge 2$, be defined by*

$$f_n(x) = \begin{cases} 1, & x = 1 \\ 1/(k+1), & 1/(k+1) \le x < 1/k, \ 1 \le k < n \\ 0, & 0 \le x < 1/n \end{cases}$$

Find the set of discontinuities of f_n and compute u and v for f_n.

Exercise 3.2.19 *Let f be the pointwise limit of the sequence of functions (f_n) for $n \ge 2$ defined by*

$$f_n(x) = \begin{cases} 1, & x = 1 \\ 1/(k+1), & 1/(k+1) \le x < 1/k, \ 1 \le k < n \\ 0, & 0 \le x < 1/n \end{cases}$$

- *Prove f is monotonic.*

- *Find the set of discontinuities of f and compute u and v for f.*

3.3 Functions of Bounded Variation

The next important topic for us is to consider the class of functions of *bounded variation*. We will develop this classically here, but in later chapters, we will define similar concepts using abstract measures. We are going to find out that functions of bounded variation can also be represented as the difference of two increasing functions and that their classical derivative exists everywhere, except a set of measure zero (yes, that idea is not defined yet, but we believe in teasers!). Let's get on with it.

Definition 3.3.1 Functions of Bounded Variation

> *Let $f : [a, b] \to \Re$ and let $\pi \in \Pi[a, b]$ be given by $\pi = \{x_0 = a, x_1, \ldots, x_p = b\}$. Define $\Delta f_j = f(x_j) - f(x_{j-1})$ for $1 \leq j \leq p$. If there exists a positive real number, M, such that $\sum_\pi |\Delta f_j| \leq M$ for all $\pi \in \Pi[a, b]$, then we say that f is of **bounded variation** on $[a, b]$. The set of all functions of bounded variation on the interval $[a, b]$ is denoted by the symbol $BV([a, b])$.*

Comment 3.3.1

1. *Note, saying a function f is of bounded variation is equivalent to saying the set $\{\sum_\pi |\Delta f_j| : \pi \in \Pi[a, b]\}$ is bounded, and, therefore, has a supremum.*

2. *Also, if f is of bounded variation on $[a, b]$, then, for any $x \in (a, b)$, the set $\{a, x, b\}$ is a partition of $[a, b]$. Hence, there exists $M > 0$ such that $|f(x) - f(a)| + |f(b) - f(x)| \leq M$. But this implies*

$$|f(x)| - |f(a)| \leq |f(x) - f(a)| + |f(b) - f(x)| \leq M$$

This tells us that $|f(x)| \leq |f(a)| + M$. Since our choice of x in $[a, b]$ was arbitrary, this shows that f is bounded, i.e. $\|f\|_\infty < \infty$.

We can state the comments above formally as Theorem 3.3.1.

Theorem 3.3.1 Functions of Bounded Variation are Bounded

> *If f is of bounded variation on $[a, b]$, then f is bounded on $[a, b]$.*

Theorem 3.3.2 Monotone Functions are of Bounded Variation

> *If f is monotone on $[a, b]$, then $f \in BV([a, b])$.*

Proof 3.3.1
As usual, we assume, for concreteness, that f is monotone increasing. Let $\pi \in \Pi[a, b]$. Hence, we can write $\pi = \{x_0 = a, x_1, \ldots, x_{p-1}, x_p = b\}$. Then

$$\sum_\pi |\Delta f_j| = \sum_\pi |f(x_j) - f(x_{j-1})|$$

Since f is monotone increasing, the absolute value signs are unnecessary, so that

$$\sum_\pi |\Delta f_j| = \sum_\pi \Delta f_j = \sum_\pi (f(x_j) - f(x_{j-1}))$$

But this is a telescoping sum, so

$$\sum_{\pi} \Delta f_j \;=\; f(x_p) - f(x_0) = f(b) - f(a)$$

Since the partition π was arbitrary, it follows that $\sum_{\pi} \Delta f_j \leq f(b) - f(a)$ for all $\pi \in \Pi[a,b]$. This implies that $f \in BV([a,b])$, for if $f(b) > f(a)$, then we can simply let $M = f(b) - f(a)$. If $f(b) = f(a)$, then f must be constant, and we can let $M = f(b) - f(a) + 1 = 1$. In either case, $f \in BV([a,b])$. ∎

Theorem 3.3.3 Bounded Differentiable Implies Bounded Variation

Suppose $f \in C([a,b])$, f is differentiable on (a,b), and $\| f' \|_{\infty} < \infty$. Then $f \in BV([a,b])$.

Proof 3.3.2
Let $\pi \in \Pi[a,b]$ so that $\pi = \{x_0 = a, x_1, \ldots, x_p = b\}$. On each subinterval $[x_{j-1}, x_j]$, for $1 \leq j \leq p$, the hypotheses of the Mean Value Theorem are satisfied. Hence, there is a point, $y_j \in (x_{j-1}, x_j)$, with $\Delta f_j = f(x_j) - f(x_{j-1}) = f'(y_j)\Delta x_j$. So, we have

$$|\Delta f_j| \;=\; |f'(y_j)|\,\Delta x_j \leq B\Delta x_j$$

where B is the bound on f' that we assume exists by hypothesis. Thus, for any $\pi \in \Pi[a,b]$, we have

$$\sum_{\pi} |\Delta f_j| \leq B \sum_{\pi} \Delta x_j \;=\; B(b-a) < \infty$$

Therefore, $f \in BV([a,b])$. ∎

Definition 3.3.2 The Total Variation of a Function of Bounded Variation

Let $f \in BV([a,b])$. The real number

$$V(f;a,b) \;=\; \sup\left\{ \sum_{\pi} |\Delta f_j| : \pi \in \Pi[a,b] \right\}$$

*is called the **Total Variation** of f on $[a,b]$.*

Note that this number always exists if $f \in BV([a,b])$.

Comment 3.3.2 *For any $f \in BV([a,b])$, we clearly have $V(f;a,b) = V(-f;a,b)$ and $V(f;a,b) \geq 0$. Moreover, we also see that $V(f;a,b) = 0$ if and only if f is constant on $[a,b]$.*

Theorem 3.3.4 Functions of Bounded Variation are Closed under Addition

If f and g are in $BV([a,b])$, then so are $f \pm g$, and $V(f \pm g;a,b) \leq V(f;a,b) + V(g;a,b)$.

Proof 3.3.3
Let $\pi \in \Pi[a,b]$, so that $\pi = \{x_0 = a, x_1, \ldots, x_p = b\}$. Consider $f + g$ first. We have, for each $1 \leq j \leq p$,

$$|\Delta(f+g)_j| \;=\; |(f+g)(x_j) - (f+g)(x_{j-1})|$$

$$\leq \ |f(x_j) - f(x_{j-1})| + |g(x_j) - g(x_{j-1})|$$
$$\leq \ |\Delta f_j| + |\Delta g_j|$$

This implies that, for any $\pi \in \Pi[a,b]$,

$$\sum_\pi |\Delta(f+g)_j| \ \leq \ \sum_\pi |\Delta f_j| + \sum_\pi |\Delta g_j|$$

Both quantities on the right-hand side are bounded by $V(f;a,b)$ and $V(g;a,b)$, respectively. Since $\pi \in \Pi[a,b]$ was arbitrary, we have

$$V(f+g;a,b) \ \leq \ V(f;a,b) + V(g;a,b)$$

This shows that $f+g \in BV([a,b])$ and proves the desired inequality for that case. Since $V(-g;a,b) = V(g;a,b)$, we also have

$$V(f-g;a,b) \leq V(f;a,b) + V(-g;a,b) \ = \ V(f;a,b) + V(g;a,b)$$

which proves that $f - g \in BV([a,b])$. ∎

Theorem 3.3.5 Products of Functions of Bounded Variation are of Bounded Variation

> *If $f,g \in BV([a,b])$, then $fg \in BV([a,b])$ and $V(fg;a,b) \leq \|g\|_\infty V(f;a,b) + \|f\|_\infty V(g;a,b)$.*

Proof 3.3.4

By Theorem 3.3.1, we know that f and g are bounded. Hence, the numbers $\|f\|_\infty$ and $\|g\|_\infty$ exist and are finite. Let $h = fg$, and let $\pi = \{x_0 = a, x_1, \ldots, x_p = b\}$ be any partition. Then

$$\begin{aligned}
|\Delta h_j| &= |f(x_j)g(x_j) - f(x_{j-1})g(x_{j-1})| \\
&= |f(x_j)g(x_j) - g(x_j)f(x_{j-1}) + g(x_j)f(x_{j-1}) - f(x_{j-1})g(x_{j-1})| \\
&\leq |g(x_j)||\Delta f_j| + |f(x_{j-1})||\Delta g_j| \\
&\leq \|g\|_\infty |\Delta f_j| + \|f\|_\infty |\Delta g_j|
\end{aligned}$$

Thus,

$$\begin{aligned}
\sum_\pi |\Delta h_j| &\leq \|g\|_\infty \sum_\pi |\Delta f_j| + \|f\|_\infty \sum_\pi |\Delta g_j| \\
&\leq \|g\|_\infty V(f;a,b) + \|f\|_\infty V(g;a,b)
\end{aligned}$$

Since π was arbitrary, we see the right-hand side is an upper bound for all the partition sums and hence, the supremum of all these sums must also be less than or equal to the right-hand side. Thus,

$$V(fg;a,b) \ \leq \ \|g\|_\infty V(f;a,b) + \|f\|_\infty V(g;a,b)$$

∎

Comment 3.3.3 *Note that we have verified that $BV([a,b])$ is a commutative algebra (i.e. a ring) of functions with an identity, since the constant function $f = 1$ is of bounded variation.*

It is natural to ask, then, what the units are in this algebra. That is, what functions have multiplicative inverses?

BOUNDED VARIATION

Theorem 3.3.6 Inverses of Functions of Bounded Variation

Let f be in $BV([a,b])$, and assume that there is a positive m such that $|f(x)| \geq m > 0$ for all $x \in [a,b]$. Then $1/f \in BV([a,b])$ and $V(1/f;a,b) \leq (1/m^2)V(f;a,b)$.

Proof 3.3.5

Let $\pi = \{x_0 = a, x_1, \ldots, x_p\}$ be any partition. Then

$$
\begin{aligned}
\left| \Delta\left(\frac{1}{f}\right)_j \right| &= \left| \frac{1}{f(x_j)} - \frac{1}{f(x_{j-1})} \right| \\
&= \left| \frac{f(x_{j-1}) - f(x_j)}{f(x_j)f(x_{j-1})} \right| \\
&= \frac{|\Delta f_j|}{|f(x_j)||f(x_{j-1})|} \\
&\leq \frac{\Delta f_j}{m^2}
\end{aligned}
$$

Thus, we have

$$
\sum_\pi \left| \Delta\left(\frac{1}{f}\right)_j \right| \leq \frac{1}{m^2} \sum_\pi |\Delta f_j|
$$

implying that $V(1/f;a,b) \leq (1/m^2)V(f;a,b)$. ∎

Comment 3.3.4

1. *Any polynomial, p, is in $BV([a,b])$, and p is a unit if none of its zeros occur in the interval.*

2. *Any rational function p/q where p and q are of bounded variation on $[a,b]$, is in $BV([a,b])$ as long as none of the zeros of q occur in the interval.*

3. *$e^x \in BV([a,b])$. In fact, $e^{u(x)} \in BV([a,b])$ if $u(x)$ is monotone or has a bounded derivative.*

4. *$\sin x$ and $\cos x$ are in $BV([a,b])$ by Theorem 3.3.3.*

5. *$\tan x \in BV([a,b])$ if $[a,b]$ does not contain any point of the form $(2k+1)\pi/2$ for $k \in \mathbf{Z}$, by Theorem 3.3.6.*

6. *The function*

$$
f(x) = \begin{cases} \sin\frac{1}{x}, & 0 < x \leq 1 \\ 0, & x = 0 \end{cases}
$$

is not in $BV([0,1])$. To see this, choose partition points $\{x_0, \ldots, x_p\}$ by $x_0 = 0$, $x_p = 1$, and

$$
x_j = \frac{2}{\pi(2p - 2j + 1)}, \qquad 1 \leq j \leq p - 1
$$

Then

$$
\begin{aligned}
\Delta f_1 &= \sin\left(\frac{\pi(2p-1)}{2}\right) = \pm 1 \\
\Delta f_2 &= \sin\left(\frac{\pi(2p-3)}{2}\right) - \sin\left(\frac{\pi(2p-1)}{2}\right) = \pm 2
\end{aligned}
$$

48   BASIC ANALYSIS IV: MEASURE THEORY AND INTEGRATION

and continuing, we find

$$\begin{aligned}
\Delta f_{p-1} &= \sin\left(\frac{\pi(2p - 2(p-1) + 1)}{2}\right) - \sin\left(\frac{\pi(2p - 2(p-2) + 1)}{2}\right) = \pm 2 \\
\Delta f_p &= \sin 1 - \sin(3\pi/2) = \sin 1 + 1
\end{aligned}$$

Thus,

$$\begin{aligned}
\sum_{\pi} &= |\Delta f_1| + \sum_{j=2}^{p-1} |\Delta f_j| + \sin 1 + 1 \\
&= 2(p-1) + \sin 1
\end{aligned}$$

Hence, we can make the value of this sum as large as we desire and so this function is not of bounded variation.

Homework

Exercise 3.3.1 *Prove that if f is of bounded variation on the finite interval $[a, b]$, then αf is also of bounded variation for any scalar α. Do this proof using the partition approach.*

Exercise 3.3.2 *Prove that if f and g are of bounded variation on the finite interval $[a, b]$, then $\alpha f + \beta g$ is also of bounded variation for any scalars α and β. Do this proof using the partition approach. Note, these two exercises essentially show $BV([a, b])$ is a vector space.*

Exercise 3.3.3 *Prove $BV([a, b])$ is a complete normed linear space with norm $|| \cdot ||$ defined by*

$$|| f || = | f(a) | + V(f, a, b)$$

Exercise 3.3.4 *Define f on $[0, 1]$ by*

$$f(x) = \begin{cases} x^2 \cos(x^{-2}) & x \neq 0 \in [0, 1] \\ 0 & x = 0 \end{cases}$$

Prove that f is differentiable on $[0, 1]$ but is not of bounded variation. This is a nice example of something we will see later. This f is a function which is continuous but not absolutely continuous.

3.4   The Total Variation Function

Theorem 3.4.1 The Total Variation is Additive on Intervals

> *If $f \in BV([a, b])$ and $c \in [a, b]$, then $f \in BV([a, c])$, $f \in BV([c, b])$, and $V(f; a, b) = V(f; a, c) + V(f; c, b)$. That is, the total variation, V, is additive on intervals.*

Proof 3.4.1

The case $c = a$ or $c = b$ is easy, so we assume $c \in (a, b)$. Let $\pi_1 \in \Pi[a, c]$ and $\pi_2 \in \Pi[c, b]$ with $\pi_1 = \{x_0 = a, x_1, \ldots, x_p = c\}$ and $\pi_2 = \{y_0 = c, y_1, \ldots, y_q = b\}$. Then $\pi_1 \vee \pi_2$ is a partition of $[a, b]$ and we know

$$\sum_{\pi_1 \vee \pi_2} |\Delta f_j| = \sum_{\pi_1} |\Delta f_j| + \sum_{\pi_2} |\Delta f_j| \leq V(f; a, b)$$

BOUNDED VARIATION

Dropping the π_2 term, and noting that $\pi_1 \in \Pi[a,c]$ was arbitrary, we see that

$$\sup_{\pi_1 \in \Pi[a,c]} \sum_{\pi_1} |\Delta f_j| \leq V(f;a,b)$$

which implies that $V(f;a,c) \leq V(f;a,b) < \infty$. Thus, $f \in BV([a,c])$. A similar argument shows that $V(f;c,b) \leq V(f;a,b)$, so $f \in BV([c,b])$.

Finally, since both π_1 and π_2 were arbitrary and we know that

$$\sum_{\pi_1} |\Delta f_j| + \sum_{\pi_2} |\Delta f_j| \leq V(f;a,b)$$

we see that $V(f;a,c) + V(f;c,b) \leq V(f;a,b)$.

Now we will establish the reverse inequality. Let $\pi \in \Pi[a,b]$, so that $\pi = \{x_0 = a, x_1, \ldots, x_p = b\}$. First, assume that c is a partition point of π, so that $c = x_{k_0}$ for some k_0. Thus, $\pi = \{x_0, \ldots, x_{k_0}\} \cup \{x_{k_0}, \ldots, x_p\}$. Let $\pi_1 = \{x_0, \ldots, x_{k_0}\} \in \Pi[a,c]$ and let $\pi_2 = \{x_{k_0}, \ldots, x_p\} \in \Pi[c,b]$. From the first part of our proof, we know that $f \in BV([a,c])$ and $f \in BV([c,b])$, so

$$\begin{aligned} \sum_{\pi} |\Delta f_j| &= \sum_{\pi_1} |\Delta f_j| + \sum_{\pi_2} |\Delta f_j| \\ &\leq V(f;a,c) + V(f;c,b) \end{aligned}$$

Since $\pi \in \Pi[a,b]$ was arbitrary, it follows that $V(f;a,b) \leq V(f;a,c) + V(f;c,b)$. For the other case, suppose c is not a partition point of π. Then c must lie inside one of the subintervals. That is, $c \in (x_{k_0-1}, x_{k_0})$ for some k_0. Let $\pi' = \{x_0, \ldots, x_{k_0-1}, c, x_{k_0}, \ldots, x_p\}$ be a new partition of $[a,b]$. Then π' refines π. Apply our previous argument to conclude that

$$\sum_{\pi'} |\Delta f_j| \leq V(f;a,c) + V(f;c,b)$$

Finally, we note that

$$\sum_{\pi} |\Delta f_j| \leq \sum_{\pi'} |\Delta f_j|,$$

since

$$|f(x_{k_0}) - f(x_{k_0-1})| \leq |f(x_{k_0}) - f(c)| + |f(c) - f(x_{k_0-1})|$$

Thus, we have

$$\sum_{\pi} |\Delta f_j| \leq V(f;a,c) + V(f;c,b)$$

Since π was arbitrary, it follows that $V(f;a,b) \leq V(f;a,c) + V(f;c,b)$. Combining these two inequalities, we see the result is established. ∎

Definition 3.4.1 The Variation Function of a Function f of Bounded Variation

BASIC ANALYSIS IV: MEASURE THEORY AND INTEGRATION

Let $f \in BV([a,b])$. *The* **Variation Function** *of f, or simply the* **Variation** *of f, is the function $V_f : [a,b] \to \Re$ defined by*

$$V_f(x) \;=\; \begin{cases} 0, & x = a \\ V(f;a,x), & a < x \leq b \end{cases}$$

Theorem 3.4.2 V_f and $V_f - f$ are Monotone for a Function f of Bounded Variation

If $f \in BV([a,b])$, then the functions V_f and $V_f - f$ are monotone increasing on $[a,b]$.

Proof 3.4.2

Pick $x_1, x_2 \in [a,b]$ with $x_1 < x_2$. By Theorem 3.4.1, $f \in BV([a,x_1])$ and $f \in BV([a,x_2])$. Apply this same theorem to the interval $[a,x_1] \cup [x_1,x_2]$ to conclude that $f \in BV([x_1,x_2])$. Thus

$$V_f(x_2) \;=\; V(f;a,x_2) = V(f;a,x_1) + V(f;x_1,x_2) \;=\; V_f(x_1) + V(f;x_1,x_2)$$

It follows that $V_f(x_2) - V_f(x_1) = V(f;x_1,x_2) \geq 0$, so V_f is monotone increasing. Now, consider $(V_f - f)(x_2) - (V_f - f)(x_1)$. We have

$$\begin{aligned} (V_f - f)(x_2) - (V_f - f)(x_1) &= V_f(x_2) - V_f(x_1) - (f(x_2) - f(x_1)) \\ &= V(f;a,x_2) - V(f;a,x_1) - (f(x_2) - f(x_1)) \\ &= V(f;x_1,x_2) - (f(x_2) - f(x_1)) \end{aligned}$$

But $\{x_1, x_2\}$ is the trivial partition of $[x_1, x_2]$, so

$$\sum_{\{x_1,x_2\}} |\, \Delta f_j \,| \;\leq\; \sup_{\pi \in \Pi[x_1,x_2]} \sum_{\pi} |\, \Delta f_j \,| = V(f;x_1,x_2)$$

Thus, $V(f;x_1,x_2) - (f(x_2) - f(x_1)) \geq 0$, implying that $V_f - f$ is monotone increasing. ∎

Theorem 3.4.3 A Function of Bounded Variation is the Difference of Two Increasing Functions

Every $f \in BV([a,b])$ can be written as the difference of two monotone increasing functions on $[a,b]$. In other words,

$$BV([a,b]) = \{f : [a,b] \to \Re \mid \exists u,v : [a,b] \to \Re, \; u,v \text{ monotone increasing}, \; f = u - v\}$$

Proof 3.4.3

If $f = u - v$, where u and v are monotone increasing, then u and v are of bounded variation. Since $BV([a,b])$ is an algebra, it follows that $f \in BV([a,b])$.

Conversely, suppose $f \in BV([a,b])$, and let $u = V_f$ and $v = V_f - f$. Then u and v are monotone increasing and $u - v = f$. ∎

Comment 3.4.1 *Theorem 3.4.3 tells us if g is of bounded variation on $[a,b]$, then $g = u - v$ where u and v are monotone increasing. Thus, we can also use the Saltus decomposition of u and v to conclude*

$$f \;=\; (u_c + S_u) \;-\; (v_c + S_v)$$

BOUNDED VARIATION 51

$$= (u_c - v_c) + (S_u - S_v)$$

The first term is the difference of two continuous functions of bounded variation and the second term is the difference of Saltus functions. This is essentially another form of decomposition theorem for a function of bounded variation.

Homework

Exercise 3.4.1 *Let $f(x) = 2x + 5$. Compute the total variation of f on $[-1.5]$.*

Exercise 3.4.2 *Let $f(x) = 2x^2 + 5$. Compute the total variation of f on $[1, 8]$.*

Exercise 3.4.3 *Let f be defined by*

$$f(x) = \begin{cases} 2x, & 0 \leq x < 1 \\ 3, & 1 = x \\ 5x^2 + 1, & 1 < x \leq 2 \end{cases}$$

Compute the total variation of f.

Exercise 3.4.4 *Let f be defined by*

$$f(x) = \begin{cases} 2, & x = -1 \\ x + 4, & -1 < x < 1 \\ 6, & 1 = x \\ x^2 + 7, & 1 < x < 2 \\ 12, & x = 2 \\ 2x^3, & 2 < x \leq 3 \end{cases}$$

Compute the total variation of f.

Exercise 3.4.5 *Let f_n, for $n \geq 2$, be defined by*

$$f_n(x) = \begin{cases} 1, & x = 1 \\ 1/(k+1), & 1/(k+1) \leq x < 1/k, 1 \leq k < n \\ 0, & 0 \leq x < 1/n \end{cases}$$

Compute the total variation of f_n.

Exercise 3.4.6 *Let f be the pointwise limit of the sequence of functions (f_n) for $n \geq 2$ defined by*

$$f_n(x) = \begin{cases} 1, & x = 1 \\ 1/(k+1), & 1/(k+1) \leq x < 1/k, 1 \leq k < n \\ 0, & 0 \leq x < 1/n \end{cases}$$

Compute the total variation of f.

Exercise 3.4.7 *Let $f(x) = 2x^2 + 5$. Compute the total variation of f on $[-5, 10]$ and find the standard decomposition of f as a difference of monotone increasing functions.*

3.5 Continuous Functions of Bounded Variation

Theorem 3.5.1 Functions of Bounded Variation Always Possess Right- and Left-Hand Limits

52 BASIC ANALYSIS IV: MEASURE THEORY AND INTEGRATION

> *Let $f \in BV([a,b])$. Then the limit $f(x^+)$ exists for all $x \in [a,b)$ and the limit $f(x^-)$ exists for all $x \in (a,b]$.*

Proof 3.5.1
By Theorem 3.4.2, V_f and $V_f - f$ are monotone increasing. So $V_f(x^+)$ and $(V_f - f)(x^+)$ both exist. Hence,

$$
\begin{aligned}
f(x^+) &= \lim_{x \to x^+} f(x) = \lim_{x \to x^+} [V_f(x) - (V_f - f)(x)] \\
&= \lim_{x \to x^+} V_f(x) + \lim_{x \to x^+} (V_f - f)(x) = V_f(x^+) + (V_f - f)(x^+)
\end{aligned}
$$

So, $f(x^+)$ exists. A similar argument shows that $f(x^-)$ exists. ∎

Theorem 3.5.2 Functions of Bounded Variation Have Countable Discontinuity Sets

> *If $f \in BV([a,b])$, then the set of discontinuities of f is countable.*

Proof 3.5.2
$f = u - v$ where u and v are monotone increasing. By Theorem 3.4.3, $S_1 = \{x \in [a,b] \mid u$ is not continuous at $x\}$ and $S_2 = \{x \in [a,b] \mid v$ is not continuous at $x\}$ are countable. The union of these sets is the set of all the points of possible discontinuity of f, so the set of discontinuities of f is countable. ∎

Theorem 3.5.3 A Function of Bounded Variation Continuous if and only if V_f is Continuous

> *Let $f \in BV([a,b])$. Then f is continuous at $c \in [a,b]$ if and only if V_f is continuous at c.*

Proof 3.5.3
The case where $c = a$ and $c = b$ are easier, so we will only prove the case where $c \in (a,b)$. First, suppose f is continuous at c. We will prove separately that V_f is continuous from the right at c and from the left at c.

Let $\epsilon > 0$ be given. Since f is continuous at c, there is a positive δ such that if x is in $(c - \delta, c + \delta) \subset [a,b]$, then $|f(x) - f(c)| < \epsilon/2$. Now,

$$
V(f; c, b) = \sup_{\pi \in \Pi[c,b]} \sum_{\pi} |\Delta f_j|
$$

So, there is a partition π_0 such that

$$
V(f; c, b) - \frac{\epsilon}{2} < \sum_{\pi_0} |\Delta f_j| \le V(f; c, b) \tag{$*$}
$$

If $\pi_0{}'$ is any refinement of π_0, we see that

$$
\sum_{\pi_0} |\Delta f_j| \le \sum_{\pi_0{}'} |\Delta f_j|
$$

since adding points to π_0 simply increases the sum. Thus,

$$
V(f; c, b) - \frac{\epsilon}{2} < \sum_{\pi_0} |\Delta f_j| \le \sum_{\pi_0{}'} |\Delta f_j| \le V(f; c, b)
$$

BOUNDED VARIATION

for any refinement π_0' of π_0. Now, choose a partition, π_1 which refines π_0 and satisfies $\| \pi_1 \| < \delta$. Then

$$V(f;c,b) - \frac{\epsilon}{2} \; < \; \sum_{\pi_1} | \Delta f_j | \leq V(f;c,b) \qquad (**)$$

So, if $\pi_1 = \{x_0 = c, x_1, \dots, x_p\}$, then $| x_1 - x_0 | < \delta$. Thus, we have $| x_1 - c | < \delta$. It follows that $| f(x_1) - f(c) | < \epsilon/2$. From Equation $**$, we then have

$$
\begin{aligned}
V(f;c,b) - \frac{\epsilon}{2} \; &< \; \sum_{\pi_1} | \Delta f_j | \\
&= \; | f(x_1) - f(c) | + \sum_{\text{rest of } \pi_1} | \Delta f_j | \\
&< \; \frac{\epsilon}{2} + \sum_{\text{rest of } \pi_1} | \Delta f_j | \\
&< \; \frac{\epsilon}{2} + V(f;x_1,b)
\end{aligned}
$$

So, we see that

$$V(f;c,b) - \frac{\epsilon}{2} \; < \; \frac{\epsilon}{2} + V(f;x_1,b)$$

which implies that

$$V(f;c,b) - V(f;x_1,b) \; < \; \frac{\epsilon}{2} + \frac{\epsilon}{2} \; = \; \epsilon$$

But $V(f;c,b) - V(f;x_1,b) = V(f;c,x_1)$ which is the same as $V_f(x_1) - V_f(c)$. Thus, we have

$$V_f(x_1) - V_f(c) \; < \; \epsilon$$

Now V_f is monotone and hence we have shown that if $x \in (c, x_1)$,

$$V_f(x) - V_f(c) \; \leq \; V_f(x_1) - V_f(c) \; < \; \epsilon$$

Since $\epsilon > 0$ was arbitrary, this verifies the right continuity of V_f at c.

The argument for left continuity is similar. We can find a partition π_1 of $[a, c]$ with partition points $\{x_0 = a, x_1, \dots, x_{p-1}, x_p = c\}$ such that $\| \pi_1 \| < \delta$ and

$$
\begin{aligned}
V(f;a,c) - \frac{\epsilon}{2} \; &< \; | f(c) - f(x_{p-1}) | + \sum_{\text{rest of } \pi_1} | \Delta f_j | \\
&\leq \; | f(c) - f(x_{p-1}) | + V(f;a,x_{p-1})
\end{aligned}
$$

Since $\| \pi_1 \| < \delta$, we see as before that $| f(c) - f(x_{p-1}) | < \epsilon/2$. Thus,

$$V(f;a,c) - \frac{\epsilon}{2} \; < \; \frac{\epsilon}{2} + V(f;a,x_{p-1})$$

and it follows that

$$V(f;a,c) - V(f;a,x_{p-1}) \; < \; \epsilon$$

or

$$V_f(c) - V_f(x_{p-1}) < \epsilon$$

Since V_f is monotone, we then have for any x in (x_{p-1}, c) that

$$V_f(c) - V_f(x) < V_f(c) - V_f(x_{p-1}) < \epsilon$$

which shows the left continuity of V_f at c. Hence, V_f is continuous at c.

Conversely, suppose V_f is continuous at $c \in (a, b)$. Given $\epsilon > 0$, there is a positive δ such that $(c - \delta, c + \delta) \subset [a, b]$ and $| V_f(x) - V_f(c) | < \epsilon$ for all $x \in (c - \delta, c + \delta)$. Pick any $x \in (c, c + \delta)$. Then $\{c, x\}$ is a trivial partition of $[c, x]$. Hence

$$0 \leq | f(x) - f(c) | \leq V(f; c, x) \quad = \quad V(f; a, x) - V(f; a, c)$$

or

$$0 \leq | f(x) - f(c) | \quad \leq \quad V_f(x) - V_f(c) < \epsilon$$

Hence, it follows that f is continuous from the right. A similar argument shows that f is continuous from the left. ∎

We immediately have this corollary.

Theorem 3.5.4 f Continuous and Bounded Variation if and Only if V_f and $V_f - f$ are Continuous and Increasing

> *$f \in C([a, b]) \cap BV([a, b])$ if and only if V_f and $V_f - f$ are monotone increasing and continuous.*

Homework

Exercise 3.5.1 *If f is Lipschitz on $[a, b]$ with Lipschitz constant $L > 0$, prove f is of bounded variation and the total variation is bounded by $L(b - a)$.*

Exercise 3.5.2 *If f'' is bounded on $[a, b]$, prove f' is of bounded variation on $[a, b]$.*

Exercise 3.5.3 *Prove functions of bounded variation are continuous a.e.*

Exercise 3.5.4 *If f is continuously differentiable on $[a, b]$, prove $V(f; a, b) \geq \int_a^b |f(x)| dx$.*

Chapter 4

The Theory of Riemann Integration

We will now develop the theory of the Riemann Integral for a bounded function f on the interval $[a, b]$. We have done this carefully in (Peterson (8) 2020) also, but now that we are going to handle integration more abstractly, it is good to have a review of the proofs before we move on to the extension to functions of bounded variation. In Chapter 2, we gave you a roadmap of these kinds of results without the necessary arguments but now we will show the details.

4.1 Defining the Riemann Integral

Definition 4.1.1 The Riemann Sum

*Let $f \in B([a, b])$, and let $\pi \in \Pi[a, b]$ be given by $\pi = \{x_0 = a, x_1, \ldots, x_p = b\}$. Define $\sigma = \{s_1, \ldots, s_p\}$, where $s_j \in [x_{j-1}, x_j]$ for $1 \leq j \leq p$. We call σ an evaluation set, and we denote this by $\sigma \subset \pi$. The **Riemann Sum** determined by the partition π and the evaluation set σ is defined by*

$$S(f, \pi, \sigma) = \sum_\pi f(s_j) \Delta x_j$$

Definition 4.1.2 Riemann Integrability of a Bounded f

*We say $f \in B([a, b])$ is **Riemann Integrable** on $[a, b]$ if there exists a real number, I, such that for every $\epsilon > 0$ there is a partition, $\pi_0 \in \Pi[a, b]$ such that*

$$\mid S(f, \pi, \sigma) - I \mid < \epsilon$$

for any refinement, π, of π_0 and any evaluation set, $\sigma \subset \pi$. We denote this value, I, by

$$I \equiv RI(f; a, b)$$

We denote the set of Riemann integrable functions on $[a, b]$ by $RI([a, b])$. Also, it is readily seen that the number $RI(f; a, b)$ in the definition above, when it exists, is unique. So we can speak of *Riemann Integral* of a function, f. We also have the following conventions.

1. $RI(f; a, b) = -RI(f; b, a)$

2. $RI(f; a, a) = 0$

3. f is called the **integrand**.

Theorem 4.1.1 $RI([a,b])$ **is a Vector Space and** $RI(f; a, b)$ **is a Linear Mapping**

$RI([a,b])$ *is a vector space over* $\Re$ *and the mapping* $I_R : RI([a,b]) \to \Re$ *defined by*

$$I_R(f) = RI(f; a, b)$$

is a linear mapping.

Proof 4.1.1
Let $f_1, f_2 \in RI([a,b])$, *and let* $\alpha, \beta \in \Re$. *For any* $\boldsymbol{\pi} \in \boldsymbol{\Pi}[a,b]$ *and* $\boldsymbol{\sigma} \subset \boldsymbol{\pi}$, *we have*

$$
\begin{aligned}
S(\alpha f_1 + \beta f_2, \boldsymbol{\pi}, \boldsymbol{\sigma}) &= \sum_{\boldsymbol{\pi}} (\alpha f_1 + \beta f_2)(s_j) \Delta x_j \\
&= \alpha \sum_{\boldsymbol{\pi}} f_1(s_j) \Delta x_j + \beta \sum_{\boldsymbol{\pi}} f_2(s_j) \Delta x_j \\
&= \alpha S(f_1, \boldsymbol{\pi}, \boldsymbol{\sigma}) + \beta S(f_2, \boldsymbol{\pi}, \boldsymbol{\sigma})
\end{aligned}
$$

Since f_1 *is Riemann integrable, given* $\epsilon > 0$, *there is a real number* $I_1 = RI(f_1, a, b)$ *and a partition* $\boldsymbol{\pi}_1 \in \boldsymbol{\Pi}[a,b]$ *such that*

$$| S(f_1, \boldsymbol{\pi}, \boldsymbol{\sigma}) - I_1 | < \frac{\epsilon}{2(| \alpha | + 1)} \qquad (*)$$

for all refinements, $\boldsymbol{\pi}$, *of* $\boldsymbol{\pi}_1$, *and all* $\boldsymbol{\sigma} \subset \boldsymbol{\pi}$.

Likewise, since f_2 *is Riemann integrable, there is a real number* $I_2 = RE(f_2; a, b)$ *and a partition* $\boldsymbol{\pi}_2 \in \boldsymbol{\Pi}[a,b]$ *such that*

$$| S(f_2, \boldsymbol{\pi}, \boldsymbol{\sigma}) - I_2 | < \frac{\epsilon}{2(| \beta | + 1)} \qquad (**)$$

for all refinements, $\boldsymbol{\pi}$, *of* $\boldsymbol{\pi}_2$, *and all* $\boldsymbol{\sigma} \subset \boldsymbol{\pi}$.

Let $\boldsymbol{\pi}_0 = \boldsymbol{\pi}_1 \vee \boldsymbol{\pi}_2$. *Then* $\boldsymbol{\pi}_0$ *is a refinement of both* $\boldsymbol{\pi}_1$ *and* $\boldsymbol{\pi}_2$. *So, for any refinement,* $\boldsymbol{\pi}$, *of* $\boldsymbol{\pi}_0$, *and any* $\boldsymbol{\sigma} \subset \boldsymbol{\pi}$, *we have Equation* $*$ *and Equation* $**$ *are valid. Hence,*

$$
\begin{aligned}
| S(f_1, \boldsymbol{\pi}, \boldsymbol{\sigma}) - I_1 | &< \frac{\epsilon}{2(| \alpha | + 1)} \\
| S(f_2, \boldsymbol{\pi}, \boldsymbol{\sigma}) - I_2 | &< \frac{\epsilon}{2(| \beta | + 1)}
\end{aligned}
$$

Thus, for any refinement $\boldsymbol{\pi}$ *of* $\boldsymbol{\pi}_0$ *and any* $\boldsymbol{\sigma} \subset \boldsymbol{\pi}$, *it follows that*

$$
\begin{aligned}
| S(\alpha f_1 + \beta f_2, \boldsymbol{\pi}, \boldsymbol{\sigma}) - (\alpha I_1 + \beta I_2) | &= | \alpha S(f_1, \boldsymbol{\pi}, \boldsymbol{\sigma}) + \beta S(f_2, \boldsymbol{\pi}, \boldsymbol{\sigma}) - \alpha I_1 - \beta I_2 | \\
&\leq | \alpha | \| S(f_1, \boldsymbol{\pi}, \boldsymbol{\sigma}) - I_1 | + | \beta | \| S(f_2, \boldsymbol{\pi}, \boldsymbol{\sigma}) - I_2 | \\
&< | \alpha | \frac{\epsilon}{2(| \alpha | + 1)} + | \beta | \frac{\epsilon}{2(| \beta | + 1)} \\
&< \epsilon
\end{aligned}
$$

RIEMANN INTEGRATION

This shows that $\alpha f_1 + \beta f_2$ is Riemann integrable and that the value of the integral $RI(\alpha f_1 + \beta f_2; a, b)$ is given by $\alpha RI(f_1; a, b) + \beta RI(f_2; a, b)$. It then follows immediately that I_R is a linear mapping. ■

Homework

The following are standard sorts of exercise you should have been exposed to already, but it is a good idea to get back into proofs using partition arguments.

Exercise 4.1.1 *If f and g are Riemann Integrable on $[a, b]$, prove that $5f - 8g$ is also Riemann Integrable on $[a, b]$.*

Exercise 4.1.2 *If f is a continuous function on $[a, b]$, prove that $\phi(f) = \int_a^b f(s)ds$ defines a bounded linear functional on $C([a, b])$ and compute an estimate of its bound.*

Exercise 4.1.3 *If $f \in RI([a, b])$, prove that $\phi(f) = \int_a^b f(s)ds$ defines a bounded linear functional on $RI([a, b])$ and compute an estimate of its bound.*

Theorem 4.1.2 Fundamental Riemann Integral Estimates

> *Let $f \in RI([a, b])$. Let $m = \inf_x f(x)$ and let $M = \sup_x f(x)$. Then*
>
> $$m(b - a) \le RI(f; a, b) \le M(b - a)$$

Proof 4.1.2
If $\pi \in \Pi[a, b]$, then for all $\sigma \subset \pi$, we see that

$$\sum_\pi m\Delta x_j \le \sum_\pi f(s_j)\Delta x_j \le \sum_\pi M\Delta x_j$$

But $\sum_\pi \Delta x_j = b - a$, so

$$m(b - a) \le \sum_\pi f(s_j)\Delta x_j \le M(b - a)$$

or

$$m(b - a) \le S(f, \pi, \sigma) \le M(b - a)$$

for any partition π and any $\sigma \subset \pi$.

Now, let $\epsilon > 0$ be given. Then there exist $\pi_0 \in \Pi[a, b]$ such that for any refinement, π, of π_0 and any $\sigma \subset \pi$,

$$RI(f; a, b) - \epsilon < S(f, \pi, \sigma) < RI(f; a, b) + \epsilon$$

Hence, for any such refinement, π, and any $\sigma \subset \pi$, we have

$$m(b - a) \le S(f, \pi, \sigma) < RI(f; a, b) + \epsilon$$

and

$$M(b - a) \ge S(f, \pi, \sigma) > RI(f; a, b) - \epsilon$$

Since $\epsilon > 0$ is arbitrary, it follows that

$$m(b - a) \le RI(f; a, b) \le M(b - a)$$

58   BASIC ANALYSIS IV: MEASURE THEORY AND INTEGRATION

Theorem 4.1.3 The Riemann Integral is Order Preserving

> *The Riemann integral is order preserving. That is, if $f, f_1, f_2 \in RI([a, b])$, then*
>
> *(i)*
> $$f \geq 0 \Rightarrow RI(f; a, b) \geq 0$$
>
> *(ii)*
> $$f_1 \leq f_2 \Rightarrow RI(f_1; a, b) \leq RI(f_2; a, b)$$

Proof 4.1.3

If $f \geq 0$ on $[a, b]$, then $\inf_x f(x) = m \geq 0$. Hence, by Theorem 4.1.2

$$\int_a^b f(x)dx \geq m(b - a) \geq 0$$

This proves the first assertion. To prove (ii), let $f = f_2 - f_1$. Then $f \geq 0$, and the second result follows from the first. ∎

Homework

Exercise 4.1.4 *Let $R > 0$ and let $B = \{f \in C([a, b]) \,\|f\|_\infty \leq R\}$. Let $\Omega = \{\int_a^b f(s)ds \, f \in B\}$. Prove Ω is bounded by $R(b - a)$.*

Exercise 4.1.5 *Assume x is a continuously differentiable function which satisfies $x'(t) = f(t, x(t))$ for $0 \leq t \leq T$ for some $T > 0$ and f is continuous on $[0, T] \times [-R, R]$ for some $R > 0$. Also, assume f satisfies the growth condition $|f(t, y)| \leq e^{-\alpha t}B$ for some $\alpha > 0$ and $B > 0$ for all $(t, y) \in [0, T] \times [-R, R]$. Prove $\|x\|_\infty \leq |x(0)| + \frac{B}{\alpha}e^{-\alpha a}$.*

Exercise 4.1.6 *Using the setup of the previous exercise, recall the Laplace transform of x is $X(s) = \int_0^\infty e^{-st}x(t)dt$. Prove the Laplace Transform of x exists and determine for what values of s this is true. We assume you remember how the improper integral is defined here although we have not formally discussed it.*

4.2   The Existence of the Riemann Integral

Although we have a definition for what it means for a bounded function to be Riemann integrable, we still do not actually know that $RI([a, b])$ is nonempty! In this section, we will show how we prove that the set of Riemann integrable functions is quite rich and varied.

Definition 4.2.1 Darboux Upper and Lower Sums

Let $f \in B([a,b])$. Let $\pi \in \Pi[a,b]$ be given by $\pi = \{x_0 = a, x_1, \ldots, x_p = b\}$. Define

$$m_j = \inf_{x_{j-1} \le x \le x_j} f(x) \qquad 1 \le j \le p$$

and

$$M_j = \sup_{x_{j-1} \le x \le x_j} f(x) \qquad 1 \le j \le p$$

We define the **Lower Darboux Sum** by

$$L(f,\pi) = \sum_{\pi} m_j \Delta x_j$$

and the **Upper Darboux Sum** by

$$U(f,\pi) = \sum_{\pi} M_j \Delta x_j$$

Comment 4.2.1

1. It is straightforward to see that

$$L(f,\pi) \quad \le \quad S(f,\pi,\sigma) \quad \le \quad U(f,\pi)$$

   for all $\pi \in \Pi[a,b]$.

2. We also have

$$U(f,\pi) - L(f,\pi) \quad = \quad \sum_{\pi}(M_j - m_j)\Delta x_j$$

Theorem 4.2.1 $\pi \preceq \pi'$ **Implies** $L(f,\pi) \le L(f,\pi')$ **and** $U(f,\pi) \ge U(f,\pi')$

If $\pi \preceq \pi'$, that is, if π' refines π, then $L(f,\pi) \le L(f,\pi')$ and $U(f,\pi) \ge U(f,\pi')$.

Proof 4.2.1
The general result is established by induction on the number of points added. It is actually quite an involved induction. Here are some of the details:

Step 1 We prove the proposition for inserting points $\{z_1, \ldots, z_q\}$ into one subinterval of π. The argument consists of

1. The Basis Step where we prove the proposition for the insertion of a single point into one subinterval.

2. The Induction Step where we assume the proposition holds for the insertion of q points into one subinterval and then we show the proposition still holds if an additional point is inserted.

3. With the Induction Step verified, the Principle of Mathematical Induction then tells us that the proposition is true for any refinement of π which places points into one subinterval of π.

Basis:

Proof *Let $\pi \in \Pi[a,b]$ be given by $\{x_0 = a, x_1, \ldots, x_p = b\}$. Suppose we form the refinement, π', by adding a single point x' to π into the interior of the subinterval $[x_{k_0-1}, x_{k_0}]$. Let*

$$
\begin{aligned}
m' &= \inf_{[x_{k_0-1}, x']} f(x) \\
m'' &= \inf_{[x', x_{k_0}]} f(x)
\end{aligned}
$$

Note that $m_{k_0} = \min\{m', m''\}$ and

$$
\begin{aligned}
m_{k_0} \Delta x_{k_0} &= m_{k_0}(x_{k_0} - x_{k_0-1}) \\
&= m_{k_0}(x_{k_0} - x') + m_{k_0}(x' - x_{k_0-1}) \\
&\leq m''(x_{k_0} - x') + m'(x' - x_{k_0-1}) \\
&\leq m'' \Delta x'' + m' \Delta x'
\end{aligned}
$$

where $\Delta x'' = x_{k_0} - x'$ and $\Delta x' = x' - x_{k_0-1}$. It follows that

$$
\begin{aligned}
L(f, \pi') &= \sum_{j \neq k_0} m_j \Delta x_j + m' \Delta x' + m'' \Delta x'' \\
&\geq \sum_{j \neq k_0} m_j \Delta x_j + m_{k_0} \Delta x_{k_0} \\
&\geq L(f, \pi)
\end{aligned}
$$

$\square$

Induction:

Proof *We assume that q points $\{z_1, \ldots, z_q\}$ have been inserted into the subinterval $[x_{k_0-1}, x_{k_0}]$. Let π' denote the resulting refinement of π. We assume that*

$$
L(f, \pi) \leq L(f, \pi')
$$

Let the additional point added to this subinterval be called x' and call π'' the resulting refinement of π'. We know that π' has broken $[x_{k_0-1}, x_{k_0}]$ into $q + 1$ pieces. For convenience of notation, let's label these $q + 1$ subintervals as $[y_{j-1}, y_j]$ where y_0 is x_{k_0-1} and y_{q+1} is x_{k_0} and the y_j values in between are the original z_i points for appropriate indices. The new point x' is thus added to one of these $q + 1$ pieces, call it $[y_{j_0-1}, y_{j_0}]$ for some index j_0. This interval plays the role of the original subinterval in the proof of the Basis *Step. An argument similar to that in the proof of the* Basis *Step then shows us that*

$$
L(f, \pi') \leq L(f, \pi'')
$$

Combining with the first inequality from the Induction *hypothesis, we establish the result. Thus, the* Induction *Step is proved.* $\square$

Step 2 *Next, we allow the insertion of a finite number of points into a finite number of subintervals of π. The induction is now on the number of subintervals.*

1. *The* Basis *Step where we prove the proposition for the insertion of points into one subinterval.*

RIEMANN INTEGRATION

2. *The* Induction *Step where we assume the proposition holds for the insertion of points into q subintervals and then we show the proposition still holds if an additional subinterval has points inserted.*

3. *With the* Induction *Step verified, the Principle of Mathematical Induction then tells us that the proposition is true for any refinement of π which places points into any number of subintervals of π.*

Basis

Proof *Step 1 above gives us the* Basis *Step for this proposition.* □

Induction

Proof *We assume the results hold for p subintervals and show it also holds when one more subinterval is added. Specifically, let π' be the refinement that results from adding points to p subintervals of π. Then the* Induction *hypothesis tells us that*

$$L(f, \pi) \leq L(f, \pi')$$

Let π'' denote the new refinement of π which results from adding more points into one more subinterval of π. Then π'' is also a refinement of π' where all the new points are added to one subinterval of π'. Thus, Step 1 holds for the pair (π', π''). We see

$$L(f, \pi') \leq L(f, \pi'')$$

and the desired result follows immediately. □

A similar argument establishes the result for upper sums. ∎

Theorem 4.2.2 $L(f, \pi_1) \leq U(f, \pi_2)$

> *Let π_1 and π_2 be any two partitions in $\mathbf{\Pi}[a, b]$. Then $L(f, \pi_1) \leq U(f, \pi_2)$.*

Proof 4.2.6
Let $\pi = \pi_1 \vee \pi_2$ be the common refinement of π_1 and π_2. Then, by the previous result, we have

$$L(f, \pi_1) \leq L(f, \pi) \leq U(f, \pi) \leq U(f, \pi_2)$$

∎

Theorem 4.2.2 then allows us to define a new type of integrability for the bounded function f. We begin by looking at the infimum of the upper sums and the supremum of the lower sums for a given bounded function f.

Theorem 4.2.3 The Upper and Lower Darboux Integrals are Finite

> *Let $f \in B([a, b])$. Let $\mathscr{L} = \{L(f, \pi) \mid \pi \in \mathbf{\Pi}[a, b]\}$ and $\mathscr{U} = \{U(f, \pi) \mid \pi \in \mathbf{\Pi}[a, b]\}$. Define $L(f) = \sup \mathscr{L}$, and $U(f) = \inf \mathscr{U}$. Then $L(f)$ and $U(f)$ are both finite. Moreover, $L(f) \leq U(f)$.*

Proof 4.2.7
By Theorem 4.2.2, the set $\mathscr{L}$ is bounded above by any upper sum for f. Hence, it has a finite supremum and so $\sup \mathscr{L}$ is finite. Also, again by Theorem 4.2.2, the set $\mathscr{U}$ is bounded below by any

62   BASIC ANALYSIS IV: MEASURE THEORY AND INTEGRATION

lower sum for f. Hence, $\inf \mathscr{U}$ is finite. Finally, since $L(f) \leq U(f, \pi)$ and $U(f) \geq L(f, \pi)$ for all π, by definition of the infimum and supremum of a set of numbers, we must have $L(f) \leq U(f)$. ∎

Definition 4.2.2 Darboux Lower and Upper Integrals

*Let f be in $B([a,b])$. The **Lower Darboux Integral** of f is defined to be the finite number $L(f) = \sup \mathscr{L}$, and the **Upper Darboux Integral** of f is the finite number $U(f) = \inf \mathscr{U}$.*

We can then define what is meant by a bounded function being *Darboux Integrable* on $[a, b]$.

Definition 4.2.3 Darboux Integrability

Let f be in $B([a,b])$. We say f is Darboux Integrable on $[a, b]$ if $L(f) = U(f)$. The common value is then called the Darboux Integral of f on $[a, b]$ and is denoted by the symbol $DI(f; a, b)$.

Comment 4.2.2 *Not all bounded functions are Darboux Integrable. Consider the function $f : [0, 1] \to \Re$ defined by*

$$f(t) \;=\; \begin{cases} 1 & t \in [0, 1] \text{ and is rational} \\ -1 & t \in [0, 1] \text{ and is irrational} \end{cases}$$

You should be able to see that for any partition of $[0, 1]$, the infimum of f on any subinterval is always -1 as any subinterval contains irrational numbers. Similarly, any subinterval contains rational numbers and so the supremum of f on a subinterval is 1. Thus $U(f, \pi) = 1$ and $L(f, \pi) = -1$ for any partition π of $[0, 1]$. It follows that $L(f) = -1$ and $U(f) = 1$. Thus, f is bounded but not Darboux Integrable.

Homework

Exercise 4.2.1 *Consider the function $f : [0, 1] \to \Re$ defined by*

$$f(t) \;=\; \begin{cases} 2 & t \in [0, 1] \text{ and is rational} \\ -3 & t \in [0, 1] \text{ and is irrational} \end{cases}$$

Prove f is not Riemann Integrable on $[0, 1]$.

Exercise 4.2.2 *Consider the function $f : [0, 1] \to \Re$ defined by*

$$f(t) \;=\; \begin{cases} t & t \in [0, 1] \text{ and is rational} \\ -t & t \in [0, 1] \text{ and is irrational} \end{cases}$$

Using uniform partitions of $[0, 1]$, prove $L(f) \geq -\frac{1}{2}$ and $U(f) \leq \frac{1}{2}$.

Exercise 4.2.3 *Consider the function $f : [0, 1] \to \Re$ defined by*

$$f(t) \;=\; \begin{cases} 2t & t \in [0, 1] \text{ and is rational} \\ -2t & t \in [0, 1] \text{ and is irrational} \end{cases}$$

Using uniform partitions of $[0, 1]$, prove $L(f) \geq -1$ and $U(f) \leq 1$.

Definition 4.2.4 Riemann's Criterion for Integrability

RIEMANN INTEGRATION

> Let $f \in B([a,b])$. We say that **Riemann's Criterion** holds for f if for every positive ϵ there exists a $\pi_0 \in \Pi[a,b]$ such that $U(f,\pi) - L(f,\pi) < \epsilon$ for any refinement, π, of π_0.

Theorem 4.2.4 The Riemann Integral Equivalence Theorem

> Let $f \in B([a,b])$. Then the following are equivalent.
>
> (i) $f \in RI([a,b])$.
>
> (ii) f satisfies Riemann's Criterion.
>
> (iii) f is Darboux Integrable, i.e., $L(f) = U(f)$, and $RI(f;a,b) = DI(f;a,b)$.

Proof 4.2.8

$(i) \Rightarrow (ii)$

Proof *Assume $f \in RI([a,b])$, and let $\epsilon > 0$ be given. Let IR be the Riemann integral of f over $[a,b]$. Choose $\pi_0 \in \Pi[a,b]$ such that $|\, S(f,\pi,\sigma) - IR \,| < \epsilon/3$ for any refinement, π, of π_0 and any $\sigma \subset \pi$. Let π be any such refinement, denoted by $\pi = \{x_0 = a, x_1, \ldots, x_p = b\}$, and let m_j, M_j be defined as usual. Using the Infimum and Supremum Tolerance Lemmas, we can conclude that, for each $j = 1, \ldots, p$, there exist $s_j, t_j \in [x_{j-1}, x_j]$ such that*

$$M_j - \frac{\epsilon}{6(b-a)} < f(s_j) \leq M_j$$

$$m_j \leq f(t_j) < m_j + \frac{\epsilon}{6(b-a)}$$

It follows that

$$f(s_j) - f(t_j) > M_j - \frac{\epsilon}{6(b-a)} - m_j - \frac{\epsilon}{6(b-a)}$$

Thus, we have

$$M_j - m_j - \frac{\epsilon}{3(b-a)} < f(s_j) - f(t_j)$$

Multiply this inequality by Δx_j to obtain

$$(M_j - m_j)\Delta x_j - \frac{\epsilon}{3(b-a)}\Delta x_j < \big(f(s_j) - f(t_j)\big)\Delta x_j$$

Now, sum over π to obtain

$$
\begin{aligned}
U(f,\pi) - L(f,\pi) &= \sum_{\pi}(M_j - m_j)\Delta x_j \\
&< \frac{\epsilon}{3(b-a)}\sum_{\pi}\Delta x_j + \sum_{\pi}\big(f(s_j) - f(t_j)\big)\Delta x_j
\end{aligned}
$$

This simplifies to

$$\sum_{\pi}(M_j - m_j)\Delta x_j - \frac{\epsilon}{3} < \sum_{\pi}\big(f(s_j) - f(t_j)\big)\Delta x_j \tag{$*$}$$

Now, we have

$$| \sum_\pi \big(f(s_j) - f(t_j)\big)\Delta x_j \,| \;=\; | \sum_\pi f(s_j)\Delta x_j - \sum_\pi f(t_j)\Delta x_j \,|$$

$$=\; | \sum_\pi f(s_j)\Delta x_j - IR + IR - \sum_\pi f(t_j)\Delta x_j \,|$$

$$\leq\; | \sum_\pi f(s_j)\Delta x_j - IR \,| + | \sum_\pi f(t_j)\Delta x_j - IR \,|$$

$$=\; | \, S(f,\pi,\sigma_s) - IR \,| + | \, S(f,\pi,\sigma_t) - IR \,|$$

where $\sigma_s = \{s_1,\ldots,s_p\}$ and $\sigma_t = \{t_1,\ldots,t_p\}$ are evaluation sets of π. Now, by our choice of partition π, we know

$$| \, S(f,\pi,\sigma_s) - IR \,| \;<\; \frac{\epsilon}{3}$$

$$| \, S(f,\pi,\sigma_t) - IR \,| \;<\; \frac{\epsilon}{3}$$

Thus, we can conclude that

$$| \sum_\pi \big(f(s_j) - f(t_j)\big)\Delta x_j \,| < \frac{2\epsilon}{3}$$

Applying this to the inequality in Equation $*$, we obtain

$$\sum_\pi (M_j - m_j)\Delta x_j < \epsilon$$

Now, π was an arbitrary refinement of π_0, and $\epsilon > 0$ was also arbitrary. So this shows that f satisfies Riemann's condition. □

$(ii) \Rightarrow (iii)$

Proof *Now, assume that f satisfies Riemann's criteria, and let $\epsilon > 0$ be given. Then there is a partition, $\pi_0 \in \mathbf{\Pi}[a,b]$ such that $U(f,\pi) - L(f,\pi) < \epsilon$ for any refinement, π, of π_0. Thus, by the definition of the upper and lower Darboux integrals, we have*

$$U(f) \leq U(f,\pi) < L(f,\pi) + \epsilon \leq L(f) + \epsilon$$

Since ϵ is arbitrary, this shows that $U(f) \leq L(f)$. The reverse inequality has already been established. Thus, we see that $U(f) = L(f)$. □

$(iii) \Rightarrow (i)$

Proof *Finally, assume f is Darboux integral which means $L(f) = U(f)$. Let ID denote the value of the Darboux integral. We will show that f is also Riemann integrable according to the definition and that the value of the integral is ID.*

Let $\epsilon > 0$ be given. Now, recall that

$$ID \;=\; L(f) \;=\; \sup_\pi L(f,\pi)$$

$$=\; U(f) \;=\; \inf_\pi U(f,\pi)$$

Hence, by the Supremum Tolerance Lemma, there exists $\pi_1 \in \mathbf{\Pi}[a,b]$ such that

$$ID - \epsilon = L(f) - \epsilon < L(f,\pi_1) \leq L(f) = ID$$

RIEMANN INTEGRATION

and by the Infimum Tolerance Lemma, there exists $\pi_2 \in \Pi[a,b]$ such that

$$ID = U(f) \le U(f, \pi_2) < U(f) + \epsilon = ID + \epsilon$$

Let $\pi_0 = \pi_1 \vee \pi_2$ be the common refinement of π_1 and π_2. Now, let π be any refinement of π_0, and let $\sigma \subset \pi$ be any evaluation set. Then we have

$$\begin{aligned} ID - \epsilon \quad &< \quad L(f, \pi_1) \le L(f, \pi_0) \le L(f, \pi) \le S(f, \pi, \sigma) \\ &\le \quad U(f, \pi) \le U(f, \pi_0) \le U(f, \pi_2) < ID + \epsilon \end{aligned}$$

Thus, it follows that

$$ID - \epsilon \quad < \quad S(f, \pi, \sigma) \quad < \quad ID + \epsilon$$

Since the refinement, π, of π_0 was arbitrary, as were the evaluation set, σ, and the tolerance ϵ, it follows that for any refinement, π, of π_0 and any $\epsilon > 0$, we have

$$\mid S(f, \pi, \sigma) - ID \mid < \epsilon$$

This shows that f is Riemann Integrable and the value of the integral is ID. $\qquad\square$

$\blacksquare$

Comment 4.2.3 *By Theorem 4.2.4, we now know that the Darboux and Riemann integrals are equivalent. Hence, it is no longer necessary to use a different notation for these two different approaches to what we call integration. From now on, we will use this notation*

$$RI(f; a, b) \equiv DI(f; a, b) \equiv \int f(t)\, dt$$

where the (t) in the new integration symbol refers to the name we wish to use for the independent variable and dt is a mnemonic to remind us that the $\|\pi\|$ is approaching zero as we choose progressively finer partitions of $[a, b]$. This is, of course, not very rigorous notation. A better notation would be

$$RI(f; a, b) \equiv DI(f; a, b) \equiv I(f; a, b)$$

where the symbol I denotes that we are interested in computing the integral of f using the equivalent approach of Riemann or Darboux. Indeed, the notation $I(f; a, b)$ does not require the uncomfortable lack of rigor that the symbol dt implies. However, for historical reasons, the symbol $\int f(t)\, dt$ will be used.

Also, the use of the $\int f(t)\, dt$ allows us to very efficiently apply the integration techniques of substitution and so forth as we have shown in Chapter 2.

Homework

Exercise 4.2.4 *Consider the function $f : [0, 1] \to \Re$ defined by*

$$f(t) \quad = \quad \begin{cases} t & t \in [0, 1] \text{ and is rational} \\ -t & t \in [0, 1] \text{ and is irrational} \end{cases}$$

We have proven in an earlier exercise that using uniform partitions of $[0, 1]$, $L(f) \ge -\frac{1}{2}$ and $U(f) \le \frac{1}{2}$. We can use that to show f is not Riemann Integrable. If f was, we can use the Riemann Criterion

66 BASIC ANALYSIS IV: MEASURE THEORY AND INTEGRATION

to show that given $\epsilon > 0$, there is a partition P_0 so that $U(f, P) - L(f, P) < \epsilon$ for any refinement of P_0. Hence

$$U(f, P_n \vee P_0) - L(f, P_n \vee P_0) \quad \leq \quad U(f, P_0) - L(f, P_0) < \epsilon$$

However, you also have an estimate for $U(f, P_n) - L(f, P_n)$ which you can use to derive a contradiction. Good luck!

Exercise 4.2.5 *Consider the function $f : [0, 1] \to \Re$ defined by*

$$f(t) \quad = \quad \begin{cases} 2t & t \in [0, 1] \text{ and is rational} \\ -2t & t \in [0, 1] \text{ and is irrational} \end{cases}$$

Prove f is not Riemann Integrable using the ideas from the previous exercise.

4.3 Properties of the Riemann Integral

We can now prove a series of properties of the Riemann integral. Let's start with a lemma about infimums and supremums.

Lemma 4.3.1 Fundamental Infimum and Supremum Equalities

If f is a bounded function on the finite interval $[a, b]$, then

1. $\sup_{a \leq x \leq b} f(x) \quad = \quad -\inf_{a \leq x \leq b}(-f(x)) \quad$ *and* $\quad -\sup_{a \leq x \leq b}(-f(x)) \quad = \quad \inf_{a \leq x \leq b}(f(x))$

2.

$$\sup_{x, y \in [a,b]} (f(x) - f(y)) \quad = \quad \sup_{y \in [a,b]} \sup_{y \in [a,b]} (f(x) - f(y))$$

$$= \quad \sup_{x \in [a,b]} \sup_{y \in [a,b]} (f(x) - f(y)) = M - m$$

where $M = \sup_{a \leq x \leq b} f(x)$ and $m = \inf_{a \leq x \leq b} f(x)$.

3. $\sup_{x, y \in [a,b]} | f(x) - f(y) | = M - m$

Proof 4.3.1

First let $Q = \sup_{a \leq x \leq b}(-f)$ and $q = \inf_{a \leq x \leq b}(-f)$.
(1):
Let $(f(x_n))$ be a sequence which converges to M. Then since $-f(x_n) \geq q$ for all n, letting $n \to \infty$, we find $-M \geq q$.

Now let $(-f(z_n))$ be a sequence which converges to q. Then, we have $f(z_n) \leq M$ for all n and letting $n \to \infty$, we have $-q \leq M$ or $q \geq -M$.

Combining, we see $-q = M$ which is the first part of the statement; i.e. $\sup_{a \leq x \leq b} f(x) = -\inf_{a \leq x \leq b}(-f(x))$. Now just replace all the f's by $-f$'s in this to get $\sup_{a \leq x \leq b}(-f(x)) = -\inf_{a \leq x \leq b}(--f(x))$ or $-\sup_{a \leq x \leq b}(-f(x)) = \inf_{a \leq x \leq b}(f(x))$ which is the other identity.

RIEMANN INTEGRATION

(2):
We know

$$
\begin{aligned}
f(x) - f(y) &\leq \sup_{a \leq x \leq b} f(x) - f(y) \leq \sup_{x \in [a,b]} - \inf_{a \leq y \leq b} f(y) = M - m \\
f(x) - f(y) &\leq f(x) - \inf_{a \leq y \leq b} f(y) \leq \sup_{x \in [a,b]} - \inf_{a \leq y \leq b} f(y) = M - m
\end{aligned}
$$

Thus,

$$
\sup_{x,y \in [a,b]} (f(x) - f(y)) \leq M - m
$$

So one side of the inequality is clear. Now let $f(x_n)$ be a sequence converging to M and $f(y_n)$ be a sequence converging to m. Then, we have

$$
f(x_n) - f(y_n) \leq \sup_{x,y \in [a,b]} (f(x) - f(y))
$$

Letting $n \to \infty$, we see

$$
M - m \leq \sup_{x,y \in [a,b]} (f(x) - f(y))
$$

This is the other side of the inequality. We have thus shown that the equality is valid.
(3):
Note

$$
| f(x) - f(y) | = \begin{cases} f(x) - f(y), & f(x) \geq f(y) \\ f(y) - f(x), & f(x) < f(y) \end{cases}
$$

In either case, we have $| f(x) - f(y) | \leq M_j - m_j$ for all x, y using Part (2) implying that $\sup_{x,y} | f(x) - f(y) | \leq M_j - m_j$.

To see the reverse inequality holds, we first note that if $M_j = m_j$, we see the reverse inequality holds trivially as $\sup_{x,y} | f(x) - f(y) | \geq 0 = M_j - m_j$. Hence, we may assume without loss of generality that the gap $M_j - m_j$ is positive.

Then, given $0 < \epsilon < (1/2)(M_j - m_j)$, there exist $s_j, t_j \in [x_{j-1}, x_j]$ such that $M_j - \epsilon/2 < f(s_j)$ and $m_j + \epsilon/2 > f(t_j)$, so that $f(s_j) - f(t_j) > M_j - m_j - \epsilon$. By our choice of ϵ, these terms are positive and so we also have $| f(s_j) - f(t_j) | > M_j - m_j - \epsilon$.

It follows that

$$
\sup_{x,y \in [x_{j-1}, x_j]} | f(x) - f(y) | \geq | f(s_j) - f(t_j) | > M_j - m_j - \epsilon |
$$

Since we can make ϵ arbitrarily small, this implies that

$$
\sup_{x,y \in [x_{j-1}, x_j]} | f(x) - f(y) | \geq M_j - m_j
$$

This establishes the reverse inequality and proves the claim. ∎

Theorem 4.3.2 Properties of the Riemann Integral

> Let $f, g \in RI([a,b])$. Then
>
> (i) $|f| \in RI([a,b])$;
>
> (ii)
> $$\left| \int_a^b f(x)dx \right| \le \int_a^b |f|\, dx$$
>
> (iii) $f^+ = \max\{f, 0\} \in RI([a,b])$;
>
> (iv) $f^- = \max\{-f, 0\} \in RI([a,b])$;
>
> (v)
> $$\int_a^b f(x)dx = \int_a^b [f^+(x) - f^-(x)]dx = \int_a^b f^+(x)dx - \int_a^b f^-(x)dx$$
> $$\int_a^b |f(x)|\, dx = \int_a^b [f^+(x) + f^-(x)]dx = \int_a^b f^+(x)dx + \int_a^b f^-(x)dx$$
>
> (vi) $f^2 \in RI([a,b])$;
>
> (vii) $fg \in RI([a,b])$;
>
> (viii) If there exists m, M such that $0 < m \le |f| \le M$, then $1/f \in RI([a,b])$.

Proof 4.3.2

(i):

Proof *Note given a partition $\pi = \{x_0 = a, x_1, \ldots, x_p = b\}$, for each $j = 1, \ldots, p$, from Lemma 4.3.1, we know*

$$\sup_{x,y \in [x_{j-1}, x_j]} (f(x) - f(y)) = M_j - m_j$$

Now, let m_j' and M_j' be defined by

$$m_j' = \inf_{[x_{j-1}, x_j]} |f(x)|, \quad M_j' = \sup_{[x_{j-1}, x_j]} |f(x)|$$

Then, applying Lemma 4.3.1 to $|f|$, we have

$$M_j' - m_j' = \sup_{x,y \in [x_{j-1}, x_j]} |f(x)| - |f(y)|$$

Thus, for each $j = 1, \ldots, p$, we have

$$M_j - m_j = \sup_{x,y \in [x_{j-1}, x_j]} |f(x) - f(y)|$$

So, since $|f(x)| - |f(y)| \le |f(x) - f(y)|$ for all x, y, it follows that $M_j' - m_j' \le M_j - m_j$, implying that $\sum_\pi (M_j' - m_j')\Delta x_j \le \sum_\pi (M_j - m_j)\Delta x_j$. This means $U(|f|, \pi) - L(|f|, \pi) \le U(f, \pi) - L(f, \pi)$ for the chosen π. Since f is integrable by hypothesis, by Theorem 4.2.4, we know the Riemann Criterion must also hold for f. Thus, given $\epsilon > 0$, there is a partition π_0 so that

$U(f, \pi) - L(f, \pi) < \epsilon$ *for any refinement π of π_0. Therefore $|f|$ also satisfies the Riemann Criterion and so $|f|$ is Riemann integrable.* $\square$

The other results now follow easily.
(ii):

Proof *We have $f \leq |f|$ and $f \geq -|f|$, so that*

$$\int_a^b f(x)dx \leq \int_a^b |f(x)| \, dx$$
$$\int_a^b f(x)dx \geq -\int_a^b |f(x)| \, dx$$

from which it follows that

$$-\int_a^b |f(x)| \, dx \leq \int_a^b f(x)dx \leq \int_a^b |f(x)| \, dx$$

and so

$$\left| \int_a^b f \right| \leq \int_a^b |f|$$

$\square$

(iii) *and* (iv):

Proof *This follows from the facts that $f^+ = \frac{1}{2}(|f| + f)$ and $f^- = \frac{1}{2}(|f| - f)$ and the Riemann integral is a linear mapping.* $\square$

(v):

Proof *This follows from the facts that $f = f^+ - f^-$ and $|f| = f^+ + f^-$ and the linearity of the integral.* $\square$

(vi):

Proof *Note that, since f is bounded, there exists $K > 0$ such that $|f(x)| \leq K$ for all $x \in [a, b]$. Then, applying Lemma 4.3.1 to f^2, we have*

$$\sup_{x,y \in [x_{j-1}, x_j]} (f^2(x) - f^2(y)) = M_j(f^2) - m_j(f^2)$$

where $[x_{j-1}, x_j]$ is a subinterval of a given partition π and $M_j(f^2) = \sup_{x \in [x_{j-1}, x_j]} f^2(x)$ and $m_j(f^2) = \inf_{x \in [x_{j-1}, x_j]} f^2(x)$. Thus, for this partition, we have

$$U(f^2, \pi) - L(f^2, \pi) = \sum_{\pi} (M_j(f^2) - m_j(f^2)) \, \Delta x_j$$

But we also know

$$\sup_{x,y \in [x_{j-1}, x_j]} (f^2(x) - f^2(y)) = \sup_{x,y \in [x_{j-1}, x_j]} (f(x) + f(y))(f(x) - f(y))$$
$$\leq 2K \sup_{x,y \in [x_{j-1}, x_j]} (f(x) - f(y)) = M_j - m_j.$$

70   BASIC ANALYSIS IV: MEASURE THEORY AND INTEGRATION

Thus,

$$\begin{aligned}
U(f^2, \boldsymbol{\pi}) - L(f^2, \boldsymbol{\pi}) &= \sum_{\boldsymbol{\pi}} \left(M_j(f^2) - m_j(f^2) \right) \Delta x_j \\
&\leq 2K \sum_{\boldsymbol{\pi}} \left(M_j - m_j \right) \Delta x_j = U(f, \boldsymbol{\pi}) - L(f\boldsymbol{\pi})
\end{aligned}$$

Now since f is Riemann Integrable, it satisfies the Riemann Criterion and so given $\epsilon > 0$, there is a partition $\boldsymbol{\pi_0}$ so that $U(f, \boldsymbol{\pi}) - L(f\boldsymbol{\pi}) < \epsilon/(2K)$ for any refinement $\boldsymbol{\pi}$ of $\boldsymbol{\pi_0}$. Thus, f^2 satisfies the Riemann Criterion too and so it is integrable. $\qquad\square$

(vii):

Proof *To prove that fg is integrable when f and g are, simply note that*

$$fg = (1/2)\left((f+g)^2 - f^2 - g^2 \right)$$

Property (vi) and the linearity of the integral then imply fg is integrable. $\qquad\square$

$(viii)$:

Proof *Suppose $f \in RI([a,b])$ and there exist $M, m > 0$ such that $m \leq | f(x) | \leq M$ for all $x \in [a,b]$. Note that*

$$\frac{1}{f(x)} - \frac{1}{f(y)} = \frac{f(y) - f(x)}{f(x)f(y)}$$

Let $\boldsymbol{\pi} = \{x_0 = a, x_1, \ldots, x_p = b\}$ be a partition of $[a,b]$, and define

$$\begin{aligned}
M_j' &= \sup_{[x_{j-1}, x_j]} \frac{1}{f(x)} \\
m_j' &= \inf_{[x_{j-1}, x_j]} \frac{1}{f(x)}
\end{aligned}$$

Then we have

$$\begin{aligned}
M_j' - m_j' &= \sup_{x,y \in [x_{j-1}, x_j]} \frac{f(y) - f(x)}{f(x)f(y)} \\
&\leq \sup_{x,y \in [x_{j-1}, x_j]} \frac{| f(y) - f(x) |}{| f(x) || f(y) |} \\
&\leq \frac{1}{m^2} \sup_{x,y \in [x_{j-1}, x_j]} | f(y) - f(x) | \\
&\leq \frac{M_j - m_j}{m^2}
\end{aligned}$$

Since $f \in RI([a,b])$, given $\epsilon > 0$, there is a partition $\boldsymbol{\pi_0}$ such that $U(f, \boldsymbol{\pi}) - L(f, \boldsymbol{\pi}) < m^2\epsilon$ for any refinement, pi, of $\boldsymbol{\pi_0}$. Hence, the previous inequality implies that, for any such refinement, we have

$$\begin{aligned}
U\left(\frac{1}{f}, \boldsymbol{\pi}\right) - L\left(\frac{1}{f}, \boldsymbol{\pi}\right) &= \sum_{\boldsymbol{\pi}} (M_j' - m_j')\Delta x_j \\
&\leq \frac{1}{m^2} \sum_{\boldsymbol{\pi}} (M_j - m_j)\Delta x_j
\end{aligned}$$

$$\leq \frac{1}{m^2}\Big(U(f,\boldsymbol{\pi}) - L(f,\boldsymbol{\pi})\Big)$$
$$< \frac{m^2\epsilon}{m^2} = \epsilon$$

Thus $1/f$ satisfies the Riemann Criterion and hence it is integrable. □

■

Homework

Exercise 4.3.1 *For $\alpha > 0$ and any β, let $f(t) = e^{-\alpha t}\cos(\beta t + \delta)$. Prove*

$$-1 + \frac{1}{\alpha}e^{-\alpha L} \;\leq\; \int_0^L f(t)dt \leq 1 - \frac{1}{\alpha}e^{-\alpha L}$$

Can you bound the improper integral $\int_0^\infty f(t)dt$ also?

Exercise 4.3.2 *For $\alpha > 0$ and any β, let $f(t) = e^{\alpha t}\cos(\beta t + \delta)$. Prove*

$$1 - \frac{1}{\alpha}e^{\alpha L} \;\leq\; \int_0^L f(t)dt \leq -1 + \frac{1}{\alpha}e^{\alpha L}$$

Can you bound the improper integral $\int_0^\infty f(t)dt$ also?

Exercise 4.3.3 *Use an induction argument to prove if f and g are Riemann Integrable on $[a, b]$, then $f^p g^q$ is also Riemann Integrable for all positive integers p and q.*

Exercise 4.3.4 *Use an induction argument to prove f is bounded below on $[a, b]$ and g is bounded below on $[a, b]$, then $(1/f)^p(1/g)^q$ is also Riemann Integrable for all positive integers p and q.*

Exercise 4.3.5 *Use an induction argument to prove if f and g are Riemann Integrable on $[a, b]$, then $(f^+)^p(g^-)^q$ is also Riemann Integrable for all positive integers p and q.*

4.4 What Functions are Riemann Integrable?

Now we need to show that the set $RI([a, b])$ is nonempty. We begin by showing that all continuous functions on $[a, b]$ will be Riemann Integrable.

Theorem 4.4.1 Continuous Implies Riemann Integrable

> *If $f \in C([a, b])$, then $f \in RI([a, b])$.*

Proof 4.4.1
Since f is continuous on a compact set, it is uniformly continuous. Hence, given $\epsilon > 0$, there is a $\delta > 0$ such that $x, y \in [a, b]$, $\mid x - y \mid < \delta \Rightarrow \mid f(x) - f(y) \mid < \epsilon/(b - a)$. Let $\boldsymbol{\pi}_0$ be a partition such that $\|\boldsymbol{\pi}_0\| < \delta$, and let $\boldsymbol{\pi} = \{x_0 = a, x_1, \ldots, x_p = b\}$ be any refinement of $\boldsymbol{\pi}_0$. Then $\boldsymbol{\pi}$ also satisfies $\|\boldsymbol{\pi}\| < \delta$. Since f is continuous on each subinterval $[x_{j-1}, x_j]$, f attains its supremum, M_j, and infimum, m_j, at points s_j and t_j, respectively. That is, $f(s_j) = M_j$ and $f(t_j) = m_j$ for each $j = 1, \ldots, p$. Thus, the uniform continuity of f on each subinterval implies that, for each j,

$$M_j - m_j = \mid f(s_j) - f(t_j) \mid < \frac{\epsilon}{b - a}$$

72 BASIC ANALYSIS IV: MEASURE THEORY AND INTEGRATION

Thus, we have

$$U(f, \boldsymbol{\pi}) - L(f, \boldsymbol{\pi}) = \sum_{\boldsymbol{\pi}} (M_j - m_j)\Delta x_j < \frac{\epsilon}{b-a} \sum_{\boldsymbol{\pi}} \Delta x_j = \epsilon.$$

Since $\boldsymbol{\pi}$ was an arbitrary refinement of $\boldsymbol{\pi}_0$, it follows that f satisfies Riemann's criterion. Hence, $f \in RI([a, b])$. ∎

Theorem 4.4.2 Constant Functions Are Riemann Integrable

> *If $f : [a, b] \to \Re$ is a constant function, $f(t) = c$ for all t in $[a, b]$, then f is Riemann Integrable on $[a, b]$ and $\int_a^b f(t)dt = c(b - a)$.*

Proof 4.4.2

For any partition $\boldsymbol{\pi}$ of $[a, b]$, since f is a constant, all the individual m_j's and M_j's associated with $\boldsymbol{\pi}$ take on the value c. Hence, $U(f, \boldsymbol{\pi}) - U(f, \boldsymbol{\pi}) = 0$ always. It follows immediately that f satisfies the Riemann Criterion and hence is Riemann Integrable. Finally, since f is integrable, by Theorem 4.1.2, we have

$$c(b - a) \leq RI(f; a, b) \leq c(b - a)$$

Thus, $\int_a^b f(t)dt = c(b - a)$. ∎

Theorem 4.4.3 Monotone Implies Riemann Integrable

> *If f is monotone on $[a, b]$, then $f \in RI([a, b])$.*

Proof 4.4.3

As usual, for concreteness, we assume that f is monotone increasing. We also assume $f(b) > f(a)$, for if not, then f is constant and must be integrable by Theorem 4.4.2. Let $\epsilon > 0$ be given, and let $\boldsymbol{\pi}_0$ be a partition of $[a, b]$ such that $\|\boldsymbol{\pi}_0\| < \epsilon/(f(b) - f(a))$. Let $\boldsymbol{\pi} = \{x_0 = a, x_1, \ldots, x_p = b\}$ be any refinement of $\boldsymbol{\pi}_0$. Then $\boldsymbol{\pi}$ also satisfies $\|\boldsymbol{\pi}\| < \epsilon/(f(b) - f(a))$. Thus, for each $j = 1, \ldots, p$, we have

$$\Delta x_j < \frac{\epsilon}{f(b) - f(a)}$$

Since f is increasing, we also know that $M_j = f(x_j)$ and $m_j = f(x_{j-1})$ for each j. Hence,

$$
\begin{aligned}
U(f, \boldsymbol{\pi}) - L(f, \boldsymbol{\pi}) &= \sum_{\boldsymbol{\pi}} (M_j - m_j)\Delta x_j \\
&= \sum_{\boldsymbol{\pi}} [f(x_j) - f(x_{j-1})]\Delta x_j \\
&< \frac{\epsilon}{f(b) - f(a)} \sum_{\boldsymbol{\pi}} [f(x_j) - f(x_{j-1})]
\end{aligned}
$$

But this last sum is telescoping and sums to $f(b) - f(a)$. So, we have

$$U(f, \boldsymbol{\pi}) - L(f, \boldsymbol{\pi}) < \frac{\epsilon}{f(b) - f(a)}(f(b) - f(a)) = \epsilon$$

Thus, f satisfies Riemann's criterion. ∎

Theorem 4.4.4 Bounded Variation Implies Riemann Integrable

RIEMANN INTEGRATION 73

> *If $f \in BV([a,b])$, then $f \in RI([a,b])$.*

Proof 4.4.4
Since f is of bounded variation, there are functions u and v, defined on $[a,b]$ and both monotone increasing, such that $f = u - v$. Hence, by the linearity of the integral and the previous theorem, $f \in RI([a,b])$. ∎

Homework

Exercise 4.4.1 *Let f be defined by*

$$f(x) \;=\; \begin{cases} 2x, & 0 \le x < 1 \\ 3, & 1 = x \\ 5x^2 + 1, & 1 < x \le 2 \end{cases}$$

Explain why f is Riemann Integrable.

Exercise 4.4.2 *Let f be defined by*

$$f(x) \;=\; \begin{cases} 2, & x = -1 \\ x + 4, & -1 < x < 1 \\ 6, & 1 = x \\ x^2 + 7, & 1 < x < 2 \\ 12, & x = 2 \\ 2x^3, & 2 < x \le 3 \end{cases}$$

Explain why f is Riemann Integrable.

Exercise 4.4.3 *Let f_n, for $n \ge 2$, be defined by*

$$f_n(x) \;=\; \begin{cases} 1, & x = 1 \\ 1/(k+1), & 1/(k+1) \le x < 1/k,\ 1 \le k < n \\ 0, & 0 \le x < 1/n \end{cases}$$

Explain why f_n is Riemann Integrable.

Exercise 4.4.4 *Let f be the pointwise limit of the sequence of functions (f_n) for $n \ge 2$ defined by*

$$f_n(x) \;=\; \begin{cases} 1, & x = 1 \\ 1/(k+1), & 1/(k+1) \le x < 1/k,\ 1 \le k < n \\ 0, & 0 \le x < 1/n \end{cases}$$

Explain why f_n is Riemann Integrable.

4.5 Further Properties of the Riemann Integral

We first want to establish the familiar summation property of the Riemann integral over an interval $[a, b] = [a, c] \cup [c, b]$. Most of the technical work for this result is done in the following lemma.

Lemma 4.5.1 The Upper and Lower Darboux Integral is Additive on Intervals

> Let $f \in B([a,b])$ and let $c \in (a,b)$. Let
>
> $$\underline{\int_a^b} f(x)\,dx = L(f) \text{ and } \overline{\int_a^b} f(x)\,dx = U(f)$$
>
> denote the lower and upper Darboux integrals of f on $[a,b]$, respectively. Then we have
>
> $$\overline{\int_a^b} f(x)dx = \overline{\int_a^c} f(x)dx + \overline{\int_c^b} f(x)dx$$
>
> $$\underline{\int_a^b} f(x)dx = \underline{\int_a^c} f(x)dx + \underline{\int_c^b} f(x)dx$$

Proof 4.5.1

We prove the result for the upper integrals as the lower integral case is similar. Let π be given by $\pi = \{x_0 = a, x_1, \ldots, x_p = b\}$. We first assume that c is a partition point of π. Thus, there is some index $1 \le k_0 \le p-1$ such that $x_{k_0} = c$. For any interval $[\alpha, \beta]$, let $U_\alpha^\beta(f, \pi)$ denote the upper sum of f for the partition π over $[\alpha, \beta]$. Now, we can rewrite π as $\pi = \{x_0, x_1, \ldots, x_{k_0}\} \cup \{x_{k_0}, x_{k_0+1}, \ldots, x_p\}$. Let $\pi_1 = \{x_0, \ldots, x_{k_0}\}$ and $\pi_2 = \{x_{k_0}, \ldots, x_p\}$. Then $\pi_1 \in \Pi[a,c]$, $\pi_2 \in \Pi[c,b]$, and

$$
\begin{aligned}
U_a^b(f, \pi) &= U_a^c(f, \pi_1) + U_c^b(f, \pi_2) \\
&\ge \overline{\int_a^c} f(x)dx + \overline{\int_c^b} f(x)dx
\end{aligned}
$$

by the definition of the upper sum. Now, if c is not in π, then we can refine π by adding c, obtaining the partition $\pi' = \{x_0, x_1, \ldots, x_{k_0}, c, x_{k_0+1}, \ldots, x_p\}$. Splitting up π' at c as we did before into π_1 and π_2, we see that $\pi' = \pi_1 \vee \pi_2$ where $\pi_1 = \{x_0, \ldots, x_{k_0}, c\}$ and $\pi_2 = \{c, x_{k_0+1}, \ldots, x_p\}$. Thus, by our properties of upper sums, we see that

$$U_a^b(f, \pi) \ge U_a^b(f, \pi') = U_a^c(f, \pi_1) + U_c^b(f, \pi_2) \ge \overline{\int_a^c} f(x)dx + \overline{\int_c^b} f(x)dx$$

Combining both cases, we can conclude that for any partition $\pi \in \Pi[a,b]$, we have

$$U_a^b(f, \pi) \ge \overline{\int_a^c} f(x)dx + \overline{\int_c^b} f(x)dx$$

which implies that

$$\overline{\int_a^b} f(x)dx \ge \overline{\int_a^c} f(x)dx + \overline{\int_c^b} f(x)dx$$

Now we want to show the reverse inequality. Let $\epsilon > 0$ be given. By the definition of the upper integral, there exists $\pi_1 \in \Pi[a,c]$ and $\pi_2 \in \Pi[c,b]$ such that

$$
\begin{aligned}
U_a^c(f, \pi_1) &< \overline{\int_a^c} f(x)dx + \frac{\epsilon}{2} \\
U_c^b(f, \pi_2) &< \overline{\int_c^b} f(x)dx + \frac{\epsilon}{2}
\end{aligned}
$$

RIEMANN INTEGRATION

Let $\pi = \pi_1 \cup \pi_2 \in \Pi[a,b]$. It follows that

$$U_a^b(f, \pi) = U_a^c(f, \pi_1) + U_c^b(f, \pi_2) < \overline{\int_a^c} f(x)dx + \overline{\int_c^b} f(x)dx + \epsilon$$

But, by definition, we have

$$\overline{\int_a^b} f(x)dx \leq U_a^b(f, \pi)$$

for all π. Hence, we see that

$$\overline{\int_a^b} f(x)dx < \overline{\int_a^c} f(x)dx + \overline{\int_c^b} f(x)dx + \epsilon$$

Since ϵ was arbitrary, this proves the reverse inequality we wanted. We can conclude, then, that

$$\overline{\int_a^b} f(x)dx = \overline{\int_a^c} f(x)dx + \overline{\int_c^b} f(x)dx$$

$\blacksquare$

Theorem 4.5.2 The Riemann Integral Exists on Subintervals

If $f \in RI([a,b])$ and $c \in (a,b)$, then $f \in RI([a,c])$ and $f \in RI([c,b])$.

Proof 4.5.2
Let $\epsilon > 0$ be given. Then there is a partition $\pi_0 \in \Pi[a,b]$ such that $U_a^b(f, \pi) - L_a^b(f, \pi) < \epsilon$ for any refinement, π, of π_0. Let π_0 be given by $\pi_0 = \{x_0 = a, x_1, \ldots, x_p = b\}$. Define $\pi_0' = \pi_0 \cup \{c\}$, so there is some index k_0 such that $x_{k_0} \leq c \leq x_{k_0+1}$. Let $\pi_1 = \{x_0, \ldots, x_{k_0}, c\}$ and $\pi_2 = \{c, x_{k_0+1}, \ldots, x_p\}$. Then $\pi_1 \in \Pi[a,c]$ and $\pi_2 \in \Pi[c,b]$. Let π_1' be a refinement of π_1. Then $\pi_1' \cup \pi_2$ is a refinement of π_0, and it follows that

$$
\begin{aligned}
U_a^c(f, \pi_1') - L_a^c(f, \pi_1') &= \sum_{\pi_1'} (M_j - m_j)\Delta x_j \\
&\leq \sum_{\pi_1' \cup \pi_2} (M_j - m_j)\Delta x_j \\
&\leq U_a^b(f, \pi_1' \cup \pi_2) - L_a^b(f, \pi_1' \cup \pi_2)
\end{aligned}
$$

But, since $\pi_1' \cup \pi_2$ refines π_0, we have

$$U_a^b(f, \pi_1' \cup \pi_2) - L_a^b(f, \pi_1' \cup \pi_2) < \epsilon$$

implying that

$$U_a^c(f, \pi_1') - L_a^c(f, \pi_1') < \epsilon$$

for all refinements, π_1', of π_1. Thus, f satisfies Riemann's criterion on $[a,c]$, and $f \in RI([a,c])$. The proof on $[c,b]$ is done in exactly the same way. $\blacksquare$

Theorem 4.5.3 The Riemann Integral is Additive on Subintervals

76 BASIC ANALYSIS IV: MEASURE THEORY AND INTEGRATION

If $f \in RI([a,b])$ and $c \in (a,b)$, then

$$\int_a^b f(x)dx = \int_a^c f(x)dx + \int_c^b f(x)dx$$

Proof 4.5.3

Since $f \in RI([a,b])$, we know that

$$\overline{\int_a^b} f(x)dx = \underline{\int_a^b} f(x)dx$$

Further, we also know that $f \in RI([a,c])$ and $f \in RI([c,b])$ for any $c \in (a,b)$. Thus,

$$\overline{\int_a^c} f(x)dx = \underline{\int_a^c} f(x)dx$$

$$\overline{\int_c^b} f(x)dx = \underline{\int_c^b} f(x)dx$$

So, applying Lemma 4.5.1, we conclude that, for any $c \in (a,b)$,

$$\int_a^b f(x)dx = \overline{\int_a^b} f(x)\,dx = \overline{\int_a^c} f(x)\,dx + \overline{\int_c^b} f(x)\,dx = \int_a^c f(x)\,dx + \int_c^b f(x)\,dx$$

∎

Homework

Exercise 4.5.1 Let f be defined by

$$f(x) = \begin{cases} 2x, & 0 \le x < 1 \\ 3, & 1 = x \\ 5x^2 + 1, & 1 < x \le 2 \end{cases}$$

Write down $\int_0^2 f(t)dt$ as a sum of terms very carefully. For example,

$$\int_0^1 f(t)dt = \int_0^1 \left\{ \begin{array}{ll} 2t, & 0 \le t < 1 \\ 3, & 1 = x \end{array} \right\} dt$$

and at the moment we don't know that

$$\int_0^1 f(t)dt = \int_0^1 \left\{ \begin{array}{ll} 2t, & 0 \le t < 1 \\ 3, & 1 = x \end{array} \right\} dt = \int_0^1 2t\,dt$$

as the two integrands match at all but one point.

Exercise **4.5.2** *Let f be defined by*

$$f(x) \;=\; \begin{cases} 2, & x = -1 \\ x + 4, & -1 < x < 1 \\ 6, & 1 = x \\ x^2 + 7, & 1 < x < 2 \\ 12, & x = 2 \\ 2x^3, & 2 < x \le 3 \end{cases}$$

Write down $\int_{-1}^{3} f(t)dt$ as a sum of terms very carefully.

Exercise **4.5.3** *Let f_n, for $n \ge 2$, be defined by*

$$f_n(x) \;=\; \begin{cases} 1, & x = 1 \\ 1/(k+1), & 1/(k+1) \le x < 1/k, \; 1 \le k < n \\ 0, & 0 \le x < 1/n \end{cases}$$

Write down $\int_{0}^{1} f_n(t)dt$ as a sum of terms very carefully.

Exercise **4.5.4** *Let $I_A(x) = 1$ if $x \in A$ and 0 if not. This is the indicator function of the set A. Define f on $[0,2]$ by*

$$f(x) = I_{\mathbb{Q} \cap [0,1]}(x) - I_{\mathbb{I} \cap [0,1]}(x) + 2x I_{\mathbb{Q} \cap (1,2]}(x) - 3x I_{\mathbb{I} \cap (1,2]}$$

Compute $\overline{\int_0^2} f(x)dx$ and $\underline{\int_0^2} f(x)dx$. From previous problems you know how to compute the necessary upper and lower integrals.

Exercise **4.5.5** *This uses the notation from the previous problem. Define f on $[0,2]$ by*

$$f(x) = 4 I_{\mathbb{Q} \cap [0,1]}(x) - 4 I_{\mathbb{I} \cap [0,1]}(x) + 3x I_{\mathbb{Q} \cap (1,2]}(x) - 7x I_{\mathbb{I} \cap (1,2]}$$

Compute $\overline{\int_0^2} f(x)\,dx$ and $\underline{\int_0^2} f(x)\,dx$.

4.6 The Fundamental Theorem of Calculus

The next result is the well-known **Fundamental of Theorem of Calculus**.

Theorem 4.6.1 The Fundamental Theorem of Calculus

Let $f \in RI([a,b])$. Define $F : [a,b] \to \Re$ by

$$F(x) = \int_a^x f(t)dt$$

Then

(i) $F \in BV([a,b])$;

(ii) $F \in C([a,b])$;

(iii) if f is continuous at $c \in [a,b]$, then F is differentiable at c and $F'(c) = f(c)$.

BASIC ANALYSIS IV: MEASURE THEORY AND INTEGRATION

Proof 4.6.1

First, note that $f \in RI([a,b]) \Rightarrow f \in R[a,x]$ for all $x \in [a,b]$, by our previous results. Hence, F is well-defined. We will prove the results in order.

(i) $F \in BV([a,b])$:

Proof *Let $\pi \in \Pi[a,b]$ be given by $\pi = \{x_0 = a, x_1, \ldots, x_p = b\}$. Then the fact that $f \in R[a, x_j]$ implies that $f \in R[x_{j-1}, x_j]$ for each $j = 1, \ldots, p$. Thus, we have*

$$m_j \Delta x_j \leq \int_{x_{j-1}}^{x_j} f(t)dt \leq M_j \Delta x_j$$

This implies that, for each j, we have

$$\left| \int_{x_{j-1}}^{x_j} f(t)dt \right| \leq \|f\|_\infty \Delta x_j$$

Thus,

$$
\begin{aligned}
|\Delta F_j| &= |F(x_j) - F(x_{j-1})| \\
&= \left| \int_a^{x_j} f(t)dt - \int_a^{x_{j-1}} f(t)dt \right| \\
&= \left| \int_{x_{j-1}}^{x_j} f(t)dt \right| \\
&\leq \|f\|_\infty \Delta x_j
\end{aligned}
$$

Summing over π, we obtain

$$\sum_\pi |\Delta F_j| \leq \|f\|_\infty \sum_\pi \Delta x_j = (b-a)\|f\|_\infty < \infty$$

Since the partition π was arbitrary, we conclude that $F \in BV([a,b])$. $\square$

(ii) $F \in C([a,b])$:

Proof *Now, let $x, y \in [a,b]$ be such that $x < y$. Then*

$$\inf_{[x,y]} f(t)\,(y-x) \leq \int_x^y f(t)dt \leq \sup_{[x,y]} f(t)\,(y-x)$$

which implies that

$$|F(y) - F(x)| = \left| \int_x^y f(t)dt \right| \leq \|f\|_\infty \,(y-x)$$

A similar argument shows that if $y, x \in [a,b]$ satisfy $y < x$, then

$$|F(y) - F(x)| = \left| \int_x^y f(t)dt \right| \leq \|f\|_\infty \,(x-y)$$

Let $\epsilon > 0$ be given. Then if

$$|x - y| < \frac{\epsilon}{\|f\|_\infty + 1}$$

we have

$$| F(y) - F(x) | \leq \|f\|_\infty \, | y - x | < \frac{\|f\|_\infty}{\|f\|_\infty + 1} \, \epsilon < \epsilon$$

Thus, F is continuous at x and, consequently, on $[a,b]$. $\square$

(iii) if f is continuous at $c \in [a, b]$, then F is differentiable at c and $F'(c) = f(c)$:

Proof *Finally, assume f is continuous at $c \in [a, b]$, and let $\epsilon > 0$ be given. Then there exists $\delta > 0$ such that $x \in (c - \delta, c + \delta) \cap [a, b]$ implies $| f(x) - f(c) | < \epsilon/2$. Pick $h \in \Re$ such that $0 < | h | < \delta$ and $c + h \in [a, b]$. Let's assume, for concreteness, that $h > 0$. Define*

$$m = \inf_{[c, c+h]} f(t) \qquad \text{and} \qquad M = \sup_{[c, c+h]} f(t)$$

If $c < x < c + h$, then we have $x \in (c - \delta, c + \delta) \cap [a, b]$ and $-\epsilon/2 < f(x) - f(c) < \epsilon/2$. That is,

$$f(c) - \frac{\epsilon}{2} < f(x) < f(c) + \frac{\epsilon}{2} \qquad \forall x \in [c, c + h]$$

Hence, $m \geq f(c) - \epsilon/2$ and $M \leq f(c) + \epsilon/2$. Now, we also know that

$$mh \leq \int_c^{c+h} f(t)dt \ \leq \ Mh$$

Thus, we have

$$\frac{F(c + h) - F(c)}{h} = \frac{\int_a^{c+h} f(t)dt - \int_a^c f(t)dt}{h} = \frac{\int_c^{c+h} f(t)dt}{h}$$

Combining inequalities, we find

$$f(c) - \frac{\epsilon}{2} \leq m \leq \frac{F(c + h) - F(c)}{h} \leq M \leq f(c) + \frac{\epsilon}{2}$$

yielding

$$\left| \frac{F(c + h) - F(c)}{h} - f(c) \right| \leq \frac{\epsilon}{2} < \epsilon$$

if $x \in [c, c + h]$.

The case where $h < 0$ is handled in exactly the same way. Thus, since ϵ was arbitrary, this shows that F is differentiable at c and $F'(c) = f(c)$. Note that if $c = a$ or $c = b$, we need only consider the definition of the derivative from one side. $\square$

■

Comment 4.6.1 *We call $F(x)$ the indefinite integral of f. F is always better behaved than f, since integration is a smoothing operation. We can see that f need not be continuous, but, as long as it is integrable, F is always continuous.*

Homework

Exercise 4.6.1 *Let f be bounded on $[a, b]$. We are going to see if we can relax the assumptions used in the Fundamental Theorem of Calculus. We only assume here f is bounded. Define $F : [a, b] \to \Re$*

80　　　　*BASIC ANALYSIS IV: MEASURE THEORY AND INTEGRATION*

by

$$F(x) = \overline{\int_a^x} f(t)dt$$

Then

 (i) $F \in BV([a,b])$.

 (ii) $F \in C([a,b])$.

 (iii) *if f is continuous at $c \in [a,b]$, then F is differentiable at c and $F'(c) = f(c)$.*

Can you prove all of these assertions with this relaxed assumption?

Exercise 4.6.2 *Let f be bounded on $[a,b]$. We are going to see if we can relax the assumptions used in the Fundamental Theorem of Calculus. We only assume here f is bounded. Define $F : [a,b] \to \Re$ by*

$$F(x) = \underline{\int_a^x} f(t)dt$$

Then

 (i) $F \in BV([a,b])$.

 (ii) $F \in C([a,b])$.

 (iii) *if f is continuous at $c \in [a,b]$, then F is differentiable at c and $F'(c) = f(c)$.*

Can you prove all of these assertions with this relaxed assumption?

The next result is one of the many mean value theorems in the theory of integration. It is a more general form of the standard mean value theorem given in beginning calculus classes.

Theorem 4.6.2 The Mean Value Theorem for Riemann Integrals

> *Let $f \in C([a,b])$, and let $g \geq 0$ be integrable on $[a,b]$. Then there is a point, $c \in [a,b]$, such that*
> $$\int_a^b f(x)g(x)dx = f(c) \int_a^b g(x)dx$$

Proof 4.6.5
Since f is continuous, it is also integrable. Hence, fg is integrable. Let m and M denote the lower and upper bounds of f on $[a,b]$, respectively. Then $mg(x) \leq f(x)g(x) \leq Mg(x)$ for all $x \in [a,b]$. Since the integral preserves order, we have

$$m \int_a^b g(x)dx \leq \int_a^b f(x)g(x)dx \leq M \int_a^b g(x)dx$$

If the integral of g on $[a,b]$ is 0, then this shows that the integral of fg will also be 0. Hence, in this case, we can choose any $c \in [a,b]$ and the desired result will follow. If the integral of g is not 0, then it must be positive, since $g \geq 0$. Hence, we have, in this case,

$$m \leq \frac{\int_a^b f(x)g(x)dx}{\int_a^b g(x)dx} \leq M$$

RIEMANN INTEGRATION

Now, f is continuous, so it attains the values M and m at some points. Hence, by the intermediate value theorem, there must be some $c \in [a, b]$ such that

$$f(c) = \frac{\int_a^b f(x)g(x)dx}{\int_a^b g(x)dx}$$

This implies the desired result. ∎

The next result is another standard mean value theorem from basic calculus. It is a direct consequence of the previous theorem, by simply letting $g(x) = 1$ for all $x \in [a, b]$. This result can be interpreted as stating that integration is an averaging process.

Theorem 4.6.3 Average Value for Riemann Integrals

If $f \in C([a, b])$, then there is a point $c \in [a, b]$ such that

$$\frac{1}{b-a} \int_a^b f(x)dx = f(c)$$

The next result is the standard means for calculating definite integrals in basic calculus. We start with a definition.

Definition 4.6.1 The Antiderivative of f

*Let $f : [a, b] \to \Re$ be a bounded function. Let $G : [a, b] \to \Re$ be such that G' exists on $[a, b]$ and $G'(x) = f(x)$ for all $x \in [a, b]$. Such a function is called an **antiderivative** or a **primitive** of f.*

Comment 4.6.2 *The idea of an antiderivative is intellectually distinct from the Riemann integral of a bounded function f. Consider the following function f defined on $[-1, 1]$.*

$$f(x) = \begin{cases} x^2 \sin(1/x^2), & x \neq 0, \ x \in [-1, 1] \\ 0, & x = 0 \end{cases}$$

It is easy to see that this function has a removable discontinuity at 0. Moreover, f is even differentiable on $[-1, 1]$ with derivative

$$f'(x) = \begin{cases} 2x \sin(1/x^2) - (2/x) \cos(1/x^2), & x \neq 0, \ x \in [-1, 1] \\ 0, & x = 0 \end{cases}$$

Note f' is not bounded on $[-1, 1]$ and hence it cannot be Riemann Integrable. Now to connect this to the idea of antiderivatives, just relabel the functions. Let g be defined by

$$g(x) = \begin{cases} 2x \sin(1/x^2) - (2/x) \cos(1/x^2), & x \neq 0, \ x \in [-1, 1] \\ 0, & x = 0 \end{cases}$$

then define G by

$$G(x) = \begin{cases} x^2 \sin(1/x^2), & x \neq 0, \ x \in [-1, 1] \\ 0, & x = 0 \end{cases}$$

We see that G is the antiderivative of g even though g itself does not have a Riemann integral. Note upper and lower integrals don't work here either as g is not bounded. Again, the point is that the idea

82   BASIC ANALYSIS IV: MEASURE THEORY AND INTEGRATION

of the antiderivative of a function is intellectually distinct from that of being Riemann integrable.

Theorem 4.6.4 Cauchy's Fundamental Theorem

> *Let $f : [a,b] \to \Re$ be integrable. Let $G : [a,b] \to \Re$ be any antiderivative of f. Then*
>
> $$\int_a^b f(t)dt = G(t)\Big|_a^b = G(b) - G(a)$$

Proof 4.6.6

Since G' exists on $[a,b]$, G must be continuous on $[a,b]$. Let $\epsilon > 0$ be given. Since f is integrable, there is a partition $\pi_0 \in \Pi[a,b]$ such that for any refinement, π, of π_0 and any $\sigma \subset \pi$, we have

$$\left| S(f, \pi, \sigma) - \int_a^b f(x)dx \right| < \epsilon$$

Let π be any refinement of π_0, given by $\pi = \{x_0 = a, x_1, \ldots, x_p = b\}$. The Mean Value Theorem for differentiable functions then tells us that there is an $s_j \in (x_{j-1}, x_j)$ such that $G(x_j) - G(x_{j-1}) = G'(s_j)\Delta x_j$. Since $G' = f$, we have $G(x_j) - G(x_{j-1}) = f(s_j)\Delta x_j$ for each $j = 1, \ldots, p$. The set of points, $\{s_1, \ldots, s_p\}$, is thus an evaluation set associated with π. Hence,

$$\sum_\pi [G(x_j) - G(x_{j-1})] = \sum_\pi G'(s_j)\Delta x_j = \sum_\pi f(s_j)\Delta x_j$$

The first sum on the left is a collapsing sum, hence we have

$$\Rightarrow G(b) - G(a) = S(f, \pi, \{s_1, \ldots, s_p\})$$

We conclude

$$\left| G(b) - G(a) - \int_a^b f(x)dx \right| < \epsilon$$

Since ϵ was arbitrary, this implies the desired result.   ∎

Comment 4.6.3 *Not all functions (in fact, most functions) will have closed form, or analytically obtainable, antiderivatives. So, the previous theorem will not work in such cases.*

Theorem 4.6.5 The Recapture Theorem

> *If f is differentiable on $[a,b]$, and if $f' \in RI([a,b])$, then*
>
> $$\int_a^x f'(t)dt = f(x) - f(a)$$

Proof 4.6.7

f is an antiderivative of f. Now apply Cauchy's Fundamental Theorem 4.6.4.   ∎

Homework

RIEMANN INTEGRATION

Exercise 4.6.3 *Can you prove this result? Let f be bounded on $[a,b]$, and let $g \geq 0$ be integrable on $[a,b]$. Then there is a point, $c \in [a,b]$, such that*

$$\overline{\int_a^b} f(x)g(x)dx = f(c)\overline{\int_a^b} g(x)dx$$

Exercise 4.6.4 *Can you prove this result? Let f be bounded on $[a,b]$, and let $g \geq 0$ be integrable on $[a,b]$. Then there is a point, $c \in [a,b]$, such that*

$$\underline{\int_a^b} f(x)g(x)dx = f(c)\overline{\int_a^b} g(x)dx$$

Exercise 4.6.5 *Let $I_A(x) = 1$ if $x \in A$ and 0 if not. This is the indicator function of the set A. Define f on $[0,2]$ by*

$$f(x) = 4xI_{\mathbb{Q} \cap [1,2]}(x) - 5xI_{\mathbb{I} \cap [1,2]}$$

Compute $F(x) = \overline{\int_1^x} f(x)\,dx$ and $\underline{\int_1^2} f(x)\,dx$.

Exercise 4.6.6 *We use the notation of the previous exercise. Define f on $[0,2]$ by*

$$f(x) = 4I_{\mathbb{Q} \cap [0,2]}(x) - 2I_{\mathbb{I} \cap [0,2]}$$

Compute $F(x) = \overline{\int_0^x} f(x)\,dx$ and $\underline{\int_1^2} f(x)\,dx$.

Another way to evaluate Riemann integrals is to directly approximate them using an appropriate sequence of partitions. Theorem 4.6.6 is a fundamental tool that tells us when and why such approximations will work.

Theorem 4.6.6 Approximation of the Riemann Integral

If $f \in RI([a,b])$, then given any sequence of partitions $\{\pi_n\}$ with any associated sequence of evaluation sets $\{\sigma_n\}$ that satisfies $\|\pi_n\| \to 0$, we have

$$\lim_{n \to \infty} S(f, \pi_n, \sigma_n) = \int_a^b f(x)\,dx$$

Proof 4.6.8
Since f is integrable, given a positive ϵ, there is a partition π_0 so that

$$\left| S(f, \pi, \sigma) - \int_a^b f(x)dx \right| \quad < \quad \epsilon/2, \ \pi_0 \preceq \pi, \ \sigma \subseteq \pi \qquad (*)$$

Let the partition π_0 be $\{x_0, x_1, \ldots, x_P\}$ and let ξ be defined to be the smallest Δx_j from π_0. Then since the norm of the partitions π_n goes to zero, there is a positive integer N so that

$$\|\pi_n\| \quad < \quad \min\left(\xi, \epsilon/(4P\|f\|_\infty)\right) \qquad (*)$$

Now pick any $n > N$ and label the points of π_n as $\{y_0, y_1, \ldots, y_Q\}$. We see that the points in π_n are close enough together so that at most one point of π_0 lies in any subinterval $[y_{j-1}, y_j]$ from π_n. This follows from our choice of ξ. So the intervals of π_n split into two pieces: those containing a

84   BASIC ANALYSIS IV: MEASURE THEORY AND INTEGRATION

point of π_0 and those that do not have a π_0 inside. Let $\mathscr{A}$ be the first collection of intervals and $\mathscr{B}$, the second. Note there are P points in π_0 and so there are P subintervals in $\mathscr{B}$. Now consider the common refinement $\pi_n \vee \pi_0$. The points in the common refinement match π_n except on the subintervals from $\mathscr{B}$. Let $[y_{j-1}, y_j]$ be such a subinterval and let γ_j denote the point from π_0 which is in this subinterval. Let's define an evaluation set σ for this refinement $\pi_n \vee \pi_0$ as follows.

1. *If we are in the subintervals labeled $\mathscr{A}$, we choose as our evaluation point, the evaluation point s_j that is already in this subinterval since $\sigma_n \subseteq \pi_n$. Here, the length of the subinterval will be denoted by $\delta_j(\mathscr{A})$, which equals $y_j - y_{j-1}$ for appropriate indices.*

2. *If we are in the subintervals labeled $\mathscr{B}$, we have two intervals to consider as $[y_{j-1}, y_j] = [y_{j-1}, \gamma_j] \cup [\gamma_j, y_j]$. Choose the evaluation point γ_j for both $[y_{j-1}, \gamma]$ and $[\gamma, y_j]$. Here, the length of the subintervals will be denoted by $\delta_j(\mathscr{B})$. Note that $\delta_j(\mathscr{B}) = \gamma_j - y_{j-1}$ or $y_j - \gamma_j$.*

Then we have

$$
\begin{aligned}
S(f, \pi_n \vee \pi_0, \sigma) &= \sum_{\mathscr{A}} f(s_j)\delta_j(\mathscr{A}) + \sum_{\mathscr{B}} f(\gamma_j)\delta_j(\mathscr{A}) \\
&= \sum_{\mathscr{A}} f(s_j)(y_j - y_{j-1}) + \sum_{\mathscr{B}} (f(\gamma_j)(y_j - \gamma_j) + f(\gamma_j)(\gamma_j - y_{j-1})) \\
&= \sum_{\mathscr{A}} f(s_j)(y_j - y_{j-1}) + \sum_{\mathscr{B}} (f(\gamma_j)(y_j - y_{j-1}))
\end{aligned}
$$

Thus, since the Riemann sums over π_n and $\pi_n \vee \pi_0$ with these choices of evaluation sets match on $\mathscr{A}$, we have, using Equation $*$, that

$$
\begin{aligned}
|\, S(f, \pi_n, \sigma_n) - S(f, \pi_n \vee \pi_0, \sigma)\, | &= |\sum_{\mathscr{A}} (f(s_j) - f(\gamma_j))(y_j - y_{j-1})\,| \\
&\leq \sum_{\mathscr{A}} (|\,f(s_j)\,| + |\,f(\gamma_j)\,|)\,(y_j - y_{j-1})\,| \\
&\leq P\,2\|f\|_\infty\, \|\pi_n\| \\
&< P\,2\|f\|_\infty\, \frac{\epsilon}{4P\|f\|_\infty} \\
&= \epsilon/2
\end{aligned}
$$

We conclude that for our special evaluation set σ for the refinement $\pi_n \vee \pi_0$ that

$$
\begin{aligned}
\|S(f, \pi_n, \sigma_n) - \int_a^b f(x)dx\,| &= |\,S(f, \pi_n, \sigma_n) - S(f, \pi_n \vee \pi_0, \sigma) \\
&\quad + S(f, \pi_n \vee \pi_0, \sigma) - \int_a^b f(x)dx\,| \\
&\leq |\,S(f, \pi_n, \sigma_n) - S(f, \pi_n \vee \pi_0, \sigma)\,| \\
&\quad + |\,S(f, \pi_n \vee \pi_0, \sigma) - \int_a^b f(x)dx\,| \\
&< \epsilon/2 + \epsilon/2 = \epsilon
\end{aligned}
$$

using Equation $*$ as $\pi_n \vee \pi_0$ refines π_0. Since we can do this analysis for any $n > N$, we see we have shown the desired result. ∎

Homework

RIEMANN INTEGRATION

Exercise 4.6.7 *Let* $f(x) = 3x^2 + 3x - 8$ *on the interval* $[-4, 8]$. *Use Theorem 4.6.6 to compute* $\int_{-4}^{8} f(x)dx$.

Exercise 4.6.8 *Let* $f(x) = -4x^2 + 5x - 7$ *on the interval* $[1, 3]$. *Use Theorem 4.6.6 to compute* $\int_{1}^{3} f(x)dx$.

Exercise 4.6.9 *Let* f *be a continuously differentiable function on* $[a, b]$. *Use Theorem 4.6.6 to show that the arc length of the curve* $\{(x, f(x)) | a \leq x \leq b\}$ *can be defined as* $\int_{a}^{b} \sqrt{1 + (f'(x))^2}dx$. *This is a standard problem discussed in earlier volumes, so it is good to review the process.*

4.7 Substitution Type Results

Using the Fundamental Theorem of Calculus, we can derive many useful tools. This discussion is a precursor for more complicated discussions of such things using integrals defined by what are called measures and integrals extending Riemann integration to what is called Riemann - Stieltjes integration.

Theorem 4.7.1 Integration by Parts

Assume $u : [a, b] \to \Re$ *and* $v : [a, b] \to \Re$ *are differentiable on* $[a, b]$ *and* u' *and* v' *are integrable. Then*

$$\int_{a}^{x} u(t)v'(t) \, dt = u(t)v(t) \Big|_{a}^{x} - \int_{a}^{x} v(t)u'(t) \, dt$$

Proof 4.7.1
Since u *and* v *are differentiable on* $[a, b]$, *they are also continuous and hence, integrable. Now apply the product rule for differentiation to obtain*

$$(u(t)v(t))' = u'(t)v(t) + u(t)v'(t)$$

By Theorem 4.3.2, we know products of integrable functions are integrable. Also, the integral is linear. Hence, the integral of both sides of the equation above is defined. We obtain

$$\int_{a}^{x} (u(t)v(t))' \, dt = \int_{a}^{x} u'(t)v(t) \, dt + \int_{a}^{x} u(t)v'(t) \, dt$$

Since $(uv)'$ *is integrable, we can apply the Recapture Theorem to see*

$$u(t)v(t) \Big|_{a}^{x} = \int_{a}^{x} u'(t)v(t) \, dt + \int_{a}^{x} u(t)v'(t) \, dt$$

This is the desired result. ∎

Theorem 4.7.2 Substitution in Riemann Integration

Let f be continuous on $[c,d]$ and u be continuously differentiable on $[a,b]$ with $u(a) = c$ and $u(b) = d$. Then

$$\int_c^d f(u)\,du \;=\; \int_a^b f(u(t))\,u'(t)\,dt$$

Proof 4.7.2

Let F be defined on $[c,d]$ by $F(u) = \int_c^u f(t)dt$. Then since f is continuous, F is continuous and differentiable on $[c,d]$ by the Fundamental Theorem of Calculus. We know $F'(u) = f(u)$ and so

$$F'(u(t)) = f(u(t)), \; a \leq t \leq b$$

implying

$$F'(u(t))\,u'(t) = f(u(t))u'(t)\,, \; a \leq t \leq b$$

By the Chain Rule for differentiation, we also know

$$(F \circ u)'(t) = F(u(t))u'(t)\,, \; a \leq t \leq b$$

and hence $(F \circ u)'(t) = f(u(t))u'(t)$ on $[a,b]$.

Now define g on $[a,b]$ by

$$g(t) \;=\; (f \circ u)(t)\,u'(t) \;=\; f(u(t))\,u'(t) = (F \circ u)'(t)$$

Since g is continuous, g is integrable on $[a,b]$. Now define G on $[a,b]$ by $G(t) = (F \circ u)(t)$. Then $G'(t) = f(u(t))u'(t) = g(t)$ on $[a,b]$ and G' is integrable. Now, apply the Cauchy Fundamental Theorem of Calculus to G to find

$$\int_a^b g(t)\,dt \;=\; G(b) - G(a)$$

or

$$\int_a^b f(u(t))\,u'(t)\,dt \;=\; F(u(b)) - F(u(a))$$

$$=\; \int_c^{u(b)=d} f(t)dt - \int_c^{u(a)=c} f(t)dt$$

$$=\; \int_c^d f(t)dt.$$

$\blacksquare$

Theorem 4.7.3 Leibnitz's Rule

Let f be continuous on $[a,b]$, $u : [c,d] \to [a,b]$ be differentiable on $[c,d]$ and $v : [c,d] \to [a,b]$ be differentiable on $[c,d]$. Then

$$\left(\int_{u(x)}^{v(x)} f(t)\,dt\right)' \;=\; f(v(x))v'(x) - f(u(x))u'(x)$$

RIEMANN INTEGRATION

Proof 4.7.3

Let F be defined on $[a,b]$ by $F(y) = \int_a^y f(t)dt$. Since f is continuous, F is also continuous and moreover, F is differentiable with $F'(y) = f(y)$. Since v is differentiable on $[c,d]$, we can use the Chain Rule to find

$$\begin{aligned} (F \circ v)'(x) &= F'(v(x))\, v'(x) \\ &= f(v(x))\, v'(x) \end{aligned}$$

This says

$$\left(\int_a^{v(x)} f(t)\, dt \right)' = f(v(x))v'(x)$$

Next, define G on $[a,b]$ by $G(y) = \int_y^b f(t)dt = \int_a^b f(t) - \int_a^y f(t)dt$. Apply the Fundamental Theorem of Calculus to conclude

$$G'(y) = -\left(\int_a^y f(t)dt \right) = -f(y)$$

Again, apply the Chain Rule to see

$$\begin{aligned} (G \circ u)'(x) &= G'(u(x))\, u'(x) \\ &= -f(u(x))\, u'(x). \end{aligned}$$

We conclude

$$\left(\int_{u(x)}^b f(t)\, dt \right)' = -f(u(x))u'(x)$$

Now combine these results as follows:

$$\int_a^b f(t)dt = \int_a^{v(x)} f(t)dt + \int_{v(x)}^{u(x)} f(t)dt + \int_{u(x)}^b f(t)dt$$

or

$$\begin{aligned} (F \circ v)(x) + (G \circ u)(x) - \int_a^b f(t)dt &= -\int_{v(x)}^{u(x)} f(t)dt \\ &= \int_{u(x)}^{v(x)} f(t)dt \end{aligned}$$

Then, differentiate both sides to obtain

$$\begin{aligned} (F \circ v)'(x) + (G \circ u)'(x) &= f(v(x))v'(x) - f(u(x))u'(x) \\ &= \left(\int_{u(x)}^{v(x)} f(t)dt \right)' \end{aligned}$$

which is the desired result. ∎

Homework

Exercise 4.7.1 *Let*

$$k(t,s) \quad = \quad \begin{cases} \cos(s)\cos(10-t), & 0 \le s \le t \\ \cos(t)\cos(10-s), & t < s \le 10 \end{cases}$$

and let

$$x(s) \quad = \quad \int_0^{10} k(t,s)f(s)ds = \int_0^t \cos(s)\cos(10-t)f(s)ds + \int_t^{10} \cos(t)\cos(10-s)f(s)ds$$

for any continuous f on $[0,10]$. *Compute* x' *and* x''.

Exercise 4.7.2 *Let*

$$k(t,s) \quad = \quad \begin{cases} \sin(s)\sin(10-t), & 0 \le s \le t \\ \sin(t)\sin 10 - s), & t < s \le 4 \end{cases}$$

and let

$$x(s) \quad = \quad \int_0^4 k(t,s)f(s)ds = \int_0^t \cos(s)\cos(10-t)f(s)ds + \int_t^4 \cos(t)\cos(10-s)f(s)ds$$

for any continuous f on $[0,4]$. *Compute* x' *and* x''.

4.8 When Do Two Functions Have the Same Integral?

The last results in this chapter seek to find conditions under which the integrals of two functions, f and g, are equal.

Lemma 4.8.1 f **Zero on** (a,b) **Implies Zero Riemann Integral**

> *Let* $f \in B([a,b])$, *with* $f(x) = 0$ *on* (a,b). *Then* f *is integrable on* $[a,b]$ *and*
>
> $$\int_a^b f(x)dx = 0$$

Proof 4.8.1
If f is identically 0, *then the result follows easily. Now, assume* $f(a) \ne 0$ *and* $f(x) = 0$ *on* $(a,b]$. *Let* $\epsilon > 0$ *be given, and let* $\delta > 0$ *satisfy*

$$\delta < \frac{\epsilon}{|f(a)|}$$

Let $\pi_0 \in \Pi[a,b]$ *be any partition such that* $\|\pi_0\| < \delta$. *Let* $\pi = \{x_0 = a, x_1, \ldots, x_p\}$ *be any refinement of* π_0. *Then* $U(f,\pi) = \max(f(a),0)\Delta x_1$ *and* $L(f,\pi) = \min(f(a),0)\Delta x_1$. *Hence, we have*

$$U(f,\pi) - L(f,\pi) = [\max(f(a),0) - \min(f(a),0)]\Delta x_1 = |f(a)| \Delta x_1$$

But

$$|f(a)| \Delta x_1 < |f(a)| \delta < |f(a)| \frac{\epsilon}{|f(a)|} = \epsilon$$

Hence, if π is any refinement of π_0, we have $U(f,\pi) - L(f,\pi) < \epsilon$. This shows that $f \in RI([a,b])$. Further, we have

$$U(f,\pi) = \max(f(a),0)\Delta x_1 \Rightarrow U(f) = \inf_{\pi} U(f,\pi) = 0$$

since we can make Δx_1 as small as we wish. Likewise, we also see that $L(f) = \sup_{\pi} L(f,\pi) = 0$, implying that

$$U(f) = L(f) = \int_a^b f(x)dx = 0$$

The case where $f(b) \neq 0$ and $f(x) = 0$ on $[a,b)$ is handled in the same way. So, assume that $f(a), f(b) \neq 0$ and $f(x) = 0$ for $x \in (a,b)$. Let $\epsilon > 0$ be given, and choose $\delta > 0$ such that

$$\delta < \frac{\epsilon}{2\max\{|f(a)|,|f(b)|\}}$$

Let π_0 be a partition of $[a,b]$ such that $|\pi_0| < \delta$, and let π be any refinement of π_0. Then

$$U(f,\pi) = \max(f(a),0)\Delta x_1 + \max(f(b),0)\Delta x_p$$

$$L(f,\pi) = \min(f(a),0)\Delta x_1 + \min(f(b),0)\Delta x_p$$

It follows that

$$\begin{aligned}
U(f,\pi) - L(f,pi) &= [\max(f(a),0) - \min(f(a),0)]\Delta x_1 + [\max(f(b),0) - \min(f(a),0)]\Delta x_p \\
&= |f(a)|\Delta x_1 + |f(b)|\Delta x_p \\
&< |f(a)|\delta + |f(b)|\delta < \epsilon
\end{aligned}$$

Since we can make Δx_1 and Δx_p as small as we wish, we see $\int_a^b f(x)dx = 0$. ■

Lemma 4.8.2 Riemann Integrable f and g on $[a,b]$ Match on (a,b) Implies Riemann Integrals Match

> *Let $f,g \in RI([a,b])$ with $f(x) = g(x)$ on (a,b). Then*
>
> $$\int_a^b f(x)dx = \int_a^b g(x)dx$$

Proof 4.8.2
Let $h = f - g$, and apply the previous lemma. ■

Theorem 4.8.3 Two Riemann Integrable Functions Match at all but Finitely Many Points Implies Integrals Match

> *Let $f,g \in RI([a,b])$, and assume that $f = g$ except at finitely many points $c_1,\ldots,c_k$. Then*
>
> $$\int_a^b f(x)dx = \int_a^b g(x)dx$$

90 BASIC ANALYSIS IV: MEASURE THEORY AND INTEGRATION

Proof 4.8.3

We may re-index the points $\{c_1, \ldots, c_k\}$, if necessary, so that $c_1 < c_2 < \cdots < c_k$. Then apply Lemma 4.8.2 on the intervals (c_{j-1}, c_j) for all allowable j. This shows

$$\int_{c_{j-1}}^{c_j} f(t)dt = \int_{c_{j-1}}^{c_j} g(t)dt$$

Then, since

$$\int_a^b f(t)dt = \sum_{j=1}^k \int_{c_{j-1}}^{c_j} f(t)dt$$

the results follows. ∎

Theorem 4.8.4 f Bounded and Continuous at all but One Point Implies f is Riemann Integrable

> *If f is bounded on $[a, b]$ and continuous except at one point c in $[a, b]$, then f is Riemann integrable.*

Proof 4.8.4

For convenience, we will assume that c is an interior point, i.e. c is in (a, b). We will show that f satisfies the Riemann Criterion and so it is Riemann integrable. Let $\epsilon > 0$ be given. Since f is bounded on $[a, b]$, there is a real number M so that $f(x) < M$ for all x in $[a, b]$. We know f is continuous on $[a, c - \epsilon/(6M)]$ and f is continuous on $[c + \epsilon/(6M), b]$. Thus, f is integrable on both of these intervals and f satisfies the Riemann Criterion on both intervals. For this ϵ there is a partition π_0 of $[a, c - \epsilon/(6M)]$ so that

$$U(f, \boldsymbol{P}) - L(f, \boldsymbol{P}) \quad < \quad \epsilon/3, \ \text{if } \pi_0 \preceq \boldsymbol{P}$$

and there is a partition π_1 of $[c + \epsilon/(6M), b]$ so that

$$U(f, \boldsymbol{Q}) - L(f, \boldsymbol{Q}) \quad < \quad \epsilon/3, \ \text{if } \pi_0 \preceq \boldsymbol{Q}$$

Let π_2 be the partition we get by combining π_0 with the points $\{c - \epsilon/(6M), c + \epsilon/(6M)\}$ and π_1. Then, we see

$$
\begin{aligned}
U(f, \pi_2) - L(f, \pi_2) &= U(f, \pi_0) - L(f, \pi_0) \\
&+ \left(\sup_{x \in [c - \epsilon/(6M), c + \epsilon/(6M)]} f(x) \right) \epsilon/3 + U(f, \pi_1) - L(f, \pi_1) \\
&< \epsilon/3 + M\epsilon/(3M) + \epsilon/3 = \epsilon
\end{aligned}
$$

Then if $\pi_2 \preceq \pi$ on $[a, b]$, we have

$$U(f, \pi) - L(f, \pi) \quad < \quad \epsilon$$

This shows f satisfies the Riemann criterion and hence is integrable if the discontinuity c is interior to $[a, b]$. The argument at $c = a$ and $c = b$ is similar but a bit simpler as it only needs to be done from one side. Hence, we conclude f is integrable on $[a, b]$ in all cases. ∎

Homework

We can now compute these integrals.

RIEMANN INTEGRATION 91

Exercise 4.8.1 *Let f be defined by*

$$f(x) = \begin{cases} 2x, & 0 \le x < 1 \\ 3, & 1 = x \\ 5x^2 + 1, & 1 < x \le 2 \end{cases}$$

Compute $\int_0^2 f(t)dt$ explaining all steps very carefully.

Exercise 4.8.2 *Let f be defined by*

$$f(x) = \begin{cases} 2, & x = -1 \\ x + 4, & -1 < x < 1 \\ 6, & 1 = x \\ x^2 + 7, & 1 < x < 2 \\ 12, & x = 2 \\ 2x^3, & 2 < x \le 3 \end{cases}$$

Compute $\int_{-1}^3 f(t)dt$ explaining all steps very carefully.

It is then easy to extend this result to a function f which is bounded and continuous on $[a, b]$ except at a finite number of points $\{x_1, x_2, \ldots, x_k\}$ for some positive integer k. We state this as Theorem 4.8.5.

Theorem 4.8.5 f Bounded and Continuous at all but Finitely Many Points Implies f is Riemann Integrable

> *If f is bounded on $[a, b]$ and continuous except at finitely many points $\{x_1, x_2, \ldots, x_k\}$ in $[a, b]$, then f is Riemann integrable.*

Proof 4.8.5
We may assume without loss of generality that the points of discontinuity are ordered as $a < x_1 < x_2 < \ldots < x_k < b$. Then f is continuous except at x_1 on $[a, x_1]$ and hence by Theorem 4.8.4 f is integrable on $[a, x_1]$. Now apply this argument on each of the subintervals $[x_{k-1}, x_k]$ in turn. ∎

Homework

Exercise 4.8.3 *Let f_n, for $n \ge 2$, be defined by*

$$f_n(x) = \begin{cases} 1, & x = 1 \\ 1/(k+1), & 1/(k+1) \le x < 1/k, \ 1 \le k < n \\ 0, & 0 \le x < 1/n \end{cases}$$

Compute $\int_0^1 f_n(t)dt$ explaining all steps very carefully.

Exercise 4.8.4 *Let f_n, for $n \ge 2$, be defined by*

$$f_n(x) = \begin{cases} 1, & x = 1 \\ 1/(k+1), & 1/(k+1) \le x < 1/k, \ 1 \le k < n \\ 0, & 0 \le x < 1/n \end{cases}$$

and let f be the pointwise limit. *Compute* $\lim n \to \infty \int_0^1 f_n(t)dt$ *explaining all steps very carefully. Can you prove this is the same as* $\int_0^1 f(t)dt$?

Chapter 5

Further Riemann Integration Results

In this chapter, we will explore certain aspects of Riemann Integration that are more subtle. We begin with a limit interchange theorem.

5.1 The Limit Interchange Theorem for Riemann Integration

Suppose you knew that the sequence of functions $\{x_n\}$ contained in $RI([a, b])$ converged uniformly to the function x on $[a, b]$. Is it true that $\int_a^b x(t)dt = \lim_{n\to\infty} x_n(t)dt$? The answer to this question is *Yes!* and it is our Theorem 5.1.1.

Theorem 5.1.1 The Riemann Integral Limit Interchange Theorem

> *Let $\{x_n\}$ be a sequence of Riemann Integrable functions on $[a, b]$ which converge uniformly to the function x on $[a, b]$. Then x is also Riemann Integrable on $[a, b]$ and*
>
> $$\int_a^b x(t)dt = \lim_{n\to\infty} \int_a^b x_n(t)dt$$

Proof 5.1.1
First, we show that x is Riemann integrable on $[a, b]$. Let ϵ be given. Then since x_n converges uniformly to x on $[a, b]$,

$$\exists\, \delta > 0 \ni \mid x_n(t) - x(t) \mid \ < \ \frac{\epsilon}{5(b-a)} \, \forall\, n > N,\, t \in [a, b] \tag{α}$$

Fix any $n_1 > N$. Then since x_{n_1} is integrable,

$$\exists\, \pi_0 \in \Pi[a, b] \ni U(x_{n_1}, \pi) - L(x_{n_1}, \pi) \ < \ frac\epsilon5 \,\forall\, \pi_0 \preceq \pi \tag{β}$$

Since x_n converges uniformly to x on $[a, b]$, you should be able to show that x is bounded on $[a, b]$. Hence, we can define

$$M_j \ = \ \sup_{[x_{j-1}, x_j]} x(t), \ M_j^1 \ = \ \sup_{[x_{j-1}, x_j]} x_{n_1}(t)$$

$$m_j \ = \ \inf_{[x_{j-1}, x_j]} x(t), \ m_j^1 \ = \ \inf_{[x_{j-1}, x_j]} x_{n_1}(t)$$

93

94   BASIC ANALYSIS IV: MEASURE THEORY AND INTEGRATION

Using the Infimum and Supremum Tolerance Lemma, there are points s_j and t_j in $[x_{j-1}, x_j]$ so that

$$M_j - \frac{\epsilon}{5(b-a)} \; < \; x(s_j) \leq M_j \tag{γ}$$

and

$$m_j \; \leq \; x(t_j) < m_j + \frac{\epsilon}{5(b-a)} \tag{ξ}$$

Thus,

$$U(x, \boldsymbol{\pi}) \; - \; L(x, \boldsymbol{\pi}) \; = \; \sum_{\boldsymbol{\pi}} (M_j - m_j)\Delta x_j$$

The term on the right hand side can be rewritten using the standard add and subtract trick *as*

$$\sum_{\boldsymbol{\pi}} \left(M_j - x(s_j) + x(s_j) - x_{n_1}(s_j) + x_{n_1}(s_j) - x_{n_1}(t_j) + x_{n_1}(t_j) - x(t_j) + x(t_j) - m_j \right)\Delta x_j$$

We can then overestimate this term using the triangle inequality to find

$$
\begin{aligned}
U(x, \boldsymbol{\pi}) \; - \; L(x, \boldsymbol{\pi}) \; \leq \; & \sum_{\boldsymbol{\pi}}(M_j - x(s_j))\Delta x_j \; + \; \sum_{\boldsymbol{\pi}}(x(s_j) - x_{n_1}(s_j))\Delta x_j \\
& + \; \sum_{\boldsymbol{\pi}}(x_{n_1}(s_j) - x_{n_1}(t_j))\Delta x_j \\
& + \; \sum_{\boldsymbol{\pi}}(x_{n_1}(t_j) - x(t_j))\Delta x_j \; + \; \sum_{\boldsymbol{\pi}}(x(t_j) - m_j)\Delta x_j
\end{aligned}
$$

The first term can be estimated by Equation γ and the fifth term by Equation ξ to give

$$
\begin{aligned}
U(x, \boldsymbol{\pi}) \; - \; L(x, \boldsymbol{\pi}) \; < \; & \frac{\epsilon}{5(b-a)} \sum_{\boldsymbol{\pi}} \Delta x_j \; + \; \sum_{\boldsymbol{\pi}}(x(s_j) - x_{n_1}(s_j))\Delta x_j \\
& + \; \sum_{\boldsymbol{\pi}}(x_{n_1}(s_j) - x_{n_1}(t_j))\Delta x_j \\
& + \; \sum_{\boldsymbol{\pi}}(x_{n_1}(t_j) - x(t_j))\Delta x_j \; + \; \frac{\epsilon}{5(b-a)} \sum_{\boldsymbol{\pi}} \Delta x_j
\end{aligned}
$$

Thus,

$$
\begin{aligned}
U(x, \boldsymbol{\pi}) \; - \; L(x, \boldsymbol{\pi}) \; < \; & 2\frac{\epsilon}{5} \; + \; \sum_{\boldsymbol{\pi}}(x(s_j) - x_{n_1}(s_j))\Delta x_j \\
& + \; \sum_{\boldsymbol{\pi}}(x_{n_1}(s_j) - x_{n_1}(t_j))\Delta x_j \; + \; \sum_{\boldsymbol{\pi}}(x_{n_1}(t_j) - x(t_j))\Delta x_j
\end{aligned}
$$

Now apply the estimate from Equation α to the first and third terms of the equation above to conclude

$$U(x, \boldsymbol{\pi}) \; - \; L(x, \boldsymbol{\pi}) \; < \; 4\frac{\epsilon}{5} \; + \; \sum_{\boldsymbol{\pi}}(x_{n_1}(s_j) - x_{n_1}(t_j))\Delta x_j$$

Finally, note

$$|\, x_{n_1}(s_j) - x_{n_1}(t_j) \,| \; \leq \; M_j^1 - m_j^1$$

and so

$$\sum_{\pi}(x_{n_1}(s_j) - x_{n_1}(t_j))\Delta x_j \;\leq\; \sum_{\pi}(M_j^1 - m_j^1)\Delta x_j$$
$$< \;\; \epsilon/5$$

by Equation β. Thus, $U(x, \pi) - L(x, \pi) < \epsilon$. Since the partition π refining π_0 was arbitrary, we see x satisfies the Riemann Criterion and hence, is Riemann integrable on $[a, b]$.

It remains to show the limit interchange portion of the theorem. Since x_n converges uniformly to x, given a positive ϵ, there is an integer N so that

$$\sup_{a \leq t \leq b} \mid x_n(t) - x(t) \mid \;\; < \;\; \epsilon/(b - a), \; if \, n > N \tag{ζ}$$

Now for any $n > N$, we have

$$\mid \int_a^b x(t)dt \;-\; \int_a^b x_n(t)dt \mid \;\; = \;\; \mid \int_a^b \left(x(t) - x_n(t) \right) dt \mid$$
$$\leq \;\; \int_a^b \left| x(t) - x_n(t) \right| dt$$
$$\leq \;\; \int_a^b \sup_{a \leq t \leq b} \mid x_n(t) - x(t) \mid dt$$
$$< \;\; \int_a^b \epsilon/(b - a)dt$$
$$= \;\; \epsilon$$

using Equation ζ. This says $\lim \int_a^b x_n(t)dt = \int_a^b x(t)dt$. ∎

Homework

Exercise 5.1.1 *Let f_n, for $n \geq 2$, be defined by*

$$f_n(x) \;\; = \;\; \begin{cases} 1, & x = 1 \\ 1/(k+1), & 1/(k+1) \leq x < 1/k, \; 1 \leq k < n \\ 0, & 0 \leq x < 1/n \end{cases}$$

and let f be the pointwise limit. Compute $\lim n \to \infty \int_0^1 f_n(t)dt$ explaining all steps very carefully. It is clear the limit function is not continuous so Theorem 5.1.1 does not apply. We asked this question before, but let's do it again. Prove $\int_0^1 f(t)dt = \lim_{n \to \infty} \int_0^1 f_n(t)dt$ so the results of the theorem can hold even if the convergence is not uniform. There are many other examples of this behavior.

Exercise 5.1.2 *If (f_n) and (g_n) converge uniformly on $[a, b]$ to f and g, respectively, modify the proof above to show $\int_a^b (\alpha f_n(t) + \beta g_n(t))dt \to \int_a^b (\alpha f(t) + \beta g(t))dt$.*

Exercise 5.1.3 *Assume g is Riemann integrable on $[a, b]$ and (f_n) is a sequence of Riemann integrable functions which converges uniformly to f on $[a, b]$. Prove $\int_a^b g(t)f_n(t)dt \to \int_a^b g(t)f(t)dt$.*

Exercise 5.1.4 *Assume (y_n') is a sequence of continuously differentiable functions which converge uniformly to y on $[a, b]$. Prove $y_n \overset{unif}{\to} F$ where $F(x) = \int_a^x y(s)ds$.*

The next result is indispensable in modern analysis. Fundamentally, it states that a continuous real-valued function defined on a compact set can be uniformly approximated by a smooth function. This is used throughout analysis to prove results about various functions. We can often verify a property of a continuous function, f, by proving an analogous property of a smooth function that is uniformly close to f. We will only prove the result for a closed finite interval in $\Re$. The general result for a compact subset of a more general set called a *Topological Space* is a modification of this proof which is actually not that more difficult, but that is another story. We follow the development of (Simmons (12) 1963) for this proof.

Theorem 5.1.2 The Weierstrass Approximation Theorem

> *Let f be a continuous real-valued function defined on $[0,1]$. For any $\epsilon > 0$, there is a polynomial, p, such that $|f(t) - p(t)| < \epsilon$ for all $t \in [0,1]$, that is $\|p - f\|_\infty < \epsilon$*

Proof 5.1.2
We first derive some equalities. We will denote the interval $[0,1]$ by I. By the binomial theorem, for any $x \in I$, we have

$$\sum_{k=0}^{n} \binom{n}{k} x^k (1-x)^{n-k} = (x + 1 - x)^n = 1 \qquad (\alpha)$$

Differentiating both sides of Equation α, we get

$$
\begin{aligned}
0 &= \sum_{k=0}^{n} \binom{n}{k} \left(k x^{k-1}(1-x)^{n-k} - x^k(n-k)(1-x)^{n-k-1} \right) \\
&= \sum_{k=0}^{n} \binom{n}{k} x^{k-1}(1-x)^{n-k-1} \left(k(1-x) - x(n-k) \right) \\
&= \sum_{k=0}^{n} \binom{n}{k} x^{k-1}(1-x)^{n-k-1} \left(k - nx \right)
\end{aligned}
$$

Now, multiply through by $x(1-x)$, to find

$$0 = \sum_{k=0}^{n} \binom{n}{k} x^k (1-x)^{n-k}(k - nx)$$

Differentiating again, we obtain

$$0 = \sum_{k=0}^{n} \binom{n}{k} \frac{d}{dx} \left(x^k(1-x)^{n-k}(k - nx) \right)$$

This leads to a series of simplifications. It is pretty messy and many texts do not show the details, but we think it is instructive.

$$
\begin{aligned}
0 &= \sum_{k=0}^{n} \binom{n}{k} \left[-nx^k(1-x)^{n-k} + (k - nx)\left((k-n)x^k(1-x)^{n-k-1} + kx^{k-1}(1-x)^{n-k} \right) \right] \\
&= \sum_{k=0}^{n} \binom{n}{k} \left[-nx^k(1-x)^{n-k} + (k - nx)(1-x)^{n-k-1}x^{k-1}\left((k-n)x + k(1-x) \right) \right] \\
&= \sum_{k=0}^{n} \binom{n}{k} \left(-nx^k(1-x)^{n-k} + (k - nx)^2(1-x)^{n-k-1}x^{k-1} \right)
\end{aligned}
$$

$$= -n \sum_{k=0}^{n} \binom{n}{k} x^k (1-x)^{n-k} + \sum_{k=0}^{n} \binom{n}{k} (k-nx)^2 x^{k-1}(1-x)^{n-k-1}$$

Thus, since the first sum is 1, we have

$$n = \sum_{k=0}^{n} \binom{n}{k} (k-nx)^2 x^{k-1}(1-x)^{n-k-1}$$

and multiplying through by $x(1-x)$, we have

$$nx(1-x) = \sum_{k=0}^{n} \binom{n}{k} (k-nx)^2 x^k (1-x)^{n-k}$$

$$\frac{x(1-x)}{n} = \sum_{k=0}^{n} \binom{n}{k} \left(\frac{k-nx}{n}\right)^2 x^k (1-x)^{n-k}$$

This last equality then leads to the

$$\sum_{k=0}^{n} \binom{n}{k} \left(x - \frac{k}{n}\right)^2 x^k (1-x)^{n-k} = \frac{x(1-x)}{n} \qquad (\beta)$$

We now define the nth order Bernstein Polynomial *associated with f by*

$$B_n(x) = \sum_{k=0}^{n} \binom{n}{k} x^k (1-x)^{n-k} f\left(\frac{k}{n}\right)$$

Note that

$$f(x) - B_n(x) = \sum_{k=0}^{n} \binom{n}{k} x^k (1-x)^{n-k} \left[f(x) - f\left(\frac{k}{n}\right)\right]$$

Also note that $f(0) - B_n(0) = f(1) - B_n(1) = 0$, so f and B_n match at the endpoints. It follows that

$$| f(x) - B_n(x) | \le \sum_{k=0}^{n} \binom{n}{k} x^k (1-x)^{n-k} \left| f(x) - f\left(\frac{k}{n}\right)\right| \qquad (\gamma)$$

Now, f is uniformly continuous on I since it is continuous. So, given $\epsilon > 0$, there is a $\delta > 0$ such that $|x - \frac{k}{n}| < \delta \Rightarrow |f(x) - f(\frac{k}{n})| < \frac{\epsilon}{2}$. Consider x to be fixed in $[0,1]$. The sum in Equation γ has only $n+1$ terms, so we can split this sum up as follows. Let $\{K_1, K_2\}$ be a partition of the index set $\{0, 1, ..., n\}$ such that $k \in K_1 \Rightarrow |x - \frac{k}{n}| < \delta$ and $k \in K_2 \Rightarrow |x - \frac{k}{n}| \ge \delta$. Then

$$| f(x) - B_n(x) | \le \sum_{k \in K_1} \binom{n}{k} x^k (1-x)^{n-k} \left| f(x) - f\left(\frac{k}{n}\right)\right| + \sum_{k \in K_2} \binom{n}{k} x^k (1-x)^{n-k} \left| f(x) - f\left(\frac{k}{n}\right)\right|$$

which implies

$$|f(x) - B_n(x)| \le \frac{\epsilon}{2} \sum_{k \in K_1} \binom{n}{k} x^k (1-x)^{n-k} + \sum_{k \in K_2} \binom{n}{k} x^k (1-x)^{n-k} \left| f(x) - f\left(\frac{k}{n}\right)\right|$$

$$= \frac{\epsilon}{2} + \sum_{k \in K_2} \binom{n}{k} x^k (1-x)^{n-k} \left| f(x) - f\left(\frac{k}{n}\right)\right|$$

BASIC ANALYSIS IV: MEASURE THEORY AND INTEGRATION

Now, f is bounded on I, so there is a real number $M > 0$ such that $|f(x)| \leq M$ for all $x \in I$. Hence

$$\sum_{k \in K_2} \binom{n}{k} x^k (1-x)^{n-k} \left| f(x) - f\left(\frac{k}{n}\right) \right| \leq 2M \sum_{k \in K_2} \binom{n}{k} x^k (1-x)^{n-k}$$

Since $k \in K_2 \Rightarrow |x - \frac{k}{n}| \geq \delta$, using Equation β, we have

$$\delta^2 \sum_{k \in K_2} \binom{n}{k} x^k (1-x)^{n-k} \leq \sum_{k \in K_2} \binom{n}{k} \left(x - \frac{k}{n}\right)^2 x^k (1-x)^{n-k} \leq \frac{x(1-x)}{n}$$

This implies that

$$\sum_{k \in K_2} \binom{n}{k} x^k (1-x)^{n-k} \leq \frac{x(1-x)}{\delta^2 n}$$

and so combining inequalities

$$2M \sum_{k \in K_2} \binom{n}{k} x^k (1-x)^{n-k} \leq \frac{2Mx(1-x)}{\delta^2 n}$$

We conclude then that

$$\sum_{k \in K_2} \binom{n}{k} x^k (1-x)^{n-k} \left| f(x) - f\left(\frac{k}{n}\right) \right| \leq \frac{2Mx(1-x)}{\delta^2 n}$$

Now, the maximum value of $x(1-x)$ on I is $\frac{1}{4}$, so

$$\sum_{k \in K_2} \binom{n}{k} x^k (1-x)^{n-k} \left| f(x) - f\left(\frac{k}{n}\right) \right| \leq \frac{M}{2\delta^2 n}$$

Finally, choose n so that $n > \frac{M}{\delta^2 \epsilon}$. Then $\frac{M}{n\delta^2} < \epsilon$ implies $\frac{M}{2n\delta^2} < \frac{\epsilon}{2}$. So, Equation γ becomes

$$| f(x) - B_n(x) | \leq \frac{\epsilon}{2} + \frac{\epsilon}{2} = \epsilon$$

Note that the polynomial B_n does not depend on $x \in I$, since n only depends on M, δ, and ϵ, all of which, in turn, are independent of $x \in I$. So, B_n is the desired polynomial, as it is uniformly within ϵ of f. ■

Comment 5.1.1 *A change of variable translates this result to any closed interval $[a, b]$.*

Homework

Exercise 5.1.5 *Prove the extension of the Weierstrass Approximation Theorem to $[a, b]$ by an appropriate change of variables.*

Exercise 5.1.6 *If f is continuous on $[a, b]$, prove there is a sequence of continuously differentiable functions (p_n) which converge in $\| \cdot \|_2$ norm to f.*

Exercise 5.1.7 *This uses ideas from (Peterson (9) 2020) concerning the Hilbert space $\mathbb{L}_2([a, b])$. We know the continuous functions on $[a, b]$ are dense in $\mathbb{L}_2([a, b])$. Prove the set of polynomials in dense in $\mathbb{L}_2([a, b])$ as well.*

Exercise 5.1.8 *Consider the ODE $u'' + 4u = f$, with $u(a) = u(b) = 0$ on $[a, b]$ where f is simply*

FURTHER RIEMANN RESULTS

Riemann integrable with $f(a) = f(b) = 0$ Now let's use ideas about Stürm - Liouville Operators from (Peterson (9) 2020). We know there is a sequence of polynomials on $[a, b]$, (g_n) which converge in $\| \cdot \|_2$ to f.

- *Prove the differential operator is invertible and so a suitable kernel continuous kernel function k on $[a, b] \times [a, b]$ we can write the solution as*

$$u_n(t) = A_n \cos(2t) + B_n \sin(2t) + \int_a^t k(t, s) g_n(s) ds$$

where

$$\begin{bmatrix} \cos(2A_n) & \sin(2A_n) \\ \cos(2B_n) & \sin(2B_n) \end{bmatrix} \begin{bmatrix} 0 \\ -\int_a^b k(t, s) g_n(s) ds \end{bmatrix}$$

always has a unique solution.

- *Prove the sequence $(\int_a^b k(t, s) g_n(s) ds)$ converges in $\| \cdot \|_2$ to $\int_a^b k(t, s) f(s) ds$.*

5.2 Showing Functions are Riemann Integrable

We already know that continuous functions, monotone functions and functions of bounded variation are classes of functions which are Riemann Integrable on the interval $[a, b]$. A good reference for some of the material in this section is (Douglas (2) 1996), although it is mostly in problems and not in the text! Hence, since $f(x) = \sqrt{x}$ is continuous on $[0, M]$ for any positive M, we know f is Riemann integrable on this interval. What about the composition $\sqrt{g}$ where g is just known to be non-negative and Riemann integrable on $[a, b]$? If g were continuous, since compositions of continuous functions are also continuous, we would have immediately that $\sqrt{g}$ is Riemann Integrable. However, it is not so easy to handle this case. Let's try this approach. Using Theorem 5.1.2, we know given a finite interval $[c, d]$, there is a sequence of polynomials $\{p_n(x)\}$ which converge uniformly to $\sqrt{x}$ on $[c, d]$. To apply this here, note that since g is Riemann Integrable on $[a, b]$, g must be bounded. Since we assume g is non-negative, we know that there is a positive number M so that $g(x)$ is in $[0, M]$ for all x in $[a, b]$. Thus, there is a sequence of polynomials $\{p_n\}$ which converge uniformly to $\sqrt{\cdot}$ on $[0, M]$.

Next, using Theorem 4.3.2, we know a polynomial in g is also Riemann integrable on $[a, b]$ ($f^2 = f \cdot f$ so it is integrable and so on). Hence, $p_n(f)$ is Riemann integrable on $[a, b]$. Then given $\epsilon > 0$, we know there is a positive N so that

$$| p_n(u) - \sqrt{u} | < \epsilon, \text{ if } n > N \text{ and } u \in [0, M]$$

Thus, in particular, since $g(x) \in [0, M]$, we have

$$| p_n(g(x)) - \sqrt{g(x)} | < \epsilon, \text{ if } n > N \text{ and } x \in [a, b]$$

We have therefore proved that $p_n \circ g$ converges uniformly to $\sqrt{g}$ on $[0, M]$. Then by Theorem 5.1.1, we see $\sqrt{g}$ is Riemann integrable on $[0, M]$.

Note this type of argument would work for any f which is continuous and g that is Riemann integrable. We state this as Theorem 5.2.1.

Theorem 5.2.1 f Continuous and g Riemann Integrable Implies $f \circ g$ is Riemann Integrable

> *If f is continuous on $g([a,b])$ where g is Riemann Integrable on $[a,b]$, then $f \circ g$ is Riemann Integrable on $[a,b]$.*

Proof 5.2.1
This proof is for you. ∎

In general, the composition of Riemann Integrable functions is not Riemann integrable. Here is the standard counterexample. This great example comes from (Douglas (2) 1996). Define f on $[0,1]$ by

$$f(y) = \begin{cases} 1 & \text{if } y = 0 \\ 0 & \text{if } 0 < y \le 1 \end{cases}$$

and g on $[0,1]$ by

$$g(x) = \begin{cases} 1 & \text{if } x = 0 \\ 1/p & \text{if } x = p/q, (p,q) = 1, x \in (0,1] \text{ and } x \text{ is rational} \\ 0 & \text{if } x \in (0,1] \text{ and } x \text{ is irrational} \end{cases}$$

We see immediately that f is integrable on $[0,1]$ by Theorem 4.8.4. We can show that g is also Riemann integrable on $[0,1]$, but we will leave this as an exercise.

Now $f \circ g$ becomes

$$f(g(x)) = \begin{cases} f(1) & \text{if } x = 0 \\ f(1/p) & \text{if } x = p/q, (p,q) = 1, x \in (0,1] \text{ and } x \text{ rational} \\ f(0) & \text{if } 0 < x \le 1 \text{ and } x \text{ irrational} \end{cases}$$
$$= \begin{cases} 1 & \text{if } x = 0 \\ 0 & \text{if if } x \text{ rational} \in (0,1] \\ 1 & \text{if if } x \text{ irrational} \in (0,1] \end{cases}$$

The function $f \circ g$ above is not Riemann integrable as $U(f \circ g) = 1$ and $L(f \circ g) = 0$. Thus, we have found two Riemann integrable functions whose composition is not Riemann integrable!
Homework

Exercise 5.2.1 *Show g as defined above is*

  1. *Continuous at each irrational points in $[0,1]$ and discontinuous at all rational points in $[0,1]$.*

  2. *Riemann integrable on $[0,1]$ with value $\int_0^1 g(x)dx = 0$.*

Exercise 5.2.2 *Prove Theorem 5.2.1.*

Exercise 5.2.3 *Prove $f(x) = x^{1/3}$ is Riemann integrable using the arguments in the proof of Theorem 5.2.1.*

Exercise 5.2.4 *Prove $f(x) = exp(x^{1/3})$ is Riemann integrable using the arguments in the proof of Theorem 5.2.1.*

5.3 Sets of Content Zero

We already know the length of the finite interval $[a,b]$ is $b - a$ and we exploit this to develop the Riemann integral when we compute lower, upper and Riemann sums for a given partition. We also

FURTHER RIEMANN RESULTS

know that the set of discontinuities of a monotone function is countable. We have seen that continuous functions with a finite number of discontinuities are integrable and in the last section, we saw a function which was discontinuous on a countably infinite set and still was integrable! Hence, we know that a function is integrable should imply something about its discontinuity set. However, the concept of length doesn't seem to apply as there are no intervals in these discontinuity sets. With that in mind, let's introduce a new notion: the *content* of a set. We will follow the development of a set of content zero as it is done in (Sagan (11) 1974).

Definition 5.3.1 Sets of Content Zero

> *A subset S of $\Re$ is said to have content zero if and only if given any positive ϵ we can find a sequence of bounded open intervals $\{J_n^\epsilon = (a_n, b_n)\}$ either finite in number or infinite so that*
> $$S \subseteq \cup J_n,$$
> *with the total length*
> $$\sum (b_n - a_n) < \epsilon$$
> *If the sequence only has a finite number of intervals, the union and sum are written from 1 to N where N is the number of intervals and if there are infinitely many intervals, the sum and union are written from 1 to ∞.*

Comment 5.3.1

1. *A single point c in $\Re$ has content zero because $c \in (c - \epsilon/2, c + \epsilon/2)$ for all positive ϵ.*

2. *A finite number of points $S = \{c_1, \ldots, c_k\}$ in $\Re$ has content zero because $B_i = c_i \in (c_i - \epsilon/(2k), c_i + \epsilon/(2k))$ for all positive ϵ. Thus, $S \subseteq \cup_{i=1}^k B_i$ and the total length of these intervals is smaller than ϵ.*

3. *The rational numbers have content zero also. Let $\{c_i\}$ be any enumeration of the rationals. Let $B_i = (c_i - \epsilon/(2^i), c_i + \epsilon/(2^i))$ for any positive ϵ. The Q is contained in the union of these intervals and the length is smaller than $\epsilon \sum_{i=1}^\infty 1/2^i = \epsilon$.*

4. *Finite unions of sets of content zero also have content zero.*

5. *Subsets of sets of content zero also have content zero.*

Hence, the function g above is continuous on $[0, 1]$ except on a set of content zero. We make this more formal with a definition.

Definition 5.3.2 Continuous Almost Everywhere

> *The function f defined on the interval $[a, b]$ is said to be continuous almost everywhere if the set of discontinuities of f has content zero. We abbreviate the phrase almost everywhere by writing a.e.*

We are now ready to prove an important theorem which is known as the **Riemann - Lebesgue Lemma.** This is also called **Lebesgue's Criterion for the Riemann Integrability of Bounded Functions.** We follow the proof given in (Sagan (11) 1974).

Homework

102 BASIC ANALYSIS IV: MEASURE THEORY AND INTEGRATION

Exercise 5.3.1 *Let f be defined by*

$$f(x) \;=\; \begin{cases} 2x, & 0 \leq x < 1 \\ 3, & 1 = x \\ 5x^2 + 1, & 1 < x \leq 2 \end{cases}$$

Prove explicitly the set of discontinuities of f is content zero.

Exercise 5.3.2 *Let f be defined by*

$$f(x) \;=\; \begin{cases} 2, & x = -1 \\ x + 4, & -1 < x < 1 \\ 6, & 1 = x \\ x^2 + 7, & 1 < x < 2 \\ 12, & x = 2 \\ 2x^3, & 2 < x \leq 3 \end{cases}$$

Prove explicitly the set of discontinuities of f is content zero.

Exercise 5.3.3 *Let f_n, for $n \geq 2$, be defined by*

$$f_n(x) \;=\; \begin{cases} 1, & x = 1 \\ 1/(k+1), & 1/(k+1) \leq x < 1/k,\, 1 \leq k < n \\ 0, & 0 \leq x < 1/n \end{cases}$$

Prove explicitly the set of discontinuities of f is content zero.

Exercise 5.3.4 *Let f be the pointwise limit of the sequence of functions (f_n) for $n \geq 2$ defined by*

$$f_n(x) \;=\; \begin{cases} 1, & x = 1 \\ 1/(k+1), & 1/(k+1) \leq x < 1/k,\, 1 \leq k < n \\ 0, & 0 \leq x < 1/n \end{cases}$$

Prove explicitly the set of discontinuities of f is content zero.

Theorem 5.3.1 The Riemann - Lebesgue Lemma

(i) $f \in B([a,b])$ *and continuous a.e. implies* $f \in RI([a,b])$.

(ii) $f \in RI([a,b])$ *implies f is continuous a.e.*

Proof 5.3.1
The proof of this result is fairly complicated. So grab a cup of coffee, a pencil and prepare for a long battle!
(i):

Proof *We will prove this by showing that for any positive ϵ, we can find a partition π_0 so that the Riemann Criterion is satisfied. First, since f is bounded, there are numbers m and M so that $m \leq f(x) \leq M$ for all x in $[a,b]$. If m and M were the same, then f would be constant and it would therefore be continuous. In this case, we know f is integrable. So we can assume without loss of generality that $M - m > 0$. Let D denote the set of points in $[a,b]$ where f is not continuous. By assumption, the content of D is zero. Hence, given a positive ϵ there is a sequence of bounded open*

FURTHER RIEMANN RESULTS

intervals $J_n = (a_n, b_n)$ (we will assume without loss of generality that there are infinitely many such intervals) so that

$$D \subseteq \cup J_n, \ \sum (b_n - a_n) < \epsilon/(2(M-m))$$

Now if x is from $[a,b]$, x is either in D or in the complement of D, D^C. Of course, if $x \in D^C$, then f is continuous at x. The set

$$E = [a,b] \cap \left(\cup J_n\right)^C$$

is compact and so f must be uniformly continuous on E. Hence, for the ϵ chosen, there is a $\delta > 0$ so that

$$\mid f(y) - f(x) \mid < \epsilon/(8(b-a)) \tag{$*$}$$

if $y \in (x - \delta, x + \delta) \cap E$. Next, note that

$$\boldsymbol{O} = \{J_n, B_{\delta/2}(x) \mid x \in E\}$$

is an open cover of $[a,b]$ and hence must have a finite sub cover. Call this finite sub cover $\boldsymbol{O}'$ and label its members as follows:

$$\boldsymbol{O}' = \{J_{n_1}, \ldots, J_{n_r}, B_{\delta/2}(x_1), \ldots, B_{\delta/2}(x_s)\}$$

Then it is also true that we know that

$$[a,b] \subseteq \boldsymbol{O}'' = \{J_{n_1}, \ldots, J_{n_r}, B_{\delta/2}(x_1) \cap E, \ldots, B_{\delta/2}(x_s) \cap E\}$$

All of the intervals in $\boldsymbol{O}''$ have endpoints. Throw out any duplicates and arrange these endpoints in increasing order in $[a,b]$ and label them as $y_1, \ldots, y_{p-1}$. Then, let

$$\boldsymbol{\pi_0} = \{y_0 = a, y_1, y_2, \ldots, y_{p-1}, y_p = b\}$$

be the partition formed by these points. Recall where the points y_j come from. The endpoints of the $B_{\delta/2}(x_i) \cap E$ sets are not in any of the intervals J_{n_k}. So suppose for two successive points y_{j-1} and y_j, y_{j-1} was in an interval J_{n_k} and the next point y_j was an endpoint of a $B_{\delta/2}(x_i) \cap E$ set which is also inside J_{n_k}. By our construction, this cannot happen as all of the $B_{\delta/2}(x_i) \cap E$ are disjoint from the J_{n_k} sets. Hence, the next point y_j either must be in the set J_{n_k} also or it must be outside. If y_{j-1} is inside and y_j is outside, this is also a contradiction as this would give us a third point, call it z temporarily, so that

$$y_{j-1} < z < y_j$$

with z a new distinct endpoint of the finite cover $\boldsymbol{O}''$. Since we have already ordered these points, this third point is not a possibility. Thus, we see (y_{j-1}, y_j) is in some J_{n_k} or neither of the points is in any J_{n_k}. Hence, we have shown that given the way the points y_j were chosen, either (y_{j-1}, y_j) is inside some interval J_{n_k} or its closure $[y_{j-1}, y_j]$ lies in none of the J_{n_q} for any $1 \leq q \leq r$. But that means (y_{j-1}, y_j) lies in some $\hat{B}_{\delta/2}(x_i)$. Note this set uses the radius $\delta/2$ and so we can say the closed interval $[y_{j-1}, y_j]$ must be contained in some $\hat{B}_{\delta}(x_i)$.

Now we separate the index set $\{1, 2, \ldots, p\}$ into two disjoint sets. We define A_1 to be the set of all indices j so that (y_{j-1}, y_j) is contained in some J_{n_k}. Then we set A_2 to be the complement of A_1 in the entire index set, i.e. $A_2 = \{1, 2, \ldots, p\} \setminus A_1$. Note, by our earlier remarks, if j is in A_2,

104   BASIC ANALYSIS IV: MEASURE THEORY AND INTEGRATION

$[y_{j-1}, y_j]$ is contained in some $B_\delta(x_i) \cap E$. Thus,

$$
\begin{aligned}
U(f, \boldsymbol{\pi_0}) - L(f, \boldsymbol{\pi_0}) &= \sum_{j=1}^{n} \left(M_j - m_j \right) \Delta y_j \\
&= \sum_{j \in A_1} \left(M_j - m_j \right) \Delta y_j + \sum_{j \in A_2} \left(M_j - m_j \right) \Delta y_j
\end{aligned}
$$

Let's work with the first sum: we have

$$
\begin{aligned}
\sum_{j \in A_1} \left(M_j - m_j \right) \Delta y_j &\leq \left(M - m \right) \sum_{j \in A_1} \Delta y_j \\
&< (M - m)\, \epsilon/(2(M - m)) = \epsilon/2
\end{aligned}
$$

Now if j is in A_2, then $[y_{j-1}, y_j]$ is contained in some $B_\delta(x_i) \cap E$. So any two points u and v in $[y_{j-1}, y_j]$ satisfy $|\, u - x_i \,| < \delta$ and $|\, v - x_i \,| < \delta$. Since these points are this close, the uniform continuity condition, Equation $*$, holds. Therefore

$$
|\, f(u) - f(v) \,| \;\leq\; |\, f(u) - f(x_i) \,| + |\, f(v) - f(x_i) \,| < \epsilon/(4(b - a))
$$

This holds for any u and v in $[y_{j-1}, y_j]$. In particular, we can use the Supremum and Infimum Tolerance Lemma to choose u_j and v_j so that

$$
M_j - \epsilon/(8(b - a)) < f(u_j), \;\; m_j + \epsilon/(8(b - a)) > f(v_j)
$$

It then follows that

$$
M_j - m_j < f(u_j) - f(v_j) + \epsilon/(4(b - a))
$$

Now, we can finally estimate the second summation term. We have

$$
\begin{aligned}
\sum_{j \in A_2} \left(M_j - m_j \right) \Delta y_j &< \sum_{j \in A_2} \left(|\, f(u_j) - f(v_j) \,| + \epsilon/(4(b - a)) \right) \Delta y_j \\
&< \sum_{j \in A_2} \left(|\, f(u_j) - f(v_j) \,| \right) \Delta y_j + \epsilon/(4(b - a)) \sum_{j \in A_2} \Delta y_j \\
&< \epsilon/(4(b - a)) \sum_{j \in A_2} \Delta y_j + \epsilon/(4(b - a)) \sum_{j \in A_2} \Delta y_j \\
&< \epsilon/2
\end{aligned}
$$

Combining our estimates, we have

$$
\begin{aligned}
U(f, \boldsymbol{\pi_0}) - L(f, \boldsymbol{\pi_0}) &= \sum_{j \in A_1} \left(M_j - m_j \right) \Delta y_j + \sum_{j \in A_2} \left(M_j - m_j \right) \Delta y_j \\
&< \epsilon/2 + \epsilon/2 = \epsilon
\end{aligned}
$$

Any partition $\boldsymbol{\pi}$ that refines $\boldsymbol{\pi_0}$ will also satisfy $U(f, \boldsymbol{\pi}) - L(f, \boldsymbol{\pi}) < \epsilon$. Hence, f satisfies the Riemann Criterion and so f is integrable. $\qquad\square$

(ii):

FURTHER RIEMANN RESULTS

Proof *We begin by noting that if f is discontinuous at a point x in $[a, b]$, if and only if there is a positive integer m so that*

$$\forall \delta > 0,\ \exists y \in (x - \delta, x + \delta) \cap [a, b] \ \ni \ |f(y) - f(x)| \geq 1/m$$

This allows us to define some interesting sets. Define the set E_m by

$$E_m \ = \ \{x \in [a, b] \mid \forall \delta > 0\ \exists y \in (x - \delta, x + \delta) \cap [a, b] \ \ni \ |f(y) - f(x)| \geq 1/m\}$$

Then, the set of discontinuities of f, $\mathbf{D}$ can be expressed as $\mathbf{D} = \cup_{j=1}^{\infty} E_m$.

Now let $\boldsymbol{\pi} = \{x_0, x_1, \ldots, x_n\}$ be any partition of $[a, b]$. Then, given any positive integer m, the open subinterval $[x_{k-1}, x_k]$ either intersects E_m or it does not. Define

$$A_1 \ = \ \left\{ k \in \{1, \ldots, n\} \mid (x_{k-1}, x_k) \cap E_m \neq \emptyset \right\}$$

$$A_2 \ = \ \left\{ k \in \{1, \ldots, n\} \mid (x_{k-1}, x_k) \cap E_m = \emptyset \right\}$$

By construction, we have $A_1 \cap A_2 = \emptyset$ and $A_1 \cup A_2 = \{1, \ldots, n\}$.

We assume f is integrable on $[a, b]$. So, by the Riemann Criterion, given $\epsilon > 0$, and a positive integer m, there is a partition $\boldsymbol{\pi_0}$ such that

$$U(f, \boldsymbol{\pi}) - L(f, \boldsymbol{\pi}) < \epsilon/(2m),\ \forall \boldsymbol{\pi_0} \preceq \boldsymbol{\pi} \tag{$**$}$$

It follows that if $\boldsymbol{\pi_0} = \{y_0, y_1, \ldots, y_n\}$, then

$$
\begin{aligned}
U(f, \boldsymbol{\pi_0}) - L(f, \boldsymbol{\pi_0}) \ &= \ \sum_{k=1}^{n} (M_k - m_k)\Delta y_k \\
&= \ \sum_{k \in A_1} (M_k - m_k)\Delta y_k \ + \ \sum_{k \in A_2} (M_k - m_k)\Delta y_k
\end{aligned}
$$

If k is in A_1, then by definition, there is a point u_k in E_m and a point v_k in (y_{k-1}, y_k) so that $|f(u_k) - f(v_k)| \geq 1/m$. Also, since u_k and v_k are both in (y_{k-1}, y_k),

$$M_k - m_k \geq |f(u_k) - f(v_k)|$$

Thus,

$$\sum_{k \in A_1} (M_k - m_k)\Delta y_k \ \geq \ \sum_{k \in A_1} |f(u_k) - f(v_k)| \, \Delta y_k \ \geq \ (1/m) \sum_{k \in A_1} \Delta y_k$$

Also, the second term, $\sum_{k \in A_2} (M_k - m_k)\Delta y_k$ is non-negative and so using Equation $**$, we find

$$\epsilon/(2m) \ > \ U(f, \boldsymbol{\pi_0}) - L(f, \boldsymbol{\pi_0}) \ \geq \ (1/m) \sum_{k \in A_1} \Delta y_k$$

which implies $\sum_{k \in A_1} \Delta y_k < \epsilon/2$.

106   BASIC ANALYSIS IV: MEASURE THEORY AND INTEGRATION

The partition π_0 divides $[a, b]$ as follows:

$$[a, b] = \left(\cup_{k \in A_1} (y_{k-1}, y_k) \right) \cup \left(\cup_{k \in A_2} (y_{k-1}, y_k) \right) \cup \left(\{y_0, \ldots, y_n\} \right)$$
$$= C_1 \cup C_2 \cup \pi_0$$

By the way we constructed the sets E_m, we know E_m does not intersect C_2. Hence, we can say

$$E_m = \left(C_1 \cap E_m \right) \cup \left(E_m \cap \pi_0 \right)$$

Therefore, we have $C_1 \cap E_m \subseteq \cup_{k \in A_1} (y_{k-1}, y_k)$ with $\sum_{k \in A_1} \Delta y_k < \epsilon/2$. Since ϵ is arbitrary, we see $C_1 \cap E_m$ has content zero. The other set $E_m \cap \pi_0$ consists of finitely many points and so it also has content zero by the comments at the end of Definition 5.3.1. This shows that E_m has content zero since it is the union of two sets of content zero. We finish by noting $D = \cup E_m$ also has content zero. The proof of this we leave as an exercise. $\square$

    ■

Let's look at some standard results about functions that differ on sets of content zero.

Theorem 5.3.2 If f is Continuous and Zero a.e., the Riemann Integral of f is zero

> *If f is Riemann Integrable on $[a, b]$ and is 0 a.e., then $\int_a^b f(t)dt = 0$.*

Proof 5.3.4
If f is Riemann Integrable and 0 a.e. on $[a, b]$, we know f^+ and f^- are also Riemann integrable and 0 a.e. Let's start with f^+. Since it is Riemann integrable, given $\epsilon > 0$, there is a partition π_0 so that $U(f^+, \pi) - L(f^+, \pi) < \epsilon$ for any refinement π of π_0. Since $f^+ = 0$ a.e., all the lower sums are zero telling us the lower Darboux integral is zero. We then have $U(f^+, \pi) < \epsilon$ for all refinements π of π_0. Hence, the upper Darboux integral is zero too and since they match, $\int_a^b f^+(t)dt = 0$.

A similar argument shows $\int_a^b f^-(t)dt = 0$ too. Combining we have $0 = \int_a^b (f^+(t) - f^-(t))dt = \int_a^b f(t)dt$.   ■

What if the integral of a non-negative function is zero? What does that say about the function?

Theorem 5.3.3 If f is Non-Negative and Riemann Integrable with Zero Value, then f is Zero a.e.

> *If $f \geq 0$ is Riemann Integrable on $[a, b]$ and $\int_a^b |f(s)|ds = 0$ then $f = 0$ a.e.*

Proof 5.3.5
Since $|f|$ is integrable, $|f|$ is continuous a.e. Suppose x is a point where $|f|$ is continuous with $|f(x)| > 0$. Then there is an $r > 0$ so that on $[x - r, x + r]$, $|f(s)| > \frac{|f(x)|}{2}$. Thus

$$0 = \int_a^b |f(s)|ds = \int_a^{x-r} |f(s)|ds + \int_{x-r}^{x+r} |f(s)|ds + \int_{x=r}^b |f(s)|ds$$
$$\geq \int_{x-r}^{x+r} |f(s)|ds > r|f(x)| > 0$$

FURTHER RIEMANN RESULTS

This is not possible, so at each point where $|f|$ is continuous, $|f(x)| = 0$. Hence $|f|$ is zero a.e. ■

It then immediately follows that integrable functions that match a.e. must have the same integrals.

Theorem 5.3.4 Riemann Integrable f and g equal a.e. have the Same Integral

> *If f and g are both Riemann Integrable on $[a,b]$ and $f = g$ a.e., then $\int_a^b f(t)dt = \int_a^b g(t)dt.$*

Proof 5.3.6
Let $h = f - g$. Then h is Riemann Integrable and $h = 0$ a.e. Thus, $0 = \int_a^b h(t)dt = \int_a^b f(t)dt - \int_a^b g(t)dt$. This gives the result. ■

This tells us if $\|f - g\|_1 = 0$, then $f = g$ a.e.

Theorem 5.3.5 Riemann Integrable f and g and $d_1(f,g) = 0$ Implies f is g a.e.

> *If f and g are both Riemann Integrable on $[a,b]$ and $\int_a^b |f(s) - g(s)|ds = 0$ then $f = g$ a.e.*

Proof 5.3.7
Let $h = f - g$. Then h is Riemann Integrable and $\int_a^b |h| = 0$. Thus $|f - g| = 0$ a.e.. This gives the result. ■

Comment 5.3.2 *However, just because a Riemann integrable function f equals another function g a.e. does not mean g is also Riemann integrable. Here is an easy example: let $f(x) = 1$ on $[0,1]$ and $g(x) = 1$ on $\mathbb{I} \cap [0,1]$ and -1 on the complement, $\mathbb{Q} \cap [0,1]$. Then g is not Riemann integrable even though it is equal to the Riemann integrable f a.e.*

Homework

Exercise 5.3.5 *Prove that if $F_n \subseteq [a,b]$ has content zero for all n, then $F = \cup F_n$ also has content zero.*

Exercise 5.3.6 *Let $f(x) = I_{\mathbb{Q} \cap [0,1]} - 2I_{\mathbb{I} \cap [0,1]}$.*

- *Prove f is continuous nowhere on $[0,1]$.*

- *Prove f is not Riemann Integrable.*

- *This is an important example: this function is equal a.e. to the continuous constant function -2, yet it is not integrable. Hence, in general, f equal to a continuous function a.e. is not enough to show f is integrable. The Riemann - Lebesgue does not say f equal to a continuous function a.e. must be integrable.*

Exercise 5.3.7 *Let $f(x) = 3xI_{\mathbb{Q} \cap [0,1]} - 12xI_{\mathbb{I} \cap [0,1]}$.*

- *Prove f is continuous only at 0 on $[0,1]$.*

- *Prove f is not Riemann Integrable.*

- *This is also an important example: this function is equal a.e. to the continuous function $-12x$, yet it is not integrable. The Riemann - Lebesgue does not say f equal to a continuous function a.e. must be integrable.*

BASIC ANALYSIS IV: MEASURE THEORY AND INTEGRATION

Exercise 5.3.8 *Let $f(x) = I_{\mathbb{Q} \cap [0,1]}$.*

- *Prove f is continuous nowhere on $[0,1]$.*

- *Prove f is not Riemann Integrable.*

- *This is another important example: this function is equal a.e. to the continuous constant function 0, yet it is not integrable. Hence, just because a function is zero a.e. does not mean it is integrable. The Riemann - Lebesgue **does not say** f equal to a continuous function a.e. must be integrable. However, if f is continuous and equal to zero a.e. then f is integrable and it is easy to show the integral of f is 0; see (Peterson (8) 2020).*

Exercise 5.3.9 *Recall Dirichlet's Function: $f : [0,1] \to \Re$ defined by*

$$\hat{f}(x) = \begin{cases} 0, & x = 0 \\ \frac{1}{q}, & x = \frac{p}{q}, (p,q) = 1, p, q \in \mathbb{N} \\ 0, & x \in \mathbb{I} \end{cases}$$

- *Go back to (Peterson (8) 2020) and review how to prove f is continuous a.e. and then write it down as carefully as possible. Note we did not use this language back then.*

- *Since f is continuous a.e. and bounded, f is Riemann Integrable. Compute its integral. This is done in (Peterson (8) 2020), so look up that proof also.*

Part III

Riemann - Stieltjes Integration

Chapter 6

The Riemann - Stieltjes Integral

In classical analysis, the Riemann - Stieltjes integral was the first attempt to generalize the idea of the size, or measure, of a subset of the real numbers. Instead of simply using the length of an interval as a measure, we can use any function that satisfies the same properties as the length function. By the way, we discussed functions of bounded variation and Riemann - Stieltjes integrals in (Peterson (9) 2020) when we were talking about game theory but we were not as careful as we are being here.

Let f and g be any bounded functions on the finite interval $[a, b]$. If π is any partition of $[a, b]$ and σ is any evaluation set, we can extend the notion of the Riemann sum $S(f, \pi, \sigma)$ to the more general **Riemann - Stieltjes** sum as follows:

Definition 6.0.1 The Riemann - Stieltjes Sum for Two Bounded Functions

> *Let $f, g \in B([a, b])$, $\pi \in \Pi[a, b]$ and $\sigma \subseteq \pi$. Let the partition points in π be $\{x_0, x_1, \ldots, x_p\}$ and the evaluation points be $\{s_1, s_2, \ldots, s_p\}$ as usual. Define*
>
> $$\Delta g_j = g(x_j) - g(x_j - i), \ 1 \leq j \leq p$$
>
> *and the Riemann - Stieltjes sum for **integrand** f and **integrator** g for partition π and evaluation set π by*
>
> $$S(f, g, \pi, \sigma) = \sum_{j \in \pi} f(s_j) \, \Delta g_j$$
>
> *This is also called the **Riemann - Stieltjes sum for the function** f **with respect to the function** g for partition π and evaluation set σ.*

Of course, you should compare this definition to Definition 4.1.1 to see the differences! We can then define the **Riemann - Stieltjes** integral of f with respect to g using language very similar to that of Definition 4.1.2.

Definition 6.0.2 The Riemann - Stieltjes Integral for Two Bounded Functions

> Let $f, g \in B([a, b])$. If there is a real number I so that for all positive ϵ, there is a partition $\pi_0 \in \Pi[a, b]$ so that
> $$|S(f, g, \pi, \sigma) - I| < \epsilon$$
> for all partitions π that refine π_0 and evaluation sets σ from π, then we say f is Riemann - Stieltjes integrable with respect to g on $[a, b]$. We call the value I the Riemann - Stieltjes integral of f with respect to g on $[a, b]$. We use the symbol
> $$I = RS(f, g; a, b)$$
> to denote this value. We call f the **integrand** and g the **integrator**.

As usual, there is the question of what pairs of functions (f, g) will turn out to have a finite Riemann - Stieltjes integral. The collection of the functions f from $B([a, b])$ that are Riemann - Stieltjes integrable with respect to a given integrator g from $B([a, b])$ is denoted by $RS([g, a, b])$.

Comment 6.0.1 *If $g(x) = x$ on $[a, b]$, then $RS([g, a, b]) = RI([a, b])$ and $RS(f, g; a, b) = \int_a^b f(x)dx$.*

Comment 6.0.2 *We will use the standard conventions: $RS(f, g; a, b) = -RS(f, g; b, a)$ and also $RS(f, g; a; a) = 0$.*

Homework

Exercise 6.0.1 *Let $P = \{1, 1.2, 1.5, 1.8, 2.0\}$ be a partition of $[1, 2]$, $e = \{1.1, 1.4, 1.6, 1.85\}$ be an in-between set and let $f(x) = |\sin(x)|$ and $g(x) = x^2$. Compute $S(f, g, P, E)$.*

Exercise 6.0.2 *Let $P = \{-1, 0.2, 1.5, 1.8, 3.0\}$ be a partition of $[-1, 2]$, $e = \{-0.6, 1.1, 1.7, 2.85\}$ be an in-between set and let $f(x) = |\cos(x)|$ and $g(x) = x^2$. Compute $S(f, g, P, E)$.*

Exercise 6.0.3 *Write MATLAB code to compute $S(f, g, P, E)$ for a given function f and g on the interval $[a, b]$ given a partition P and in-between set E. You can modify code given in (Peterson (8) 2020).*

6.1 Standard Properties of the Riemann - Stieltjes Integral

We can easily prove the usual properties that we expect an *integration* type mapping to have.

Theorem 6.1.1 The Linearity of the Riemann - Stieltjes Integral

> If f_1 and f_2 are in $RS([g, a, b])$, then
>
> (i)
> $$c_1 f_1 + c_2 f_2 \in RS([g, a, b]), \ \forall c_1, c_2 \in \Re$$
>
> (ii)
> $$RS(c_1 f_1 + c_2 f_2, g; a, b) = c_1 RS(f_1, g; a, b) + c_2 RS(f_2, g; a, b)$$
>
> If $f \in RS([g_1, a, b])$ and $f \in RS([g_2, a, b])$ then
>
> (i)
> $$f \in RS([c_1 g_1 + c_2 g_2, a, b]), \ \forall c_1, c_2 \in \Re$$
>
> (ii)
> $$RS(f, c_1 g_1 + c_2 g_2; a, b) = c_1 RS(f, g_1; a, b) + c_2 RS(f, g_2; a, b)$$

THE RIEMANN - STIELTJES INTEGRAL

Proof 6.1.1

We leave these proofs to you as an exercise. The proof of these statements is quite similar in spirit to those of Theorem 4.1.1. You should compare the techniques! ■

Homework

Exercise 6.1.1 *Prove all the standard properties of the Riemann - Stieltjes integral as described in Theorem 6.1.1 in general. These are $\epsilon - \pi$ proofs.*

To make sure these ideas sink in, let's prove the properties for specific choices.

Exercise 6.1.2 *Prove if (f_1, g) and (f_2, g) have Riemann - Stieltjes integrals then $2f_1 + 3f_2 \in RS(g, a, b)$ and $RS(2f_1 + 3f_2, g; a, b) = 2RS(f_1, g; a, b) + 3RS(f_2, g; a, b)$. This is an $\epsilon - \pi$ proof.*

Exercise 6.1.3 *Prove if (f, g_1) and (f, g_2) have Riemann - Stieltjes integrals then $f \in RS(3g_1 - 5g_2, a, b)$ and $RS(f, 3g_1 - 5g_2; a, b) = 3RS(f, g_1; a, b) - 5RS(f, g_2; a, b)$. This is an $\epsilon - \pi$ proof.*

Exercise 6.1.4 *If $f \in RS(g, a, b)$, let $T(f) = RS(f, g; a, b)$. Prove T is a linear functional on the set of bounded functions on $[a, b]$.*

Exercise 6.1.5 *Prove $RS(g, a, b)$ is a vector space over $\Re$.*

To give you a feel for the kind of partition arguments we use for Riemann - Stieltjes proofs (you will no doubt enjoy working out these details for yourselves in various exercises), we will go through the proof of the standard **Integration by Parts** formula in this context.

Theorem 6.1.2 Riemann - Stieltjes Integration by Parts

> *If $f \in RS([g, a, b])$, then $g \in RS([f, a, b])$ and*
>
> $$RS(g, f; a, b) = f(x)g(s)|_a^b - RS(f, g; a, b)$$

Proof 6.1.2

Since $f \in RS([g, a, b])$, there is a number $I_f = RS(f, g; a, b)$ so that given a positive ϵ, there is a partition π_0 such that

$$\left| S(f, g, \pi, \sigma) - I_f \right| \; < \; \epsilon, \; \pi_0 \preceq \pi, \; \sigma \subseteq \pi \tag{α}$$

For such a partition π and evaluation set $\sigma \subseteq \pi$, we have

$$\begin{aligned} \pi &= \{x_0, x_1, \ldots, x_p\}, \\ \sigma &= \{s_1, \ldots, s_p\} \end{aligned}$$

and

$$S(g, f, \pi, \sigma) \; = \; \sum_{\pi} g(s_j)\Delta f_j$$

We can rewrite this as

$$S(g, f, \pi, \sigma) \; = \; \sum_{\pi} g(s_j)f(x_j) - \sum_{\pi} g(s_j)f(x_{j-1}) \tag{β}$$

114   BASIC ANALYSIS IV: MEASURE THEORY AND INTEGRATION

Also, we have the identity (it is a collapsing sum)

$$\sum_{\pi} \Big(f(x_j)g(x_j) - f(x_{j-1})g(x_{j-1}) \Big) = f(b)g(b) - f(a)g(a). \qquad (\gamma)$$

Thus, using Equation β and Equation γ, we have

$$
\begin{aligned}
f(b)g(b) - f(a)g(a) \, - \, S(g, f, \pi, \sigma) \;\; &= \;\; \sum_{\pi} f(x_j) \Big(g(x_j) - g(s_j) \Big) \qquad (\xi) \\
&+ \;\; \sum_{\pi} f(x_{j-1}) \Big(g(s_j) - g(x_{j-1}) \Big)
\end{aligned}
$$

Since $\sigma \subseteq \pi$, we have the ordering

$$a = x_0 \leq s_1 \leq x_1 \leq s_2 \leq x_2 \leq \ldots \leq x_{p-1} \leq s_p \leq x_p = b$$

Hence, the points above are a refinement of π we will call π'. Relabel the points of π' as

$$\pi' = \{y_0, y_1, \ldots, y_q\}$$

and note that the original points of π now form an evaluation set σ' of π'. We can therefore rewrite Equation ξ as

$$f(b)g(b) - f(a)g(a) \, - \, S(g, f, \pi, \sigma) = \sum_{\pi} f(y_j)\Delta g_j \;=\; S(f, g, \pi', \sigma')$$

Let $I_g = f(b)g(b) - f(a)g(a) - I_f$. Then since $\pi_0 \preceq \pi \preceq \pi'$, we can apply Equation α to conclude

$$
\begin{aligned}
\epsilon \;\; &> \;\; \Big| S(f, g, \pi', \sigma') - I_f \Big| \\
&= \;\; \Big| f(b)g(b) - f(a)g(a) \, - \, S(g, f, \pi, \sigma) \, - \, I_f \Big| \\
&= \;\; \Big| S(g, f, \pi, \sigma) \, - \, I_g \Big|
\end{aligned}
$$

Since our choice of refinement π of π_0 and evaluation set σ was arbitrary, we have shown that $g \in RS([f, a, b])$ with value

$$RS(g, f, a, b) = f(x)g(x)\big|_a^b - RS(f, g, a, b)$$

$\blacksquare$

Homework

Exercise 6.1.6 *Assume $f \in RS(x^2, a, b)$ where we abuse notation and just place the definition of $g(x) = x^2$ directly inside the formula. Follow the proof we did above and prove $g \in RS(f, a, b)$.*

Exercise 6.1.7 *Assume $f \in RS(x^3, a, b)$. Follow the proof we did above and prove $g \in RS(f, a, b)$.*

THE RIEMANN - STIELTJES INTEGRAL   115

6.2   Step Function Integrators

We now turn our attention to the question of what pairs of functions might have a Riemann - Stieltjes integral. All we know so far is that if $g(x) = x$ on $[a, b]$ is labeled as $g = id$, then $RS([f, id, a, b]) = RI([f, a, b])$.

First, we need to define what we mean by a **Step Function**.

Definition 6.2.1 Step Function

> *We say $g \in B([a, b])$ is a Step Function if g only has finitely many jump discontinuities on $[a, b]$ and g is constant on the intervals between the jump discontinuities. Thus, we may assume there is a non-negative integer p so that the jump discontinuities are ordered and labeled as*
> $$c_0 < c_1 < c_2 < \ldots < c_p$$
> *and g is constant on each subinterval (c_{k-1}, c_k) for $1 \leq k \leq p$.*

Comment 6.2.1 *We can see $g(c_k^-)$ and $g(c_k^+)$ both exist and are finite with $g(c_k^-)$ the value g has on (c_{k-1}, c_k) and $g(c_k^+)$ the value g has on (c_k, c_{k+1}). At the endpoints, $g(a^+)$ and $g(b^-)$ are also defined. The actual finite values g takes on at the points c_j are completely arbitrary.*

We can prove a variety of results about Riemann - Stieltjes integrals with step function integrations.

Lemma 6.2.1 One Jump Step Functions as Integrators: One

> *Let $g \in B([a, b])$ be a step function having only one jump at some c in $[a, b]$. Let $f \in B([a, b])$. Then $f \in C([a, b])$ implies $f \in RS([g, a, b])$ and*
>
> - *If $c \in (a, b)$, then $RS(f, g; a, b) = f(c)[g(c^+) - g(c^-)]$.*
> - *If $c = a$, then $RS(f, g; a, b) = f(a)[g(a^+) - g(a)]$.*
> - *If $c = b$, then $RS(f, g; a, b) = f(b)[g(b) - g(b^-)]$.*

Proof 6.2.1
Let π be any partition of $[a, b]$. We will assume that c is a partition point of π because if not, we can use the argument we have used before to construct an appropriate refinement as done, for example, in the proof of Lemma 4.5.1. Letting the partition points be

$$\pi = \{x_0, x_1, \ldots, x_p\}$$

we see there is a partition point $x_{k_0} = c$ with $k_0 \neq 0$ or p. Hence, on $[x_{k_0-1}, x_{k_0}] = [x_{k_0-1}, c]$, $\Delta g_{k_0} = g(c) - g(x_{k_0-1})$. However, since g has a single jump at c, we see that the value $g(x_{k_0-1})$ must be $g(c^-)$. Thus, $\Delta g_{k_0} = g(c) - g(c^-)$. A similar argument shows that $\Delta g_{k_0} = g(c^+) - g(c)$. Further, since g has only one jump, all the other terms Δg_k are zero. Hence, for any evaluation set σ in π, we have $\sigma = \{s_1, \ldots, s_p\}$ and

$$
\begin{aligned}
S(f, g, \pi, \sigma) &= f(s_{k_0})\Delta g_{k_0} + f(s_{k_0+1})\Delta g_{k_0+1} \\
&= f(s_{k_0})\bigg(g(c) - g(c^-)\bigg) + f(s_{k_0+1})\bigg(g(c^+) - g(c)\bigg) \\
&= \bigg(f(s_{k_0}) - f(c) + f(c)\bigg)\bigg(g(c) - g(c^-)\bigg)
\end{aligned}
$$

$$+ \left(f(s_{k_0+1}) - f(c) + f(c)\right)\left(g(c^+) - g(c)\right)$$

Thus, we obtain

$$
\begin{aligned}
S(f,g,\boldsymbol{\pi},\boldsymbol{\sigma}) \;=\;& \left(f(s_{k_0}) - f(c)\right)\left(g(c) - g(c^-)\right) \\[2mm]
+\;& \left(f(s_{k_0+1}) - f(c)\right)\left(g(c^+) - g(c)\right) \qquad (\boldsymbol{\alpha}) \\[2mm]
+\;& f(c)\left(g(c^+) - g(c^-)\right)
\end{aligned}
$$

We know f is continuous at c. Let $A = \max\left(\mid g(c) - g(c^-)\mid,\mid g(c^+) - g(c)\mid\right)$. Then $A > 0$ because g has a jump at c. Since f is continuous at c, given $\epsilon > 0$, there is a $\delta > 0$, so that

$$\mid f(x) - f(c)\mid \; < \; \epsilon/(2A),\; x \in (c - \delta, c + \delta) \cap [a,b]. \qquad (\boldsymbol{\beta})$$

In fact, since c is an interior point of $[a,b]$, we can choose δ so small that $(c - \delta, c + \delta) \subseteq [a,b]$. Now, if $\boldsymbol{\pi_0}$ is any partition with $\|\boldsymbol{\pi_0}\| < \delta$ containing c as a partition point, we can argue as we did in the prefatory remarks to this proof. Thus, there is an index k_0 so that

$$[x_{k_0-1}, x_{k_0} = c] \subseteq (c - \delta, c],\; [c = x_{k_0}, x_{k_0+1}] \subseteq [c, c + \delta)$$

This implies that

$$[x_{k_0-1}, x_{k_0+1}] \subseteq (c - \delta, c + \delta)$$

and so the evaluation points, labeled as usual, s_{k_0} and s_{k_0+1} are also in $(c - \delta, c + \delta)$. Applying Equation β, we have

$$\mid f(s_{k_0}) - f(c)\mid < \epsilon/(2A),\; \mid f(s_{k_0+1}) - f(c)\mid < \epsilon/(2A)$$

From Equation $\boldsymbol{\alpha}$, we then have

$$
\begin{aligned}
\left| S(f,g,\boldsymbol{\pi},\boldsymbol{\sigma}) - f(c)\left(g(c^+) - g(c^-)\right)\right| \;\le\;& \left|\left(f(s_{k_0}) - f(c)\right)\left(g(c) - g(c^-)\right)\right| \\[2mm]
+\;& \left|\left(f(s_{k_0+1}) - f(c)\right)\left(g(c^+) - g(c)\right)\right| \\[2mm]
<\;& \epsilon/(2A)\left|g(c) - g(c^-)\right| + \epsilon/(2A)\left|g(c^+) - g(c)\right| \\[2mm]
<\;& \epsilon
\end{aligned}
$$

Finally, if $\boldsymbol{\pi_0} \preceq \boldsymbol{\pi}$, then $\|\boldsymbol{\pi}\| < \delta$ also and the same argument shows that for any evaluation set $\boldsymbol{\sigma} \subseteq \boldsymbol{\pi}$, we have

$$\left| S(f,g,\boldsymbol{\pi},\boldsymbol{\sigma}) - f(c)\left(g(c^+) - g(c^-)\right)\right| \; < \; \epsilon$$

THE RIEMANN - STIELTJES INTEGRAL   117

This proves that $f \in RS([g, a, b])$ and $RS(f, g; a, b) = f(c)\left(g(c^+) - g(c^-)\right)$. Now, if $c = a$ or $c = b$, the arguments are quite similar, except one-sided and we find

$$RS(f, g; a, b) = f(a)\left(g(a^+) - g(a)\right) \text{ or } RS(f, g; a, b) = f(b)\left(g(b) - g(b^-)\right)$$

■

Homework

Here f is continuous at the jump and g simply has a jump.

Exercise 6.2.1 *Let*

$$f(x) = x^2, \quad \text{and } g(x) = \begin{cases} 5, 1 \leq x < 2 \\ 10, & x = 2 \\ 7, & 2 < x \leq 4 \end{cases}$$

Compute $RS(f, g; 1, 4)$.

Exercise 6.2.2 *Let*

$$f(x) = 5x^2 + 3x_2, \quad \text{and } g(x) = \begin{cases} 5, & 1 \leq x \leq 3 \\ -2, & 3 < x \leq 5 \end{cases}$$

Compute $RS(f, g; 1, 5)$.

Exercise 6.2.3 *Let*

$$f(x) = 4x^2, \quad \text{and } g(x) = \begin{cases} -4, & -1 \leq x < 2 \\ -2, & 2 \leq x \leq 3 \end{cases}$$

Compute $RS(f, g; -1, 3)$.

Lemma 6.2.2 One Jump Step Functions as Integrators: Two

Let $g \in B([a, b])$ be a step function having only one jump at some c in $[a, b]$. Let $f \in B([a, b])$. If $c \in (a, b)$, $f(c^-) = f(c)$ and $g(c^+) = g(c)$, then $f \in RS([g, a, b])$. We can rephrase this as: if c is an interior point, f is continuous from the left at c and g is continuous from the right at c, then $f \in RS([g, a, b])$ and

- *If $c \in (a, b)$, then $RS(f, g; a, b) = f(c)[g(c) - g(c^-)]$.*

- *If $c = a$, then $RS(f, g; a, b) = f(a)[g(a) - g(a)] = 0$.*

- *If $c = b$, then $RS(f, g; a, b) = f(b)[g(b) - g(b^-)]$.*

Proof 6.2.2

To prove this result, we first use the initial arguments of Lemma 6.2.1 and then we note in this case, f is continuous from the left at c so $f(c^-) = f(c)$. Further, g is continuous from the right; hence,

118   BASIC ANALYSIS IV: MEASURE THEORY AND INTEGRATION

$g(c) = g(c^+)$. *Thus, Equation* α *reduces to*

$$S(f,g,\pi,\sigma) = \left(f(s_{k_0}) - f(c)\right)\left(g(c) - g(c^-)\right) + f(c)\left(g(c) - g(c^-)\right) \qquad (\alpha')$$

Let $L = | \, g(c) - g(c^-) \, |$. *Then, given* $\epsilon > 0$, *since* f *is continuous from the left, there is a* $\delta > 0$ *so that*

$$| \, f(x) - f(c) \, | < \epsilon/L, \quad x \in (c - \delta, c] \subseteq [a, b]$$

As usual, we can restrict our attention to partitions that contain the point c. *We continue to use* x_i*'s and* s_j*'s to represent points in these partitions and associated evaluation sets. Let* π *be such a partition with* $x_{k_0} = c$ *and* $\|\pi\| < \delta$. *Let* σ *be any evaluation set of* π. *Then, we have*

$$[x_{k_0-1}, x_{k_0}] \subseteq (c - \delta, c]$$

and thus

$$| \, f(s_{k_0}) - f(c) \, | < \epsilon/L$$

Hence,

$$\left| S(f,g,\pi,\sigma) - f(c)\left(g(c^+) - g(c)\right) \right| = \left| f(s_{k_0}) - f(c) \right| \left| g(c) - g(c^-) \right|$$

$$< \epsilon$$

Finally, just as in the previous proof, if $\pi_0 \preceq \pi$, *then* $\|\pi\| < \delta$ *also and the same argument shows that for any evaluation set* $\sigma \subseteq \pi$, *we have*

$$\left| S(f,g,\pi,\sigma) - f(c)\left(g(c) - g(c^-)\right) \right| < \epsilon$$

This proves that $f \in RS([g, a, b])$ *and* $RS(f, g; a, b) = f(c)\left(g(c) - g(c^-)\right)$. *Now, if* $c = a$ *or* $c = b$, *the arguments are again similar, except one-sided. For* $c = a$, *we find that* $RS(f, g; a, b) = f(a)(g(a) - g(a)) = 0$ *and for* $c = b$, *we find* $RS(f, g; a, b) = f(b)(g(b) - g(b^-))$. ∎

Homework

Here f is continuous from the left at the jump and g is continuous from the right at the jump.

Exercise 6.2.4 *Let*

$$f(x) \begin{cases} 12, & 1 \le x \le 2 \\ -2, & 2 < x \le 4 \end{cases} \quad and \quad g(x) = \begin{cases} 5, & 1 \le x < 2 \\ 7, & 2 \le x \le 4 \end{cases}$$

Compute $RS(f, g; 1, 4)$.

Exercise 6.2.5 *Let*

$$f(x) \begin{cases} x^2, & 1 \le x \le 2 \\ -2x^2, & 2 < x \le 4 \end{cases} \quad and \quad g(x) = \begin{cases} 1, & 1 \le x < 2 \\ 8, & 2 \le x \le 4 \end{cases}$$

Compute $RS(f, g; 1, 4)$.

THE RIEMANN - STIELTJES INTEGRAL 119

Exercise 6.2.6 *Let*

$$f(x) \begin{cases} 1 + 2x - x^2, & -11 \leq x \leq 3 \\ -2x^2, & 3 < x \leq 6 \end{cases} \quad and \; g(x) = \begin{cases} -2 & , -11 \leq x < 3 \\ 9 & , 3 \leq x \leq 6 \end{cases}$$

Compute $RS(f, g; -11, 6)$.

Lemma 6.2.3 One Jump Step Functions as Integrators: Three

Let $g \in B([a, b])$ *be a step function having only one jump at some c in* $[a, b]$. *Let* $f \in B([a, b])$. *If* $c \in (a, b)$, $f(c^+) = f(c)$ *and* $g(c^-) = g(c)$, *then* $f \in RS([g, a, b])$. *We can rephrase this as: if c is an interior point, f is continuous from the right at c and g is continuous from the left at c, then* $f \in RS([g, a, b])$ *and*

- *If* $c \in (a, b)$, *then* $RS(f, g; a, b) = f(c)[g(c^+) - g(c)]$.

- *If* $c = a$, *then* $RS(f, g; a, b) = f(a)[g(a^+) - g(a)]$.

- *If* $c = b$, *then* $RS(f, g; a, b) = f(b)[g(b) - g(b)] = 0$.

Proof 6.2.3
This is quite similar to the argument presented for Lemma 6.2.2. We find $f \in RS([g, a, b])$ *and* $RS(f, g; a, b) = f(c)(g(c^+) - g(c))$. *Now, if* $c = a$ *or the arguments are again similar, except one-sided and we find* $RS(f, g; a, b) = f(a)(g(a^+) - g(a)) = 0$. *At* $c = b$, *we have* $RS(f, g; a, b) = f(b)(g(b) - g(b)) = 0$. ∎

Homework

Here f is continuous from the right at the jump and g is continuous from the left at the jump.

Exercise 6.2.7 *Let*

$$f(x) \begin{cases} 12, & 1 \leq x < 2 \\ -2, & 2 \leq x \leq 4 \end{cases} \quad and \; g(x) = \begin{cases} 5, & 1 \leq x \leq 2 \\ 7, & 2 < x \leq 4 \end{cases}$$

Compute $RS(f, g; 1, 4)$.

Exercise 6.2.8 *Let*

$$f(x) \begin{cases} x^2, & 1 \leq x < 2 \\ -2x^2 & , 2 \leq x \leq 4 \end{cases} \quad and \; g(x) = \begin{cases} 1, & 1 \leq x \leq 2 \\ 8, & 2 < x \leq 4 \end{cases}$$

Compute $RS(f, g; 1, 4)$.

Exercise 6.2.9 *Let*

$$f(x) \begin{cases} 1 + 2x - x^2, & -11 \leq x < 3 \\ -2x^2, & 3 \leq x \leq 6 \end{cases} \quad and \; g(x) = \begin{cases} -2, & -11 \leq x \leq 3 \\ 9, & 3 < x \leq 6 \end{cases}$$

Compute $RS(f, g; -11, 6)$.

Exercise 6.2.10 *Let*

$$f(x) \begin{cases} 1 + 2x - x^2, & -11 \leq x < 3 \\ -2x^2, & 3 \leq x \leq 6 \end{cases} \quad and \; g(x) = \begin{cases} 5, & = 11 = x \\ -2, & -11 < x \leq 6 \end{cases}$$

Compute $RS(f, g; -11, 6)$.

Exercise 6.2.11 *Let*

$$f(x) \begin{cases} 1 + 2x - x^2, & -11 \le x < 3 \\ -2x^2, & 3 \le x \le 6 \end{cases} \quad \text{and } g(x) = \begin{cases} -2, & -11 \le x < 6 \\ 10, & 6 = x \end{cases}$$

Compute $RS(f, g; -11, 6)$.

We can then generalize to a finite number of jumps.

Lemma 6.2.4 Finite Jump Step Functions as Integrators

Let g be a step function on $[a, b]$ with jump discontinuities at

$$\{a \le c_0, c_1, \ldots, c_{k-1}, c_k \le b\}$$

Assume $f \in B([a, b])$. Then, if

    (i) f is continuous at c_j, or

    (ii) f is left continuous at c_j and g is right continuous at c_j, or

    (iii) f is right continuous at c_j and g is left continuous at c_j,

then, $f \in RS([f, g, a, b])$ and

$$RS(f, g, a, b) = f(a)\Big(g(a^+) - g(a)\Big) + \sum_{j=0}^{k} f(c_j)\Big(g(c_j^+) - g(c_j^-)\Big) + f(b)\Big(g(b) - g(b^-)\Big)$$

Proof 6.2.4
Use Lemma 6.2.1, Lemma 6.2.2 and Lemma 6.2.3 repeatedly. ∎

Homework

Exercise 6.2.12 *Let $f(x) = x^2$ on $[0, 10]$ and let g be defined by*

$$g(x) = \begin{cases} 2, & 0 \le x < 1 \\ 5, & 1 \le x < 2 \\ -3, & 2 \le x < 4 \\ 12, & 4 \le x \le 10 \end{cases}$$

Compute $RS(f, g; 0, 10)$.

Exercise 6.2.13 *Let $f(x) = x^3$ on $[-1, 3]$ and let g be defined by*

$$g(x) = \begin{cases} 4, & -1 \le x < 1 \\ 7, & 1 \le x < 2 \\ 3, & 2 \le x \le 3 \end{cases}$$

Compute $RS(f, g; 0, 10)$.

THE RIEMANN - STIELTJES INTEGRAL
121

Exercise 6.2.14 *Let $f(x) = x^2$ on $[0, 10]$ and let g be defined by*

$$g(x) \;=\; \begin{cases} 2, & 0 \le x \le 1 \\ 5, & 1 < x <\le 2 \\ -3, & 2 < x \le 4 \\ 12, & 4 < x \le 10 \end{cases}$$

Compute $RS(f, g; 0, 10)$.

Exercise 6.2.15 *Let $f(x) = x^3$ on $[-1, 3]$ and let g be defined by*

$$g(x) \;=\; \begin{cases} 4, & -1 \le x \le 1 \\ 7, & 1 < x \le 2 \\ 3, & 2 < x \le 3 \end{cases}$$

Compute $RS(f, g; 0, 10)$.

Exercise 6.2.16 *Let $f(x) = x^4$ on $[0, 1]$ and let g_n, for $n \ge 2$, be defined by*

$$g_n(x) \;=\; \begin{cases} 1, & x = 1 \\ 1/(k+1), & 1/(k+1) \le x < 1/k, \; 1 \le k < n \\ 0, & 0 \le x < 1/n \end{cases}$$

Compute $RS(f, g; 0, 1)$.

6.3 Monotone Integrators

The next step is to learn how to deal with integrators that are monotone functions. To do this, we extend the notion of Darboux Upper and Lower Sums in the obvious way.

Definition 6.3.1 Upper and Lower Riemann - Stieltjes Darboux Sums

> *Let $f \in B([a, b])$ and $g \in B([a, b])$ be monotone increasing. Let π be any partition of $[a, b]$ with partition points*
> $$\pi = \{x_0, x_1, \ldots, x_p\}$$
> *as usual. Define*
> $$M_j = \sup_{x \in [x_{j-1}, x_j]} f(x), \;\; m_j = \inf_{x \in [x_{j-1}, x_j]} f(x)$$
> *The Lower Riemann - Stieltjes Darboux Sum for f with respect to g on $[a, b]$ for the partition π is*
> $$L(f, g, \pi) = \sum_{\pi} m_j \Delta g_j$$
> *and the Upper Riemann - Stieltjes Darboux Sum for f with respect to g on $[a, b]$ for the partition π is*
> $$U(f, g, \pi) = \sum_{\pi} M_j \Delta g_j$$

Comment 6.3.1 *It is clear that for any partition π and associated evaluation set σ, we have the usual inequality chain:*
$$L(f, g, \pi) \;\le\; S(f, g, \pi, \sigma \;\le\; U(f, g, \pi)$$

122 BASIC ANALYSIS IV: MEASURE THEORY AND INTEGRATION

The following theorems have proofs very similar to the ones we did for Theorem 4.2.1 and Theorem 4.2.2.

Theorem 6.3.1 $\pi \preceq \pi'$ **Implies** $L(f, g, \pi) \leq L(f, g, \pi')$ **and** $U(f, g, \pi) \geq U(f, g, \pi')$

Assume g is a bounded monotone increasing function on $[a, b]$ and $f \in B([a, b])$. Then if $\pi \preceq \pi'$, then $L(f, g, \pi) \leq L(f, g, \pi')$ and $U(f, g, \pi) \geq U(f, g, \pi')$.

Theorem 6.3.2 $L(f, g, \pi_1) \leq U(f, g, \pi_2)$

Let π_1 and π_2 be any two partitions in $\mathbf{\Pi}[a, b]$. Then $L(f, g, \pi_1) \leq U(f, g, \pi_2)$.

These two theorems allow us to prove the following

Theorem 6.3.3 The Upper and Lower Riemann - Stieltjes Darboux Integrals are Finite

Let $f \in B([a, b])$ and let g be a bounded monotone increasing function on $[a, b]$. Let $\mathscr{U} = \{L(f, g, \pi) \,|\, \pi \in \mathbf{\Pi}[a, b]\}$ and $\mathscr{V} = \{U(f, g, \pi) \,|\, \pi \in \mathbf{\Pi}[a, b]\}$. Define $L(f, g) = \sup \mathscr{U}$, and $U(f, g) = \inf \mathscr{V}$. Then $L(f, g)$ and $U(f, g)$ are both finite. Moreover, $L(f, g) \leq U(f, g)$.

We can then define upper and lower Riemann - Stieltjes integrals analogous to the way we defined the upper and lower Riemann integrals.

Definition 6.3.2 Upper and Lower Riemann - Stieltjes Integrals

Let $f \in B([a, b])$ and g be a bounded, monotone increasing function on $[a, b]$. The Upper and Lower Riemann - Stieltjes integrals of f with respect to g are $U(f, g)$ and $L(f, g)$, respectively.

Thus, we can define the Riemann - Stieltjes Darboux integral of $f \in B([a, b])$ with respect to the bounded monotone increasing integrator g.

Definition 6.3.3 The Riemann - Stieltjes Darboux Integral

Let $f \in B([a, b])$ and g be a bounded, monotone increasing function on $[a, b]$. We say f is Riemann - Stieltjes Darboux integrable with respect to the integrator g if $U(f, g) = L(f, g)$. We denote this common value by $RSD(f, g, a, b)$.

Homework

Exercise 6.3.1 *Let g be defined by*

$$g(x) = \begin{cases} 2x, & 0 \leq x < 1 \\ 3, & 1 = x \\ 5x^2 + 1, & 1 < x \leq 2 \end{cases}$$

- *Prove g is a monotone function. However, it is not a step function.*

- *Let $P = \{0.0, 1.0, 1.5, 2.0\}$ be a partition of the interval $[0.0, 2.0]$ and compute $L(f, g, P)$ and $U(f, g, P)$ for the function $f(x) = x^2$.*

THE RIEMANN - STIELTJES INTEGRAL 123

Exercise 6.3.2 *Let g be defined by*

$$g(x) = \begin{cases} 2, & x = -1 \\ x + 4, & -1 < x < 1 \\ 6, & 1 = x \\ x^2 + 7, & 1 < x < 2 \\ 12, & x = 2 \\ 2x^3, & 2 < x \leq 3 \end{cases}$$

- *Prove g is a monotone function. However, it is not a step function.*

- *Let $P = \{-1.0, 0.0.0.4.1.3, 2.0, 2.5, 3.0\}$ be a partition of the interval $[-1.0, 3.0]$ and compute $L(f, g, P)$ and $U(f, g, P)$ for the function $f(x) = 2x^2 + 3x_2$.*

Exercise 6.3.3 *Let g_n, for $n \geq 2$, be defined by*

$$g_n(x) = \begin{cases} 1, & x = 1 \\ 1/(k+1), & 1/(k+1) \leq x < 1/k, \ 1 \leq k < n \\ 0, & 0 \leq x < 1/n \end{cases}$$

- *Prove g_n is a monotone function. However, it is not a step function.*

- *Let $P = \{0.0.0.1, 0.3, 0.5, 0.7, 1.0\}$ be a partition of the interval $[0.0, 1.0]$ and compute $L(f, g_n, P)$ and $U(f, g_n, P)$ for the function $f(x) = (2x - 1)^2$.*

Exercise 6.3.4 *Let g be the pointwise limit of the sequence of functions (g_n) for $n \geq 2$ defined by*

$$g_n(x) = \begin{cases} 1, & x = 1 \\ 1/(k+1), & 1/(k+1) \leq x < 1/k, \ 1 \leq k < n \\ 0, & 0 \leq x < 1/n \end{cases}$$

- *Prove g is a monotone function. However, it is not a step function.*

- *Let $P = \{0.0.0.1, 0.3, 0.5, 0.7, 1.0\}$ be a partition of the interval $[0.0, 1.0]$ and compute $L(f, g, P)$ and $U(f, g, P)$ for the function $f(x) = (2x - 1)^2$.*

6.4 The Riemann - Stieltjes Equivalence Theorem

The connection between the Riemann - Stieltjes and Riemann - Stieltjes Darboux integrals is obtained using an analog of the familiar Riemann Condition we have seen before in Definition 4.2.4.

Definition 6.4.1 The Riemann - Stieltjes Criterion for Integrability

Let $f \in B([a, b])$ and g be a bounded monotone increasing function on $[a, b]$. We say the Riemann - Stieltjes Condition or Criterion holds for f with respect to g if there is a partition of $[a, b]$, π_0 so that

$$U(f, g, \pi) - L(f, g, \pi) < \epsilon, \ \pi_0 \preceq \pi$$

We can then prove an equivalence theorem for Riemann - Stieltjes and Riemann - Stieltjes Darboux integrability.

Theorem 6.4.1 The Riemann - Stieltjes Integral Equivalence Theorem

> Let $f \in B([a,b])$ and g be a bounded monotone increasing function on $[a,b]$. Then the following are equivalent.
>
> (i) $f \in RS([g,a,b])$.
>
> (ii) The Riemann - Stieltjes Criterion holds for f with respect to g.
>
> (iii) f is Riemann - Stieltjes Darboux Integrable, i.e., $L(f,g) = U(f,g)$, and $RS(f,g;a,b) = RSD(f,g;a,b)$.

Proof 6.4.1
The arguments are essentially the same as presented in the proof of Theorem 4.2.4 and hence, you will be asked to go through the original proof and replace occurrences of Δx_j with Δg_j and $b - a$ with $g(b) - g(a)$. ∎

Comment 6.4.1 *We have been very careful to distinguish between Riemann - Stieltjes and Riemann - Stieltjes Darboux integrability. Since we now know they are equivalent, we can begin to use a common notation. Recall, the common notation for the Riemann integral is $\int_a^b f(x)dx$. We will now begin using the notation $\int_a^b f(x)dg(x)$ to denote the common value $RS(f,g;a,b) = RSD(f,g;a,b)$. We thus know $int_a^b f(x)dx$ is equivalent to the Riemann - Stieltjes integral of f with respect to the integrator $g(x) = x$. Hence, in this case, we could write $g(x) = \mathbf{id}(x) = x$, where $\mathbf{id}$ is the identity function. We could then use the notation $\int_a^b f(x)dx = \int_a^b f(x)d\mathbf{id}$. However, that is cumbersome. We can easily remember that the identity mapping is simply x itself. So replace $d\mathbf{id}$ by dx to obtain $\int_a^b f(x)dx$. The use of the (x) in these notations has always been helpful as it allows us to handle substitution type rules, but it is certainly somewhat awkward. A reasonable change of notation would be to go to using boldface for the f and g in these integrals and write $\int_a^b \mathbf{f}d\mathbf{g}$ giving $\int_a^b \mathbf{f}d\mathbf{x}$ for the simpler Riemann integral.*

Comment 6.4.2 *You can see no matter what we do the symbolism becomes awkward. For example, suppose $f(x) = sin(x^2)$ on $[0,\pi]$ and $g(x) = x^2$. Then, how do we write $\int_0^\pi \mathbf{f}d\mathbf{g}$? We will usually abuse our integral notation and write $\int_0^\pi sin(x^2)d(x^2)$.*

6.5 Further Properties of the Riemann - Stieltjes Integral

We can prove the following useful collection of facts about Riemann - Stieltjes integrals.

Theorem 6.5.1 Properties of the Riemann - Stieltjes Integral

Let the integrator g be bounded and monotone increasing on $[a, b]$. Assume f_1, f_2 and f_3 are in $RS([g, a, b])$. Then

(i) $|f| \in RS([g, a, b])$;

(ii)

$$\left| \int_a^b f(x) dg(x) \right| \leq \int_a^b |f| \, dg(x)$$

(iii) $f^+ = \max\{f, 0\} \in RS([g, a, b])$;

(iv) $f^- = \max\{-f, 0\} \in RS([g, a, b])$;

(v)

$$\int_a^b f(x) dg(x) = \int_a^b [f^+(x) - f^-(x)] dg(x)$$

$$= \int_a^b f^+(x) dg(x) - \int_a^b f^-(x) dg(x)$$

$$\int_a^b |f(x)| \, dg(x) = \int_a^b [f^+(x) + f^-(x)] dg(x)$$

$$= \int_a^b f^+(x) dg(x) + \int_a^b f^-(x) dg(x)$$

(vi) $f^2 \in RS([g, a, b])$;

(vii) $f_1 f_2 \in RS([g, a, b])$;

(viii) If there exists m such that $0 < m \leq f(x)$ for all x in $[a, b]$, then $1/f \in RS([g, a, b])$.

Proof 6.5.1
The arguments are straightforward modifications of the proof of Theorem 4.3.2 replacing $b - a$ with $g(b) - g(a)$ and replacing Δx_j with Δg_j. ∎

Homework

Exercise 6.5.1 *Use an induction argument to prove if f_1 and f_2 are in $RS(g, a, b)$ for a bounded monotone increasing g, then $f^p g^q$ is also in $RS(g, a, b)$ for all positive integers p and q.*

Exercise 6.5.2 *Use an induction argument to prove f is bounded below on $[a, b]$ and g is bounded below on $[a, b]$ with f and g in $RS(h, a, b)$ for a bounded monotone increasing h, then $(1/f)^p (1/g)^q$ is also in $RS(h, a, b)$ for all positive integers p and q.*

Exercise 6.5.3 *Use an induction argument to prove if f and g are in $RS(h, a, b)$ for a bounded monotone increasing h, then $(f^+)^p (g^-)^q$ is also in $RS(h, a, b)$ for all positive integers p and q.*

We can also easily prove the following fundamental estimate.

Theorem 6.5.2 Fundamental Riemann - Stieltjes Integral Estimates

> *Let g be bounded and monotone increasing on $[a,b]$ and let $f \in RS([g,a,b])$. Let $m = \inf_x f(x)$ and let $M = \sup_x f(x)$. Then*
>
> $$m(g(b) - g(a)) \leq \int_a^b f(x)dg(x) \leq M(g(b) - g(a))$$

In addition, Riemann - Stieltjes integrals are also order preserving as we can modify the proof of Theorem 4.1.3 quite easily.

Theorem 6.5.3 The Riemann - Stieltjes Integral is Order Preserving

> *Let g be bounded and monotone increasing on $[a,b]$ and $f, f_1, f_2 \in RS([g,a,b])$ with $f_1 \leq f_2$ on $[a,b]$. Then the Riemann - Stieltjes integral is order preserving in the sense that*
>
> *(i)*
> $$f \geq 0 \Rightarrow \int_a^b f(x)dg(x) \geq 0$$
>
> *(ii)*
> $$f_1 \leq f_2 \Rightarrow \int_a^b f_1(x)dg(x) \leq \int_a^b f_2(x)dg(x)$$

We also want to establish the familiar summation property of the Riemann - Stieltjes integral over an interval $[a,b] = [a,c] \cup [c,b]$. We can modify the proof of the corresponding result in Lemma 4.5.1 as usual to obtain Lemma 6.5.4.

Lemma 6.5.4 The Upper and Lower Riemann - Stieltjes Darboux Integral is Additive on Intervals

> *Let g be bounded and monotone increasing on $[a,b]$ and $f \in B([a,b])$. Let $c \in (a,b)$. Define*
>
> $$\int_{\underline{a}}^b f(x)\, dg(x) = L(f,g) \text{ and } \overline{\int_a^b} f(x)\, dg(x) = U(f,g)$$
>
> *Denote the lower and upper Riemann - Stieltjes Darboux integrals of f on with respect to g on $[a,b]$, respectively. Then we have*
>
> $$\overline{\int_a^b} f(x)dg(x) = \overline{\int_a^c} f(x)dg(x) + \overline{\int_c^b} f(x)dg(x)$$
>
> $$\int_{\underline{a}}^b f(x)dg(x) = \int_{\underline{a}}^c f(x)dg(x) + \int_{\underline{c}}^b f(x)dg(x).$$

Lemma 6.5.4 allows us to prove the existence of the Riemann - Stieltjes on $[a,b]$, implies it also exists on subintervals of $[a,b]$ and the Riemann - Stieltjes value is additive. The proofs are obvious modifications of the proofs of Theorem 4.5.2 and Theorem 4.5.3, respectively.

Theorem 6.5.5 The Riemann - Stieltjes Integral Exists on Subintervals

> *Let g be bounded and monotone increasing on $[a,b]$. If $f \in RS([g,a,b])$ and $c \in (a,b)$, then $f \in RS([g,a,c])$ and $f \in RS([g,c,b])$.*

THE RIEMANN - STIELTJES INTEGRAL   127

Theorem 6.5.6 The Riemann - Stieltjes Integral is Additive on Subintervals

> If $f \in RS([g, a, b])$ and $c \in (a, b)$, then
>
> $$\int_a^b f(x)dg(x) = \int_a^c f(x)dg(x) + \int_c^b f(x)dg(x)$$

6.6   Bounded Variation Integrators

We now turn our attention to integrators which are of bounded variation. By Theorem 3.4.3, we know that if $g \in BV([a, b])$, then we can write $g = u - v$ where u and v are monotone increasing on $[a, b]$. Note if h is any other monotone increasing function on $[a, b]$, we could also use the decomposition

$$g = (u + h) - (v + h)$$

as well, so this representation is certainly not unique. We must be very careful when we extend the Riemann - Stieltjes integral to bounded variation integrators. For example, even if $f \in RS([g, a, b])$ it does not always follow that $f \in RS([u, a, b])$ and /or $f \in RS([v, a, b])$! However, we can prove that this statement is true if we use a particular decomposition of f. Let $u(x) = V_g(x)$ and $v(x) = V_g(x) - g(x)$ be our decomposition of g. Then, we will be able to show $f \in RS([g, a, b])$ implies $f \in RS([V_g, a, b])$ and $f \in RS([V_g - g, a, b])$.

Theorem 6.6.1 f Riemann - Stieltjes Integrable with Respect to g of Bounded Variation Implies Integrable with respect to V_g and $V_g - g$

> Let $g \in BV([a, b])$ and $f \in RS([g, a, b])$. Then $f \in RS([V_g, a, b])$ and $f \in RS([V_g - g, a, b])$.

Proof 6.6.1

For convenience of notation, let $u = V_g$ and $v = V_g - g$. First, we show that $f \in RS([u, a, b])$ by showing the Riemann - Stieltjes Criterion holds for f with respect to u on $[a, b]$. Fix a positive ϵ. Then there is a partition π_0 so that

$$\left| S(f, g, \pi, \sigma) - \int_a^b f(x)dg(x) \right| < \epsilon$$

for all refinements π of π_0 and evaluation sets σ of π. Thus, given two such evaluation sets σ_1 and σ_2 of a refinement π, we have

$$
\begin{aligned}
\left| S(f, g, \pi, \sigma_1) - S(f, g, \pi, \sigma_2) \right| \ \leq \ & \left| S(f, g, \pi, \sigma_1) - \int_a^b f(x)dg(x) \right| \\
+ \ & \left| S(f, g, \pi, \sigma_2) - \int_a^b f(x)dg(x) \right| \\
< \ & 2\epsilon
\end{aligned}
$$

Hence, we know for $\sigma_1 = \{s_1, \dots, s_p\}$ and $\sigma_2 = \{s_1', \dots, s_p'\}$, that

$$\left| S(f, g, \pi, \sigma_1) - S(f, g, \pi, \sigma_2) \right| \ < \ 2\epsilon \qquad (\alpha)$$

Now, $u(b) = V_g(b) = \sup_\pi \sum_\pi |\Delta g_j|$. Thus, by the Supremum Tolerance Lemma, there is a partition π_1 so that

$$u(b) - \epsilon < \sum_{\pi_1} |\Delta g_j| \le u(b)$$

Then if π refines π_1, we have

$$u(b) - \epsilon < \sum_{\pi_1} |\Delta g_j| \le \sum_{\pi} |\Delta g_j| \le u(b)$$

and so for all $\pi_1 \preceq \pi$,

$$u(b) - \epsilon < \sum_{\pi} |\Delta g_j| \le u(b). \qquad (\beta)$$

Now let $\pi_2 = \pi_0 \vee \pi_1$ and choose any partition π that refines π_2. Then,

$$\sum_{\pi} \left((M_j - m_j) |\Delta u_j| - |\Delta g_j| \right) \le \sum_{\pi} \left((M_j + m_j)\Delta u_j - |\Delta g_j| \right)$$

$$\le 2M \sum_{\pi} |\Delta u_j| - \sum_{\pi} |\Delta g_j|$$

where $M = \|f\|_\infty$. But the term $\sum_\pi \Delta u_j$ is a collapsing sum which becomes $u(b) - u(a) = u(b)$ as $u(a) = 0$. We conclude

$$\sum_{\pi} \left((M_j - m_j) |\Delta u_j| - |\Delta g_j| \right) \le 2M \left(u(b) - \sum_{\pi} |\Delta g_j| \right)$$

Now by Equation α, for all refinements of π_2, we have

$$u(b) - \sum_{\pi} |\Delta g_j| < \epsilon$$

Hence,

$$\sum_{\pi} \left((M_j - m_j) |\Delta u_j| - |\Delta g_j| \right) \le 2M \epsilon. \qquad (\gamma)$$

Next, for any refinement of π of π_2, let the partition points be $\{x_0, \dots, x_n\}$ as usual and define

$$J^+(\pi) = \{j \in \pi | \Delta g_j \ge 0\}, \quad J^-(\pi) = \{j \in \pi | \Delta g_j < 0\}$$

By the Infimum and Supremum Tolerance Lemma, if $j \in J^+(\pi)$,

$$\exists s'_j \in [x_{j-1}, x_j] \ni m_j \le f(s'_j) < m_j + \epsilon/2, \ \exists s_j \in [x_{j-1}, x_j] \ni M_j - \epsilon/2 < f(s_j) \le M_j$$

It follows

$$f(s_j) - f(s'_j) > M_j - m_j - \epsilon, \ j \in J^+(\pi). \qquad (\xi)$$

On the other hand, if $j \in J^-(\pi)$, we can find s_j and s'_j in $[x_{j-1}, x_j]$ so that

$$\exists s'_j \in [x_{j-1}, x_j] \ni m_j \le f(s_j) < m_j + \epsilon/2, \ \exists s_j \in [x_{j-1}, x_j] \ni M_j - \epsilon/2 < f(s'_j) \le M_j$$

THE RIEMANN - STIELTJES INTEGRAL

This leads to

$$f(s_j') - f(s_j) > M_j - m_j - \epsilon, \; j \in J^-(\boldsymbol{\pi}). \tag{ζ}$$

Thus,

$$
\begin{aligned}
\sum_{\boldsymbol{\pi}} \left(M_j - m_j \right) \mid \Delta g_j \mid \; &= \; \sum_{j \in J^+(\boldsymbol{\pi})} \left(M_j - m_j \right) \Delta g_j \; + \; \sum_{j \in J^-(\boldsymbol{\pi})} \left(M_j - m_j \right) \left(-\Delta g_j \right) \\
&< \; \sum_{j \in J^+(\boldsymbol{\pi})} \left(f(s_j) - f(s_j') \right) \Delta g_j \; + \; \epsilon \sum_{j \in J^+(\boldsymbol{\pi})} \Delta g_j \\
&+ \; \sum_{j \in J^-(\boldsymbol{\pi})} \left(f(s_j') - f(s_j) \right) \left(-\Delta g_j \right) \; + \; \epsilon \sum_{j \in J^+(\boldsymbol{\pi})} \left(-\Delta g_j \right) \\
&= \; \sum_{j \in \boldsymbol{\pi}} \left(f(s_j) - f(s_j') \right) \Delta g_j \; + \; \epsilon \sum_{j \in \boldsymbol{\pi}} \mid \Delta g_j \mid
\end{aligned}
$$

Also, by the definition of the variation function of g, we have

$$\sum_{j \in \boldsymbol{\pi}} \mid \Delta g_j \mid \; \leq u(b) = V_g(b)$$

Since the points $\{s_1, \ldots, s_n\}$ and $\{s_1', \ldots, s_n'\}$ are evaluation sets of $\boldsymbol{\pi}$, we can apply Equation $\boldsymbol{\alpha}$ to conclude

$$
\begin{aligned}
\mid S(f, g, \boldsymbol{\pi}, \boldsymbol{\sigma_1}) - S(f, g, \boldsymbol{\pi}, \boldsymbol{\sigma_2}) \mid \; &= \; \sum_{j \in \boldsymbol{\pi}} \left(f(s_j) - f(s_j') \right) \Delta g_j \\
&< \; 2\epsilon
\end{aligned}
$$

Hence,

$$\sum_{\boldsymbol{\pi}} \left(M_j - m_j \right) \mid \Delta g_j \mid \; < \; 2\epsilon + \epsilon u(b) = (2 + u(b))\epsilon \tag{$\boldsymbol{\theta}$}$$

Then, using Equation $\boldsymbol{\gamma}$ and Equation $\boldsymbol{\theta}$, we find

$$
\begin{aligned}
\sum_{\boldsymbol{\pi}} \left(M_j - m_j \right) \Delta u_j \; &= \; \sum_{\boldsymbol{\pi}} \left(M_j - m_j \right) \left(\Delta u_j - \mid \Delta g_j \mid \right) \; + \; \sum_{\boldsymbol{\pi}} \left(M_j - m_j \right) \left(\mid \Delta g_j \mid \right) \\
&< \; 2M\epsilon \; + \; (2 + u(b))\epsilon = (2M + 2 + u(b))\epsilon
\end{aligned}
$$

Letting $A = 2M + 2 + u(b)$, and recalling that $u = V_g$, we have

$$U(f, V_g, \boldsymbol{\pi}) - L(f, V_g, \boldsymbol{\pi}) < A\epsilon$$

for any refinement $\boldsymbol{\pi}$ of $\boldsymbol{\pi_2}$. Hence, f satisfies the Riemann - Stieltjes Criterion with respect to V_g on $[a, b]$. We conclude $f \in RS([V_g, a, b])$.

Thus, $f \in RS([g, a, b])$ and $f \in RS([V_g, a, b])$ and by Theorem 6.1.1, we have $f \in RS([V_g - g, a, b])$ also. ∎

Homework

130 BASIC ANALYSIS IV: MEASURE THEORY AND INTEGRATION

Exercise 6.6.1 *Prove the set of all functions $f \in RS(g, a, b)$ for a bounded variation g on $[a, b]$ is a vector space over $\Re$.*

Exercise 6.6.2 *Define $T(f) = \int_a^b f(x)dg(x)$. Prove T is a linear functional on the set of all functions $f \in RS(g, a, b)$ for a bounded variation g on $[a, b]$ and find an upper bound for the set of all possible $|T(f)|$.*

Theorem 6.6.2 Products and Reciprocals of Functions Riemann - Stieltjes Integrable with Respect to g of Bounded Variation are also Integrable

> *Let $g \in BV([a, b])$ and $f, f_1, f_2 \in RS([g, a, b])$. Then*
>
> *(i) $f^2 \in RS([g, a, b])$.*
>
> *(ii) $f_1 f_2 \in RS([g, a, b])$.*
>
> *(iii) If there is a positive constant m, so that $|f(x)| > m$ for all x in $[a, b]$, then $1/f \in Rs([g, a, b])$.*

Proof 6.6.2

(i)

Proof *Since $f \in RS([g, a, b])$, $f \in RS([V_g, a, b])$ and $f \in RS([V_g - g, a, b])$ by Theorem 6.6.1. Hence, by Theorem 6.5.1, $f^2 \in RS([V_g, a, b])$ and $f^2 \in RS([V_g - g, a, b])$. Then, by the linearity of the Riemann - Stieltjes integral for monotone integrators, Theorem 6.1.1, we have $f^2 \in RS([V_g - (V_g - g) = g, a, b])$.* $\square$

(ii)

Proof *$f_1, f_2 \in RS([g, a, b])$ implies $f_1, f_2 \in RS([V_g, a, b])$ and $f_1, f_2 \in RS([V_g - g, a, b])$. Thus, using reasoning just like that in Part (i), we have $f_1 f_2 \in RS([g, a, b])$.* $\square$

(iii)

Proof *By our assumptions, we know $1/f \in RS([V_g, a, b])$ and $1/f \in RS([V_g - g, a, b])$. Thus, by the linearity of the Riemann - Stieltjes integral with respect to monotone integrators, $1/f \in RS([g, a, b])$.* $\square$

∎

Theorem 6.6.3 The Riemann - Stieltjes Integral is Additive on Subintervals

> *Let $g \in BV([a, b])$ and $f \in RS([g, a, b])$. Then, if $a \le c \le b$,*
>
> $$\int_a^b f(x)dg(x) = \int_a^c f(x)dg(x) + \int_c^b f(x)dg(x)$$

Proof 6.6.6
From Theorem 6.5.6, we know

$$\int_a^b f(x)dV_g(x) = \int_a^c f(x)dV_g(x) + \int_c^b f(x)dV_g(x)$$

THE RIEMANN - STIELTJES INTEGRAL

$$\int_a^b f(x)d(V_g - g)(x) = \int_a^c f(x)d(V_g - g)(x) + \int_c^b f(x)d(V_g - g)(x)$$

Also, we know

$$\int_a^b f(x)dg(x) = \int_a^b f(x)dV_g(x) - \int_c^b f(x)d(V_g - g)(x)$$

and so the result follows. ∎

Homework

Exercise 6.6.3 *Use an induction argument to prove if f_1 and f_2 are in $RS(g, a, b)$ for a bounded variation g, then $f^p g^q$ is also in $RS(g, a, b)$ for all positive integers p and q.*

Exercise 6.6.4 *Use an induction argument to prove f is bounded below on $[a, b]$ and g is bounded below on $[a, b]$ with f and g in $RS(h, a, b)$ for a bounded variation h, then $(1/f)^p(1/g)^q$ is also in $RS(h, a, b)$ for all positive integers p and q.*

Exercise 6.6.5 *Use an induction argument to prove if f and g are in $RS(h, a, b)$ for a bounded variation h, then $(f^+)^p(g^-)^q$ is also in $RS(h, a, b)$ for all positive integers p and q.*

Chapter 7

Further Riemann - Stieltjes Results

We know quite a bit about the Riemann - Stieltjes integral in theory. However, we do not know how to compute a Riemann - Stieltjes integral and we only know that Riemann - Stieltjes integrals exist for a few types of integrators: those that are bounded with a finite number of jumps and the identity integrator $g(x) = x$. It is time to learn more.

7.1 The Riemann - Stieltjes Fundamental Theorem of Calculus

As you might expect, we can prove a Riemann - Stieltjes variant of the Fundamental Theorem of Calculus.

Theorem 7.1.1 Riemann - Stieltjes Fundamental Theorem of Calculus

> *Let $g \in BV([a,b])$ and $f \in RS([g,a,b])$. Define $F : [a,b] \to \Re$ by*
>
> $$F(x) = \int_a^x f(t) dg(t)$$
>
> *Then*
>
>   *(i) $F \in BV([a,b])$*
>
>   *(ii) If g is continuous at c in $[a,b]$, then F is continuous at c*
>
>   *(iii) If g is monotone and if at c in $[a,b]$, $g'(c)$ exists and f is continuous at c, then $F'(c)$ exists with*
>
> $$F'(c) = f(c) \, g'(c)$$

Proof 7.1.1
First, assume g is monotone increasing and $g(a) < g(b)$. Let π be a partition of $[a,b]$. Then, we immediately have the fundamental estimates

$$m(g(b) - g(a)) \leq L(f,g) \leq U(f,g) \leq M(g(b) - g(a))$$

where m and M are the infimum and supremum of f on $[a,b]$ respectively. Since $f \in RS([g,a,b])$, we then have

$$m(g(b) - g(a)) \leq \int_a^b f dg \leq M(g(b) - g(a))$$

133

or

$$m \le \frac{\int_a^b fdg}{g(b) - g(a)} \le M$$

Let $K(a,b) = \int_a^b fdg/(g(b)-g(a))$. Then, $m \le K(a,b) \le M$ and $\int_a^b fdg = K(a,b)(g(b)-g(a))$.

Now assume $x < y$ in $[a,b]$. Since $f \in RS([g,a,b])$, by Theorem 6.5.5, $f \in RS([g,x,y])$. By the argument just presented, we can show there is a number $K(x,y)$ so that

$$\begin{aligned} K(x,y) &= \int_x^y fdg/(g(y) - g(x)), \\ m \le \inf_{t \in [x,y]} f(t) &\le K(x,y) \le \sup_{t \in [x,y]} f(t) \le M \qquad \qquad (\boldsymbol{\alpha}) \\ \int_x^y fdg &= K(x,y)(g(y) - g(x)) \end{aligned}$$

(i)

Proof *We show $f \in BV([a,b])$. Let π be a partition of $[a,b]$. Then, labeling the partition points in the usual way,*

$$\begin{aligned} \sum_\pi |\Delta F_j| &= \sum_\pi |\Delta F(x_j) - F(x_{j-1})| \\ &= \sum_\pi \left| \int_{x_{j-1}}^{x_j} fdg \right| \\ &= \sum_\pi |K(x_{j-1},x_j)| \, |g(x_j) - g(x_{j-1})| = \sum_\pi |K(x_{j-1},x_j)| \, |\Delta g_j| \end{aligned}$$

using Equation $\boldsymbol{\alpha}$ on each subinterval $[x_{j-1}, x_j]$. However, we know each $m \le K(x_{j-1}, x_j) \le M$ and so

$$\begin{aligned} \sum_\pi |\Delta F_j| &\le \|f\|_\infty \sum_\pi |\Delta g_j| \\ &= \|f\|_\infty \, (g(b) - g(a)) \end{aligned}$$

as g is monotone increasing. Since this inequality holds for all partitions of $[a,b]$, we see

$$V(F;a,b) \le \|f\|_\infty \, (g(b) - g(a))$$

implying $F \in BV([a,b])$. $\qquad\qquad\qquad\qquad\qquad\qquad\qquad\qquad\qquad\qquad\square$

(ii)

Proof *Let g be continuous at c. Then given a positive ϵ, there is a $\delta > 0$, so that*

$$|g(c) - g(y)| < \epsilon/(1 + \|f\|_\infty), \;\; |y - c| < \delta, \;\; y \in [a,b]$$

For any such y, apply Equation $\boldsymbol{\alpha}$ to the interval $[c,y]$ or $[y,c]$ depending on whether $y > c$ or vice-versa. For concreteness, let's look at the case $y > c$. Then, there is a $K(c,y)$ so that $m \le K(c,y) \le M$ and $\int_c^y f(t)dg(t) = K(c,y)(g(y) - g(c))$. Thus, since y is within δ of c, we have

$$\left| \int_c^y f(t)dg(t) \right| = |K(c,y)| \, |g(y) - g(c)| \le \|f\|_\infty \, \epsilon/(1 + \|f\|_\infty) < \epsilon$$

FURTHER RIEMANN - STIELTJES RESULTS

We conclude that if $y \in [c, c+\delta)$, then $\left| \int_c^y f(t)dg(t) \right| < \epsilon$. A similar argument holds for $y \in (c-\delta, c]$. Combining, we see $y \in (c - \delta, c + \delta)$ and in $[a, b]$ implies

$$\left| F(y) - F(c) \right| = \left| \int_c^y f(t)dg(t) \right| < \epsilon$$

So F is continuous at c. $\quad\square$

(iii)

Proof *If $c \in [a, b]$, $g'(c)$ exists and f is continuous at c, we must show that $F'(c) = f(c)g'(c)$. Let a positive ϵ be given. Then,*

$$\exists \delta_1 \ni \left| \frac{g(y) - g(c)}{y - c} - g'(c) \right| \ <\ \epsilon, \ 0 <| \, y - c \,| < \delta_1, \ y \in [a, b]. \tag{β}$$

and

$$\exists \delta_2 \ni \left| f(y) - f(c) \, \right| \ <\ \epsilon, \ | \, y - c \, | < \delta_2, \ y \in [a, b]. \tag{γ}$$

Choose any $\delta < \min(\delta_1, \delta_2)$. Let y be in $\left((c - \delta, c) \cup (c, c + \delta) \right) \cap [a, b]$. We are interested in the interval I with endpoints c and y which is either of the form $[c, y]$ or vice-versa. Apply Equation α to this interval. We find there is a $K(I)$ that satisfies

$$\inf_{t \in I} f(t) \le K(I) \le \sup_{t \in I} f(t)$$

and

$$\int_c^y f(t)dg(t) = K([c, y])(g(y) - g(c)), \ y > c$$

or

$$\int_y^c f(t)dg(t) \ = \ K([y, c])(g(c) - g(y)), \ y < c$$

or

$$-\int_c^y f(t)dg(t) \ = \ K([y, c])(g(c) - g(y)), \ y < c$$

which gives

$$\int_c^y f(t)dg(t) = K([y, c])(g(y) - g(c)), \ y < c$$

So we conclude we can write

$$\int_c^y f(t)dg(t) = K(I)(g(y) - g(c))$$

where $K(I)$ denotes $K([c, y])$ or $K([y, c])$ depending on where y is relative to c. Next, since $\delta < \min(\delta_1, \delta_2)$, both Equation α and Equation β hold. Thus,

$$f(c) - \epsilon < f(t) < f(c) + \epsilon, \ y \in \left((c - \delta, c) \cup (c, c + \delta) \right) \cap [a, b]$$

BASIC ANALYSIS IV: MEASURE THEORY AND INTEGRATION

This tells us that $\sup_{t \in I} f(t) \leq f(c) + \epsilon$ *and* $\inf_{t \in I} f(t) \geq f(c) - \epsilon$. *Thus,*

$$f(c) - \epsilon \leq K([c, y]), K([y, c]) \leq f(c) + \epsilon$$

or $\mid K([c, y]) - f(c) \mid < \epsilon$ *and* $\mid K([y, c]) - f(c) \mid < \epsilon$. *Finally, consider*

$$
\begin{aligned}
\left| \frac{F(y) - F(c)}{y - c} - f(c)g'(c) \right| &= \left| \frac{K(I)(g(y) - g(c))}{y - c} - f(c)g'(c) \right| \\
&= \left| \frac{K(I)(g(y) - g(c))}{y - c} - f(c)g'(c) + K(I)g'(c) - K(I)g'(c) \right| \\
&\leq \mid K(I) \mid \left| \frac{g(y) - g(c)}{y - c} - g'(c) \right| + \left| K(I) - f(c) \right| \mid g'(c) \mid \\
&< \parallel f \parallel_\infty \epsilon + \mid g'(c) \mid \epsilon
\end{aligned}
$$

Since ϵ is arbitrary, this shows F is differentiable at c with value $f(c)g'(c)$. $\square$

This proves the proposition for the case that g is monotone. To finish the proof, we note if $g \in BV([a, b])$, then $g = V_g - (V_g - g)$ is the standard decomposition of g into the difference of two monotone increasing functions. Let $F_1(x) = \int_a^x f(t)d(V_g)(t)$ and $F_2(x) = \int_a^x f(t)d(V_g - g)(t)$. From Part (i), we see $F = F_1 - F_2$ is of bounded variation. Next, if g is continuous at c, so is V_g and $V_g - g$ by Theorem 3.5.3. So by Part (ii), F_1 and F_2 are continuous at c. This implies F is continuous at c. ■

7.2 Existence Results

We begin by looking at continuous integrands.

Theorem 7.2.1 Integrand Continuous and Integrator of Bounded Variation Implies Riemann - Stieltjes Integral Exists

> *If $f \in C([a, b])$ and $g \in BV([a, b])$, then $f \in RS([g, a, b])$.*

Proof 7.2.1

Let's begin by assuming g is monotone increasing. We may assume without loss of generality that $g(a) < g(b)$. Let $K = g(b) - g(a) > 0$. Since f is continuous on $[a, b]$, f is uniformly continuous on $[a, b]$. Hence, given a positive ϵ, there is a positive δ so that

$$\mid f(s) - f(t) \mid < \epsilon/K, \ \mid t - s \mid < \delta, \ t, s \in [a, b]$$

Now, repeat the proof of Theorem 4.4.1 which shows that if f is continuous on $[a, b]$, then $f \in RI([a, b])$, but replace all the Δx_j by Δg_j. This shows that f satisfies the Riemann - Stieltjes Criterion for integrability. Thus, by the equivalence theorem, $f \in RS([g, a, b])$.

Next, let $g \in BV([a, b])$. Then $g = V_g - (V_g - g)$ as usual. Since V_g and $V_g - g$ are monotone increasing, we can apply our first argument to conclude $f \in RS([V_g, a, b])$ and $f \in RS([V_g - g, a, b])$. Then, by the linearity of the Riemann - Stieltjes integral with respect to the integrator, Theorem 6.1.1, we have $f \in RS([g, a, b])$ with

$$\int_a^b f dg = \int_a^b f dv_g - \int_a^b f d(V_g - g)$$

FURTHER RIEMANN - STIELTJES RESULTS

Next, we let the integrand be of bounded variation.

Theorem 7.2.2 Integrand Bounded Variation and Integrator Continuous Implies Riemann - Stieltjes Integral Exists

If $f \in BV([a,b])$ and $g \in C([a,b])$, then $f \in RS([g,a,b])$.

Proof 7.2.2
If $f \in BV([a,b])$ and $g \in C([a,b])$, then by the previous theorem, Theorem 7.2.1, $g \in RS([f,a,b])$. Now apply integration by parts, Theorem 6.1.2, to conclude $f \in RS([g,a,b])$. ∎

What if the integrator is differentiable?

Theorem 7.2.3 Integrand Continuous and Integrator Continuously Differentiable Implies Riemann - Stieltjes Integrable

Let $f \in C([a,b])$ and $g \in C^1([a,b])$. Then $f \in RS([g,a,b])$, $fg' \in RI([a,b])$ and

$$\int_a^b f(x)dg(x) = \int_a^b f(x)g'(x)dx$$

where the integral on the left side is a traditional Riemann integral.

Proof 7.2.3
Pick an arbitrary positive ϵ. Since g' is continuous on $[a,b]$, g' is uniformly continuous on $[a,b]$. Thus, there is a positive δ so that

$$| g'(s) - g'(t) | \quad < \quad \epsilon, \ | s - t | < \delta, \ s,t \in [a,b]. \tag{α}$$

Since g' is continuous on $[a,b]$, there is a number M so that $| g(x) | \leq M$ for all x in $[a,b]$. We conclude that $g \in BV([a,b])$ by Theorem 3.3.3. Now apply Theorem 7.2.1, to conclude $f \in RS([g,a,b])$. Thus, there is a partition $\boldsymbol{\pi_0}$ of $[a,b]$, so that

$$\left| S(f,g,\boldsymbol{\pi},\boldsymbol{\sigma}) - \int_a^b fdg \right| \quad < \quad \epsilon, \ \boldsymbol{\pi_0} \preceq \boldsymbol{\pi}, \ \boldsymbol{\sigma} \subseteq \boldsymbol{\pi}. \tag{β}$$

Further, since fg' is continuous on $[a,b]$, $fg' \in RI([a,b])$ and so $\int_a^b fg'$ exists also.

Now let $\boldsymbol{\pi_1}$ be a refinement of $\boldsymbol{\pi_0}$ with $\|\boldsymbol{\pi_1}\| < \delta$. Then we can apply Equation β to conclude

$$\left| S(f,g,\boldsymbol{\pi},\boldsymbol{\sigma}) - \int_a^b fdg \right| \quad < \quad \epsilon, \ \boldsymbol{\pi_1} \preceq \boldsymbol{\pi}, \ \boldsymbol{\sigma} \subseteq \boldsymbol{\pi}. \tag{γ}$$

Next, apply the Mean Value Theorem to g on the subintervals $[x_{j-1}, x_j]$ from partition $\boldsymbol{\pi}$ for which Equation γ holds. Then, $\Delta g_j = g'(t_j)(x_j - x_{j-1})$ for some t_j in (x_{j-1}, x_j). Hence,

$$S(f,g,\boldsymbol{\pi},\boldsymbol{\sigma}) = \sum_{\boldsymbol{\pi}} f(s_j)\Delta g_j = \sum_{\boldsymbol{\pi}} f(s_j)g'(t_j)\Delta x_j$$

Also, we see

$$S(fg',\boldsymbol{\pi},\boldsymbol{\sigma}) = \sum_{\boldsymbol{\pi}} f(s_j)g'(s_j)\Delta x_j$$

138 BASIC ANALYSIS IV: MEASURE THEORY AND INTEGRATION

Thus, we can compute

$$\left| S(f, g, \boldsymbol{\pi}, \boldsymbol{\sigma}) - S(fg', \boldsymbol{\pi}, \boldsymbol{\sigma}) \right| = \left| \sum_{\boldsymbol{\pi}} f(s_j) \Big(g'(t_j) - g'(s_j) \Delta x_j \Big) \right|$$

$$\leq \|f\|_\infty \sum_{\boldsymbol{\pi}} \left| g'(t_j) - g'(s_j) \Delta x_j \right|$$

By Equation $\boldsymbol{\alpha}$, since $\|\boldsymbol{\pi}\| < \delta$, $|t_j - s_j| < \delta$ and so $|g'(t_j) - g'(s_j)| < \epsilon$. We conclude

$$\left| S(f, g, \boldsymbol{\pi}, \boldsymbol{\sigma}) - S(fg', \boldsymbol{\pi}, \boldsymbol{\sigma}) \right| < \epsilon \|f\|_\infty \sum_{\boldsymbol{\pi}} \Delta x_j = \epsilon \|f\|_\infty (b - a). \tag{$\boldsymbol{\xi}$}$$

Thus,

$$\left| S(fg', \boldsymbol{\pi}, \boldsymbol{\sigma}) - \int_a^b f \, dg \right| \leq \left| S(f, g, \boldsymbol{\pi}, \boldsymbol{\sigma}) - S(fg', \boldsymbol{\pi}, \boldsymbol{\sigma}) \right| + \left| S(f, g, \boldsymbol{\pi}, \boldsymbol{\sigma}) - \int_a^b f \, dg \right|$$

$$< \epsilon \|f\|_\infty (b - a) + \epsilon$$

by Equation $\boldsymbol{\gamma}$ and Equation $\boldsymbol{\xi}$. This proves the desired result. ∎

Homework

Exercise 7.2.1 *Let $f(x) = x^2$ and $g(x) = \tanh(x)$ on $[-6, 5]$.*

- *Express $\int_{-6}^{5} f(x) dg(x) dx$ as a Riemann integral.*

- *Compute $\int_{-6}^{5} f(x) dg(x) dx$ approximately using Riemann sums in MATLAB.*

Exercise 7.2.2 *Let $f(x) = x^2$ and $g(x) = x_2^3 x + 10$ on $[0, 7]$.*

- *Express $\int_{0}^{7} f(x) dg(x) dx$ as a Riemann integral.*

- *Evaluate $\int_{0}^{7} f(x) dg(x) dx$.*

Exercise 7.2.3 *Let $f(x) = \sin(5x)$ and $g(x) = \cos(3x)$ on $[0, 5]$.*

- *Express $\int_{0}^{5} f(x) dg(x) dx$ as a Riemann integral.*

- *Evaluate $\int_{0}^{5} f(x) dg(x) dx$.*

It should be easy to see that the assumptions of Theorem 7.2.3 can be relaxed.

Theorem 7.2.4 Integrand Riemann Integrable and Integrator Continuously Differentiable Implies Riemann - Stieltjes Integrable

Let $f \in RI([a, b])$ and $g \in C^1([a, b])$. Then $f \in RS([g, a, b])$, $fg' \in RI([a, b])$ and

$$\int_a^b f(x) dg(x) = \int_a^b f(x) g'(x) dx$$

where the integral on the left side is a traditional Riemann integral.

FURTHER RIEMANN - STIELTJES RESULTS

Proof 7.2.4
We never use the continuity of f in the proof given for Theorem 7.2.3. All we use is the fact that f is Riemann integrable. Hence, we can use the proof of Theorem 7.2.3 without change to find

$$\left| S(fg', \pi, \sigma) - \int_a^b f dg \right| \; < \; \epsilon \|f\|_\infty (b - a) + \epsilon$$

This tells us that fg' is Riemann integrable on $[a, b]$ with value $\int_a^b f dg$. ∎

Homework

Exercise 7.2.4 *Let f be defined by*

$$f(x) \;=\; \begin{cases} 2x, & 0 \le x < 1 \\ 3, & 1 = x \\ 5x^2 + 1, & 1 < x \le 2 \end{cases}$$

and let $g(x) = x^2$. Compute $\int_0^2 f(x) dg(x)$.

Exercise 7.2.5 *Let f be defined by*

$$f(x) \;=\; \begin{cases} 2, & x = -1 \\ x + 4, & -1 < x < 1 \\ 6, & 1 = x \\ x^2 + 7, & 1 < x < 2 \\ 12, & x = 2 \\ 2x^3, & 2 < x \le 3 \end{cases}$$

and let $g(x) = x^3$. Compute $\int_{-1}^3 f(x) dg(x)$.

Exercise 7.2.6 *Let f_n, for $n \ge 2$, be defined by*

$$f_n(x) \;=\; \begin{cases} 1, & x = 1 \\ 1/(k+1), & 1/(k+1) \le x < 1/k, \; 1 \le k < n \\ 0, & 0 \le x < 1/n \end{cases}$$

and let $g(x) = x^4$. Compute $\int_0^1 f(x) dg_n(x)$.

7.3 Worked Out Examples of Riemann - Stieltjes Computations

How *do* we compute a Riemann - Stieltjes integral? Let's look at some examples.

Example 7.3.1 *Let f and g be defined on $[0, 2]$ by*

$$f(x) \;=\; \begin{cases} x, & x \in Q \cap [0, 2] \\ 2 - x, & x \in Ir \cap [0, 2], \end{cases} \quad g(x) \;=\; \begin{cases} 1, & 0 \le x < 1 \\ 3, & 1 \le x \le 2. \end{cases}$$

Does $\int f dg$ exist?

Solution *We can answer this two ways so far. Method 1: We note f is continuous at 1 (you should be able to do a traditional $\epsilon - \delta$ proof of this fact!) and since g has a jump at 1, we can look at Lemma 6.2.1 and Lemma 6.2.3 to see that f is indeed Riemann - Stieltjes with respect to g. The value is* *given by*

$$\int_0^2 f dg = f(1)(g(1^+) - g(1^-)) = 1(3 - 1) = 2$$

Method 2: We can compute the integral using a partition approach. Let π be a partition of $[0, 2]$. We may assume without loss of generality that $1 \in \pi$ (recall all of our earlier arguments that allow us to make this statement!). Hence, there is an index k_0 such that $x_{k_0} = 1$. We have

$$L(f, g, \pi) = \left(\inf_{x \in [x_{k_0 - 1}, 1]} f(x)\right)\left(g(1) - g(x_{k_0 - 1})\right) + \left(\inf_{x \in [1, x_{k_0 + 1}]} f(x)\right)\left(g(x_{k_0 + 1}) - g(1)\right)$$

Now use how g is defined to see,

$$L(f, g, \pi) = \left(\inf_{x \in [x_{k_0 - 1}, 1]} f(x)\right)\left(3 - 1\right) + \left(\inf_{x \in [1, x_{k_0 + 1}]} f(x)\right)\left(3 - 3\right)$$

Hence,

$$L(f, g, \pi) = 2\left(\inf_{x \in [x_{k_0 - 1}, 1]} f(x)\right)$$

If you graphed x and $2 - x$ simultaneously on $[0, 2]$, you would see that they cross at 1 and x is below $2 - x$ before 1. This graph works well for f even though we can only use the graph of x when x is rational and the graph of $2 - x$ when x is irrational. We can see in our mind how to do the visualization. For this mental picture, you should be able to see that the infimum of f on $[x_{k_0 - 1}, 1]$ will be the value $x_{k_0 - 1}$. We have thus found that $L(f, g, \pi) = 2x_{k_0 - 1}$. A similar argument will show that $U(f, g, \pi) = 2(2 - x_{k_0 - 1})$. This immediately implies that $L(f, g) = U(f, g) = 2$.

Example 7.3.2 *Let f be any bounded function which is discontinuous from the left at 1 on $[0, 2]$. Again, let g be defined on $[0, 2]$ by*

$$g(x) = \begin{cases} 1, & 0 \leq x < 1 \\ 3, & 1 \leq x \leq 2. \end{cases}$$

Does $\int f dg$ exist?

Solution *First, since we know f is not continuous from the left at 1 and g is continuous from the right at 1, the conditions of Lemma 6.2.2 do not hold. So it is possible this integral does not exist. We will in fact show this using arguments that are similar to the previous example. Again, π is a partition which has $x_{k_0} = 1$. We find*

$$L(f, g, \pi) = \left(\inf_{x \in [x_{k_0 - 1}, 1]} f(x)\right)\left(3 - 1\right), \ U(f, g, \pi) = \left(\sup_{x \in [x_{k_0 - 1}, 1]} f(x)\right)\left(3 - 1\right)$$

Since we can choose $x_{k_0} - 1$ as close to 1 as we wish, we see

$$\inf_{x \in [x_{k_0 - 1}, 1]} f(x) \to \min(f(1^-), f(1))$$

$$\sup_{x \in [x_{k_0 - 1}, 1]} f(x) \to \max(f(1^-), f(1))$$

But f is discontinuous from the left at 1 and so $f(1^-) \neq f(1)$. For concreteness, let's assume $f(1^-) < f(1)$ (the argument the other way is very similar). We see $L(f, g) = 2f(1^-)$ and $U(f, g) = 2f(1)$. Since these values are not the same, f is not Riemann - Stieltjes integrable with respect to g by the Riemann - Stieltjes equivalence theorem, Theorem 6.4.1.

FURTHER RIEMANN - STIELTJES RESULTS

Example 7.3.3 *Let f be any bounded function which is continuous from the left at 1 on $[0,2]$. Again, let g be defined on $[0,2]$ by*

$$g(x) \;=\; \begin{cases} 1, & 0 \le x < 1 \\ 3, & 1 \le x \le 2. \end{cases}$$

Does $\int f dg$ exist?

Solution *First, since we know f is continuous from the left at 1 and g is continuous from the right at 1, the conditions of Lemma 6.2.2 do hold. So this integral does exist. Using Lemma 6.2.2, we see*

$$\int_0^2 f dg = f(1)(g(1^+) - g(1^-)) = 2f(1)$$

We can also show this using partition arguments as we have done before. Again, π is a partition which has $x_{k_0} = 1$. Again, we have

$$L(f,g,\pi) = \left(\inf_{x \in [x_{k_0-1},1]} f(x)\right)\left(3 - 1\right), \;\; U(f,g,\pi) = \left(\sup_{x \in [x_{k_0-1},1]} f(x)\right)\left(3 - 1\right)$$

Since we can choose $x_{k_0} - 1$ as close to 1 as we wish, we see

$$\inf_{x \in [x_{k_0-1},1]} f(x) \to \min(f(1^-), f(1))$$

$$\sup_{x \in [x_{k_0-1},1]} f(x) \to \max(f(1^-), f(1))$$

But f is continuous from the left at 1, $f(1^-) = f(1)$. We see $L(f,g) = 2f(1)$ and $U(f,g) = 2f(1)$. Since these values are the same, f is Riemann - Stieltjes integrable with respect to g by the Riemann - Stieltjes equivalence theorem, Theorem 6.4.1.

Example 7.3.4 *Define a step function g on $[0,12]$ by*

$$g(x) \;=\; \begin{cases} 0, & 0 \le x < 2 \\ \sum_{j=2}^{\lfloor x \rfloor} (j-1)/36, & 2 \le x < 8 \\ 21/36 + \sum_{j=8}^{\lfloor x \rfloor} (13-j)/36, & 8 \le x \le 12 \end{cases}$$

where $\lfloor x \rfloor$ is the greatest integer which is less than or equal to x. The function g is everywhere continuous from the right and represents the probability of rolling a number $j \le x$. It is called the cumulative probability distribution function of a fair pair of dice. *The Riemann - Stieltjes integral $\mu = \int_0^{12} x dg(x)$ is called the* mean *of this distribution. The* variance *of this distribution is denoted by σ^2 (unfortunate choice, isn't it as that is the letter we use to denote evaluation sets of partitions!) and defined to be*

$$\sigma^2 \;=\; \int_0^{12} (x - \mu)^2 dg(x)$$

Compute μ and σ^2.

Solution *Since $f(x) = x$ is continuous on $[0,12]$, Lemma 6.2.4 applies and we have*

$$\int_0^{12} x dg(x) = \sum_{j=2}^{12} j\left(g(j+) - g(j^-)\right)$$

142   BASIC ANALYSIS IV: MEASURE THEORY AND INTEGRATION

The evaluations are a bit messy.

$$
\begin{aligned}
36(g(2+) - g(2^-)) &= 36(g(2) - g(2^-)) = 1 - 0 = 1 \\
36(g(3+) - g(3^-)) &= 36(g(3) - g(3^-)) = 3 - 1 = 2 \\
36(g(4+) - g(4^-)) &= 36(g(4) - g(4^-)) = 6 - 3 = 3 \\
36(g(5+) - g(5^-)) &= 36(g(5) - g(5^-)) = 10 - 6 = 4 \\
36(g(6+) - g(6^-)) &= 36(g(6) - g(6^-)) = 15 - 10 = 5 \\
36(g(7+) - g(7^-)) &= 36(g(7) - g(7^-)) = 21 - 15 = 6 \\
36(g(8+) - g(8^-)) &= 36(g(8) - g(8^-)) = 26 - 21 = 5 \\
36(g(9+) - g(9^-)) &= 36(g(9) - g(9^-)) = 30 - 26 = 4 \\
36(g(10+) - g(10^-)) &= 36(g(10) - g(10^-)) = 33 - 30 = 3 \\
36(g(11+) - g(11^-)) &= 36(g(11) - g(11^-)) = 35 - 33 = 2 \\
36(g(12+) - g(12^-)) &= 36(g(12) - g(12^-)) = 36 - 35 = 1
\end{aligned}
$$

Thus,

$$
\begin{aligned}
\int_0^{12} x\,dg(x) &= \bigg(2(1) + 3(2) + 4(3) + 5(4) + 6(5) + 7(6) \\
&\qquad + 8(5) + 9(4) + 10(3) + 11(2) + 12(1) \bigg)/36 \\
&= \bigg(2 + 6 + 12 + 20 + 30 + 42 + 40 + 36 + 30 + 22 + 12 \bigg)/36 \\
&= 252/36 = 7
\end{aligned}
$$

So, the mean *or* expected value *of a single roll of a fair pair of dice is* 7. *To find the* variance, *we calculate*

$$
\begin{aligned}
\sigma^2 &= \int_0^{12} (x - 7)\,dg(x) \\
&= \sum_{j=2}^{12} (j - 7)^2 \bigg(g(j+) - g(j^-) \bigg) \\
&= \bigg(25(1) + 16(2) + 9(3) + 4(4) + 1(5) + 0(6) + 1(5) + 4(4) + 9(3) + 16(2) + 25(1) \bigg)/36 \\
&= \bigg(25 + 32 + 27 + 16 + 5 + 5 + 16 + 27 + 32 + 25 \bigg)/36 \\
&= 210/36 = 35/6
\end{aligned}
$$

Example 7.3.5 *Let $f(x) = e^x$ and let g be defined on $[0,2]$ by*

$$
g(x) = \begin{cases} x^2, & 0 \leq x \leq 1 \\ x^2 + 1, & 1 < x \leq 2. \end{cases}
$$

Show $\int f\,dg$ exists and evaluate it.

Solution *Since g is monotone, $\int_0^2 f\,dg$ exists. We can thus decompose g into its continuous and saltus part. We find*

$$g_c(x) \;=\; x^2, \;\; s_g(x) \;=\; \left\{ \begin{array}{ll} 0, & 0 \le x \le 1 \\ 1, & 1 < x \le 2. \end{array} \right.$$

The saltus integral is evaluated using Lemma 6.2.1. The integrand is continuous and the jump is at 1, so we have

$$\begin{aligned} \int_0^2 f\,ds_g &= \int_0^2 e^x \, ds_g(x) \\ &= e^1(s_g(1^+) - s_g(1^-)) \;=\; e(1-0) \;=\; e \end{aligned}$$

and for the continuous part, we can use the fact the integrator is continuously differentiable on $[0,2]$ to apply Theorem 7.2.3 to obtain

$$\int_0^2 f\,dg_c \;=\; \int_0^2 e^x \, d(x^2) \;=\; \int_0^2 e^x \, 2x\,dx \;=\; 2(e^2+1)$$

Thus,

$$\begin{aligned} \int_0^2 f\,dg &= \int_0^2 f\,dg_c + \int_0^2 f\,ds_g \\ &= 2(e^2+1) + e \end{aligned}$$

We can also do this by integration by parts, Theorem 6.1.2. Since $f \in RS([g,0,2])$, it follows that $g \in RS([f,0,2])$ and

$$\begin{aligned} \int_0^2 f(x)\,dg(x) &= \left. e^x\, g(x) \right|_0^2 - \int_0^2 g(x)\,df(x) \\ &= e^2 g(2) - g(0) - \int_0^2 g(x)\,d(e^x) \\ &= e^2 - \int_0^2 g(x) e^x\,dx \end{aligned}$$

Example 7.3.6 *Let $f(x) = e^x$ and let g be defined on $[0,2]$ by*

$$g(x) \;=\; \left\{ \begin{array}{ll} x^2, & 0 \le x < 1 \\ \sin(x), & 1 \le x \le 2. \end{array} \right.$$

Show $\int f\,dg$ exists and evaluate it.

Solution *We know that g is of bounded variation on $[1,2]$ because it is continuously differentiable with bounded derivative there. But what about on $[0,1]$? We know that the function $h(x) = x^2$ on $[0,1]$ is of bounded variation on $[0,1]$ because it is also continuously differentiable with a bounded derivative. If π is any partition of $[0,1]$ then we must have, using standard notation for the partition points of π, that*

$$\sum_\pi |\,\Delta g_j\,| \;=\; \sum_{j=0}^{p-1} |\,\Delta g_j\,| + |\,g(1) - g(x_{p-1})\,| \le V(h,0,1) + 2\|g\|_\infty$$

144 BASIC ANALYSIS IV: MEASURE THEORY AND INTEGRATION

Since the choice of partition on $[0,1]$ is arbitrary, we see $g \in BV([0,1])$. Thus, combining, we have that $g \in BV([0,2])$. It then follows that $f \in RS([g,0,2])$. Now note that on $[0,1]$, we can write $g(x) = h(x) + u(x)$ where

$$u(x) = \begin{cases} 0, & 0 \le x < 1 \\ \sin(1) - 1, & x = 1. \end{cases}$$

Then, to evaluate $\int_0^2 f dg$ we write

$$
\begin{aligned}
\int_0^2 f dg &= \int_0^1 f dg + \int_1^2 f dg \\
&= \int_0^1 f d(h + u) + \int_1^2 f d(\sin(x)) \\
&= \int_0^1 f d(h) + \int_0^1 f d(u) + \int_1^2 f \cos(x) dx \\
&= \int_0^1 e^x 2x dx + f(1)(u(1) - u(1^-)) + \int_1^2 e^x \cos(x) dx \\
&= \int_0^1 e^x 2x dx + e(\sin(1) - 1) + \int_1^2 e^x \cos(x) dx
\end{aligned}
$$

and these integrals are standard Riemann integrals that can be evaluated by parts.

Homework

Exercise 7.3.1 *Define g on $[0,2]$ by*

$$g(x) = \begin{cases} -2 & x = 0 \\ x^3 & 0 < x < 1 \\ 9/8 & x = 1 \\ x^4/4 + 1 & 1 < x < 2 \\ 7 & x = 2 \end{cases}$$

This function is from a previous exercise.

1. *Show that if $f(x) = x^4$ on $[0,2]$, then $f \in RS([g,0,2])$.*

2. *Compute $\int_0^2 f dg$.*

3. *Explain why $g \in RS([f,0,2])$.*

4. *Compute $\int_0^2 g df$.*

Exercise 7.3.2 *Define g on $[0,2]$ by*

$$g(x) = \begin{cases} -1 & x = 0 \\ x^2 & 0 < x < 1 \\ 7/4 & x = 1 \\ \sqrt{x+3} & 1 < x < 2 \\ 3 & x = 2 \end{cases}$$

This function is also from a previous exercise.

1. *Show that if $f(x) = x^2 + 5$ on $[0,2]$, then $f \in RS([g,0,2])$.*

FURTHER RIEMANN - STIELTJES RESULTS

2. Compute $\int_0^2 f dg$.

3. Explain why $g \in RS([f, 0, 2])$.

4. Compute $\int_0^2 g df$.

Exercise 7.3.3 *Let f and g be defined on $[0, 4]$ by*

$$f(x) = \begin{cases} x, & x \in Q \cap [0,4] \\ 2x, & x \in Ir \cap [0,4], \end{cases} \quad g(x) = \begin{cases} 1, & 0 \le x < 1 \\ 2, & 1 \le x < 2 \\ 3, & 2 \le x < 3 \\ 4, & 3 \le x \le 4. \end{cases}$$

Does $\int f dg$ exist and if so, what is its value?

Exercise 7.3.4 *Let $f(x) = x^3$ and let g be defined on $[0, 3]$ by*

$$g(x) = \begin{cases} x^2, & 0 \le x \le 2 \\ x^2 + 4, & 2 < x \le 3. \end{cases}$$

Show $\int f dg$ exists and evaluate it.

Exercise 7.3.5 *Let $f(x) = x^2 + 3x + 10$ and let g be defined on $[-1, 5]$ by*

$$g(x) = \begin{cases} x^3, & -1 \le x \le 2 \\ -10x^2, & 2 < x \le 5. \end{cases}$$

Show $\int f dg$ exists and evaluate it.

Exercise 7.3.6 *The following are definitions of integrands f_1, f_2 and f_3 and integrators g_1, g_2 and g_3 on $[0, 2]$. For each pair of indices i, j determine if $\int_0^2 f_i dg_j$ exists. If the integral exists, compute the value and if the integral does not exist, provide a proof of its failure to exist.*

$$f_1(x) = \begin{cases} 1, & 0 \le x < 1 \\ x - 1, & 1 \le x \le 2, \end{cases} \quad f_2(x) = \begin{cases} 1, & x = 0 \\ x, & 0 < x \le 2 \end{cases} \quad f_3(x) = \begin{cases} 2, & x = 0 \\ 1, & 0 < x < 1 \\ x - 1, & 1 \le x \le 2 \end{cases}$$

$$g_1(x) = \begin{cases} x, & 0 \le x < 1 \\ x + 1, & 1 \le x \le 2, \end{cases} \quad g_2(x) = \begin{cases} x, & 0 \le x \le 1 \\ x + 1, & 1 < x < 2 \\ 4, & x = 2, \end{cases} \quad g_3(x) = \begin{cases} -1, & x = 0 \\ x, & 0 < x \le 1 \\ x + 1, & 1 < x < 2 \\ 4, & x = 2. \end{cases}$$

Exercise 7.3.7 *Prove*

Theorem Limit Interchange Theorem for Riemann - Stieltjes Integrals

> Assume $g \in BV([a,b])$ and $\{f_n\} \subseteq RS([g,a,b])$ converges uniformly to f_0 on $[a,b]$. Then
>
> (i) $f_0 \in RS([g,a,b])$.
>
> (ii) If $F_n(x) = \int_a^x f_n(t)dg(t)$ and $F_0(x) = \int_a^x f_0(t)dg(t)$, then F_n converges uniformly to F_0 on $[a,b]$.
>
> (iii)
> $$\lim_n \int_a^b f_n(t)dg(t) = \int_a^b f_0(t)dg(t)$$

Exercise 7.3.8 *Let g be strictly monotone on $[a,b]$. For f_1, f_2 in $C([a,b])$, define $\omega : C([a,b]) \times C([a,b]) \to \Re$ by $\omega(f_1, f_2) = \int_a^b f_1(t)f_2(t)dg(t)$.*

(i) Prove that ω is an inner product on $C([a,b])$.

(ii) Prove if $\omega(f,h) = 0$ for all $h \in RS([g,a,b])$, then $f = 0$.

Part IV

Abstract Measure Theory One

Chapter 8

Measurable Functions and Spaces

If you have been looking closely at how we prove the properties of Riemann and Riemann - Stieltjes integration, you will have noted that these proofs are intimately tied to the way we use partitions to divide the function domain into small pieces. We are now going to explore a new way to associate a given bounded function with a real number which can be interpreted as the integral.

Let X be a nonempty set. In mathematics, we study sets such as X when various properties and structures have been added. For example, we might want X to have a metric d to allow us to measure an abstract version of distance between points in X. We could study sets X which have a linear or vector space structure and if this resulting vector space possessed a norm $\| \cdot \|$, we could determine an abstract version of the magnitude of objects in X. Here, we want to look at collections of subsets of the set X and impose some conditions on the structure of these collections.

Definition 8.0.1 Sigma-Algebras

Let X be a nonempty set. A family of subsets S is called a σ - algebra if

(i) $\emptyset, X \in S$.

(ii) If $A \in S$, so is A^C. We say S is closed under complementation or complements.

(iii) If $\{A_n\}_{n=1}^{\infty} \in S$, then $\cup_{n=1}^{\infty} A_n \in S$. We say S is closed under countable unions.

The pair (X, S) will be called a measurable space and if $A \in S$, we will call A an S measurable set. If the underlying σ - algebra is understood, we usually just say, A is a measurable subset of X.

A common tool we use in working with countable collections of sets is *De Morgan's Laws*.

Lemma 8.0.1 De Morgan's Laws

150

> Let X be a nonempty set and $\{A_\alpha | \alpha \in \Lambda\}$ be any collection of subsets of X. Hence, the index set Λ may be finite, countably infinite or arbitrary cardinality. Then
>
> (i)
>
> $$\left(\cup_\alpha A_\alpha\right)^C = \cap_\alpha A_\alpha^C$$
>
> (ii)
>
> $$\left(\cap_\alpha A_\alpha\right)^C = \cup_\alpha A_\alpha^C$$

Proof 8.0.1
This is a standard proof and is left to you as an exercise. ∎

Homework

Exercise 8.0.1 *Prove DeMorgan's Laws.*

Now let's work through a series of examples of σ algebras mostly via exercises for you to do.
Homework

Exercise 8.0.2 *Let X be any not empty set and let $S = \{A | A \subseteq X\}$. This is the collection of all subsets and is sometimes called the power set of X. It is often denoted by the symbol $\mathcal{P}(X)$. This collection clearly is a σ algebra. Prove $(\mathcal{P}(X), X)$ is a measurable space and all subsets of X are $\mathcal{P}(X)$ measurable.*

Exercise 8.0.3 *Let X be any set and $S = \{\emptyset, X\}$. Prove this collection is also a σ algebra, albeit not a very interesting one! With this σ algebra, X is a measurable space with only two measurable sets.*

Exercise 8.0.4 *Let X be the set of counting numbers and let $S = \{\emptyset, O, E, X\}$ where O is the odd counting numbers and E, the odd. Prove (X, S) is a measurable space.*

Let's look at a harder example. We'll work out the details for you this time. Let X be any uncountable set and let $S = \{A \subseteq X | A$ is countable or A^C is countable$\}$. It is easy to see $\emptyset$ and X itself are in S. If $A \in S$, then there are two cases: A is countable and /or A^C is countable. In both cases, it is easy to see A^C is also in S. It remains to show that S is closed under countable unions. To do this, assume we have a sequence of sets A_n from S. Consider $A = \cup_n A_n$. There are several cases to consider.

1. If all the A_n are countable, then so is the countable union implying $A \in S$.

2. If all the A_n are not countable, then each A_n^C is countable. Thus, $\cap_n A_n^C = (\cup_n A_n)^C$ is countable. Again, this tells us $A \in S$.

MEASURABILITY 151

3. If a countable number of A_n and a countable number of A_n^C are uncountable, then we have, since X is uncountable,

$$
\begin{aligned}
\left(\cup_n A_n \right)^C &= \left(\cap_n A_n^C \right) \\
&= \left(\cap_{(A_n \text{ countable})} A_n^C \right) \cap \left(\cap_{(A_n \text{ uncountable})} A_n^C \right) \\
&= \left(\cap_{(A_n^C \text{ uncountable})} A_n^C \right) \cap \left(\cap_{(A_n^C \text{ countable})} A_n^C \right)
\end{aligned}
$$

Now, for any index n, we must have $\cap_n A_n^C \subseteq A_n^C$. Thus, since some A_n^C are countable, we must have $\cap_n A_n^C$ is countable. By De Morgan's Laws, it follows that $(\cup_n A_n)^C$ is countable. This implies $A \in \mathcal{S}$.

We conclude $(X, \mathcal{S})$ is a measurable space.

Homework

Exercise 8.0.5 *Let X be any nonempty set and let $\mathcal{S}_1$ and $\mathcal{S}_2$ be two sigma-algebras of X. Let*

$$
\begin{aligned}
\mathcal{S} &= \{A \subseteq X \mid A \in \mathcal{S}_1 \text{ and } A \in \mathcal{S}_2\} \\
&\equiv \mathcal{S}_1 \cap \mathcal{S}_2
\end{aligned}
$$

Prove $(X, \mathcal{S})$ is a measurable space.

Exercise 8.0.6 *Let X be any nonempty set and let $\mathcal{S}_1$, $\mathcal{S}_2$ and $\mathcal{S}_3$ be three sigma-algebras of X. Let*

$$
\mathcal{S} = \mathcal{S}_1 \cap \mathcal{S}_2 \cap \mathcal{S}_3
$$

Prove $(X, \mathcal{S})$ is a measurable space.

Exercise 8.0.7 *Let X be any nonempty set and let $(\mathcal{S}_j)_{j=1}^N$ be N sigma-algebras of X for some $N > 1$*

$$
\mathcal{S} = \cap_{j=1}^N \mathcal{S}_j
$$

Prove $(X, \mathcal{S})$ is a measurable space.

Let X be any nonempty set. Let $\mathcal{A}$ be any nonempty collection of subsets of X. Note that $\mathcal{P}(X)$, the collection of all subsets of X, is a sigma-algebra of X and hence, $(X, \mathcal{P}(X))$ is a measurable space that contains $\mathcal{A}$. It is easy to show if $(\mathcal{S}_j)$ is a finite collection of sigma-algebras that contain $\mathcal{A}$, then $\cap_j \mathcal{S}_1$ is a new sigma-algebra that also contains $\mathcal{A}$. This suggests we search for the *smallest* sigma-algebra that contains $\mathcal{A}$.

Definition 8.0.2 The sigma-algebra Generated by Collection A

The sigma-algebra generated by a collection of subsets $\mathcal{A}$ in a nonempty set X, is denoted by $\sigma(\mathcal{A})$ and is defined by

$$
\sigma(\mathcal{A}) = \cap \{\mathcal{S} \mid \mathcal{A} \subseteq \mathcal{S}\}
$$

Since any sigma-algebra $\mathcal{S}$ that contains $\mathcal{A}$ by definition satisfies $\sigma(\mathcal{A}) \subseteq \mathcal{S}$, it is easy to see why we interpret this generated sigma-algebra as the smallest *sigma-algebra that contains the collection $\mathcal{A}$.*

8.1 The Borel Sigma-Algebra of $\Re$

We now discuss a very important sigma-algebra of subsets of the real line called the Borel sigma-algebra which is denoted by $\mathcal{B}$. Define four collections of subsets of $\Re$ as follows:

1. $\boldsymbol{A}$ is the collection of finite open intervals of the form (a, b),

2. $\boldsymbol{B}$ is the collection of finite half open intervals of the form $(a, b]$,

3. $\boldsymbol{C}$ is the collection of finite half open intervals of the form $[a, b)$ and

4. $\boldsymbol{D}$ is the collection of finite closed intervals of the form $[a, b]$.

It is possible to show that

$$\sigma(\boldsymbol{A}) \;=\; \sigma(\boldsymbol{B}) = \sigma(\boldsymbol{C}) = \sigma(\boldsymbol{D})$$

This common sigma-algebra is what we will call the Borel sigma-algebra of $\Re$. It should be evident to you that a set can be very complicated and still be in $\mathcal{B}$. Some of these equalities will be left to you as homework exercises, but we will prove that $\sigma(\boldsymbol{A}) = \sigma(\boldsymbol{D})$. Let $\mathcal{S}$ be any sigma-algebra that contains $\boldsymbol{A}$. We know that

$$\begin{aligned}
[a, b] \;&=\; (-\infty, b] \cap [a, \infty) \\
&=\; (b, \infty)^C \cap (-\infty, a)^C \\
&=\; \Big((-\infty, a) \cup (b, \infty) \Big)^C
\end{aligned}$$

In the representation of $[a, b]$ above, note we can write

$$\begin{aligned}
(-\infty, a) \;&=\; \bigcup_{-\infty}^{n=\lfloor a \rfloor} (n, a) \\
(b, \infty) \;&=\; \bigcup_{n=\lceil b \rceil}^{\infty} (b, n)
\end{aligned}$$

Since, $\mathcal{S}$ is a sigma-algebra containing $\boldsymbol{A}$, the unions on the right-hand sides in the equations above must be in $\mathcal{S}$. This immediately tells us that $[a, b]$ is also in $\mathcal{S}$. Hence, since $[a, b]$ is arbitrary, we conclude $\boldsymbol{D}$ is contained in $\mathcal{S}$ also. Further, this is true for any sigma-algebra that contains $\boldsymbol{A}$ and so we have that $\boldsymbol{D} \subseteq \sigma(\boldsymbol{A})$. Thus, by definition, we can say $\sigma(\boldsymbol{D}) \subseteq \sigma(\boldsymbol{A})$.

To show the reverse containment is quite similar. Let $\mathcal{S}$ be any sigma-algebra that contains $\boldsymbol{D}$. We know that

$$\begin{aligned}
(a, b) \;&=\; (-\infty, b) \cap (a, \infty) \\
&=\; [b, \infty)^C \cap (-\infty, a]^C \\
&=\; \Big((-\infty, a] \cup [b, \infty) \Big)^C
\end{aligned}$$

In the representation of (a, b) above, note we can write

$$(-\infty, a] \;=\; \bigcup_{\lfloor a \rfloor}^{\infty} [-n, a]$$

MEASURABILITY   153

and

$$[b, \infty) \;=\; \bigcup_{\lceil b \rceil}^{\infty} [b, n]$$

Since, $\mathcal{S}$ is a sigma-algebra containing D, the unions on the right-hand sides in the equations above must be in $\mathcal{S}$. This immediately tells us that (a, b) is also in $\mathcal{S}$. Hence, since (a, b) is arbitrary, we conclude A is contained in $\mathcal{S}$ also. Again, since this is true for any sigma-algebra that contains A, we have that $A \subseteq \sigma(D)$. Thus, by definition, we can say $\sigma(A) \subseteq \sigma(D)$. Combining, we have the equality we seek.

Homework

Exercise 8.1.1 *Prove* $\sigma(A) \;=\; \sigma(B)$.

Exercise 8.1.2 *Prove* $\sigma(A) \;=\; \sigma(C)$.

Exercise 8.1.3 *Prove* $\sigma(A) \;=\; \sigma(D)$.

Exercise 8.1.4 *Prove* $\sigma(B) \;=\; \sigma(C)$.

Exercise 8.1.5 *Prove* $\sigma(B) \;=\; \sigma(D)$.

Exercise 8.1.6 *Prove* $\sigma(C) \;=\; \sigma(D)$.

8.2   The Extended Borel Sigma-Algebra

It is often very convenient to deal with a number system that explicitly adjoins the symbols ∞ and $-\infty$ to the standard real line $\Re$. This is actually called the two - point compactification of $\Re$, but that is another story!

Definition 8.2.1 The Extended Real Number System

The extended real number system is denoted by $\overline{\Re}$ and is defined as the real numbers with two additional elements:

$$\overline{\Re} \;=\; \Re \cup \{+\infty\} \cup \{-\infty\}$$

We want arithmetic involving the new symbols $\pm\infty$ to reflect our everyday experience with limits of sequences of numbers which either grow without bound positively or negatively. Hence, we use the conventions for all real numbers x:

$$
\begin{aligned}
(\pm\infty) + (\pm\infty) \;&=\; x + (\pm\infty) = (\pm\infty) + x = \pm\infty, \\
(\pm\infty) \cdot (\pm\infty) \;&=\; \infty, \\
(\pm\infty) \cdot (\mp\infty) \;&=\; = (\mp\infty) \cdot (\pm\infty) = -\infty, \\
x \cdot (\pm\infty) \;&=\; = (\pm\infty) \cdot x = \pm\infty \; if \, x > 0, \\
x \cdot (\pm\infty) \;&=\; = (\pm\infty) \cdot x = 0 \; if \, x = 0, \\
x \cdot (\pm\infty) \;&=\; = (\pm\infty) \cdot x = \mp\infty \; if \, x < 0
\end{aligned}
$$

We cannot define the arithmetic operations $(\infty) + (-\infty)$, $(-\infty) + (\infty)$ or any of the four ratios of the form $(\pm\infty)/(\pm\infty)$.

BASIC ANALYSIS IV: MEASURE THEORY AND INTEGRATION

We can now define the Borel sigma-algebra in $\overline{\Re}$. Let E be any Borel set in $\Re$. Let

$$E_1 \;=\; E \cup \{-\infty\}, \;\; E_2 \;=\; E \cup \{+\infty\}, \;\; \text{and} \;\; E_3 \;=\; E \cup \{+\infty\} \cup \{+\infty\}$$

Then, we define

$$\overline{\mathcal{B}} \;=\; \{E, E_1, E_2, E_3 \mid E \in \mathcal{B}\}$$

We leave to you the exercise of showing that $\overline{\mathcal{B}}$ is a sigma-algebra in $\overline{\Re}$.

Homework

Exercise 8.2.1 *Prove $\overline{\mathcal{B}}$ is a sigma-algebra in $\overline{\Re}$. This is quite messy actually, so be patient.*

It is true that open intervals in $\Re$ are in $\overline{\mathcal{B}}$, but is it true that $\overline{\mathcal{B}}$ contains arbitrary open sets? To see that it does, we must prove a characterization for the open sets of $\Re$.

Theorem 8.2.1 Open Set Characterization Lemma

> *If $\mathcal{U}$ is an open set in $\Re$, then there is a countable collection of disjoint open intervals $\mathcal{C} = \{(a_n, b_n)\}$ so that $\mathcal{U} = \cup_n (a_n, b_n)$.*

Proof 8.2.1
Since $\mathcal{U}$ is open, if $p \in \mathcal{U}$, there is an $r > 0$ so that $B(p; r) \subseteq \mathcal{U}$. Hence, $(p - r, p + r) \subseteq \mathcal{U}$ implying both $(p, p + r) \subseteq \mathcal{U}$ and $(p - r, p) \subseteq \mathcal{U}$. Let

$$S_p \;=\; \{y \mid (p, y) \subseteq \mathcal{U}\} \;\; \text{and} \;\; T_p \;=\; \{x \mid (x, p) \subseteq \mathcal{U}\}$$

It is easy to see that both S_p and T_p are nonempty since $\mathcal{U}$ is open. Let $b_p = \sup S_p$ and $a_p = \inf T_p$. Clearly, b_p could be $+\infty$ and a_p could be $-\infty$.

Consider $u \in (a_p, b_p)$. From the Infimum and Supremum tolerance lemmas, we know there are points x^* and y^* so that

$$u < y^* \leq b_p \leq \infty \;\;\; \text{and} \;\;\; (p, y^*) \subseteq \mathcal{U},$$
$$-\infty \leq a_p \leq x^* < u \;\;\; \text{and} \;\;\; (x^*, p) \subseteq \mathcal{U}$$

Hence, $u \in (x^*, y^*) \subseteq \mathcal{U}$ which implies $u \in \mathcal{U}$. Thus, since u in (a_p, b_p) is arbitrary, we have $(a_p, b_p) \subseteq \mathcal{U}$. If a_p or b_p were not finite, they cannot be in $\Re$ and cannot be in $\mathcal{U}$. However, what if either one was finite? Is it possible for the point to be in $\mathcal{U}$? We will show that in this case, the points a_p and b_p still cannot lie in $\mathcal{U}$. For concreteness, let us assume that a_p is finite and in $\mathcal{U}$. Then, a_p would be an interior point of $\mathcal{U}$. Hence, there would be a radius $\rho > 0$ so that $(a_p - \rho, a_p) \subseteq \mathcal{U}$ implying $a_p - \rho \in T_p$. Thus, $\inf T_p = a_p \leq a_p - \rho$ which is not possible. Hence, $a_p \notin \mathcal{U}$. A similar argument then shows that if b_p is finite, b_p is not in $\mathcal{U}$.

Thus, we know that a_p and b_p are never in $\mathcal{U}$ and that p is always in the open interval $(a_p, b_p) \subseteq \mathcal{U}$. Let $\mathcal{F} = \{(a_p, b_p) \mid p \in \mathcal{U}\}$. We see immediately that

$$\mathcal{U} \;=\; \cup_{\mathcal{F}} (a_p, b_p)$$

Let (a, b) and (c, d) be any two intervals from $\mathcal{F}$ which overlap. From the definition of $\mathcal{F}$, we then know that a, b, c and d are not in $\mathcal{U}$. Then, if $a \geq d$, the two intervals would be disjoint; hence, we must have $a < d$. By the same sort of argument, it is also true that $c < b$. Hence, if c is in the *intersection, we have a chain of inequalities like this:*

$$a < c < b < d$$

Next, since $a \notin \mathcal{U}$, we see $a \leq c$ since $(c,d) \subseteq \mathcal{U}$. Further, since $c \notin \mathcal{U}$ and $(a,b) \subseteq \mathcal{U}$, it follows that $c \leq a$. Combining, we have $a = c$. A similar argument shows that $b = d$. Hence, $(a,b) \cap (c,d) \neq \emptyset$ implies that $(a,b) = (c,d)$. Thus, two interval I_p and I_q in $\mathcal{F}$ are either the same or disjoint. We conclude

$$\mathcal{U} = \bigcup_{(disjoint\ I_p \in \mathcal{F})} I_p$$

Let $\mathcal{F}_0$ be this collection of disjoint intervals from $\mathcal{F}$. Each I_p in $\mathcal{F}_0$ contains a rational number r_p. By definition, it then follows that if I_p and I_q are in $\mathcal{F}_0$, then $r_p \neq r_q$. The set of these rational numbers is countable and so we can label them using an enumeration r_n. Label the interval I_p which contains r_n as I_n. Then, we have

$$\mathcal{U} = \bigcup_{n=1}^{\infty} I_n$$

which is the desired result. ∎

Thus any open set can be written as a countable union of open intervals and so is in the $\overline{\mathcal{B}}$. This also tells us any closed set is in $\overline{\mathcal{B}}$ as it is the complement of an open set. Cool beans! You should think about this carefully. It is very difficult to get a handle on how to characterize what a set in $\overline{\mathcal{B}}$ looks like as $\overline{\mathcal{B}}$ is closed under countable unions, countable intersections and complementations!

8.3 Measurable Functions

Let $f : \Re \to \Re$ be a continuous function. Let $\mathcal{O}$ be an open subset of $\Re$. By Theorem 8.2.1, we know that we can write

$$\mathcal{O} = \bigcup_n (a_n, b_n)$$

where the (a_n, b_n) are mutually disjoint finite open intervals of $\Re$. It follows immediately that $\mathcal{O}$ is in the Borel sigma-algebra $\mathcal{B}$. Now consider the inverse image of $\mathcal{O}$ under f, $f^{-1}(\mathcal{O})$. If $p \in f^{-1}(\mathcal{O})$, then $f(p) \in \mathcal{O}$. Since $\mathcal{O}$ is open, $f(p)$ must be an interior point. Hence, there is a radius $r > 0$ so that $(f(p) - r, f(p) + r) \subseteq \mathcal{O}$. Since f is continuous at p, there is a $\delta > 0$ so that $f(x) \in (f(p) - r, f(p) + r)$ if $x \in (p - \delta, p + \delta)$. This tells us that $(p - \delta, p + \delta) \subseteq f^{-1}(\mathcal{O})$. Since p was arbitrarily chosen, we conclude that $f^{-1}(\mathcal{O})$ is an open set.

We see that if f is continuous on $\Re$, then $f^{-1}(\mathcal{O})$ is in the Borel sigma-algebra for any open set $\mathcal{O}$ in $\Re$. We can then say that $f^{-1}(\alpha, \infty)$ is in $\mathcal{B}$ for all $\alpha > 0$. This suggests that an interesting way to generalize the notion of *continuity* might be to look for functions f on an arbitrary nonempty set X with sigma-algebra $\mathcal{S}$ satisfying $f^{-1}(\mathcal{O}) \in \mathcal{S}$ for all open sets $\mathcal{O}$. Further, by our last remark, it should be enough to ask that $f^{-1}((\alpha, \infty)) \in \mathcal{S}$ for all $\alpha \in \Re$. This is exactly what we will do. It should be no surprise to you that functions f satisfying this new definition will not have to be continuous!

Definition 8.3.1 The Measurability of a Function

Let X be a nonempty set and $\mathcal{S}$ be a sigma-algebra of subsets of X. We say that $f : X \to \Re$ is a $\mathcal{S}$ - measurable function on X or simply $\mathcal{S}$ measurable if

$$\forall \alpha \in \Re, \ \{x \in X \mid f(x) > \alpha\} \in \mathcal{S}$$

BASIC ANALYSIS IV: MEASURE THEORY AND INTEGRATION

We can easily prove that there are equivalent ways of proving a function is measurable.

Lemma 8.3.1 Equivalent Conditions for the Measurability of a Function

> *Let X be a nonempty set and $\mathcal{S}$ be a sigma-algebra of subsets of X. The following statements are equivalent:*
>
> *(i):* $\forall \alpha \in \Re$, $A_\alpha = \{x \in X \mid f(x) > \alpha\} \in \mathcal{S}$,
>
> *(ii):* $\forall \alpha \in \Re$, $B_\alpha = \{x \in X \mid f(x) \leq \alpha\} \in \mathcal{S}$,
>
> *(iii):* $\forall \alpha \in \Re$, $C_\alpha = \{x \in X \mid f(x) \geq \alpha\} \in \mathcal{S}$,
>
> *(iv):* $\forall \alpha \in \Re$, $D_\alpha = \{x \in X \mid f(x) < \alpha\} \in \mathcal{S}$.

Proof 8.3.1

(i) $\Rightarrow$ (ii):

Proof *If $A_\alpha \in \mathcal{S}$, then its complement is in $\mathcal{S}$ also. Since $B_\alpha = A_\alpha^C$, (ii) follows.* $\qquad\square$

(ii) $\Rightarrow$ (i):

Proof *If $B_\alpha \in \mathcal{S}$, then its complement is in $\mathcal{S}$ also. Since $A_\alpha = B_\alpha^C$, (i) follows.* $\qquad\square$

(iii) $\Leftrightarrow$ (iv):

Proof *Since $C_\alpha = D_\alpha^C$ and $D_\alpha = C_\alpha^C$, arguments similar to those of the previous cases can be applied.* $\qquad\square$

Hence, if we show (i) $\Leftrightarrow$ (iii), we will be done. (i) $\Rightarrow$ (iii):

Proof *By (i), $A_{\alpha-1/n} \in \mathcal{S}$ for all n. We know*

$$C_\alpha = \bigcap_n A_{\alpha-1/n} = \bigcap_n \{x \mid f(x) > \alpha - 1/n\}$$

We also know $A_{\alpha-1/n}^C$ is measurable and so $\cup_n A_{\alpha-1/n}^C$ is also measurable. Thus, the complement of $\cup_n A_{\alpha-1/n}^C$ is also measurable. Then, by De Morgan's Laws, $C_\alpha = \cap_n A_{\alpha-1/n}$ is measurable. $\square$

(iii) $\Rightarrow$ (i):

Proof *Note, $C_{\alpha+1/n} \in \mathcal{S}$ for all n and so*

$$A_\alpha = \bigcup_n C_{\alpha+1/n} = \bigcup_n \{x \mid f(x) \geq \alpha + 1/n\}$$

is also measurable. $\qquad\square$

We conclude all four statements are equivalent. $\qquad\blacksquare$

8.3.1 Examples

Example 8.3.1 *Any constant function f on a nonempty set X with given sigma-algebra S is measurable as if $f(x) = c$ for some $c \in \Re$, then*

$$\{x | f(x) > \alpha\} \;=\; \begin{cases} \emptyset \in S & \alpha \geq c \\ X \in S & \alpha < c \end{cases}$$

Example 8.3.2 *Let X be a nonempty set X with given sigma-algebra S. Let $E \in S$ be given. Define*

$$I_E(x) \;=\; \begin{cases} 1 & \text{if } x \in E \\ 0 & \text{if } x \notin E \end{cases}$$

Then I_E is measurable. Note

$$\{x | I_E(x) > \alpha\} \;=\; \begin{cases} \emptyset \in S & \alpha \geq 1 \\ E \in S & 0 \leq \alpha < 1 \\ X \in S & \alpha < 0 \end{cases}$$

Example 8.3.3 *Let $X = \Re$ and $S = \mathcal{B}$. Then, if $f : \Re \to \Re$ is continuous, f is measurable by the arguments we made at the beginning of this section. More generally, let $f : [a, b] \to \Re$ be continuous on $[a, b]$. Then, extend f to $\Re$ as $\hat{f}$ defined by*

$$\hat{f} \;=\; \begin{cases} f(a) & x < a \\ f(x) & a \leq x \leq b \\ f(b) & x > b \end{cases}$$

Then $\hat{f}$ is continuous on $\Re$ and measurable with $\hat{f}^{-1}(\alpha, \infty) \in \mathcal{B}$ for all α. It is not hard to show that

$$\mathcal{B} \cap [a, b] \;=\; \{E \subseteq [a, b] \mid E \in \mathcal{B}\}$$

is a sigma -algebra of the set $[a, b]$. Further, the standard arguments for f continuous on $[a, b]$ show us that $f^{-1}(\alpha, \infty) \in \mathcal{B} \cap [a, b]$ for all α. Hence, a continuous f on the interval $[a, b]$ will be measurable with respect to the sigma-algebra $\mathcal{B} \cap [a, b]$.

We can argue in a similar fashion for functions continuous on intervals of the form $(a, b], [a, b)$ and (a, b) whether a and b is finite or not.

Example 8.3.4 *If $X = \Re$ and $S = \mathcal{B}$, then any monotone function is Borel measurable. To see this, note we can restrict our attention to monotone increasing functions as the argument is quite similar for monotone decreasing. It is enough to consider the cases where f takes on the value α without a jump at the point x_0 or f has a jump across the value α at x_0. In the first case, since f is monotone increasing and $f(x_0) = \alpha$, $f^{-1}(\alpha, \infty) = (x_0, \infty) \in \mathcal{B}$. On the other hand, if f has a jump at x_0 across the value α, then $f(x_0^-) \neq f(x_0^+)$ and $\alpha \in [f(x_0^-), f(x_0^+)]$. There are three possibilities:*

(i): *$f(x_0^-) = f(x_0) < f(x_0^+)$: If $\alpha = f(x_0)$, then since f is monotone, $f^{-1}(\alpha, \infty) = (x_0, \infty)$. If $f(x_0) < \alpha < f(x_0^+)$, we again have $f^{-1}(\alpha, \infty) = (x_0, \infty)$. Finally, if $\alpha = f(x_0^+)$, we have $f^{-1}(\alpha, \infty) = [x_0, \infty)$. In all cases, these inverse images are in $\mathcal{B}$.*

(ii): *$f(x_0^-) < f(x_0) < f(x_0^+)$: A similar analysis shows that all the possible inverse images are Borel sets.*

158 BASIC ANALYSIS IV: MEASURE THEORY AND INTEGRATION

(iii): $f(x_0^-) < f(x_0) = f(x_0^+)$: we handle the arguments is a similar way.

We conclude that in all cases, $f^{-1}(\alpha, \infty) \in \mathcal{B}$ and hence f is measurable.

Note, the analysis of the previous example could be employed here also to show that a monotone function defined on an interval such as $[a, b]$, (a, b) and so forth is Borel measurable with respect to the restricted sigma-algebra $\mathcal{B} \cap [a, b]$, etc.

Homework

Exercise 8.3.1 *Prove*

$$\mathcal{B} \cap [a, b] \;\; = \;\; \{E \subseteq [a, b] \mid E \in \mathcal{B}\}$$

is a sigma -algebra of the set $[a, b]$.

Exercise 8.3.2 *Prove if f is continuous on $[a, b]$ then $f^{-1}(\alpha, \infty) \in \mathcal{B} \cap [a, b]$ for all α. Thus, prove a continuous f on the interval $[a, b]$ is measurable with respect to the sigma-algebra $\mathcal{B} \cap [a, b]$.*

Exercise 8.3.3 *Using the arguments of the previous exercise with suitable changes, prove functions continuous on intervals of the form $(a, b]$ are measurable with respect to the sigma-algebra $\mathcal{B} \cap (a, b]$ whether the intervals are finite or not.*

Exercise 8.3.4 *Using the arguments of the previous exercise with suitable changes, prove functions continuous on intervals of the form $[a, b)$ are measurable with respect to the sigma-algebra $\mathcal{B} \cap [a, b)$ whether the intervals are finite or not.*

Exercise 8.3.5 *Using the arguments of the previous exercise with suitable changes, prove functions continuous on intervals of the form (a, b) are measurable with respect to the sigma-algebra $\mathcal{B} \cap (a, b)$ whether the intervals are finite or not.*

Exercise 8.3.6 *Prove functions of bounded variation on $[a, b]$ are measurable with respect to the sigma-algebra $\mathcal{B} \cap [a, b]$. Other variations of this are possible, of course, but there is no need to go through all that again!*

Exercise 8.3.7 *Let f be piecewise continuous on $[a, b]$. Prove that f is measurable with respect to the restricted Borel sigma-algebra $\mathcal{B} \cap [a, b]$. Recall, a function is piecewise continuous on $[a, b]$ if there are a finite number of points x_i in $[a, b]$ where f is not continuous.*

Comment 8.3.1 *For convenience, we will start using a more abbreviated notation for sets like*

$$\{x \in X \mid f(x) > \alpha\};$$

we will shorten this to $\{f(x) > \alpha\}$ or $(f(x) > \alpha)$ in our future discussions.

8.4 Properties of Measurable Functions

We now want to see how we can build new measurable functions from old ones we know.

Lemma 8.4.1 Properties of Measurable Functions

MEASURABILITY

Let X be a nonempty set and S a sigma-algebra on X. Then if f and g are S measurable, so are

(i): cf for all $c \in \Re$.

(ii): f^2.

(iii): $f + g$.

(iv): fg.

(v): $|f|$.

Proof 8.4.1

(i):

Proof If $c = 0$, $cf = 0$ and the result is clear. If $c > 0$, then $(cf(x) > \alpha) = (f(x) > \alpha/c)$ which is measurable as f is measurable. If $c < 0$, a similar argument holds. $\square$

(ii):

Proof If $\alpha < 0$, then $(f^2(x) > \alpha) = X$ which is in S. Otherwise, if $\alpha \geq 0$, then

$$(f^2(x) > \alpha) = (f(x) > \sqrt{\alpha}) \cup (f(x) < -\sqrt{\alpha})$$

and both of these sets are measurable since f is measurable. The conclusion follows. $\square$

(iii):

Proof If $r \in Q$, let $S_r = (f(x) > r) \cap (g(x) > \alpha - r)$ which is measurable since f and g are measurable. We claim that

$$(f(x) + g(x) > \alpha) = \bigcup_{r \in Q} S_r$$

To see this, let x satisfy $f(x) + g(x) > \alpha$. Thus, $f(x) > \alpha - g(x)$. Since the rationals are dense in $\Re$, we see there is a rational number r so that $f(x) > r > \alpha - g(x)$. This clearly implies that $f(x) > r$ and $g(x) > \alpha - r$ and so $x \in S_r$. Since our choice of x was arbitrary, we have shown that

$$(f(x) + g(x) > \alpha) \subseteq \bigcup_{r \in Q} S_r$$

The converse is easier as if $x \in S_r$, it follows immediately that $f(x) + g(x) > \alpha$.

Since S_r is measurable for each r and the rationals are countable, we see $(f(x) + g(x) > \alpha)$ is measurable. $\square$

(iv):

Proof To prove this result, note that

$$fg = (1/4)\left((f+g)^2 - (f-g)^2\right)$$

and all the individual pieces are measurable by (iii) and (i). $\square$

(v):

Proof *If $\alpha < 0$, $(f(x) > \alpha) = X$ which is measurable. On the other hand, if $\alpha \geq 0$,*

$$(|\,f\,|\,(x) > \alpha) \;=\; (f(x) > \alpha) \,\cup\, (f(x) < -\alpha)$$

which implies the measurability of $|\,f\,|$. $\qquad\qquad$ □

$\blacksquare$

We can also prove another characterization of the measurability of f.

Lemma 8.4.2 A Function is Measurable if and only if Its Positive and Negative Parts are Measurable

> *Let X be a nonempty set and S be a sigma-algebra on X. Then $f : X \to \Re$ is measurable if and only if f^+ and f^- are measurable, where*
>
> $$f^+(x) \;=\; \max\{f(x), 0\}, \quad and \;\; f^+(x) \;=\; -\min\{f(x), 0\}$$

Proof 8.4.7
We note $f = (f^+) - (f^-)$ and $|\,f\,| = (f^+(+(f^-))$. It is then easy to see $f^+ = (1/2)(|\,f\,| + f)$ and $f^- = (1/2)(|\,f\,| - f)$. Hence, if f is measurable, by Lemma 8.4.1 (i), (iii) and (v), f^+ and f^- are also measurable. Conversely, if f^+ and f^- are measurable, $f = f^+ - f^-$ is measurable as well. $\blacksquare$

Homework

Exercise 8.4.1 *Let f be defined by*

$$f(x) \;=\; \begin{cases} -8, & -1 \leq x \leq 0 \\ 2x, & 0 \leq x < 1 \\ 3, & 1 = x \\ 5x^2 + 1, & 1 < x \leq 3 \end{cases} \quad and \;\; g(x) = \begin{cases} 2, & x = -1 \\ x + 4, & -1 < x < 1 \\ 6, & 1 = x \\ x^2 + 7, & 1 < x < 2 \\ 12, & x = 2 \\ 2x^3, & 2 < x \leq 3 \end{cases}$$

- *Prove f is $\overline{B}$ measurable from first principles.*

- *Prove g is $\overline{B}$ measurable from first principles.*

- *Prove $f + g$ is $\overline{B}$ measurable from first principles.*

- *Prove $3f + 5g$ is $\overline{B}$ measurable from first principles.*

Exercise 8.4.2 *Let f be defined by*

$$f(x) \;=\; \begin{cases} 2, & x = 0 \\ x + 4, & 0 < x < 0.5 \\ 6, & 0.5 = x \\ x^2 + 7, & 0.5 < x < 1 \\ 12, & x = 1 \end{cases} \quad and \;\; \forall\, n \geq 2$$

$$g_n(x) \;=\; \begin{cases} 1, & x = 1 \\ 1/(k+1), & 1/(k+1) \leq x < 1/k,\; 1 \leq k < n \\ 0, & 0 \leq x < 1/n \end{cases}$$

MEASURABILITY

- *Prove f is $\overline{B}$ measurable from first principles.*

- *Prove g is $\overline{B}$ measurable from first principles.*

- *Prove $f + g$ is $\overline{B}$ measurable from first principles.*

- *Prove $3f + 5g$ is $\overline{B}$ measurable from first principles.*

Exercise 8.4.3 *Let X be a nonempty set and S a sigma-algebra on X. Then if f and g are S measurable, prove $f^p g^q$ is S measurable for all positive powers p and q.*

Exercise 8.4.4 *Let $f(x) = x^3$ on $[-1, 1]$. Prove, using first principles, that f, f^+ and f^- are measurable with respect to the sigma-algebra $B \cap [a, b]$.*

8.5 Extended Real-Valued Measurable Functions

We now extend these ideas to functions which are extended real-valued.

Definition 8.5.1 The Measurability of an Extended Real-Valued Function

> *Let X be a nonempty set and S be a sigma-algebra on X. Let $f : X \to \overline{\Re}$. We say f is S measurable if $(f(x) > \alpha)$ is in S for all α in $\Re$.*

Comment 8.5.1 *If the extended valued function f is measurable, then $(f(x) = +\infty) = \cap_n (f(x) > n)$ is measurable. Also, since*

$$(f(x) = -\infty) = \left(\cup_n (f(x) > -n) \right)^C,$$

it is measurable also.

We can then prove an equivalence theorem just like before.

Lemma 8.5.1 Equivalent Conditions for the Measurability of an Extended Real-Valued Function

> *Let X be a nonempty set and S be a sigma-algebra of subsets of X. The following statements are equivalent:*
>
> *(i): $\forall \alpha > 0$, $A_\alpha = \{x \in X \mid f(x) > \alpha\} \in S$,*
>
> *(ii): $\forall \alpha > 0$, $B_\alpha = \{x \in X \mid f(x) \leq \alpha\} \in S$,*
>
> *(iii): $\forall \alpha > 0$, $C_\alpha = \{x \in X \mid f(x) \geq \alpha\} \in S$,*
>
> *(iv): $\forall \alpha > 0$, $D_\alpha = \{x \in X \mid f(x) < \alpha\} \in S$.*

Proof 8.5.1
The proof follows that of Lemma 8.3.1 ∎

The collection of all extended real-valued measurable functions is important to future work. We make the following definition:

Definition 8.5.2 The Set of Extended Real-Valued Measurable Functions

BASIC ANALYSIS IV: MEASURE THEORY AND INTEGRATION

> Let X be a nonempty set and S be a sigma-algebra of subsets of X. We denote by $M(X, S)$ the set of all extended real-valued measurable functions on X. Thus,
>
> $$M(X, S) = \{f : X \to \overline{\mathbb{R}} \mid f \text{ is } S \text{ measurable}\}$$

It is also easy to prove the following equivalent definition of measurability for extended real-valued functions.

Lemma 8.5.2 Extended Real-Valued Measurability in Terms of the Finite Part of the Function

> Let X be a nonempty set and S be a sigma-algebra of subsets of X. Then $f \in M(X, S)$ if and only if (i): $(f(x) = +\infty) \in S$, (ii): $(f(x) = -\infty) \in S$ and (iii): f_1 is measurable where
>
> $$f_1(x) = \begin{cases} f(x) & x \notin (f(x) = +\infty) \cup (f(x) = -\infty), \\ 0 & x \in (f(x) = +\infty) \cup (f(x) = -\infty). \end{cases}$$

Proof 8.5.2
By Comment 8.5.1, if f is measurable, (i) and (ii) are true. Now, if $\alpha \geq 0$ is given, we see

$$(f_1(x) > \alpha) = (f(x) > \alpha) \cap (f(x) = +\infty)^C$$

which is a measurable set. On the other hand, if $\alpha < 0$, then

$$(f_1(x) > \alpha) = (f(x) > \alpha) \cup (f(x) = -\infty)$$

which is measurable as well. We conclude f_1 is measurable. Conversely, if (i), (ii) and (iii) hold, then if $\alpha \geq 0$, we have

$$(f(x) > \alpha) = (f_1(x) > \alpha) \cup (f(x) = +\infty)$$

and if $\alpha < 0$,

$$(f(x) > \alpha) = (f_1(x) > \alpha) \cap (f(x) = -\infty)^C$$

implying both sets are measurable. Thus, f is measurable. ∎

Example 8.5.1 *Let X be a nonempty set X with given sigma-algebra S. Let $E \in S$ be given. Define the extended real-value characteristic function*

$$J_E(x) = \begin{cases} \infty & \text{if } x \in E \\ 0 & \text{if } x \notin E \end{cases}$$

Then J_E is measurable. Note

$$\{x \mid I_E(x) > \alpha\} = \begin{cases} E \in S & \alpha \geq 0 \\ X \in S & \alpha < 0 \end{cases}$$

MEASURABILITY 163

Note also that if we define

$$J_E(x) \;=\; \begin{cases} \infty & \text{if } x \in E \\ -\infty & \text{if } x \notin E \end{cases}$$

Then J_E is measurable. We have

$$\{x \,|\, J_E(x) > \alpha\} \;=\; \begin{cases} E \in \mathcal{S} & \alpha \geq 0 \\ E \in \mathcal{S} & \alpha < 0 \end{cases}$$

Finally, $(J_E(x) = +\infty) = E$ and $(J_E(x) = -\infty) = E^C$ are both measurable and the f_1 type function used in Lemma 8.5.2 here is $(J_E)_1(x) = 0$ always.

Homework

Exercise 8.5.1 *Let X be a nonempty set X with given sigma-algebra $\mathcal{S}$. Let $E \in \mathcal{S}$ be given. Define*

$$I_E(x) \;=\; \begin{cases} 1 & \text{if } x \in E \\ \infty & \text{if } x \notin E \end{cases}$$

Prove I_E is measurable.

Exercise 8.5.2 *Let X be a nonempty set X with given sigma-algebra $\mathcal{S}$. Let E_1 and E_2 be in $\mathcal{S}$ be given. Define I_{E_1} and I_{E_2} are above. Prove $I_{E_1} + I_{E_2}$ is measurable. Is it equal to $I_{E_1 \cup E_2}$?*

Exercise 8.5.3 *Let X be a nonempty set X with given sigma-algebra $\mathcal{S}$. Let $E \in \mathcal{S}$ be given. Define*

$$I_E(x) \;=\; \begin{cases} 1 & \text{if } x \in E \\ -\infty & \text{if } x \notin E \end{cases}$$

Prove I_E is measurable.

Exercise 8.5.4 *Let X be a nonempty set X with given sigma-algebra $\mathcal{S}$. Let $E \in \mathcal{S}$ be given. Define*

$$I_E(x) \;=\; \begin{cases} -\infty & \text{if } x \notin E \\ 0 & \text{if } x \notin E \end{cases}$$

Prove I_E is measurable.

Exercise 8.5.5 *Let X be a nonempty set X with given sigma-algebra $\mathcal{S}$. Let $E \in \mathcal{S}$ be given and let f be measurable. Prove $f + I_E$ is measurable.*

8.6 Properties of Extended Real-Valued Measurable Functions

It is straightforward to prove these properties:

Lemma 8.6.1 Properties of Extended Real-Valued Measurable Functions

164 BASIC ANALYSIS IV: MEASURE THEORY AND INTEGRATION

Let X be a nonempty set and $\mathcal{S}$ a sigma-algebra on X. Then if f and g are in $M(X,\mathcal{S})$, so are

(i): cf for all $c \in \Re$.

(ii): f^2.

(iii): $f + g$, as long as we restrict the domain of $f + g$ to be E_{fg} where

$$E_{fg}^C = \Big((f(x) = +\infty) \cap (g(x) = -\infty) \cup (f(x) = -\infty) \cap (g(x) = +\infty) \Big)^C$$

We usually define $(f + g)(x) = 0$ on E_{fg}. Note E_{fg} is measurable since f and g are measurable functions.

(iv): $|f|$, f^+ and f^-.

Proof 8.6.1

These proofs are similar to those shown in the proof of Lemma 8.4.1. However, let's look at the details of the proof of (ii). We see that our definition of addition of the extended real-valued sum means that

$$(f + g)(x) = \Big(f + g \Big) I_{E_{fg}^C}$$

Define h by

$$h(x) = \Big(f + g \Big) I_{E_{fg}^C}(x)$$

Let α be a real number. Then

$$(h(x) > \alpha) = \begin{cases} (f(x) + g(x) > \alpha) \cap E_{fg}^C & \alpha \geq 0 \\ \Big((f(x) + g(x) > \alpha) \cap E_{fg}^C \Big) \cup E_{fg} & \alpha < 0 \end{cases}$$

Similar to what we did in Lemma 8.4.1, for $r \in Q$, let

$$S_r = (f(x) > r) \cap (g(x) > \alpha - r) \cap E_{fg}^C$$

which is measurable since f and g are measurable. We claim that

$$(f(x) + g(x) > \alpha) \cap E_{fg}^C = \bigcup_{r \in Q} S_r \tag{8.1}$$

To see this, let x be in the left-hand side of Equation 8.1. There are several cases. First, neither $f(x)$ or $g(x)$ can be $-\infty$ since α is a real number. Now, if $f(x) = \infty$, then $g(x) > -\infty$ is and it is easy to see there is a rational number r satisfying $f(x) = \infty > r > \alpha - g(x)$ and so x is in the right-hand side. If $g(x) = \infty$, then $f(x) > -\infty$ and again, we see there is a rational number so that $f(x) > r > \alpha - g(x) = -\infty$. Thus, x is in the right-hand side again. The case where both $f(x)$

MEASURABILITY

165

and $g(x)$ are finite is then handled just like we did in the proof of Lemma 8.4.1. We conclude

$$(f(x) + g(x) > \alpha) \subseteq \bigcup_{r \in Q} S_r$$

The converse is easier as if $x \in S_r$, it follows immediately that $f(x) + g(x)$ is defined and $f(x) + g(x) > \alpha$.

Since S_r is measurable for each r and the rationals are countable, we see $(f(x) + g(x) > \alpha) \cap E_{fg}^C$ is measurable. ∎

To prove that products of extended real-valued measurable functions are also measurable, we have to use a pointwise limit approach.

Lemma 8.6.2 Pointwise Infimums, Supremums, Limit Inferiors and Limit Superiors are Measurable

> *Let X be a nonempty set and S a sigma-algebra on X. Let $(f_n) \subseteq M(X, S)$. Then*
>
> *(i): If $f(x) = \inf_n f_n(x)$, then $f \in M(X, S)$.*
>
> *(ii): If $F(x) = \sup_n f_n(x)$, then $F \in M(X, S)$.*
>
> *(iii): If $f^*(x) = \liminf_n f_n(x)$, then $f^* \in M(X, S)$.*
>
> *(iv): If $F^*(x) = \limsup_n f_n(x)$, then $F^* \in M(X, S)$.*

Proof 8.6.2
It is straightforward to see that $(f(x) \geq \alpha) = \cap_n (f_n(x) \geq \alpha)$ and $(F(x) \geq \alpha) = \cup_n (f_n(x) > \alpha)$ and hence, are measurable for all α. It follows that f and F are in $M(X, S)$ and so (i) and (ii) hold. Next, recall from classical analysis that at each point x,

$$\liminf(f_n(x)) = \sup_n \inf_{k \geq n} f_k(x)$$
$$\limsup(f_n(x)) = \inf_n \sup_{k \geq n} f_k(x)$$

Now let $z_n(x) = \inf_{k \geq n} f_k(x)$ and $w_n(x) = \sup_{k \geq n} f_k(x)$. Applying (i) to z_n, we have $z_n \in M(X, S)$ and applying (ii) to w_n, we have $w_n \in M(X, S)$. Then apply (i) and (ii) to $\sup_n z_n$ and $\inf w_n$, respectively, to get the desired result. ∎

This leads to an important result.

Theorem 8.6.3 Pointwise Limits of Measurable Functions Are Measurable

> *Let X be a nonempty set and S a sigma-algebra on X. Let $(f_n) \subseteq M(X, S)$ and let $f : X \to \overline{\Re}$ be a function such that $f_n \to f$ pointwise on X. Then $f \in M(X, S)$.*

Proof 8.6.3
We know that $\liminf_n f_n(x) = \limsup_n f_n(x) = \lim_n f_n(x)$. Thus, by Lemma 8.6.2, we know that f is measurable. ∎

Comment 8.6.1 *This is a huge result. We know from classical analysis that the pointwise limit of continuous functions need not be continuous (e.g. let $f_n(t) = t^n$ on $[0, 1]$). Thus, the closure of a class of functions which satisfy a certain property (like continuity) under a limit operation is not*

166 BASIC ANALYSIS IV: MEASURE THEORY AND INTEGRATION

always guaranteed. We see that although measurable functions are certainly not as smooth *as we would like, they are well behaved enough to be closed under pointwise limits!*

Homework

Exercise 8.6.1 *Let X be a nonempty set and $\mathcal{S}$ a sigma-algebra on X. Then if f is in $M(X, \mathcal{S})$ so is f^p for positive integers p.*

Exercise 8.6.2 *Let X be a nonempty set and $\mathcal{S}$ a sigma-algebra on X. Then if f is in $M(X, \mathcal{S})$, so is $(f^+)^p$ for positive integers p.*

Exercise 8.6.3 *Let f be the pointwise limit of the sequence of functions (f_n) for $n \geq 2$ defined by*

$$
f_n(x) \;=\; \begin{cases} 1, & x = 1 \\ 1/(k+1), & 1/(k+1) \leq x < 1/k,\ 1 \leq k < n \\ 0, & 0 \leq x < 1/n \\ \infty, & x < 0\ or\ x > 1 \end{cases}
$$

Why is f measurable? There are two answers here: one is that it is monotone and the other uses the theorems of this section.

Exercise 8.6.4

$$
x_n(t) \;=\; \begin{cases} 0, & 0 \leq t \leq 1/2 \\ n(t - 1/2), & 1/2 < t \leq 1/2 + 1/n \\ 1, & 1/2 + 1/n < t \leq 1 \\ \infty, & t < 0\ or\ t > 1 \end{cases}
$$

Prove the pointwise limit function x is not continuous on $[0, 1]$ but it is measurable.

Exercise 8.6.5 *Let $f_n(x) = |x|^n$ for $n \in \mathbb{N}$. Prove $f(x) = \sup_n f_n(x)$ is given by*

$$
f(x) \;=\; \begin{cases} 0, & -1 < x < 1 \\ 1, & x = 1 \\ \infty, & |x| > 1 \end{cases}
$$

Then prove f_n is measurable for all n and $f(x) = \sup_n f_n(x)$ is also measurable.

Exercise 8.6.6 *Let(x_n) be a sequence of positive numbers which converges to x_0. Let*

$$
f_n(x) \;=\; \begin{cases} x_n, & x \in \mathbb{Q} \cap [0, 1] \\ -x_n, & x \in\in \mathbb{Q} \cap [0, 1] \\ \infty, & x \in ([0, 1])^C \end{cases}
$$

- *Prove each f_n is measurable.*
- *Prove $\underline{\lim}\, f_n$ is measurable.*
- *Prove $\overline{\lim}\, f_n$ is measurable.*
- *Compute $\underline{\lim}\, f_n$ and $\underline{\lim}\, f_n$.*

We now show that $M(X, \mathcal{S})$ is closed under multiplication.

Lemma 8.6.4 Products of Measurable Functions are Measurable

> Let X be a nonempty set and $\mathcal{S}$ a sigma-algebra on X. Let $f, g \in M(X, \mathcal{S})$. Then $fg \in M(X, \mathcal{S})$.

Proof 8.6.4
Let f_n, the truncation of f, be defined by

$$
f_n(x) \;=\;
\begin{cases}
f(x) & |\,f(x)\,| \le n \\
n & f(x) > n \\
-n & f(x) < -n
\end{cases}
$$

We define the truncation of g, g_n, in a similar way. We can easily show f_n and g_m are measurable for any n and m. We only show the argument for f_n as the argument for g_m is identical. Let α be a given real number. Then

$$
(f_n(x) > \alpha) \;=\;
\begin{cases}
\emptyset & \alpha \ge n, \\
(f(x) > n) \cup (\alpha < f(x) \le n) & 0 \le \alpha < n, \\
(f(x) > n) \cup (\alpha < f(x) \le n) & -n < \alpha < 0, \\
X & \alpha \le -n.
\end{cases}
$$

It is easy to see all of these sets are in $\mathcal{S}$ since f is measurable. Thus, each real-valued f_n is measurable.

It then follows by Lemma 8.4.1 that $f_n g_m$ is also measurable. Note we are using the definition of measurability for real-valued functions here. Next, an easy argument shows that at each x,

$$
f(x) \;=\; \lim_n f_n(x) \quad \text{and} \quad g(x) \;=\; \lim_m g_m(x)
$$

It then follows that

$$
f(x)\, g_m(x) \;=\; \lim_n \Big(f_n(x) g_m(x) \Big)
$$

Using Theorem 8.6.3, we see $f g_m$ is measurable. Then, noting

$$
f(x)\, g(x) \;=\; \lim_m \Big(f(x) g_m(x) \Big)
$$

another application of Theorem 8.6.3 establishes the result. $\blacksquare$

Homework

Exercise 8.6.7 Let X be a nonempty set and $\mathcal{S}$ a sigma-algebra on X. Then if f and g are in $M(X, \mathcal{S})$ so is $f^p g^q$ for positive integers p and q.

Exercise 8.6.8 Let X be a nonempty set and $\mathcal{S}$ a sigma-algebra on X. Then if f and g are in $M(X, \mathcal{S})$, so is $(f^+)^p (g^-)^q$ for positive integers p and q.

8.7 Continuous Functions of Measurable Functions

We wish to explore what properties the composition of a continuous function and a measurable function might have.

8.7.1 The Composition with Finite Measurable Functions

We begin with the case of finite measurable functions.

Lemma 8.7.1 Continuous Functions of Finite Measurable Functions are Measurable

> Let X be nonempty and $(X, \mathcal{S}, \mu)$ be a measure space. Let $f \in M(X, \mathcal{S})$ be finite. Let $\phi : \Re \to \Re$ be continuous. Then $\phi \circ f$ is measurable.

Proof 8.7.1
Let α be in $\Re$. We claim

$$\left(\phi \circ f\right)^{-1}(\alpha, \infty) \;=\; f^{-1}\left(\phi^{-1}(\alpha, \infty)\right)$$

First, let x be in the right-hand side. Then,

$$f(x) \in \phi^{-1}(\alpha, \infty) \;\Rightarrow\; \phi\left(f(x)\right) \in (\alpha, \infty)$$

$$\Rightarrow\; x \in \left(\phi \circ f\right)^{-1}(\alpha, \infty)$$

Conversely, if x is in the left-hand side, then

$$\left(\phi \circ f\right)(x) \in (\alpha, \infty) \;\Rightarrow\; f(x) \in \phi^{-1}(\alpha, \infty)$$

$$\Rightarrow\; x \in f^{-1}\left(\phi^{-1}(\alpha, \infty)\right)$$

Since ϕ is continuous, $G = \phi^{-1}(\alpha, \infty)$ is an open set. Finally, since f is measurable, $f^{-1}(G)$ is in $\mathcal{S}$. We conclude that $\phi \circ f$ is measurable, since our choice of α is arbitrary. ■

To handle the composition of a continuous function and an extended real-valued measurable function, we need an approximation result.

8.7.2 The Approximation of Non-Negative Measurable Functions

Theorem 8.7.2 The Approximation of Non-Negative Measurable Functions by Monotone Sequences

> Let X be a nonempty set and $\mathcal{S}$ a sigma-algebra on X. Let $f \in M(X, \mathcal{S})$ which is non-negative. Then there is a sequence $(\phi_n) \subseteq M(X, \mathcal{S})$ so that
>
> (i): $0 \leq \phi_n(x) \leq \phi_{n+1}$ for all x and for all $n \geq 1$.
>
> (ii): $\phi_n(x) \leq f(x)$ for all x and n and $f(x) = \lim_n \phi_n(x)$.
>
> (iii): Each ϕ_n has a finite range of values.

MEASURABILITY 169

Proof 8.7.2

Pick a positive integer n. Let

$$E_{k,n} = \begin{cases} \{x \in X \mid \frac{k}{2^n} \leq f(x) < \frac{k+1}{2^n}\}, & for\ 0 \leq k \leq n2^n - 1 \\ \{x \in X \mid n \leq f(x)\}, & for\ k = n2^n \end{cases}$$

You should draw some of these sets for a number of choices of non-negative functions f to get a feel for what they mean. Once you have done this, you will see that this definition slices the $[0, n]$ range of f into $n2^n$ slices each of height 2^{-n}. The last set, $E_{n2^n,n}$ is the set of all points where $f(x)$ exceeds n. This gives us a total of $n2^n + 1$ sets. It is clear that $X = \cup_k E_{k,n}$ and that each of these sets are disjoint from the others. Now define the functions ϕ_n by

$$\phi_n(x) = \frac{k}{2^n},\ x \in E_{k,n}.$$

It is evident that ϕ_n only takes on a finite number of values and so (iii) is established. Also, since f is measurable, we know each $E_{k,n}$ is measurable. Then, given any real number α, the set $(\phi_n(x) > \alpha)$ is either empty or consists of a union of the finite number of sets $E_{k,n}$ with the property that $\alpha > (k/2^n)$. Thus, $(\phi_n(x) > \alpha)$ is measurable for all α. We conclude each ϕ_n is measurable. If $f(x) = +\infty$, then by definition, $\phi_n(x) = n$ for all n and we have $f(x) = \lim_n \phi_n(x)$. Note, the ϕ_n values are strictly monotonically increasing which shows (i) and (ii) both hold in this case.

On the other hand, if $f(x)$ is finite, let n_0 be the first integer with $n_0 - 1 \leq f(x) < n_0$. Then, we must have $\phi_1(x) = 1$, $\phi_2(x) = 2$ and so forth until we have $\phi_{n_0-1} = n_0 - 1$. These first values are monotone increasing. We also know from the definition of ϕ_{n_0} that there is a k_0 so that

$$\frac{k_0}{2^{n_0}} \leq f(x) < \frac{k_0 + 1}{2^{n_0}}$$

Thus, $0 \leq f(x) - \phi_{n_0}(x) < 2^{-n_0}$. Now consider the function ϕ_{n_0+1}. We know

$$\begin{aligned} f(x) &\in \left[\frac{k_0}{2^{n_0}}, \frac{k_0 + 1}{2^{n_0}}\right) \\ &= \left[\frac{2k_0}{2^{n_0+1}}, \frac{2k_0 + 1}{2^{n_0+1}}\right) \cup \left[\frac{2k_0 + 1}{2^{n_0+1}}, \frac{2k_0 + 2}{2^{n_0+1}}\right) \end{aligned}$$

If $f(x)$ lands in the first interval above, we have

$$\phi_{n_0+1}(x) = \frac{2k_0}{2^{n_0+1}} = \frac{k_0}{2^{n_0}} = \phi_{n_0}(x)$$

and if $f(x)$ is in the second interval, we have

$$\phi_{n_0+1}(x) = \frac{2k_0 + 1}{2^{n_0+1}} > \frac{k_0}{2^{n_0}} = \phi_{n_0}(x).$$

In both cases, we have $\phi_{n_0}(x) \leq \phi_{n_0+1}(x)$. We also have immediately that $0 \leq f(x) - \phi_{n_0+1}(x) < 2^{-n_0-1}$.

The argument for $n_0 + 2$ and so on is quite similar and is omitted. This establishes (i) for this case. In general, we have $0 \leq f(x) - \phi_k(x) < 2^{-k}$ for all $k \geq n_0$. This implies that $f(x) = \lim_n \phi_n(x)$ which establishes (ii). ∎

Homework

Exercise 8.7.1 *Let* $f(x) = x^2$ *on* $\Re$. *Compute* E_{nk} *for this function and draw the functions* ϕ_n.

Exercise 8.7.2 *Let* $f(x) = \tanh(x)$ *on* $\Re$. *Compute* E_{nk} *for this function and draw the functions* ϕ_n.

Finally, we can handle the case of the composition of a continuous function and an extended real-valued measurable function.

8.7.3 Continuous Functions of Extended Real-Valued Measurable Functions

Lemma 8.7.3 Continuous Functions of Measurable Functions are Measurable

> *Let* X *be nonempty and* $(X, \mathcal{S}, \mu)$ *be a measure space. Let* $f \in M(X, \mathcal{S})$. *Let* $\phi : \Re \to \Re$ *be continuous and assume that* $\lim_n \phi(n)$ *and* $\lim_n \phi(-n)$ *are well-defined extended real-value numbers. Then* $\phi \circ f$ *is measurable.*

Proof 8.7.3
Assume first that f *is non-negative. Then by Theorem 8.7.2, there is a sequence of finite non-negative increasing functions* (f_n) *which are measurable and satisfy* $f_n \uparrow f$. *Let* E *be the set of points where* f *is infinite. Then,* $f_n(x) = n$ *when* x *is in* E *and*

$$\lim_n f_n(x) = \begin{cases} f(x) & x \in E^C \\ \infty & x \in E. \end{cases}$$

Thus, since ϕ *is continuous on* E^C *and* $\lim_n \phi(n)$ *is a well-defined extended valued real number, we have,*

$$\lim_n \phi\Big(f_n(x)\Big) = \begin{cases} \phi(f(x)) & x \in E^C \\ \lim_n \phi(n) & x \in E. \end{cases}$$

Let $\lim_n \phi(n) = \beta$ *in* $[\infty, \infty]$. *Thus, if* β *is finite, we have*

$$\lim_n \phi\Big(f_n(x)\Big) = \phi\Big(f\, I_{E^C}\Big) + \beta\, I_E$$

which is measurable since the first part is measurable by Lemma 8.7.1 and the second part is measurable since E *is a measurable set by Lemma 8.5.2. If* $\beta = \infty$, *we have*

$$\lim_n \phi\Big(f_n(x)\Big) = \begin{cases} \phi(f(x)) & x \in E^C \\ \infty & x \in E. \end{cases}$$

Now apply Lemma 8.5.2. Since E *is measurable and* f_1 *defined by*

$$f_1(x) = \begin{cases} \phi(f(x)) & x \in E^C \\ 0 & x \in E, \end{cases}$$

is measurable, we see $\lim_n \phi\Big(f_n(x)\Big)$ *is measurable. A similar argument holds if* $\beta = -\infty$. *We conclude that if* f *is non-negative,* $\phi \circ f$ *interpreted as above is a measurable function.*

Thus, if f *is arbitrary, the argument above shows that* $\phi \circ f^+$ *and* $\phi \circ f^-$ *are measurable. This implies that* $\phi \circ f = \phi \circ (f^+ - f^-)$ *is measurable when interpreted right.* ∎

Homework

MEASURABILITY

Exercise 8.7.3 *If a, b and c are real numbers, define the value in the middle, $mid(a,b,c)$ by*

$$mid(a,b,c) \quad = \quad \inf\{\,\sup\{a,b\},\ \sup\{a,c\},\ \sup\{b,c\}\,\}$$

Let X be a nonempty set and $\mathcal{S}$ a sigma-algebra on X. Let $f_1, f_2, f_3 \in M(X, \mathcal{S})$. Prove the function h defined pointwise by $h(x) = mid(f_1(x), f_2(x), f_3(x))$ is measurable.

Exercise 8.7.4 *Let X be a nonempty set and $\mathcal{S}$ a sigma-algebra on X. Let $f \in M(X, \mathcal{S})$ and $A > 0$. Define f_A by*

$$f_A(x) \quad = \quad \left\{ \begin{array}{ll} f(x), & \mid f(x) \mid \leq A \\ A, & f(x) > A \\ -A, & f(x) < -A \end{array} \right.$$

Prove f_A is measurable.

Exercise 8.7.5 *Let X be a nonempty set and $\mathcal{S}$ a sigma-algebra on X. Let $f \in M(X, \mathcal{S})$ and assume there is a positive K so that $0 \leq f(x) \leq K$ for all x. Prove the sequence ϕ_n of functions given in Theorem 8.7.2 converges uniformly to f on X.*

Exercise 8.7.6 *Let X and Y be nonempty sets and let $f : X \to Y$ be given. Prove that if $\mathcal{T}$ is a sigma-algebra of subsets of Y, then $\{f^{-1}(E) \mid E \in \mathcal{T}\}$ is a sigma-algebra of subsets of X.*

Exercise 8.7.7 *Let f be measurable on $\mathcal{B}$ and let $\phi(x) = \sin(x)$.*

- *If $f(x) = x^2$ is $\phi \circ f$ measurable?*
- *If $f(x) = \frac{1}{1+x^2}$, is $\phi \circ f$ measurable?*

Chapter 9

Measure and Integration

Once we have a nonempty set X with a given sigma-algebra $\mathcal{S}$, we can develop an abstract version of integration. To motivate this, consider the Borel sigma-algebra on $\overline{\mathfrak{R}}$, $\overline{\mathcal{B}}$. We know how to develop and use an integration theory that is based on finite intervals of the form $[a, b]$ for bounded functions. Hence, we have learned to understand and perform integrations of the form $\int_a^b f(t)dt$ for the standard Riemann integral. We could also write this as

$$\int_a^b f(t)dt \;=\; \int_{[a,b]} f(t)dt$$

and we have learned that

$$\int_{[a,b]} f(t)dt \;=\; \int_{(a,b)} f(t)dt = \int_{(a,b]} f(t)dt = \int_{[a,b)} f(t)dt$$

Note that we can thus say that we can compute $\int_E f(t)dt$ for $E \in \overline{\mathcal{B}}$ for sets E which are finite and have the form $[a, b]$, $(a, b]$, $[a, b)$ and (a, b). We can extend this easily to finite unions of disjoint intervals of the form E as given above by taking advantage of Theorem 4.5.3 to see

$$\int_{\cup_n E_n} f(t)dt \;=\; \sum_n \int_{E_n} f(t)dt$$

However, the development of the Riemann integral is closely tied to the interval $[a, b]$ and so it is difficult to extend these integrals to arbitrary elements F of $\overline{\mathcal{B}}$. Still, we can see that the Riemann integral is defined on some subset of the sigma-algebra $\overline{\mathcal{B}}$.

From our discussions of the Riemann - Stieltjes integral, we know that the Riemann integral can be interpreted as a Riemann - Stieltjes integral with the integrator given by the identity function $id(x) = x$. Let's switch to a new notation. Define the function $\mu(x) = x$. Then for our allowable E, we can write $\int_E f(t)dt = \int_E f(t)d\mu(t)$ which we can further simplify to $\int_E f d\mu$ as usual. Note that μ is a function which assigns a real value which we interpret as *length* to all of the allowable sets E we have been discussing. In fact, note μ is a mapping which satisfies

(i): If E is the empty set, then the length of E is 0; i.e. $\mu(\emptyset) = 0$.

(ii): If E is the finite interval $[a, b]$, $(a, b]$, $[a, b)$ or (a, b), $\mu(E) = b - a$.

(iii): If (E_n) is a finite collection of disjoint intervals, then the length of the union is clearly the sum of the individual lengths; i.e. $\mu(\cup_n E_n) = \sum_n \mu(E_n)$.

173

However, μ is not defined on the entire sigma -algebra. Also, it seems that we would probably like to extend (iii) above to countable disjoint unions as it is easy to see how that would arise in practice. If we could find a way to extend the usual length calculation of an interval to the full sigma -algebra, we could then try to extend the notion of integration as well.

It turns out we can do all of these things but we cannot do it by reusing our development process from Riemann integration. Instead, we must focus on developing a theory that can handle integrators which are mappings μ defined on a full sigma-algebra. It is time to precisely define what we mean by such a mapping.

9.1 Measures

Definition 9.1.1 Measures

> Let X be a nonempty set and S a sigma-algebra of subsets in X. We say $\mu : S \to \overline{\Re}$ is a measure on S if
>
> (i): $\mu(\emptyset) = 0$,
>
> (ii): $\mu(E) \geq 0$, for all $E \in S$,
>
> (iii): μ is countably additive on S; i.e. if $(E_n) \subseteq S$ is a countable collection of disjoint sets, then $\mu(\cup_n E_n) = \sum_n \mu(E_n)$.
>
> We also say (X, S, μ) is a measure space. If $\mu(X)$ is finite, we say μ is a finite measure. Also, even if $\mu(X) = \infty$, the measure μ is "almost finite" if we can find a collection of measurable sets (F_n) so that $X = \cup_n F_n$ with $\mu(F_n)$ finite for all n. In this case, we say the measure μ is σ-finite.

We can drop the requirement that the mapping μ be non-negative. The resulting mapping is called a *charge* instead of a measure. This will be important later. You can go back and read our discussion of charges in the context of our development of games in (Peterson (9) 2020) also.

Definition 9.1.2 Charges

> Let X be a nonempty set and S a sigma-algebra of subsets in X. We say $\nu : S \to \overline{\Re}$ is a charge on S if
>
> (i): $\nu(\emptyset) = 0$,
>
> (ii): ν is countably additive on S; i.e. if $(E_n) \subseteq S$ is a countable collection of disjoint sets, then $\nu(\cup_n E_n) = \sum_n \nu(E_n)$.
>
> Note that we want the value of the charge to be finite on all members of S as otherwise we could potentially have trouble with subsets having value ∞ and $-\infty$ inside a given set. That would then lead to undefined $\infty - \infty$ operations.

Let's look at some examples using exercises to help it sink in.

Homework

Exercise 9.1.1 *Let X be any nonempty set and let the sigma-algebra be $S = \mathcal{P}(X)$, the power set of X. Define μ_1 on S by $\mu_1(E) = 0$ for all E. Prove μ_1 is a very uninteresting measure, albeit not* *very interesting! Another non-interesting measure is defined by* $\mu_2(E) = \infty$ *if* E *is not empty and* 0 *if* $E = \emptyset$.

Exercise 9.1.2 *Let* X *be any nonempty set and let the sigma-algebra be* $\mathcal{S} = \mathcal{P}(X)$, *the power set of* X. *Define* $\mu_2(E) = \infty$ *if* E *is not empty and* 0 *if* $E = \emptyset$. *Prove this is another uninteresting measure!*

Exercise 9.1.3 *Let* X *be any set and again let* $\mathcal{S} = \mathcal{P}(X)$. *Pick any element* p *in* X. *Define* μ *by* $\mu(E) = 0$ *if* $p \notin E$ *and* 1 *if* $p \in E$. *Prove* μ *is a measure.*

Exercise 9.1.4 *Let* X *be the counting numbers,* $\mathbf{N}$, *and* $\mathcal{S} = \mathcal{P}(\mathbf{N})$. *Define* μ *by* $\mu(E)$ *is the cardinality of* E *if* E *is a finite set and* ∞ *otherwise. Prove* μ *is a measure. This measure is called the* **counting measure**. *Note that* $\mathbf{N} = \cup_n \{1, \ldots, n\}$ *for all* n *and* $\mu(\{1, \ldots, n\}) = n$, *which implies* μ *is a* σ-*finite measure.*

To help you see where we are going, here are some look ahead's to future material we will be covering. Let $\overline{\mathcal{B}}$ be the extended Borel sigma-algebra. We will show later there is a measure $\lambda : \overline{\mathcal{B}} \to \overline{\Re}$ that extends the usual idea of the length of an interval. That is, if E is a finite interval of the form (a, b), $[a, b)$, $(a, b]$ or $[a, b]$, then the length of E is $b - a$ and $\lambda(E) = b - a$. Further, if the interval has infinite length, (for example, E is $(-\infty, a)$), then $\lambda(E) = \infty$ also. The measure λ will be called **Borel measure** and since $\overline{\Re} = \cup_n [-n, n]$, we see Borel measure is a σ-finite measure. The sets in $\overline{\mathcal{B}}$ are called Borel measurable sets.

We will also be able to show that there is a larger sigma-algebra $\mathcal{M}$ of subsets of $\overline{\Re}$ and a measure μ defined on $\mathcal{M}$ which also returns the usual length of intervals. Hence, $\overline{\mathcal{B}} \subseteq \mathcal{M}$ strictly (i.e. there are sets in $\mathcal{M}$ not in $\overline{\mathcal{B}}$) with $\mu = \lambda$ on $\overline{\mathcal{B}}$. This measure will be called **Lebesgue measure** and the sets in $\mathcal{M}$ will be called Lebesgue measurable sets. There is a lot of abstraction going on here. Specifically,

- The proof that there are Lebesgue measurable sets that are not Borel measurable sets will require a non-constructive argument using the Axiom of Choice (which, of course, is equivalent to Zorn's Lemma, but we digress!)

- Further, we will be able to show that the Lebesgue sigma-algebra is not the entire power set as there are non-Lebesgue measurable sets. The proof that such sets exist requires the use of the interesting functions built using Cantor sets discussed in Chapter 13.

In the setting of Borel measure on $\overline{\Re}$, we will be able to show that if g is a continuous and monotone increasing function of $\Re$, then there is a measure, λ_g defined on $\overline{\mathcal{B}}$ which satisfies

$$\lambda_g(E) = \int_E dg$$

for any finite interval E. Here, $\int_E dg$ is the usual Riemann - Stieltjes integral. These are really important measures as it is easy to construct them using the base measure generated by g.

9.2 Some Basic Properties of Measures

Lemma 9.2.1 Monotonicity

Let $(X, \mathcal{S}, \mu)$ *be a measure space. If* $E, F \in \mathcal{S}$ *with* $E \subseteq F$, *then* $\mu(E) \leq \mu(F)$. *Moreover, if* $\mu(E)$ *is finite, then* $\mu(F \setminus E) = \mu(F) - \mu(E)$.

BASIC ANALYSIS IV: MEASURE THEORY AND INTEGRATION

Proof 9.2.1

We know $F = E \cup (F \setminus E)$ is a disjoint decomposition of F. By the countable additivity of μ, it follows immediately that $\mu(F) = \mu(E) + \mu(F \setminus E)$. Since μ is non-negative, we see $\mu(F) \geq \mu(E)$. Finally, if $\mu(E)$ is finite, then subtraction is allowed in $\mu(F) = \mu(E) + \mu(F \setminus E)$ which leads to $\mu(F \setminus E) = \mu(F) - \mu(E)$. ∎

Lemma 9.2.2 The Measure of Monotonic Sequences of Sets

Let $(X, \mathcal{S}, \mu)$ be a measure space.

(i): If (E_n) is an increasing sequence of sets in $\mathcal{S}$ (i.e. $E_n \subseteq E_{n+1}$ for all n), then $\mu(\cup_n E_n) = \lim_n \mu(E_n)$.

(ii): If (F_n) is a decreasing sequence of sets in $\mathcal{S}$ (i.e. $F_{n+1} \subseteq F_n$ for all n) and $\mu(F_1)$ is finite, then $\mu(\cap_n F_n) = \lim_n \mu(F_n)$.

Proof 9.2.2

To prove (i), if there is an index n_0 where $\mu(E_{n_0})$ is infinite, then by the monotonicity of μ, we must have $\infty = \mu(E_{n_0}) \leq \mu(\cup_n E_n)$. Hence, $\mu(\cup_n E_n) = \infty$. However, since $E_{n_0} \subseteq E_n$ for all $n \geq n_0$, again by monotonicity, $n \geq n_0$ implies $\mu(E_n) = \infty$. Thus, $\lim_n \mu(E_n) = \mu(\cup_n E_n) = \infty$. On the other hand, if $\mu(E_n)$ is finite for all n, define the disjoint sequence of set (A_n) as follows:

$$
\begin{aligned}
A_1 &= E_1 \\
A_2 &= E_2 \setminus E_1 \\
A_3 &= E_3 \setminus E_2 \\
&\vdots \quad \vdots \quad \vdots \\
A_n &= E_n \setminus E_{n-1}
\end{aligned}
$$

We see $\cup_n A_n = \cup_n E_n$ and since μ is countably additive, we must have $\mu(\cup_n A_n) = \sum_n \mu(A_n)$. Since by assumption $\mu(E_n)$ is finite in this case, we know $\mu(A_n) = \mu(E_n) - \mu(E_{n-1})$. It follows that

$$
\begin{aligned}
\sum_{k=1}^n \mu(A_k) &= \mu(E_1) + \sum_{k=2}^n \left(\mu(E_k) - \mu(E_{k-1}) \right) \\
&= \mu(E_1) + \mu(E_n) - \mu(E_1) \\
&= \mu(E_n)
\end{aligned}
$$

We conclude

$$
\begin{aligned}
\mu(\cup_n E_n) &= \mu(\cup_n A_n) \\
&= \lim_n \sum_{k=1}^n \mu(A_k) \\
&= \lim_n \mu(E_n)
\end{aligned}
$$

This proves the validity of (i). Next, for (ii), construct the sequence of sets (E_n) by

$$
\begin{aligned}
E_1 &= \emptyset \\
E_2 &= F_1 \setminus F_2
\end{aligned}
$$

ABSTRACT INTEGRATION

$$E_3 = F_1 \setminus F_3$$
$$\vdots \quad \vdots \quad \vdots$$
$$E_n = F_1 \setminus F_n$$

Then (E_n) is an increasing sequence of sets and so by (i), $\mu(\cup_n E_n) = \lim_n \mu(E_n)$. Since $\mu(F_1)$ is finite, we then know that $\mu(E_n) = \mu(F_1) - \mu(F_n)$. Hence, $\mu(\cup_n E_n) = \mu(F_1) - \lim_n \mu(F_n)$. Next, note by De Morgan's Laws,

$$\mu(\cup_n E_n) = \mu\left(\cup_n F_1 \cap F_n^C\right)$$
$$= \mu\left(F_1 \cap \cup_n F_n^C\right)$$
$$= \mu\left(F_1 \cap \left(\cap_n F_n\right)^C\right)$$
$$= \mu\left(F_1 \setminus \left(\cap_n F_n\right)\right)$$

Thus, since $\mu(F_1)$ is finite and $\cap_n F_n \subseteq F_1$, we have $\mu(\cup_n E_n) = \mu(F_1) - \mu(\cap_n F_n)$. Combining these results, we have

$$\mu(F_1) - \lim_n \mu(F_n) = \mu(F_1) - \mu(\cap_n F_n)$$

The result then follows by canceling $\mu(F_1)$ from both sides which is allowed as this is a finite number. ∎

Homework

Exercise 9.2.1 *Let $E_n = [-n, n]$ define a sequence of subsets in $\Re$. Let's assume we know all about Borel measure λ so that $\lambda(E_n) = 2n$.*

- *Find the decomposition of (E_n) into disjoint sets (A_n).*

- *Compute $\underline{\lim}(E_n)$, $\lambda(\underline{\lim}(E_n))$ and $\lim_n \lambda(E_n)$.*

- *Compute $\underline{\lim}(E_n)$, $\lambda(\underline{\lim}(E_n))$ and $\lim_n \lambda(E_n)$.*

- *Compute $\overline{\lim}(E_n)$, $\lambda(\overline{\lim}(E_n))$ and $\lim_n \lambda(E_n)$.*

Exercise 9.2.2 *Let $E_n = [1/n, 1 - 1/n]$ define a sequence of subsets in $\Re$ for $n > 2$. Again assume we know all about Borel measure λ.*

- *Find the decomposition of (E_n) into disjoint sets (A_n).*

- *Compute $\underline{\lim}(E_n)$, $\lambda(\underline{\lim}(E_n))$ and $\lim_n \lambda(E_n)$.*

- *Compute $\overline{\lim}(E_n)$, $\lambda(\overline{\lim}(E_n))$ and $\lim_n \lambda(E_n)$.*

Exercise 9.2.3 *Let $F_n = [1/(n+1), 1/n]$ define a sequence of subsets in $\Re$ for $n > 1$. Again assume we know all about Borel measure λ.*

- *Find the decomposition of (F_n) into disjoint sets (E_n).*

- *Compute $\underline{\lim}(E_n)$, $\lambda(\underline{\lim}(E_n))$ and $\lim_n \lambda(E_n)$.*

- *Compute $\overline{\lim}(E_n)$, $\lambda(\overline{\lim}(E_n))$ and $\lim_n \lambda(E_n)$.*

9.3 Sequences of Sets

We will now develop a series of ideas involving sequences of sets.

Definition 9.3.1 The Limit Inferior and Superior of Sequences of Sets

> Let X be a nonempty set and (A_n) be a sequence of subsets of X. The limit inferior of (A_n) is defined to be the set
>
> $$\liminf \; = \; \underline{\lim}(A_n) \; = \; \bigcup_{m=1}^{\infty} \bigcap_{n=m}^{\infty} A_n$$
>
> while the limit superior of (A_n) is defined by
>
> $$\limsup \; = \; \overline{\lim}(A_n) \; = \; \bigcap_{m=1}^{\infty} \bigcup_{n=m}^{\infty} A_n$$

It is convenient to have a better characterization of these sets.

Lemma 9.3.1 Characterizing Limit Inferior and Limit Superior of Sequences of Sets

> Let (A_n) be a sequence of subsets of the nonempty set X. Then we have
>
> $$\liminf(A_n) \; = \; \{x \in X \mid x \in A_k \text{ for all but finitely many indices } k\} \; = \; B$$
>
> and
>
> $$\limsup(A_n) \; = \; \{x \in X \mid x \in A_k \text{ for infinitely many indices } k\} \; = \; C$$

Proof 9.3.1

We will prove the statement about $\liminf(A_n)$ *first. Let* $x \in B$. *If there are no indices k so that* $x \notin A_k$, *then* $x \in \cap_{n=1}^{\infty}$ *telling us that* $x \in \liminf(A_n)$. *On the other hand, if there are a finite number of indices k that satisfy* $x \notin A_k$, *we can label these indices as* $\{k_1, \ldots, k_p\}$ *for some positive integer p. Let k^* be the maximum index in this finite list. Then, if $k > k^*$,* $x \in \cap_{n=k}^{\infty}$. *This implies immediately that* $x \in \liminf(A_n)$. *Conversely, if* $x \in \liminf(A_n)$, *there is an index k_0 so that* $x \in \cap_{n=k_0}^{\infty}$. *This implies that x can fail to be in at most a finite number of A_k where $k < k_0$. Hence,* $x \in B$.*

Next, we prove that $\limsup(A_n) = C$. *If* $x \in C$, *then if there were an index m_0 so that* $x \notin \cup_{n=m_0}^{\infty}$, *then x would belong to only a finite number of sets A_k which contradicts the definition of the set C. Hence, there is no such index m_0 and so* $x \in \cup_{n=m}^{\infty}$ *for all m. This implies* $x \in \limsup(A_n)$. *On the other hand, if* $x \in \limsup(A_n)$, *then* $x \in \cup_{n=m}^{\infty}$ *for all m. So, if x was only in a finite number of sets A_n, there would be a largest index m^* satisfying* $x \in A_{m^*}$ *but* $x \notin A_m$ *if $m > m^*$. But this then says* $x \notin \limsup(A_n)$. *This is a contradiction. Thus, our assumption that x was only in a finite number of sets A_n is false. This implies* $x \in C$. $\blacksquare$

Lemma 9.3.2 Limit Inferior and Limit Superior of Monotone Sequences of Sets

ABSTRACT INTEGRATION

Let X be a nonempty set. Then

  (i): If (A_n) is an increasing sequence of subsets of X, then

$$\liminf(A_n) \;=\; \limsup(A_n) \;=\; \bigcup_{n=1}^{\infty} A_n$$

  (ii): If (A_n) is a decreasing sequence of subsets of X, then

$$\liminf(A_n) \;=\; \limsup(A_n) \;=\; \bigcap_{n=1}^{\infty} A_n$$

  (iii): If (A_n) is an arbitrary sequence of subsets of X, then

$$\emptyset \;\subseteq\; \liminf(A_n) \subseteq \limsup(A_n)$$

Proof 9.3.2

(i): If $x \in \limsup(A_n)$, then $x \in \cup_{n=1}^{\infty} A_n$. Conversely, if $x \in \cup_{n=1}^{\infty} A_n$, there is an index n_0 so that $x \in A_{n_0}$. But since the sequence (A_n) is increasing, this means $x \in A_n$ for all $n > n_0$ also. Hence, $x \in \cup_{n=m}^{\infty} A_n$ for all indices $m \geq n_0$. However, it is also clear that x is in any union that starts at n smaller than n_0. Thus, x must be in $\cap_{m=1}^{\infty} \cup_{n=m}^{\infty} A_n$. But this is the set $\limsup(A_n)$. We conclude $\limsup(A_n) = \cup_{n=1}^{\infty}$. Now look at the definition of $\liminf(A_n)$. Since A_n is monotone increasing, $\cap_{n=m}^{\infty} A_n = A_m$. Hence, it is immediate that $\liminf(A_n) = \cup_{n=1}^{\infty}$.

(ii): The argument for this case is similar to the argument for case (i) and is left to you.

(iii): It suffices to show that $\liminf(A_n) \subseteq \limsup(A_n)$. If $x \in \liminf(A_n)$, by Lemma 9.3.1, x belongs to all but finitely many A_n. Hence, x belongs to infinitely many A_n. Then, applying Lemma 9.3.1 again, we have the result. ∎

There will be times when it will be convenient to write an arbitrary union of sets as a countable union of disjoint sets. In the next result, we show how this is done.

Lemma 9.3.3 Disjoint Decompositions of Unions

Let X be a nonempty set and let (A_n) be a sequence of subsets of X. Then there exists a sequence of mutually disjoint set (F_n) satisfying $\cup_n A_n = \cup_n F_n$.

Proof 9.3.3
Define sets E_n and F_n as follows:

$$
\begin{aligned}
E_0 &= \emptyset, \; F_1 = A_1 \setminus E_0 = A_1 \\
E_1 &= A_1, \; F_2 = A_2 \setminus E_1 = A_2 \setminus A_1 \\
E_2 &= A_1 \bigcup A_2, \; F_3 = A_3 \setminus E_2 = A_3 \setminus \left(A_1 \bigcup A_2 \right) \\
E_3 &= \bigcup_{k=1}^{3} A_k, \; F_4 = A_4 \setminus E_3 = A_4 \setminus \left(\bigcup_{k=1}^{3} A_k \right)
\end{aligned}
$$

BASIC ANALYSIS IV: MEASURE THEORY AND INTEGRATION

$$\vdots \;\; = \;\; \vdots, \;\; \vdots$$

$$E_n \;\; = \;\; \bigcup_{k=1}^{n} A_k, \;\; F_{n+1} \;\; = \;\; A_{n+1} \setminus E_n \;\; = \;\; A_4 \setminus \left(\bigcup_{k=1}^{n} A_k \right)$$

Note that (E_n) forms a monotonically increasing sequence of sets with $cup_n A_n \; \cup_n E_n$. We claim the sets F_n are mutually disjoint and $\cup_{j=1}^{n} f_j = \cup_{j=1}^{n} A_j$. We do this by induction.

Proof *Basis: It is clear that F_1 and F_2 are disjoint and $F_1 \cup F_2 = A_1 \cup A_2$. Induction: We assume that (F_k) are mutually disjoint for $1 \leq k \leq n$ and $\cup_{j=1}^{k} f_j = \cup_{j=1}^{k} A_j$ for $1 \leq k \leq n$ as well. Then*

$$F_{n+1} \;\; = \;\; A_{n+1} \setminus E_n$$

$$= \;\; A_{n+1} \bigcap \left(\bigcup_{j=1}^{n} A_j \right)^C$$

$$= \;\; \bigcap_{j=1}^{n} \left(A_{n+1} \bigcap A_j^C \right)$$

Now, by construction, $F_j \subseteq A_j$ for all j. However, from the above expansion of F_{n+1}, we see $F_{n+1} \subseteq A_j^C$ for all $1 \leq j \leq n$. This tells us $F_{n+1} \subseteq F_j^C$ for these indices also. We conclude F_{n+1} is disjoint from all the previous F_j. This shows (F_j) is a collection of mutually disjoint sets for $1 \leq j \leq n+1$. This proves the first part of the assertion. To prove the last part, note

$$\bigcup_{j=1}^{n+1} F_j \;\; = \;\; \bigcup_{j=1}^{n} F_j \bigcup F_{n+1}$$

$$= \;\; \bigcup_{j=1}^{n} A_j \bigcup \left(A_{n+1} \setminus \left(\bigcup_{j=1}^{n} A_j \right) \right)$$

$$= \;\; \bigcup_{j=1}^{n+1} A_j$$

This completes the induction step. We conclude that this proposition holds for all n. $\qquad\square$

Since the claim holds, it is then obvious that $\cup_{j=1}^{n} f_j = \cup_{j=1}^{n} A_j$. $\qquad\blacksquare$

Homework

Exercise 9.3.1 *Let $E_n = f^{-1}([n, n+1])$ define a sequence of subsets in $\Re$ for $f(x) = \ln(x)$ for $n \geq 1$. Again assume we know all about Borel measure λ.*

- *Find the decomposition of (E_n) into disjoint sets (A_n).*

- *Compute $\underline{\lim}(E_n)$, $\lambda(\underline{\lim}(E_n))$ and $\lim_n \lambda(E_n)$*

- *Compute $\overline{\lim}(E_n)$, $\lambda(\overline{\lim}(E_n))$ and $\lim_n \lambda(E_n)$*

- *Compute $\overline{\lim}(E_n)$.*

Exercise 9.3.2 *Let $E_n = [-1/n, -1/(n+1)] \cup [n, n+1]$ define a sequence of subsets in $\Re$ for $n \geq 1$. Again assume we know all about Borel measure λ.*

- *Find the decomposition of (E_n) into disjoint sets (F_n).*

ABSTRACT INTEGRATION 181

- *Compute* $\underline{\lim}(E_n)$, $\lambda(\underline{\lim}(E_n))$ *and* $\lim_n \lambda(E_n)$

- *Compute* $\overline{\lim}(E_n)$, $\lambda(\overline{\lim}(E_n))$ *and* $\lim_n \lambda(E_n)$

- *Compute* $\overline{\lim}(E_n)$.

Exercise 9.3.3 *Let* $E_n = [-(n+1), 1/n] \cup [1 - 1/n, n]$ *define a sequence of subsets in* $\Re$ *for* $n \geq 1$. *Again assume we know all about Borel measure* λ.

- *Find the decomposition of* (E_n) *into disjoint sets* (F_n).

- *Compute* $\underline{\lim}(E_n)$, $\lambda(\underline{\lim}(E_n))$ *and* $\lim_n \lambda(E_n)$

- *Compute* $\overline{\lim}(E_n)$, $\lambda(\overline{\lim}(E_n))$ *and* $\lim_n \lambda(E_n)$

- *Compute* $\overline{\lim}(E_n)$.

To finish this section on measures, we want to discuss the idea that a property holds except on a set of measure zero. Recall, this subject came up when we discussed the content of a subset of $\Re$ earlier in Section 5.3. However, we can extend this concept of an arbitrary measure space $(X, \mathcal{S}, \mu)$ as follows.

Definition 9.3.2 Propositions Holding Almost Everywhere

> Let $(X, \mathcal{S}, \mu)$ be a measure space. We say a proposition $\mathcal{P}$ holds almost everywhere on X if $\{x \in X \mid \mathcal{P}$ does not hold $\}$ has μ measure zero. We usually say the proposition holds μ a.e. rather than writing out the phrase μ almost everywhere. Also, if the measure μ is understood from context, we usually just say the proposition hold a.e. to make it even easier to write down.

Comment 9.3.1 *Given the measure space* $(X, \mathcal{S}, \mu)$, *if* f *and* g *are extended real-valued functions on* X *which are measurable, we would say* $f = g$ μ *a.e. if* $\mu(\{x \in X \mid f(x) \neq g(x)\}) = 0$.

Comment 9.3.2 *Given the measure space* $(X, \mathcal{S}, \mu)$, *if* (f_n) *is a sequence of measurable extended real-valued functions on the* X, *and* $f : X \to \overline{\Re}$ *is another measurable function on* X, *we would say* f_n *converges pointwise a.e. to* f *if the set* $\{x \in X \mid f_n(x) \not\to f(x)\}$ *has measure* 0. *We would usually write* $f_n \to f$ *pointwise* μ *a.e.*

Homework

Exercise 9.3.4 *What are the sets of measure zero with counting measure?*

Exercise 9.3.5 *Assuming you know about Borel measure,* λ, *compute the measure of a single point in* $\Re$.

Exercise 9.3.6 *Define* f_n *for* $n \geq 2$ *by*

$$f_n(x) = \begin{cases} 1, & x = 1 \\ 1/(k+1), & 1/(k+1) \leq x < 1/k,\ 1 \leq k < n \\ 0, & 0 \leq x < 1/n \end{cases}$$

Assuming you know about Borel measure, λ, *is* f_n *continuous a.e.?*

Exercise 9.3.7 *Let* f *be the pointwise limit of the sequence of functions* (f_n) *for* $n \geq 2$ *defined by*

$$f_n(x) = \begin{cases} 1, & x = 1 \\ 1/(k+1), & 1/(k+1) \leq x < 1/k,\ 1 \leq k < n \\ 0, & 0 \leq x < 1/n \end{cases}$$

Assuming you know about Borel measure, λ, is f continuous a.e.?

Exercise 9.3.8

$$x_n(t) \;=\; \begin{cases} 0, & 0 \le t \le 1/2 \\ n(t-1/2), & 1/2 < t \le 1/2 + 1/n \\ 1, & 1/2 + 1/n < t \le 1 \end{cases}$$

Assuming you know about Borel measure, λ, is f continuous a.e.?

9.4 Integration

In this section, we will introduce an abstract notion of integration on the measure space $(X, \mathcal{S}, \mu)$. Recall that $M(X, \mathcal{S})$ denotes the class of extended real-valued measurable functions f on X. First we introduce a standard notation for some useful classes of functions. When we want to restrict our attention to the non-negative members of $M(X, \mathcal{S})$, we will use the notation that $f \in M^+(X, \mathcal{S})$.

To construct an abstract integration process on the measure space $(X, \mathcal{S}, \mu)$, we begin by defining the *integral* of a class of functions which can be used to approximate any function f in $M^+(X, \mathcal{S})$.

Definition 9.4.1 Simple Functions

Let $(X, \mathcal{S}, \mu)$ be a measure space and let $f : X \to \Re$ be a function. We say f is a simple function if the range of f is a finite set and f is $\mathcal{S}$ measurable. This implies the following standard unique representation of f. Since the range is finite, there is a positive integer N and distinct numbers a_j, $1 \le j \le N$ so that

(i): the sets $E_j = f^{-1}(a_j)$ are measurable and mutually disjoint for $1 \le j \le N$,

(ii): $X = \bigcup_{j=1}^{N} E_j$,

(iii): f has the characterization

$$f(x) \;=\; \sum_{j=1}^{N} a_j I_{E_j}(x)$$

We then define the *integral* of a simple function as follows.

Definition 9.4.2 The Integral of a Simple Function

Let $(X, \mathcal{S}, \mu)$ be a measure space and let $\phi : X \to \Re$ be a simple function. Let

$$\phi(x) \;=\; \sum_{j=1}^{N} a_j I_{E_j}(x)$$

be the standard representation of ϕ where the numbers a_j are distinct and the sets E_j are mutually disjoint, cover X, and are measurable for $1 \le j \le N$ for some positive integer N. Then the integral of ϕ with respect to the measure μ is the extended real-valued number

$$\int \phi \, d\mu \;=\; \sum_{j=1}^{N} a_j \, \mu(E_j)$$

ABSTRACT INTEGRATION

183

Comment 9.4.1 *We note that $\int \phi d\mu$ can be $+\infty$. Recall, our convention that $0 \cdot \infty = 0$. Hence, if one of the values a_j is 0, the contribution to the integral is $0\mu(E_j)$ which is 0 even if $\mu(E_j) = \infty$. Further, note the 0 function on X can be defined as $I_\emptyset$ which is a simple function. Hence, $\int 0 d\mu = 0$.*

Homework

Exercise 9.4.1 *Assume you know about Borel measure λ on $[a, b]$. Let $P = \{x_0 = a, x_1, \ldots, x_{n-1}, x_n = b\}$ be a partition of $[a, b]$. For a given evaluation set $E = \{s_1, \ldots, s_n\}$ from this partition, the Riemann sum $S(f, P, E)$ for a given bounded function f on $[a, b]$ is*

$$S(f, P, E) = \sum_{i=1}^{n} f(s_i) \, (x_{i+1} - x_i)$$

Define the new sets $\{E_1, \ldots, E_n\}$ by $E_i = [x_i, x_{i+1})$ if $1 \leq i < n$ and $E_n = [x_{n-1}, x_n]$.

- *Prove the integral of the simple function $\phi = \sum_{i=1}^{n} f(s_i) I_{E_i}$ is the same as the Riemann sum.*

- *Does it matter if we define the E_n differently?*

- *How do you handle the case where some $f(s_i)$'s match in value? Show all the details here.*

Exercise 9.4.2 *Assume you know about Borel measure λ on $[a, b]$. Let $P = \{x_0 = a, x_1, \ldots, x_{n-1}, x_n = b\}$ be a partition of $[a, b]$. The Lower Darboux sum $L(f, P)$ for a given bounded function f on $[a, b]$ is*

$$L(f, P) = \sum_{i=1}^{n} \inf_{x \in [x_i, x_{i+1}]} f(x) \, (x_{i+1} - x_i)$$

Define the new sets $\{E_1, \ldots, E_n\}$ by $E_i = [x_i, x_{i+1})$ if $1 \leq i < n$ and $E_n = [x_{n-1}, x_n]$.

- *Prove the integral of the simple function $\phi = \sum_{i=1}^{n} \inf_{x \in [x_i, x_{i+1}]} f(x) I_{E_i}$ is the same as the Riemann sum.*

- *Does it matter if we define the E_n differently?*

- *How do you handle the case where some $\inf_{x \in [x_i, x_{i+1}]} f(x)$'s match in value? Show all the details here.*

Exercise 9.4.3 *Assume you know about Borel measure λ on $[a, b]$. Let $P = \{x_0 = a, x_1, \ldots, x_{n-1}, x_n = b\}$ be a partition of $[a, b]$. The Upper Darboux sum $U(f, P)$ for a given bounded function f on $[a, b]$ is*

$$U(f, P) = \sum_{i=1}^{n} \sup_{x \in [x_i, x_{i+1}]} f(x) \, (x_{i+1} - x_i)$$

Define the new sets $\{E_1, \ldots, E_n\}$ by $E_i = [x_i, x_{i+1})$ if $1 \leq i < n$ and $E_n = [x_{n-1}, x_n]$.

- *Prove the integral of the simple function $\phi = \sum_{i=1}^{n} \sup_{x \in [x_i, x_{i+1}]} f(x) I_{E_i}$ is the same as the Riemann sum.*

- *Does it matter if we define the E_n differently?*

- *How do you handle the case where some $\sup_{x \in [x_i, x_{i+1}]} f(x)$'s match in value? Show all the details here.*

184   BASIC ANALYSIS IV: MEASURE THEORY AND INTEGRATION

Exercise 9.4.4 *Assume you know about the measure λ_g on $[a, b]$ where g is a continuous monotonic increasing function and the* length *of $[a, b]$ is $g(b) - g(a)$. Hence, $\lambda_g([a, b]) = g(b) - g(a)$. Let $P = \{x_0 = a, x_1, \ldots, x_{n-1}, x_n = b\}$ be a partition of $[a, b]$. For a given evaluation set $E = \{s_1, \ldots, s_n\}$ from this partition, the Riemann - Stieltjes sum $S(f, g, P, E)$ for a given bounded function f on $[a, b]$ is*

$$S(f, g, P, E) \quad = \quad \sum_{i=1}^{n} f(s_i) \left(g(x_{i+1}) - g(x_i) \right)$$

Define the new sets $\{E_1, \ldots, E_n\}$ by $E_i = [x_i, x_{i+1})$ if $1 \leq i < n$ and $E_n = [x_{n-1}, x_n]$.

- *Prove the integral of the simple function $\phi = \sum_{i=1}^{n} f(s_i) I_{E_i}$ is the same as the Riemann - Stieltjes sum.*

- *Does it matter if we define the E_n differently?*

- *How do you handle the case where some $f(s_i)$'s match in value? Show all the details here.*

Exercise 9.4.5 *Assume you know about the measure λ_g on $[a, b]$ where g is a continuous monotonic increasing function and the* length *of $[a, b]$ is $g(b) - g(a)$. Hence, $\lambda_g([a, b]) = g(b) - g(a)$. Let $P = \{x_0 = a, x_1, \ldots, x_{n-1}, x_n = b\}$ be a partition of $[a, b]$. The Lower Darboux sum $L(f, g, P)$ for a given bounded function f on $[a, b]$ is*

$$L(f, g, P) \quad = \quad \sum_{i=1}^{n} \inf_{x \in [x_i, x_{i+1}]} f(x) \left(g(x_{i+1}) - g(x_i) \right)$$

Define the new sets $\{E_1, \ldots, E_n\}$ by $E_i = [x_i, x_{i+1})$ if $1 \leq i < n$ and $E_n = [x_{n-1}, x_n]$.

- *Prove the integral of the simple function $\phi = \sum_{i=1}^{n} \inf_{x \in [x_i, x_{i+1}]} f(x) I_{E_i}$ is the same as the Riemann - Stieltjes sum.*

- *Does it matter if we define the E_n differently?*

- *How do you handle the case where some $\inf_{x \in [x_i, x_{i+1}]} f(x)$'s match in value? Show all the details here.*

Exercise 9.4.6 *Assume you know about the measure λ_g on $[a, b]$ where g is a continuous monotonic increasing function and the* length *of $[a, b]$ is $g(b) - g(a)$. Hence, $\lambda_g([a, b]) = g(b) - g(a)$. Let $P = \{x_0 = a, x_1, \ldots, x_{n-1}, x_n = b\}$ be a partition of $[a, b]$. The Upper Darboux sum $U(f, g, P)$ for a given bounded function f on $[a, b]$ is*

$$U(f, g, P) \quad = \quad \sum_{i=1}^{n} \sup_{x \in [x_i, x_{i+1}]} f(x) \left(g(x_{i+1}) - g(x_i) \right)$$

Define the new sets $\{E_1, \ldots, E_n\}$ by $E_i = [x_i, x_{i+1})$ if $1 \leq i < n$ and $E_n = [x_{n-1}, x_n]$.

- *Prove the integral of the simple function $\phi = \sum_{i=1}^{n} \sup_{x \in [x_i, x_{i+1}]} f(x) I_{E_i}$ is the same as the Riemann - Stieltjes sum.*

- *Does it matter if we define the E_n differently?*

- *How do you handle the case where some $\sup_{x \in [x_i, x_{i+1}]} f(x)$'s match in value? Show all the details here.*

Using this, we can define the integral of any function in $M^+(X, \mathcal{S})$.

ABSTRACT INTEGRATION

185

Definition 9.4.3 The Integral of a Non-Negative Measurable Function

Let $(X, \mathcal{S}, \mu)$ be a measure space and let $f \in M^+(X, \mathcal{S})$. For convenience of notation, let $\mathcal{F}^+$ denote the collection of all non-negative simple functions on X. Then, the integral of f with respect to the measure μ is the extended real-valued number

$$\int f \, d\mu \;=\; \sup\{ \int \phi \, d\mu \mid \phi \in \mathcal{F}^+, \; \phi \leq f \}$$

If $E \in \mathcal{S}$, we define the integral of f over E with respect to μ to be

$$\int_E f \, d\mu \;=\; \int f I_E \, d\mu$$

Homework

This continues the work we began with the previous set of exercises.

Exercise 9.4.7 *Assume you know about Borel measure λ on $[a, b]$. Let $P = \{x_0 = a, x_1, \ldots, x_{n-1}, x_n = b\}$ be a partition of $[a, b]$. The Lower Darboux sum $L(f, P)$ for a given bounded non-negative function f on $[a, b]$ is*

$$L(f, P) \;=\; \sum_{i=1}^{n} \inf_{x \in [x_i, x_{i+1}]} f(x) \, (x_{i+1} - x_i)$$

Define the new sets $\{E_1, \ldots, E_n\}$ by $E_i = [x_i, x_{i+1})$ if $1 \leq i < n$ and $E_n = [x_{n-1}, x_n]$. Prove $L(f, g) \leq \int f \, d\lambda$.

Exercise 9.4.8 *Assume you know about Borel measure λ on $[a, b]$. Let $P = \{x_0 = a, x_1, \ldots, x_{n-1}, x_n = b\}$ be a partition of $[a, b]$. The Upper Darboux sum $U(f, P)$ for a given bounded function f on $[a, b]$ is*

$$U(f, P) \;=\; \sum_{i=1}^{n} \sup_{x \in [x_i, x_{i+1}]} f(x) \, (x_{i+1} - x_i)$$

Define the new sets $\{E_1, \ldots, E_n\}$ by $E_i = [x_i, x_{i+1})$ if $1 \leq i < n$ and $E_n = [x_{n-1}, x_n]$. Prove $U(f, g) \leq \int f \, d\lambda$.

Exercise 9.4.9 *Assume you know about the measure λ_g on $[a, b]$ where g is a continuous monotonic increasing function and the length of $]a, b]$ is $g(b) - g(a)$. Hence, $\lambda_g([a, b]) = g(b) - g(a)$. Let $P = \{x_0 = a, x_1, \ldots, x_{n-1}, x_n = b\}$ be a partition of $[a, b]$. The Lower Darboux sum $L(f, g, P)$ for a given bounded function f on $[a, b]$ is*

$$L(f, g, P) \;=\; \sum_{i=1}^{n} \inf_{x \in [x_i, x_{i+1}]} f(x) \, (g(x_{i+1}) - g(x_i))$$

Define the new sets $\{E_1, \ldots, E_n\}$ by $E_i = [x_i, x_{i+1})$ if $1 \leq i < n$ and $E_n = [x_{n-1}, x_n]$. Prove $U(f, g) \leq \int f \, d\lambda_g$.

Exercise 9.4.10 *Assume you know about the measure λ_g on $[a, b]$ where g is a continuous monotonic increasing function and the length of $[a, b]$ is $g(b) - g(a)$. Hence, $\lambda_g([a, b]) = g(b) - g(a)$. Let $P = \{x_0 = a, x_1, \ldots, x_{n-1}, x_n = b\}$ be a partition of $[a, b]$. The Upper Darboux sum $U(f, g, P)$*

for a given bounded function f on $[a, b]$ is

$$U(f, g, P) = \sum_{i=1}^{n} \sup_{x \in [x_i, x_{i+1}]} f(x) \, (g(x_{i+1}) - g(x_i))$$

Define the new sets $\{E_1, \ldots, E_n\}$ by $E_i = [x_i, x_{i+1})$ if $1 \leq i < n$ and $E_n = [x_{n-1}, x_n]$. Prove $U(f, g) \leq \int f \, d\lambda_g$.

Exercise 9.4.11 *If the bounded non-negative function f is Riemann Integrable, prove $\int_a^b f(x) dx \leq \int f \, d\lambda$.*

Exercise 9.4.12 *If the bounded non-negative function f is Riemann - Stieltjes with respect to continuous monotonic increasing g prove $\int_a^b f(x) dg(x) \leq \int f \, d\lambda_g$.*

9.5 Integration Properties

It is time to prove some results about this new abstract version of integration.

Lemma 9.5.1 Properties of Simple Function Integrations

> *Let $(X, \mathcal{S}, \mu)$ be a measure space and let $\phi, \psi \in M^+(X, \mathcal{S})$ be simple functions. Then,*
>
> *(i): If $c \geq 0$ is a real number, then $c\phi$ is also a simple function and $\int c\phi \, d\mu = c \int \phi \, d\mu$.*
>
> *(ii): $\phi + \psi$ is also a simple function and $\int (\phi + \psi) \, d\mu = \int \phi \, d\mu + \int \psi \, d\mu$.*
>
> *(iii): The mapping $\lambda : \mathcal{S} \to \overline{\mathfrak{R}}$ defined by $\lambda(E) = \int_E \phi \, d\mu$ for all E in $\mathcal{S}$ is a measure.*

Proof 9.5.1
Let ϕ have the standard representation

$$\phi(x) = \sum_{j=1}^{N} a_j I_{E_j}(x)$$

where the numbers a_j are distinct, the sets E_j are mutually disjoint, cover X, and are measurable for $1 \leq j \leq N$ for some positive integer N. Similarly, let ψ have the standard representation

$$\psi(x) = \sum_{k=1}^{M} b_k I_{F_k}(x)$$

where the numbers b_k are distinct, the sets F_k are mutually disjoint, cover X, and are measurable for $1 \leq k \leq M$ for some positive integer M. Now to the proofs of the assertions:

(i):
First, if $c = 0$, $c\phi = 0$ and $\int 0 d\mu = 0 \cdot \int \phi d\mu$. Next, if $c > 0$, then it is easy to see $c\phi$ is a simple

function with representation

$$c\phi(x) \;=\; \sum_{j=1}^{N} ca_j I_{E_j}(x)$$

and hence, by the definition of the integral of a simple function

$$\int c\phi \, d\mu \;=\; \sum_{j=1}^{N} ca_j \, \mu(E_j)$$

$$=\; c\left(\sum_{j=1}^{N} a_j \, \mu(E_j)\right)$$

$$=\; c \int \phi \, d\mu$$

(ii):
This one is more interesting to prove. First, to prove $\phi + \psi$ is a simple function, all we have to do is find its standard representation. From the standard representations of ϕ and ψ, it is clear the sets $F_k \cap E_j$ are mutually disjoint and since $X = \cup E_j = \cup F_k$, we have the identities

$$F_k \;=\; \bigcup_{j=1}^{N} F_k \cap E_j, \;\; and \;\; E_j \;=\; \bigcup_{k=1}^{M} F_k \cap E_j$$

Now define $h : X \to \Re$ by

$$h(x) \;=\; \sum_{j=1}^{N} \sum_{k=1}^{M} (a_j + b_k) \, I_{F_k \cap E_j}(x)$$

Next, since $X = \cup_j \cup_k F_k \cap E_j$, given $x \in X$, there are indices k_0 and j_0 so that $x \in F_{k_0} \cap E_{j_0}$. Thus,

$$\phi(x) + \psi(x) \;=\; a_{j_0} I_{E_{j_0}} + b_{k_0} I_{F_{k_0}} = a_{j_0} + b_{k_0} = h(x)$$

From the above argument, we see $h(x) = \phi(x) + \psi(x)$ for all x in X. It follows that the range of h is finite and hence it is a measurable simple function, but we still do not know its standard representation.

To find the standard representation, let c_i, $1 \le i \le P$ be the set of distinct numbers formed by the collection $\{a_j + b_k \mid 1 \le j \le N, \, 1 \le k \le M\}$. Then let U_i be the set of index pairs (j, k) that satisfy $c_i = a_j + b_k$. Finally, let

$$G_i \;=\; \bigcup_{(j,k) \in U_i} E_j \cap F_k$$

Since the sets $F_k \cap E_j$ are mutually disjoint, we have

$$\mu(G_i) \;=\; \sum_{(j,k) \in U_i} \mu(E_j \cap F_k)$$

It follows that

$$h(x) \;=\; \sum_{i=1}^{P} c_i \, I_{G_i}$$

is the standard representation of $h = \phi + \psi$. Thus

$$\int h \, d\mu \;=\; \int (\phi + \psi) \, d\mu \;=\; \sum_{i=1}^{P} c_i \, \mu(G_i)$$

$$=\; \sum_{i=1}^{P} c_i \left(\sum_{(j,k)\in U_i} \mu(E_j \cap F_k) \right)$$

$$=\; \sum_{i=1}^{P} \sum_{(j,k)\in U_i} c_i \, \mu(E_j \cap F_k)$$

But we know that

$$\sum_{j=1}^{N} \sum_{k=1}^{M} \;=\; \sum_{i=1}^{P} \sum_{(j,k)\in U_i}$$

Hence, we can write

$$\int (\phi + \psi) \, d\mu \;=\; \sum_{j=1}^{N} \sum_{k=1}^{M} (a_j + b_k) \, \mu(E_j \cap F_k)$$

$$=\; \sum_{j=1}^{N} \sum_{k=1}^{M} a_j \, \mu(E_j \cap F_k) \;+\; \sum_{j=1}^{N} \sum_{k=1}^{M} b_k \, \mu(E_j \cap F_k)$$

This can be reorganized as

$$\sum_{j=1}^{N} a_j \sum_{k=1}^{M} \mu(E_j \cap F_k) \;+\; \sum_{k=1}^{M} b_k \sum_{j=1}^{N} \mu(E_j \cap F_k)$$

$$=\; \sum_{j=1}^{N} a_j \, \mu\!\left(\bigcup_{k=1}^{M} E_j \cap F_k \right) \;+\; \sum_{k=1}^{M} b_k \, \mu\!\left(\bigcup_{j=1}^{N} E_j \cap F_k \right)$$

$$=\; \sum_{j=1}^{N} a_j \, \mu(E_j) \;+\; \sum_{k=1}^{M} b_k \, \mu(F_k)$$

$$=\; \int \phi \, d\mu \;+\; \int \psi \, d\mu$$

(iii):
Given

$$\phi(x) \;=\; \sum_{j=1}^{N} a_j I_{E_j}(x)$$

it is easy to see that

$$\phi \, I_E(x) \;=\; \sum_{j=1}^{N} a_j I_{E \cap E_j}(x)$$

Further, it is straightforward to show that the mappings $\mu_j : (S) \to \overline{\Re}$ defined by $\mu_j(A) = \mu(A \cap E_j)$ for all A in S are measures on the sigma-algebras $S \cap E_j$ for each $1 \le j \le N$. It is also easy to see that the finite linear combination of these measures given by $\xi = \sum_{j=1}^{N} a_j \, \mu_j$ is a measure on S itself. Thus, applying part (ii) of this lemma, we see

$$\begin{aligned}
\lambda(E) \;&=\; \int \phi I_E \, d\mu \;=\; \int \phi I_{\cup_{j=1}^{N} E \cap E_j} \, d\mu \\[2mm]
&=\; \int \left(\sum_{j=1}^{N} \phi I_{E \cap E_j} \right) d\mu \;=\; \sum_{j=1}^{N} \int a_j I_{E \cap E_j} \, d\mu \\[2mm]
&=\; \sum_{j=1}^{N} a_j \mu(E \cap E_j) \;=\; \sum_{j=1}^{N} a_j \, \mu_j(E) \;=\; \xi(E)
\end{aligned}$$

We conclude $\lambda = \xi$ and λ is a measure on S. ■

Lemma 9.5.2 Monotonicity of the Abstract Integral for Non-Negative Functions

Let (X, S, μ) be a measure space and let f and g be in $M^+(X, S)$ with $f \le g$. Then, $\int f d\mu \le \int g d\mu$. Further, if $E \subseteq F$ with E and F measurable sets, then $\int_E f d\mu \le \int_F f d\mu$.

Proof 9.5.2

Let ϕ be a positive simple function which is dominated by f; i.e., $\phi \le f$. Then ϕ is also dominated by g and so by the definition of the integral of f, we have

$$\begin{aligned}
\int f d\mu \;&=\; \sup \left\{ \int \phi d\mu \,\Big|\, 0 \le \phi \le f \right\} \\[2mm]
&\le\; \sup \left\{ \int \psi d\mu \,\Big|\, 0 \le \psi \le g \right\} \\[2mm]
&=\; \int g d\mu
\end{aligned}$$

Next, if $E \subseteq F$ with E and F measurable sets, then $f I_E \le f I_F$ and from the first result, we have

$$\int f I_E d\mu \;\le\; \int f I_F d\mu$$

which implies the result we seek. ■

Homework

Exercise 9.5.1 *Let (X, S, μ) be a measure space and let $\phi, \psi \in M^+(X, S)$ be simple functions where $\phi(x) = 2 I_{E_1} + 3 I_{E_2}$ and $\psi(x) = 4 I_{F_1} + 7 I_{F_2}$. Find the standard representation of $\phi + \psi$ in detail. This means you flesh out the arguments in the proof carefully.*

190 BASIC ANALYSIS IV: MEASURE THEORY AND INTEGRATION

Exercise 9.5.2 *Let $(X, \mathcal{S}, \mu)$ be a measure space and let $\phi, \psi \in M^+(X, \mathcal{S})$ be simple functions where $\phi(x) = 2I_{E_1} + 3I_{E_2} + 8I_{E_3}$ and $\psi(x) = 4I_{F_1} + 7I_{F_2} + 3I_{F_3}$. Find the standard representation of $\phi + \psi$ in detail. This again means you flesh out the arguments in the proof carefully. This will be a bit more intense than the last exercise.*

Exercise 9.5.3 *Let $(X, \mathcal{S}, \mu)$ be a measure space and let $\phi \in M^+(X, \mathcal{S})$ be a simple function. Let E be a measurable set. Prove $\phi\, I_E(x) = \sum_{j=1}^{N} a_j I_{E \cap E_j}(x)$.*

9.6 Equality a.e. Problems

We know if a sequence of extended real-valued measurable functions (f_n) converges pointwise to a function f, then the limit function is also measurable. But what if the convergence was pointwise a.e? Is it still true that the limit function is also measurable? In general, the answer is no. We have to add an additional property to the measure. We will motivate this with an example that we are not really fully prepared for, but it should make sense anyway. The argument below will also be placed in Chapter 13 for completeness of our exposition.

Let $\boldsymbol{B} \cap [0, 1]$ denote the Borel sigma-algebra of subsets of $[0, 1]$. We will show in later chapters that there is a measure called Lebesgue measure, μ_L, defined on a sigma-algebra of subsets $\boldsymbol{L}$, the Lebesgue sigma-algebra, which extends the usual meaning of length in the following sense. If $[a, b]$ is a finite interval then the length of $[a, b]$ is the finite number $b - a$. Denote this length by $\ell([a, b])$. Then we can show that

$$\mu_L([a, b]) = \ell([a, b]) = b - a$$

We can show also that every subset in $\boldsymbol{B}$ is also in $\boldsymbol{L}$. The restriction of μ_L to $\boldsymbol{B}$ is called Borel measure and we will denote it by μ_B.

We can argue that the Borel sigma-algebra is strictly contained in the Lebesgue sigma-algebra by using the special functions we construct in Chapter 13. In that Chapter, we show if $\boldsymbol{C}$ is a Cantor set constructed from the generating sequence $(a_n) = ((1/3)^n)$ which satisfies $\lim_n 2^n a_n = 0$, the content of $\boldsymbol{C}$ is 0. Then if we let Ψ be the mapping discussed above for this $\boldsymbol{C}$ in Section 13.3, we define the mapping $g : [0, 1] \rightarrow [0, 1]$ by $g(x) = (\Psi(x) + x)/2$. The mapping g is quite nice: it is $1 - 1$, onto, strictly increasing and continuous. We also showed in the exercises in Section 13.3 that $g(\boldsymbol{C})$ is another Cantor set with $\lim 2^n a_n' = 1/2$, where (a_n') is the generating sequence for $g(\boldsymbol{C})$.

Now it turns out that the notion of content and Lebesgue measure coincide. Thus, we can say since $\boldsymbol{C}$ is a Borel set,

$$\mu_B(\boldsymbol{C}) = \mu_L(\boldsymbol{C}) = 0$$

Also, we can show that since $\lim 2^n a_n' = 1/2$,

$$\mu_B(g(\boldsymbol{C})) = \mu_L(g(\boldsymbol{C})) = 1/2$$

A nonconstructive argument we will present later using the Axiom of Choice allows us to show that any Lebesgue measurable set with positive Lebesgue measure must contain a subset which is not in the Lebesgue sigma-algebra. So since $\mu_L(g(\boldsymbol{C})) = 1/2$, there is a set $F \subseteq g(\boldsymbol{C})$ which is not in $\boldsymbol{L}$. Thus, $g^{-1}(F) \subseteq \boldsymbol{C}$ which has Lebesgue measure 0. Lebesgue measure is a measure which has the property that every subset of a set of measure 0 must be in the Lebesgue sigma-algebra. Then, using the monotonicity of μ_L, we have $\mu_L(g^{-1}(F))$ is also 0. From the above remarks, we can infer something remarkable.

Let the mapping h be defined to be g^{-1}. Then h is also continuous and hence it is measurable with respect to the Borel sigma-algebra. Note since $\boldsymbol{B} \subseteq \boldsymbol{L}$, this tells us immediately that h is also measurable with respect to the Lebesgue sigma-algebra. Thus, $h^{-1}(U)$ is in the Borel sigma-algebra for all Borel sets U. But we know $h^{-1} = g$, so this tells us $g(U)$ is in the Borel sigma -algebra if U is a Borel set. Hence, if we chose $U = g^{-1}(F)$, then $g(U) = F$ would have to be a Borel set if U is a Borel set. However, we know that F is not in $\boldsymbol{L}$ and so it is also not a Borel set. We can only conclude that $g^{-1}(F)$ cannot be a Borel set. However, $g^{-1}(F)$ is in the Lebesgue sigma-algebra. Thus, there are Lebesgue measurable sets which are not Borel! Thus, the Borel sigma-algebra is strictly contained in the Lebesgue sigma-algebra!

We can use this example to construct another remarkable thing.

Comment 9.6.1 *Using all the notations from above, note the indicator function of $\boldsymbol{C}^C$, the complement of $\boldsymbol{C}$, is defined by*

$$I_{\boldsymbol{C}^C}(x) = \begin{cases} 1 & x \in \boldsymbol{C}^C \\ 0 & x \in \boldsymbol{C}. \end{cases}$$

We see $f = I_{\boldsymbol{C}^C}$ is Borel measurable. Next, define a new mapping like this:

$$\phi(x) = \begin{cases} 1 & x \in \boldsymbol{C}^C \\ 2 & x \in \boldsymbol{C} \setminus g^{-1}(F) \\ 3 & x \in g^{-1}(F). \end{cases}$$

*Note that $\phi = f$ a.e. with respect to Borel measure. However, ϕ is **not** Borel measurable because $\phi^{-1}(3)$ is the set $g^{-1}(F)$ which is not a Borel set.*

We conclude that in this case, even though the two functions were equal a.e. with respect to Borel measure, only one was measurable! The reason this happens is that even though $\boldsymbol{C}$ has Borel measure 0, there are subsets of $\boldsymbol{C}$ which are not Borel sets!

9.7 Complete Measures

Hence, in some situations, we will have to stipulate that the measure we are working with has the property that every subset of a set of measure zero is measurable. We make this formal with a definition.

Definition 9.7.1 Complete Measure

Let X be a nonempty set and $(X, \mathcal{S}, \mu)$ be a measure space. If $E \in \mathcal{S}$ with $\mu(E) = 0$ and $F \subseteq E$ implies $F \in \mathcal{S}$, we say μ is a complete measure. Further, it follows immediately that since $\mu(F) \leq \mu(E) = 0$, that $\mu(F) = 0$ also.

Comment 9.7.1 *This example above can be used in another way. Consider the composition of the measurable function I_C and the function g defined above. For convenience, let $W = g^{-1}(F)$ which is Lebesgue measurable. Then I_W is a measurable function. Consider*

$$\left(I_W \circ g^{-1}\right)(x) = \begin{cases} 1 & g^{-1}(x) \in W \\ 0 & g^{-1}(x) \in W^C \end{cases} = \begin{cases} 1 & x \in g(W) \\ 0 & x \in g(W^C) \end{cases} = \begin{cases} 1 & x \in F \\ 0 & x \in F^C \end{cases} = I_F$$

But I_F is not a measurable function as F is not a measurable set! Hence, the composition of the measurable function I_W and the continuous function g^{-1} is not measurable. This is why we can only prove measurability with the order of the composition reversed as we did in Lemma 8.7.1.

BASIC ANALYSIS IV: MEASURE THEORY AND INTEGRATION

Theorem 9.7.1 Equality a.e. Implies Measurability if the Measure is Complete

Let X be a nonempty set and $(X, \mathcal{S}, \mu)$ be a complete measure space. Let f and g both be extended real-valued functions on X with $f = g$ a.e. Then, if f is measurable, so is g.

Proof 9.7.1

We leave it as an exercise for you to show that an equivalent condition for measurability is to prove $\{f(x) \in G\}$ is measurable for all open sets G. Then to prove this result, let G be open in $\overline{\Re}$ and let $E = (f(x) \neq g(x))$. Then, by assumption, E is measurable and $\mu(E) = 0$. Then, we claim

$$g^{-1}(G) = \left(g^{-1}(G) \cap E \right) \cup \left(f^{-1}(G) \setminus E \right)$$

If x is in $g^{-1}(G)$, then $g(x)$ is in $G \cap E$ or it is in $G \cap E^C$. Now if $g(x) \in E$, $g(x) \neq f(x)$, but if $g(x)$ is in the complement of E, g and f must match. Thus, we see x is in the right-hand side. Conversely, if x is in $g^{-1}(G) \cap E$, x is clearly in $g^{-1}(G)$. Finally, if x is in $f^{-1}(G) \setminus E$, then since x is not in E, $f(x) = g(x)$. Thus, $x \in g^{-1}(G)$ also. We conclude $x \in g^{-1}(G)$. This shows the right-hand side is contained in the left-hand side. Combining these arguments, we conclude the two sets must be equal.

Since $g^{-1}(G) \cap E$ is a subset of E, the completeness of μ implies that $g^{-1}(G) \cap E$ is measurable and has measure 0. The measurability of f tells us that $f^{-1}(G) \setminus E$ is also measurable. Hence, $g^{-1}(G)$ is measurable implying g is measurable. ∎

If the measure μ is not complete, we can still prove the following.

Theorem 9.7.2 Equality a.e. Can Imply Measurability Even if the Measure is Not Complete

Let X be a nonempty set and $(X, \mathcal{S}, \mu)$ be a measure space. Let f and g both be extended real-valued functions on X with $f = g$ on the measurable set E^C with $\mu(E) = 0$. Then, if f is measurable and g is constant on E, g is measurable.

Proof 9.7.2

We will repeat the notation of the previous theorem's proof. As before, if G is open, we can write

$$g^{-1}(G) = \left(g^{-1}(G) \cap E \right) \cup \left(f^{-1}(G) \setminus E \right)$$

Then, since g is constant on E with value say c, we have

$$g^{-1}(G) = \left(\left\{ \begin{array}{ll} E & c \in G \\ \emptyset & c \notin G \end{array} \right\} \right) \bigcup \left(f^{-1}(G) \setminus E \right) = \left\{ \begin{array}{ll} E \cup \left(f^{-1}(G) \setminus E \right) & c \in G \\ \left(f^{-1}(G) \setminus E \right) & c \notin G. \end{array} \right.$$

In both cases, the resulting set is measurable. Hence, we conclude g is measurable. ∎

Comment 9.7.2 *In Comment 9.6.1, we set*

$$\phi(x) = \left\{ \begin{array}{ll} 1, & x \in \boldsymbol{C}^C \\ 2, & x \in \boldsymbol{C} \setminus g^{-1}(F) \\ 3, & x \in g^{-1}(F). \end{array} \right.$$

ABSTRACT INTEGRATION

and since ϕ was not constant on $E = \boldsymbol{C}$, ϕ was not measurable. However, if we had defined

$$\phi(x) \;=\; \left\{ \begin{array}{ll} 1 & x \in \boldsymbol{C}^C \\ c & x \in \boldsymbol{C}, \end{array} \right.$$

then ϕ would have been measurable!

Homework

Exercise 9.7.1 *Using all the notations from above, note the indicator function of $\boldsymbol{C}^C$, the complement of $\boldsymbol{C}$, is defined by*

$$I_{\boldsymbol{C}^c}(x) \;=\; \left\{ \begin{array}{ll} 1 & x \in \boldsymbol{C}^C \\ 0 & x \in \boldsymbol{C}. \end{array} \right.$$

We see $f = I_{C^c}$ is Borel measurable. Next, define a new mapping like this:

$$\phi(x) \;=\; \left\{ \begin{array}{ll} 1 & x \in \boldsymbol{C}^C \\ 4 & x \in \boldsymbol{C} \setminus g^{-1}(F) \\ 5 & x \in g^{-1}(F). \end{array} \right.$$

- *Prove $\phi = f$ a.e. with respect to Borel measure.*

- *Prove $\phi^{-1}(5)$ is the set $g^{-1}(F)$ which is not a Borel set.*

- *Prove ϕ is **not** Borel measurable.*

- *Prove even though ϕ and f are equal a.e. with respect to Borel measure, only one was measurable.*

Exercise 9.7.2 *Using all the notations from above, note the indicator function of $\boldsymbol{C}^C$, the complement of $\boldsymbol{C}$, is defined by*

$$I_{\boldsymbol{C}^c}(x) \;=\; \left\{ \begin{array}{ll} 1 & x \in \boldsymbol{C}^C \\ 0 & x \in \boldsymbol{C}. \end{array} \right.$$

We see $f = I_{C^c}$ is Borel measurable. Next, define a new mapping like this:

$$\phi(x) \;=\; \left\{ \begin{array}{ll} 1 & x \in \boldsymbol{C}^C \\ 8 & x \in \boldsymbol{C} \setminus g^{-1}(F) \\ 6 & x \in g^{-1}(F). \end{array} \right.$$

- *Prove $\phi = f$ a.e. with respect to Borel measure.*

- *Prove $\phi^{-1}(6)$ is the set $g^{-1}(F)$ which is not a Borel set.*

- *Prove ϕ is **not** Borel measurable.*

- *Prove even though ϕ and f are equal a.e. with respect to Borel measure, only one was measurable.*

Exercise 9.7.3 *Define ϕ by*

$$\phi(x) \;=\; \left\{ \begin{array}{ll} 1, & x \in \boldsymbol{C}^C \\ \alpha, & x \in \boldsymbol{C} \setminus g^{-1}(F) \\ \beta, & x \in g^{-1}(F). \end{array} \right.$$

Prove if $\alpha = \beta$, then ϕ is measurable.

9.8 Convergence Theorems

We are now ready to look at various types of interchange theorems for abstract integrals. We will be able to generalize the results of Chapter 5 substantially. There are three basic results: (i) The Monotone Convergence Theorem, (ii) Fatou's Lemma and (iii) The Lebesgue Dominated Convergence Theorem. We will examine each in turn.

9.8.1 Monotone Convergence Theorems

Theorem 9.8.1 The Monotone Convergence Theorem

> *Let $(X, \mathcal{S}, \mu)$ be a measure space and let (f_n) be an increasing sequence of functions in $M^+(X, \mathcal{S})$. Let $f : X \to \overline{\mathbb{R}}$ be an extended real-valued function such that $f_n \to f$ pointwise on X. Then f is also in $M^+(X, \mathcal{S})$ and*
>
> $$\lim_n \int f_n d\mu = \int f d\mu.$$

Proof 9.8.1
Since f_n converges to f pointwise, we know that f is measurable by Theorem 8.6.3. Further, since $f_n \geq 0$ for all n on X, it is clear that $f \geq 0$ also. Thus, $f \in M^+(X, \mathcal{S})$. Since $f_n \leq f_{n+1} \leq f$, the monotonicity of the integral tells us

$$\int f_n d\mu \leq \int f_{n+1} d\mu \leq \int f d\mu$$

Hence, $\int f_n d\mu$ is an increasing sequence of real numbers bounded above by $\int f d\mu$. Of course, this limit could be ∞. Thus, we have the inequality

$$\int f_n d\mu \leq \int f d\mu$$

We now show the reverse inequality, $\int f d\mu \leq \int f_n d\mu$. Let α be in $(0, 1)$ and choose any non-negative simple function ϕ which is dominated by f. Let

$$A_n = \{x \mid f_n(x) \geq \alpha \, \phi(x)\}$$

We claim that $X = \cup_n A_n$. If this was not true, then there would be an x which is not in any A_n. This implies x is in $\cap_n A_n^C$. Thus, using the definition of A_n, $f_n(x) < \alpha \, \phi(x)$ for all n. Since f_n is increasing and converges pointwise to f, this tells us

$$f(x) \leq \alpha \, \phi(x) \leq \alpha \, f(x)$$

We can rewrite this as $(1 - \alpha) f(x) \leq 0$ and since $1 - \alpha$ is positive by assumption, we can conclude $f(x) \leq 0$. But f is non-negative, so combining, we see $f(x) = 0$. Since f dominates ϕ, we must have $\phi(x) = 0$ too. However, if this is true, $f_n(x)$ must be 0 also. Hence, $f_n(x) = 0 \geq \alpha\phi(x) = 0$ for all n. This says $x \in A_n$ for all n. This is a contradiction; thus, $X = \cup_n A_n$.

Next, since f and $\alpha\phi$ are measurable, so is $f - \alpha\phi$. This implies $\{x \mid f(x) - \alpha\phi(x) \geq 0\}$ is a measurable set. Therefore, A_n is measurable for all n. Further, it is easy to $A_n \subseteq A_{n+1}$ for all n; hence, (A_n) is an increasing sequence of measurable sets. Then, we know by the monotonicity of the *integral, that*

$$\int_{A_n} \alpha\phi d\mu \leq \int_{A_n} f_n d\mu \leq \int f d\mu$$

Next, we know that $\lambda(E) = \int_E \phi d\mu$ defines a measure. Thus,

$$\lim_n \lambda(A_n) = \lambda(\cup_n A_n) = \lambda(X)$$

Replacing λ by its meaning in terms of ϕ, we have

$$\int \phi d\mu = \lim_n \int_{A_n} \phi d\mu$$

Multiplying through by the positive number α, we have

$$\alpha \int \phi d\mu = \lim_n \int_{A_n} \alpha \phi d\mu \leq \lim_n \int_{A_n} f_n d\mu$$

Thus, for all $\alpha \in (0,1)$, we have

$$\alpha \int \phi d\mu \leq \lim_n \int_{A_n} f_n d\mu$$

Letting $\alpha \to 1$, we obtain

$$\int \phi d\mu \leq \lim_n \int_{A_n} f_n d\mu$$

Since the above inequality is valid for all non-negative simple functions dominated by f, we have immediately

$$\int f d\mu \leq \lim_n \int_{A_n} f_n d\mu$$

which provides the other inequality we need to prove the result. ∎

This has an immediate extension to series of non-negative functions.

Theorem 9.8.2 The Extended Monotone Convergence Theorem

Let $(X, \mathcal{S}, \mu)$ be a measure space and let (g_n) be a sequence of functions in $M^+(X, \mathcal{S}, \mu)$. Then, the sequence of partial sums,

$$S_n = \sum_{k=1}^{n} g_n$$

converges pointwise on X to the extended real-valued non-negative valued function $S = \sum_{k=1}^{\infty} g_n$. Further, S is also in $M^+(X, \mathcal{S})$ and

$$\lim_n \int S_n d\mu = \int S d\mu$$

Comment 9.8.1 *Once we establish Theorem 9.8.3 (below), we can rewrite this in series notation as*

$$\sum_{k=1}^{\infty} \int g_k d\mu = \int \sum_{k=1}^{\infty} g_k d\mu$$

196 BASIC ANALYSIS IV: MEASURE THEORY AND INTEGRATION

Proof 9.8.2

To prove this result, just apply the Monotone Convergence Theorem to the sequence of partial sums (S_n). ∎

Homework

Exercise 9.8.1 *Let f be the pointwise limit of the sequence of functions (f_n) for $n \geq 2$ defined by*

$$f_n(x) = \begin{cases} 1, & x = 1 \\ 1/(k+1), & 1/(k+1) \leq x < 1/k, \ 1 \leq k < n \\ 0, & 0 \leq x < 1/n \end{cases}$$

- *Prove (f_n) is an increasing sequence in $M^+([0,1], \mathcal{B} \cap [0,1])$ where we let λ denote the Borel measure.*

- *Prove the pointwise limit f is in $M^+([0,1], \mathcal{B} \cap [0,1])$.*

- *Compute $\int f \, d\lambda = \lim_n \int f_n \, d\lambda$.*

Exercise 9.8.2 *Let f be a non-negative bounded function on $[a, b]$. Let P_n be a sequence of partitions with $\|P_n\| \to 0$. Let (E_{ni}) be the corresponding sequence of disjoint sets which allows us to write the lower sums as simple functions.*

- *Prove $\lim_n L(f, P_n) \to L(f)$.*

- *Prove $\phi_n(x) = \sum_{i \in P_n} \inf_{x \in E_{ni}} f(x) \, I_{E_{ni}}(x)$ is increasing.*

- *If we fix x, prove $\phi_n(x)$ converges to a number a_x as it is an increasing sequence bounded above by $\|f\|_\infty$.*

- *If we define $f(x) = a_x$, then f is the pointwise limit of this sequence.*

- *Prove $\int f \, d\lambda = \lim_n \int \phi_n \, d\lambda = L(f)$.*

Exercise 9.8.3 *Let (f_n) for $n \geq 2$ be defined by $f_0(x) = I_{\{1\}}(x)$ and for $n \geq 1$*

$$f_n(x) = \frac{1}{n+1} I_{[\frac{1}{n+1}, \frac{1}{n}]}(x)$$

- *Prove $S_n = \sum_{i=0}^{n}(f_i)$ is an increasing sequence in $M^+([0,1], \mathcal{B} \cap [0,1])$ where we let λ denote the Borel measure.*

- *Prove the pointwise limit S is in $M^+([0,1], \mathcal{B} \cap [0,1])$.*

- *Compute*

$$\int S \, d\lambda = \lim_n \int S_n \, d\lambda = \lim_n \int \sum_{i=0}^{n}(f_i) \, d\lambda$$

The Monotone Convergence Theorem allows us to prove that this notion of integration is additive and linear for positive constants.

Theorem 9.8.3 Abstract Integration is Additive

Let $(X, \mathcal{S}, \mu)$ be a measure space and let f and g be functions in $M^+(X, \mathcal{S})$. Further, let α be a non-negative real number. Then

(i): αf is in $M^+(X, \mathcal{S})$ and

$$\int \alpha f d\mu \; = \; \alpha \int f d\mu$$

(ii): Also, $f + g$ is in $M^+(X, \mathcal{S})$ and

$$\int (f + g) d\mu \; = \; \int f d\mu + \int g d\mu$$

Proof 9.8.3

(i): *The case $\alpha = 0$ is clear, so we may assume without loss of generality that $\alpha > 0$. We know from Theorem 8.7.2 that there is a sequence of non-negative simple functions (ϕ_n) which are increasing and converge up to f on X. Hence, since $\alpha > 0$, we also know that $\alpha \phi_n \uparrow \alpha f$. Thus, by the Monotone Convergence Theorem, αf is in $M^+(X, \mathcal{S})$ and*

$$\lim_n \int \alpha \phi_n d\mu \; = \; \int \alpha f d\mu$$

From Lemma 9.5.1, we know that $\int \alpha \phi_n d\mu = \alpha \int \phi_n d\mu$. Thus,

$$\int \alpha f d\mu \; = \; \alpha \lim_n \int \phi_n d\mu \; = \; \alpha \int f d\mu$$

(ii): *If we apply Theorem 8.7.2 to f and g, we find two sequences of increasing simple functions (ϕ_n) and (ψ_n) so that $\phi_n \uparrow f$ and $\psi_n \uparrow g$. Thus, $(\phi_n + \psi_n) \uparrow (f + g)$. Hence, by the Monotone Convergence Theorem, $f + g$ is in $M^+(X, \mathcal{S})$ and*

$$
\begin{aligned}
\int (f + g) d\mu \; &= \; \lim_n \int (\phi_n + \psi_n) d\mu = \lim_n \int \phi_n d\mu + \lim_n \int \psi_n d\mu \\
&= \; \int f d\mu + \int g d\mu
\end{aligned}
$$

Comment 9.8.2 *The conclusion of Theorem 9.8.2 can now be restated. By Theorem 9.8.3, each S_n is measurable and*

$$\int S_n d\mu \; = \; \sum_{k=1}^{n} \int g_k \, d\mu$$

Thus, the left-hand side can be written as

$$lim_n \int S_n d\mu \; = \; \lim_n \sum_{k=1}^{n} \int g_k \, d\mu = \sum_{k=1}^{\infty} \int f_k \, d\mu$$

198   BASIC ANALYSIS IV: MEASURE THEORY AND INTEGRATION

The limit function S can clearly be written as an infinite series giving

$$\int S \, d\mu \;=\; \int \sum_{k=1}^{\infty} f_k \, d\mu$$

Combining these statements, we get the result.

Homework

Exercise 9.8.4 *Let $(X, \mathcal{S}, \mu)$ be a measure space and let f and g be functions in $M^+(X, \mathcal{S})$. Go through the details of the above proof to show that $2f + 3g$ is in $M^+(X, \mathcal{S})$ and $\int (2f + 3g) d\mu = 2 \int f d\mu + 3 \int g d\mu$.*

Exercise 9.8.5 *Let $(X, \mathcal{S}, \mu)$ be a measure space and let f and g be functions in $M^+(X, \mathcal{S})$. Go through the details of the above proof to show that $6f + 7g$ is in $M^+(X, \mathcal{S})$ and $\int (6f + 7g) d\mu = 6 \int f d\mu + 7 \int g d\mu$.*

Exercise 9.8.6 *Let $(X, \mathcal{S}, \mu)$ be a measure space and let $f_i, 1 \le i \le N$ be functions in $M^+(X, \mathcal{S})$. Use induction to prove $\sum_{i=1}^{N} f_i$ is in $M^+(X, \mathcal{S})$ and $\int \left(\sum_{i=1}^{N} \right) f_i d\mu = \sum_{i=1}^{N} \int f_i d\mu$.*

9.8.2   Fatou's Lemma

We now extend these convergence results a bit.

Theorem 9.8.4 Fatou's Lemma

Let $(X, \mathcal{S}, \mu)$ be a measure space and let (f_n) be a sequence of functions in $M^+(X, \mathcal{S})$. Then

$$\int \liminf f_n \, d\mu \;\le\; \liminf \int f_n \, d\mu$$

Proof 9.8.4
Recall

$$\liminf f_n(x) \;=\; \sup_m \inf_{n \ge m} f_n(x) = \lim_m \inf_{n \ge m} f_n(x)$$
$$\limsup f_n(x) \;=\; \inf_m \sup_{n \ge m} f_n(x) = \lim_m \sup_{n \ge m} f_n(x)$$

Further, if we define $g_m = \inf_{n \ge m} f_n(x)$, we know

$$g_m \;\uparrow\; \liminf f_n(x)$$

It follows immediately that g_m is measurable for all m and by the monotonicity of the integral

$$\int g_m \, d\mu \;\le\; \int f_n(x) \, d\mu \;\; \forall \, n \ge m$$

This implies that $\int g_m \, d\mu$ is a lower bound for the set of numbers $\{\int f_n(x) \, d\mu\}$ and so by definition of the infimum,

$$\int g_m \, d\mu \;\le\; \inf_{n \ge m} \left(\int f_n(x) \, d\mu \right)$$

ABSTRACT INTEGRATION

Let α_m denote the number $\inf_{n \geq m} \left(\int f_n(x) \, d\mu \right)$. Then, $\alpha_m \uparrow \liminf \int f_n d\mu$. We see

$$\lim_m \int g_m \, d\mu \;\; \leq \;\; \lim_m \inf_{n \geq m} \left(\int f_n(x) \, d\mu \right) = \lim_m \alpha_m \;=\; \liminf \int f_n d\mu$$

But since $g_m \uparrow \liminf f_n(x)$, this implies

$$\int \liminf f_n(x) \, d\mu \;\; \leq \;\; \liminf \int f_n \, d\mu$$

$\blacksquare$

Homework

Exercise 9.8.7 Let $(X, \mathcal{S}, \mu)$ be a measure space and let (f_n) and (g_n) be sequences of functions in $M^+(X, \mathcal{S})$. Then $\int \liminf (f_n + g_n) \, d\mu \leq \liminf \int f_n \, d\mu + \liminf \int f_n \, d\mu$.

Exercise 9.8.8 Let $(X, \mathcal{S}, \mu)$ be a measure space and let (f_n) be a sequence of functions in $M^+(X, \mathcal{S})$ Then if $\alpha > 0$, $\int \liminf (\alpha f_n) \, d\mu \leq \alpha \liminf \int f_n \, d\mu$.

Exercise 9.8.9 Let $(X, \mathcal{S}, \mu)$ be a measure space and let (f_n) and (g_n) be sequences of functions in $M^+(X, \mathcal{S})$. Then $\int \liminf (2f_n + 3g_n) \, d\mu \leq 2 \liminf \int f_n \, d\mu + 3 \liminf \int f_n \, d\mu$.

These results allow us to construct additional measures.

Theorem 9.8.5 Constructing Measures from Non-Negative Measurable Functions

Let $(X, \mathcal{S}, \mu)$ be a measure space and let f be a function in $M^+(X, \mathcal{S})$. Then $\lambda : \mathcal{S} \to \overline{\mathfrak{R}}$ defined by

$$\lambda(E) \;=\; \int_E f \, d\mu, \; E \in \mathcal{S}$$

is a measure.

Proof 9.8.5

It is clear $\lambda(\emptyset)$ is 0 and that $\lambda(E)$ is always non-negative. To show that λ is countably additive, let (E_n) be a sequence of disjoint measurable sets in $\mathcal{S}$ and let $E = \cup_n E_n$, be their union. Then E is measurable. Define

$$f_n \;=\; \sum_{k=1}^{n} f I_{E_k} \;=\; f I_{\cup_{k=1}^{n} E_k}.$$

We note that $f_n \uparrow f I_E$ and so by the Monotone Convergence Theorem,

$$\lambda(E) \;=\; \int_E f \, d\mu \;=\; \int f I_E d\mu \;=\; \lim_n \int f_n \, d\mu$$

But,

$$\int f_n \, d\mu \;=\; \sum_{k=1}^{n} \int f I_{E_k} \, d\mu = \sum_{k=1}^{n} \int_{E_k} f \, d\mu = \sum_{k=1}^{n} \lambda(E_k)$$

200 BASIC ANALYSIS IV: MEASURE THEORY AND INTEGRATION

Combining, we have

$$\lambda(E) \;=\; \lim_{n} \sum_{k=1}^{n} \lambda(E_k) \;=\; \sum_{k=1}^{\infty} \lambda(E_k)$$

which proves that λ is countably additive. ∎

Homework

Exercise 9.8.10 *Let $(X, \mathcal{S}, \mu)$ be a measure space and let f and g be functions in $M^+(X, \mathcal{S})$. Then $\lambda : \mathcal{S} \to \overline{\Re}$ defined by $\lambda(E) = \int_E (2f + 4g)\, d\mu, \ E \in \mathcal{S}$ is a measure, and if μ_f and μ_g are the measures defined by f and g, then $\lambda = 2\mu_f + 3\mu_g$.*

Exercise 9.8.11 *Let $(X, \mathcal{S}, \mu)$ be a measure space and let $f_i, 1 \le i \le N$ be functions in $M^+(X, \mathcal{S})$. Then $\lambda : \mathcal{S} \to \overline{\Re}$ defined by $\lambda(E) = \int_E \left(\sum_{i=1}^{N} f_i \right) d\mu, \ E \in \mathcal{S}$ is a measure, and if μ_{f_i} is the measure defined by f_i, then $\lambda = \sum_{i=1}^{N} \mu_{f_i}$.*

9.9 The Absolute Continuity of a Measure

Once we can construct another measure λ from a given measure μ, it is useful to think about their relationship. One useful relationship is that of *absolute continuity*.

Definition 9.9.1 The Absolute Continuity of a Measure

> *Let $(X, \mathcal{S}, \mu)$ be a measure space and let λ be another measure defined on $\mathcal{S}$. We say λ is absolutely continuous with respect to the measure μ if given E in $\mathcal{S}$ with $\mu(E) = 0$, then $\lambda(E) = 0$ also. This is written as $\lambda \ll \mu$.*

We can also now prove an important result set within the framework of functions which are equal a.e.

Lemma 9.9.1 Function f Zero a.e. if and Only if Its Integral is Zero

> *Let $(X, \mathcal{S}, \mu)$ be a measure space and let f be a function in $M^+(X, \mathcal{S})$. Then $f = 0$ a.e. if and only if $\int f d\mu = 0$.*

Proof 9.9.1

($\Leftarrow$): If $\int f d\mu = 0$, then let $E_n = (f(x) > 1/n)$. Note $E_n \subseteq E_{n+1}$ so that (E_n) is an increasing sequence. Since (E_n) is an increasing sequence, we also know $\lim_n \mu(E_n) = \mu(\cup_n E_n)$. Further,

$$\cup_n E_n \;=\; \{x \mid f(x) > 0\}$$

From the definition of E_n, we have

$$f(x) \;\ge\; \frac{1}{n} I_{E_n}$$

which implies

$$0 = \int f d\mu \;\ge\; \int \frac{1}{n} I_{E_n} \;=\; \frac{1}{n}\, \mu(E_n)$$

ABSTRACT INTEGRATION

201

We see $\mu(E_n) = 0$ for all n which implies

$$\mu(f(x) > 0) \;=\; \lim_n \mu(E_n) \;=\; 0$$

Hence, f is zero a.e.
($\Rightarrow$): If f is zero a.e., let E be the set where $f(x) > 0$. Let $f_n = nI_E$. Note that

$$\liminf f_n(x) \;=\; \sup_m \inf_{n \geq m} \left\{ \begin{array}{ll} n & f(x) > 0 \\ 0 & f(x) = 0 \end{array} \right. \;=\; \sup_m \left\{ \begin{array}{ll} m & f(x) > 0 \\ 0 & f(x) = 0 \end{array} \right. \;=\; \left\{ \begin{array}{ll} \infty & f(x) > 0 \\ 0 & f(x) = 0. \end{array} \right.$$

Clearly, $f(x) \leq \liminf f_n(x)$ which implies $\int f \, d\mu \leq \int \liminf f_n \, d\mu$. Finally, by Fatou's Lemma, we find

$$\int f \, d\mu \;\leq\; \int \liminf f_n \, d\mu \;\leq\; \liminf \int f_n \, d\mu \;=\; \liminf n\mu(E) \;=\; 0.$$

We conclude $\int f \, d\mu = 0$. ∎

Comment 9.9.1 *Given f in $M^+(X, \mathcal{S})$, Theorem 9.8.5 allows us to construct the new measure λ by $\lambda(E) = \int_E f d\mu$. If E has μ measure 0, then any simple function ϕ dominated by fI_E has integral 0 and so $\int_E f d\mu = 0$ also. Hence, $\lambda(E) = 0$ and a measure constructed in this way is absolutely continuous with respect to μ. Another way to look at it, is that the previous theorem says since $fI_E = 0\mu a.e.$, then $\int_E f d\mu = \lambda(E) = 0$ too.*

Comment 9.9.2 *Note we don't know how to handle $\int f d\lambda$ for the measures λ we construct here when all we know is f is μ measurable. However, it does look like a type of Stieltjes integral doesn't it? More on that to come.*

We can now extend the Monotone Convergence Theorem slightly. It is often difficult to know that we have pointwise convergence up to a limit function on all of X. The next theorem allows us to relax the assumption to almost everywhere convergence as long as the underlying measure is complete.

Theorem 9.9.2 The Extended Monotone Convergence Theorem Two

Let $(X, \mathcal{S}, \mu)$ be a measure space with complete measure μ and let (f_n) be an increasing sequence of functions in $M^+(X, \mathcal{S})$. Let $f : X \to \overline{\mathfrak{R}}$ be an extended real-valued function such that $f_n \to f$ pointwise a.e. on X. Then f is also in $M^+(X, \mathcal{S})$ and

$$\lim_n \int f_n d\mu \;=\; \int f d\mu$$

Proof 9.9.2
Let E be the set of points where f_n does not converge to f. Then by assumption E has measure 0 and $f_n \uparrow f$ on E^C. Thus,

$$f_n I_{E^C} \uparrow f I_{E^C}$$

and applying the Monotone Convergence Theorem, we have

$$\lim_n \int f_n I_{E^C} \;=\; \int f I_{E^C}$$

and we can say $f I_{E^C}$ is in $M^+(X, \mathcal{S})$. Now f is equal to fI_{E^C} a.e. and so although in general, f need not be measurable, since μ is a complete measure, we can invoke Theorem 9.7.1 to conclude

202   *BASIC ANALYSIS IV: MEASURE THEORY AND INTEGRATION*

that f is actually measurable. Hence, fI_E is measurable too. Since $\mu(E) = 0$, we thus know that

$$\int fI_E \, d\mu = \int fI_E \, d\mu = \int f_n I_E \, d\mu = 0$$

Therefore, we have

$$\int f \, d\mu = \int_E f \, d\mu + \int_{E^C} f \, d\mu = \int_{E^C} f \, d\mu$$

$$= \lim_n \int_{E^C} f_n \, d\mu = \lim_n \left(\int_{E^C} f_n \, d\mu + \int_E f_n \, d\mu \right) = \lim_n \int f_n \, d\mu$$

∎

Homework

Exercise 9.9.1 *Let $(X, \mathcal{S}, \mu)$ be a measure space and let f and g be functions in $M^+(X, \mathcal{S})$. Prove $\lambda : \mathcal{S} \to \overline{\mathfrak{R}}$ defined by $\lambda(E) = \int_E (2f + 4g) \, d\mu$, $E \in \mathcal{S}$ is absolutely continuous with respect to μ.*

Exercise 9.9.2 *Let $(X, \mathcal{S}, \mu)$ be a measure space and let $f_i, 1 \leq i \leq N$ be functions in $M^+(X, \mathcal{S})$. Prove $\lambda : \mathcal{S} \to \overline{\mathfrak{R}}$ defined by $\lambda(E) = \int_E \left(\sum_{i=1}^N f_i \right) \, d\mu$, $E \in \mathcal{S}$ is absolutely continuous with respect to μ.*

Exercise 9.9.3 *Let $(X, \mathcal{S}, \mu)$ be a measure space and let f be a function in $M^+(X, \mathcal{S})$ which defines the measure $\lambda(E) = \int_E f \, d\mu$, $E \in \mathcal{S}$. Let $(Y, \mathcal{S}, \lambda)$ be the resulting measure space. If (f_n) is an increasing sequence of functions in $M^+(Y, \mathcal{S})$ and $f : X \to \overline{\mathfrak{R}}$ is an extended real-valued function such that $f_n \to f$ pointwise a.e. on X, then f is also in $M^+(Y, \mathcal{S})$ and*

$$\lim_n \int f_n d\lambda = \int f d\lambda$$

9.10   Summable Functions

Now to develop the *Dominated Convergence Theorem*, we need a few more concepts.

9.11   Extending Integration to Extended Real-Valued Functions

The results of the previous sections can now be used to extend the notion of integration to general extended real-valued functions f in $M(X, \mathcal{S})$.

Definition 9.11.1 Summable Functions

ABSTRACT INTEGRATION

Let $(X, \mathcal{S}, \mu)$ be a measure space and f be in $M(X, \mathcal{S})$. We say f is summable or integrable on X if $\int f^+ d\mu$ and $\int f^- d\mu$ are both finite. In this case, we define the integral of f on X with respect to the measure μ to be

$$\int f \, d\mu = \int f^+ \, d\mu - \int f^- \, d\mu$$

Also, if E is a measurable set, we define

$$\int_E f \, d\mu = \int_E f^+ \, d\mu - \int_E f^- \, d\mu$$

We let $L_1(X, \mathcal{S}, \mu)$ be the collection of summable functions on X with respect to the measure μ.

Comment 9.11.1 *If f can be decomposed into two non-negative measurable functions f_1 and f_2 as $f = f_1 - f_2$ a.e. with $\int f_1 d\mu$ and $\int f_2 d\mu$ both finite, then note since $f = f^+ - f^-$ also, we have*

$$f_1 + f^- = f_2 + f^+$$

Thus, since all functions involved are summable,

$$\int f_1 d\mu + \int f^- d\mu = \int f_2 d\mu + \int f^+ d\mu$$

This implies that

$$\int (f_2 - f_1) d\mu = \int (f^+ - f^-) d\mu = \int f d\mu$$

Hence, the value of the integral of f is independent of the decomposition.

There are a number of results that follow right away from this definition.

Theorem 9.11.1 Summable Implies Finite a.e.

Let $(X, \mathcal{S}, \mu)$ be a measure space and f be in $L_1(X, \mathcal{S})$. Then the set of points where f is not finite has measure 0.

Proof 9.11.1
Let $E_n = (f(x) > n)$. Then it is easy to see that (E_n) is a decreasing sequence of sets and so

$$\mu\left(\bigcap_n E_n\right) = \lim_n \mu(E_n)$$

It is also clear that

$$(f(x) = \infty) = \bigcap_n E_n$$

Next, note

$$\int f^+ \, d\mu = \int_{E_n} f^+ \, d\mu + \int_{E_n^C} f^+ \, d\mu$$

$$\geq \int_{E_n} f^+ \, d\mu > n \, \mu(E_n)$$

204   *BASIC ANALYSIS IV: MEASURE THEORY AND INTEGRATION*

Thus, $\mu(E_n) < (\int f^+ d\mu)/n$. Since the integral is a finite number, this tells us that $\lim_n \mu(E_n) = 0$. This immediately implies that $\mu(E) = 0$.

A similar argument shows that the set $(f(x) = -\infty)$, which is the same as the set $(f^-(x) = \infty)$, has measure 0. ∎

Theorem 9.11.2 Summable Function Equal a.e. to Another Measurable Function Implies the Other Function is Also Summable

> *Let $(X, \mathcal{S}, \mu)$ be a measure space and f be in $L_1(X, \mathcal{S})$. Then if $g \in M(X, \mathcal{S})$ with $f = g$ a.e., g is also summable.*

Proof 9.11.2
Let E be the set of points in X where f and g are not equal. Then E has measure zero. We then have $fI_{E^c} = gI_{E^c}$ and so gI_{E^c} must be summable. Further, $f^+ I_{E^c} = g^+ I_{E^c}$ and $f^- I_{E^c} = g^- I_{E^c}$. We then note that

$$\int g^+ I_{E^c} d\mu = \int g^+ I_{E^c} d\mu + \int g^+ I_E d\mu$$

because $\int g^+ I_E d\mu = 0$ since E has measure zero. But then we see

$$\int g^+ \, d\mu = \int g^+ I_{E^c} d\mu + \int g^+ I_E d\mu = \int f^+ I_{E^c} d\mu + \int f^+ I_E d\mu = \int f^+ \, d\mu$$

Thus, we can see that $\int g^+ \, d\mu$ is finite. A similar argument shows $\int g^+ \, d\mu$ is finite and so g is summable. ∎

Theorem 9.11.3 Summable Function Equal a.e. to Another Function with Measure Complete Implies the Other Function is Also Summable

> *Let $(X, \mathcal{S}, \mu)$ be a measure space with μ complete and f be in $L_1(X, \mathcal{S})$. Then if g is a function equal a.e. to f, g is also summable.*

Proof 9.11.3
First, the completeness of μ implies that g is measurable. The argument to show g is summable is then the same as in the previous theorem's proof. ∎

Homework

Exercise 9.11.1 *Let μ be counting measure on $\mathbb{N}$. Let $f : \mathbb{N} \to \Re$ be a bounded function.*

- *Assume $f \geq 0$. Use the monotone convergence theorem to prove $\int f d\mu = \lim_{n \to \infty} \sum_{i=1}^{n} f(i) = \sum_{i=1}^{\infty} f(i)$.*

- *Assume $f \leq 0$. Use the monotone convergence theorem to prove $\int f d\mu = \lim_{n \to \infty} \sum_{i=1}^{n} f(i) = \sum_{i=1}^{\infty} f(i)$.*

- *Hence, in this theory $\sum_{i=1}^{\infty} f(i)$ converges to a finite value only if the $\sum_{i=1}^{\infty} f^+(i)$ and $\sum_{i=1}^{\infty} f^-(i)$ converge to finite values. Contrast this to our theory of infinite series where we can handle the convergence of $\sum_{i=1}^{\infty} (-1)^i \frac{1}{i}$ even though the positive and negative parts by themselves do not converge.*

Exercise 9.11.2 *Discuss the difference between this idea of summable functions and the classical idea of an improper Riemann integral.*

ABSTRACT INTEGRATION

Exercise 9.11.3 *Let $(X, \mathcal{S}, \mu)$ be a measure space and f be in $L_1(X, \mathcal{S})$. Prove the set of points where $(f(x) = -\infty)$ has measure 0.*

Exercise 9.11.4 *Let $(X, \mathcal{S}, \mu)$ be a measure space and f be in $L_1(X, \mathcal{S})$. Then if $g \in M(X, \mathcal{S})$ with $f = g$ a.e., prove $\int g^+ \, d\mu$ is finite.*

9.12 Levi's Theorem

We can extend the Monotone Convergence a bit more and actually construct a summable limit function in certain instances. This is known as Levi's Theorem.

Theorem 9.12.1 Levi's Theorem

Let $(X, \mathcal{S}, \mu)$ be a measure space and let (f_n) be a sequence of functions in $L_1(X, \mathcal{S}, \mu)$ which satisfy $f_n \leq f_{n+1}$ a.e. Further, assume

$$\lim_n \int f_n \, d\mu < \infty$$

Then, there is a summable function f on X so that $f_n \uparrow f$ a.e. and $\int f_n d\mu \uparrow \int f d\mu$.

Proof 9.12.1
Define the new sequence of functions (g_n) by $g_n = f_n - f_1$. Then, since (f_n) is increasing a.e., (g_n) is increasing and non-negative a.e. By assumption, $\lim_n \int g_n d\mu$ is then finite. Call its value I for convenience of exposition. Now define the function g pointwise on X by

$$g(x) = \lim_n g_n(x)$$

This limit always exists as an extended real-valued number in $[0, \infty]$ and since each g_n is measurable, so is g. Let $E = (g(x) = \infty)$. Note that

$$E = \bigcap_i \left(\bigcup_n \Big(g_n(x) > i \Big) \right)$$

and so we know that E is measurable.

For each non-negative measurable function g_i, there is an increasing sequence of simple functions (ϕ_n^i) such that $\phi_n^i \uparrow g_i$. For each n, define (recall the binary operator $\vee$ means a pointwise maximum)

$$\Psi_n = \phi_n^1 \vee \phi_n^2 \vee \cdots \vee \phi_n^n$$

Then it is clear that Ψ_n is measurable. Given any x in X, we have that

$$\begin{aligned}
\Psi_{n+1}(x) &= \phi_{n+1}^1 \vee \phi_{n+1}^2 \vee \cdots \vee \phi_{n+1}^{n+1} \\
&\geq \phi_{n+1}^1 \vee \phi_{n+1}^2 \vee \cdots \vee \phi_n^{n+1} \\
&\geq \phi_n^1 \vee \phi_n^2 \vee \cdots \vee \phi_n^n \\
&= \Psi_n(x)
\end{aligned}$$

Hence, (Ψ_n) is an increasing sequence. Moreover, it is straightforward to see that

$$\Psi_n(x) \leq g_1(x) \vee g_2(x) \vee \cdots \vee g_n(x) \leq g_n(x) = g(x)$$

206 BASIC ANALYSIS IV: MEASURE THEORY AND INTEGRATION

Hence, we know that $\lim_n \Psi_n(x) \leq g(x)$. If this limit was strictly less than $g(x)$, let r denote half of the gap size; i.e., $r = (1/2)(g(x) - \lim_n \Psi_n(x))$. Then, since $\Psi_n(x) \geq \phi_n^i$ where i is an index between 1 and n, we would have

$$\phi_n^i < g(x) - r, \ 1 \leq i \leq n$$

This implies that $\phi_n^n \leq g(x) - r$ for all n. In particular, fixing the index i, we see that $\phi_n^i \leq g(x) - r$ for all n. But since $\phi_n^i \uparrow g_i$, this says $g_i(x) \leq g(x) - r$. Since, we can do this for all indices i, we have $\lim_i g_i(x) \leq g(x) - r$ or $g(x) \leq g(x) - r$ which is not possible. We conclude $\lim_n \Psi_n = g$ pointwise on X.

Next, we claim $\int \Psi_n d\mu = \lim_n \int g_n d\mu$. To see this, first notice that $\int \Psi_n d\mu \geq \int \phi_n^i d\mu$ for all $1 \leq i \leq n$. In fact, for any index j, there is an index n^* so that $n^* > j$. Hence, $\int \Psi_{n^*} d\mu \geq \int \phi_n^j d\mu$. This still holds for any $n > n^*$ as well. Thus, for any index j, we can say

$$\lim_n \int \Psi_n \, d\mu \geq \lim_n \int \phi_n^j \, d\mu = \int g_j \, d\mu$$

This implies that

$$\lim_n \int \Psi_n \, d\mu \geq \sup_j \int g_j \, d\mu = \lim_j \int g_j \, d\mu = I$$

Also, since $\Psi_n \leq g_n(x)$,

$$\lim_n \int \Psi_n d\mu \leq \lim_n \int g_n d\mu = I$$

This completes the proof that $\int \Psi_n d\mu = \lim_n \int g_n d\mu$.

We now show the measure of E is zero. To do that, we start with the functions $\Psi_n \wedge k I_E$ for any positive integer k, where the wedge operation $\wedge$ is simply taking the minimum. If $g(x)$ is finite, then $I_E(x) = 0$ and since Ψ_n is non-negative, $\Psi_n \wedge k I_E = 0$. On the other hand, if $g(x) = \infty$, then $x \in E$ and so $k I_E(x) = k$. Since $\Psi_n \uparrow g$, eventually, $\Psi_n(x)$ will exceed k and we will have $\Psi_n \wedge k I_E = k$. These two cases allow us to conclude

$$\Psi_n \wedge k I_E \uparrow k I_E$$

for all x. Thus,

$$\int k I_E \, d\mu = \int \Psi_n \wedge k I_E \, d\mu \leq \int \Psi_n \, d\mu \leq \lim_n \int g_n d\mu = I$$

We conclude $k \, \mu(E) \leq I$ for all k which implies that $\mu(E) = 0$.

Finally, to construct the summable function f we need. Define $h = g I_{E^C}$. Clearly, $g_n \uparrow h$ on E^C, that is, a.e. Also, since $\Psi_n \uparrow g$ on E^C, the Monotone Convergence Theorem tells us that

$$\lim_n \int_{E^C} \Psi_n \, d\mu \uparrow \int_{E^C} g \, d\mu$$

But,

$$\int_{E^C} g \, d\mu = \int h \, d\mu$$

Hence, h is summable and so $f_1 + h$ is also summable. Define $f = f_1 + h$ on X and we have f is summable and

$$f_n \quad \uparrow \quad f_1 + h$$

$$\int f_n \, d\mu \;=\; \int f_1 \, d\mu \;+\; \int h \, d\mu$$

$$=\; \int f \, d\mu$$

∎

Homework

Exercise 9.12.1 *In the proof of Levi's Theorem, we let $E = (g(x) = \infty)$. Prove*

$$E = \bigcap_i \left(\bigcup_n \Big(g_n(x) > i \Big) \right)$$

and that E is measurable.

Exercise 9.12.2 *Define f_n for $n \geq 2$ by*

$$f_n(x) \;=\; \begin{cases} 1, & x = 1 \\ 1/(k+1), & 1/(k+1) \leq x < 1/k,\ 1 \leq k < n \\ 0, & 0 \leq x < 1/n \\ \infty, & x < 0 \ or \ x > 1 \end{cases}$$

Assuming you know about Borel measure, λ prove $\lim_n \int f_n d\lambda < \infty$ and prove there is a summable function f so that $f_n \uparrow f$ and $\int f_n d\lambda \uparrow \int f d\lambda$.

Exercise 9.12.3 *Define f_n for $n \geq 1$ by*

$$f_n(x) \;=\; \begin{cases} 0, & 0 \leq x < \frac{1}{2} \\ n(x - \frac{1}{2}), & \frac{1}{2} \leq x < \frac{1}{2} + \frac{1}{n} \\ 1, & \frac{1}{2} + \frac{1}{n} \leq x \leq 1 \\ \infty, & x < 0 \ or \ x > 1 \end{cases}$$

Assuming you know about Borel measure, λ

- *Prove f_n is summable.*

- *Prove $\lim_n \int f_n d\lambda < \infty$.*

- *Prove there is a summable function f so that $f_n \uparrow f$ and $\int f_n d\lambda \uparrow \int f d\lambda$.*

9.13 Constructing Charges

Each summable function can also be used to construct a charge.

Theorem 9.13.1 Integrals of Summable Functions Create Charges

> *Let $(X, \mathcal{S}, \mu)$ be a measure space and let f be a function in $L_1(X, \mathcal{S}, \mu)$. Then the mapping $\lambda : \mathcal{S} \to \Re$ defined by*
>
> $$\lambda(E) \;=\; \int_E f \, d\mu$$
>
> *for all E in $\mathcal{S}$ defines a charge on $\mathcal{S}$. The integral $\int_E f \, d\mu$ is also called the indefinite integral of f with respect to the measure μ.*

BASIC ANALYSIS IV: MEASURE THEORY AND INTEGRATION

Proof 9.13.1

Since f is summable, note that the mappings λ^+ and λ^- defined by

$$\lambda^+(E) = \int_E f^+ \, d\mu, \;\; \lambda^-(E) = \int_E f^- \, d\mu$$

both define measures. It then follows immediately that λ is countably additive and hence is a charge.
∎

Comment 9.13.1 *Since $\int_E f \, d\mu$ defines a charge and is countably additive, we see that if (E_n) is a collection of mutually disjoint measurable subsets, then*

$$\int_{\cup_n E_n} f \, d\mu = \sum_n \int_{E_n} f \, d\mu$$

Homework

Exercise 9.13.1 *Let $(X, \mathcal{S}, \mu)$ be a measure space and let f and g be functions in $L_1(X, \mathcal{S}, \mu)$. Prove the mapping $\lambda : \mathcal{S} \to \Re$ defined by*

$$\lambda(E) = \int_E (2f + 5g) \, d\mu$$

for all E in $\mathcal{S}$ defines a charge on $\mathcal{S}$.

Exercise 9.13.2 *Let $(X, \mathcal{S}, \mu)$ be a measure space and let f and g be functions in $L_1(X, \mathcal{S}, \mu)$. Prove the mapping $\lambda : \mathcal{S} \to \Re$ defined by*

$$\lambda(E) = \int_E (-3f + 6g) \, d\mu$$

for all E in $\mathcal{S}$ defines a charge on $\mathcal{S}$.

Exercise 9.13.3 *Let $(X, \mathcal{S}, \mu)$ be a measure space and let f and g be functions in $L_1(X, \mathcal{S}, \mu)$. Prove the mapping $\lambda : \mathcal{S} \to \Re$ defined by*

$$\lambda(E) = \int_E (8f + 2g) \, d\mu$$

for all E in $\mathcal{S}$ defines a charge on $\mathcal{S}$.

Exercise 9.13.4 *Let $(X, \mathcal{S}, \mu)$ be a measure space and let f and g be functions in $L_1(X, \mathcal{S}, \mu)$. Prove the mapping $\lambda : \mathcal{S} \to \Re$ defined by*

$$\lambda(E) = \int_E (2f - 5g) \, d\mu$$

for all E in $\mathcal{S}$ defines a charge on $\mathcal{S}$.

Exercise 9.13.5 *Let $(X, \mathcal{S}, \mu)$ be a measure space and let f and g be functions in $L_1(X, \mathcal{S}, \mu)$. Prove the mapping $\lambda : \mathcal{S} \to \Re$ defined by*

$$\lambda(E) = \int_E (4f - 3g) \, d\mu$$

for all E in $\mathcal{S}$ defines a charge on $\mathcal{S}$.

ABSTRACT INTEGRATION

Exercise 9.13.6 *Let (X, S, μ) be a measure space and let f, g and h be functions in $L_1(X, S, \mu)$. Prove the mapping $\lambda : S \to \Re$ defined by*

$$\lambda(E) \;=\; \int_E (4f - 3g + 6h)\, d\mu$$

for all E in S defines a charge on S.

Exercise 9.13.7 *Let (X, S, μ) be a measure space and let $f_i\, 1 \le i \le N$ be functions in $L_1(X, S, \mu)$. Prove the mapping $\lambda : S \to \Re$ defined by*

$$\lambda(E) \;=\; \int_E \left(\sum_{i=1}^{N} f_i \right) d\mu$$

for all E in S defines a charge on S.

9.14 Properties of Summable Functions

We need to know if $L_1(X, S, \mu)$ is a linear space under the right interpretation of scalar multiplication and addition. To do this, we need some fundamental inequalities and conditions that force summability.

Theorem 9.14.1 Fundamental Abstract Integration Inequalities

Let (X, S, μ) be a measure space.

(i): $f \in L_1(X, S, \mu)$ if and only if $\mid f \mid \in L_1(X, S, \mu)$.

(ii): $f \in L_1(X, S, \mu)$ implies $\mid \int f\, d\mu \mid \le \int \mid f \mid d\mu$.

(iii): f measurable and $g \in L_1(X, S, \mu)$ with $\mid f \mid \le \mid g \mid$ implies f is also summable and $\int \mid f \mid d\mu \le \int \mid g \mid d\mu$.

Proof 9.14.1

(i): If f is summable, f^+ and f^- are in $M^+(X, S)$ with finite integrals. Since $\mid f \mid = f^+ + f^-$, we see $\mid f \mid^+ = \mid f \mid$ and $\mid f \mid^- = 0$. Thus, $\int \mid f \mid^+ d\mu = \int (f^+ + f^-) d\mu$ which is finite. Also, since $\int \mid f \mid^- d\mu = 0$, we see that $\mid f \mid$ is summable. Conversely, if $\mid f \mid$ is summable, then $\int \mid f \mid^+ d\mu = \int (f^+ + f^-) d\mu$ is finite. This, in turn, tells us each piece is finite and hence f is summable too.

(ii): If f is summable, then

$$\mid \int f\, d\mu \mid \;=\; \mid \int f^+ d\mu - \int f^- d\mu \mid$$

$$\le\; \int f^+ d\mu + \int f^- d\mu$$

$$=\; \int (f^+ + f^-)\, d\mu \;=\; \int \mid f \mid d\mu$$

210   *BASIC ANALYSIS IV: MEASURE THEORY AND INTEGRATION*

(iii): Since g is summable, so is $|\,g\,|$ by (i). Also, because $|\,f\,| \leq |\,g\,|$, each function is in $M^+(X,\mathcal{S})$ and so $\int |\,f\,|^+ d\mu \leq \int |\,g\,|^+ d\mu$ which is finite. Hence, $|\,f\,|$ is summable. Then, also by (i), f is summable.   ∎

We can now tackle the question of the linear structure of $L_1(X,\mathcal{S},\mu)$. We have to define appropriate scalar multiply and addition operations and show these operations are well-defined.

Theorem 9.14.2 The Summable Functions Form a Linear Space

Let $(X,\mathcal{S},\mu)$ be a measure space. We define operations on $L_1(X,\mathcal{S},\mu)$ as follows:

- *scalar multiplication: for all α in $\Re$, αf is the function defined pointwise by $(\alpha f)(x) = \alpha f(x)$.*

- *addition of functions: for any two functions f and g the sum of f and g is the new function defined pointwise on E_{fg}^C by $(f+g)(x) = f(x) + g(x)$, where, recall,*

$$E_{fg} = \left((f=\infty) \cap (g=-\infty) \bigcup (f=-\infty) \cap (g=\infty) \right)$$

This is equivalent to defining $f + g$ to be the function $h = (f+g)I_{E_{fg}^C}$. This is a measurable function as we discussed in the proof of Lemma 8.6.1.

Then, we have

(i): αf is summable for all real α if f is summable and $\int \alpha f d\mu = \alpha \int f d\mu$.

(ii): $f + g$ is summable for all f and g which are summable and $\int (f+g)d\mu = \int f d\mu + \int g d\mu$.

Hence, $L_1(X,\mathcal{S},\mu)$ is a vector space over $\Re$.

Proof 9.14.2

(i): If α is 0, this is easy. Next, assume $\alpha > 0$. Then, $(\alpha f)^+ = \alpha f^+$ and $(\alpha f)^- = \alpha f^-$ and these two functions are clearly summable since f^+ and f^- are. Thus, αf is summable. Then, we have

$$
\begin{aligned}
\int \alpha f \, d\mu &= \int (\alpha f)^+ \, d\mu - \int (\alpha f)^- \, d\mu \\
&= \alpha \left(\int f^+ \, d\mu - \int f^- \, d\mu \right) \\
&= \alpha \int f \, d\mu
\end{aligned}
$$

Finally, if $\alpha < 0$, we have $(\alpha f)^+ = -\alpha f^-$ and $(\alpha f)^- = -\alpha f^+$. Now simply repeat the previous arguments making a few obvious changes.

(ii): Since f and g are summable, we know that $\mu(E_{fg}) = 0$. Further, we know $|\,f\,|$ and $|\,g\,|$ are summable. Since

$$|\,f+g\,|\,I_{E_{fg}^C} \leq \left(|\,f\,| + |\,g\,| \right) I_{E_{fg}^C} \leq |\,f\,| + |\,g\,|$$

ABSTRACT INTEGRATION

we see $\mid f + g \mid I_{E^C_{fg}}$ is summable by Theorem 9.14.1, part (iii). Hence, $(f + g) I_{E^C_{fg}}$ is summable also. Now decompose $f + g$ on E^C_{fg} as

$$f + g = (f^+ + g^+) - (f^- + g^-)$$

Then, note

$$\int_{E^C_{fg}} (f + g)\, d\mu \;=\; \int_{E^C_{fg}} (f^+ + g^+)\, d\mu \;-\; \int (f^- + g^-)\, d\mu$$

$$\;=\; \int_{E^C_{fg}} (f^+ - f^-)\, d\mu \;+\; \int (g^+ - g^-)\, d\mu$$

where we are permitted to manipulate the terms in the integrals above because all are finite in value. However, we can rewrite this as

$$\int_{E^C_{fg}} (f + g)\, d\mu \;=\; \int_{E^C_{fg}} f\, d\mu \;+\; \int g\, d\mu$$

Since we define the sum of f and g to be the function $(f + g) I_{E^C_{fg}}$, we see $f + g$ is in $L_1(X, \mathcal{S}, \mu)$. ∎

Homework

Exercise 9.14.1 Let $(X, \mathcal{S}, \mu)$ be a measure space. Let f and g be functions in $L_1(X, \mathcal{S}, \mu)$. Show all the details to prove $2f + 6g$ is in $L_1(X, \mathcal{S}, \mu)$ also and $\int (2f + 6g) d\mu = 2 \int f d\mu + 6 \int g d\mu$.

Exercise 9.14.2 Let $(X, \mathcal{S}, \mu)$ be a measure space. Let f and g be functions in $L_1(X, \mathcal{S}, \mu)$. Show all the details to prove $2f - 6g$ is in $L_1(X, \mathcal{S}, \mu)$ also and $\int (2f - 6g) d\mu = 2 \int f d\mu - 6 \int g d\mu$.

Exercise 9.14.3 Let $(X, \mathcal{S}, \mu)$ be a measure space. Let $f_i 1 \leq i \leq N$ be functions in $L_1(X, \mathcal{S}, \mu)$. Prove $\sum_{i=1}^{N} f_i$ is in $L_1(X, \mathcal{S}, \mu)$ also and $\int (\sum_{i=1}^{N} f_i) d\mu = \sum_{i=1}^{N} \int f_i d\mu$.

9.15 The Dominated Convergence Theorem

We can now complete this chapter by proving the important limit interchange called the *Lebesgue Dominated Convergence Theorem*.

Theorem 9.15.1 Lebesgue's Dominated Convergence Theorem

> Let $(X, \mathcal{S}, \mu)$ be a measure space, (f_n) be a sequence of functions in $L_1(X, \mathcal{S}, \mu)$ and $f : X \to \overline{\mathfrak{R}}$ so that $f_n \to f$ a.e. Further, assume there is a summable g so that $\mid f_n \mid \leq g$ for all n. Then, suitably defined, f is also measurable and summable with $\lim_n \int f_n d\mu = \int f d\mu$.

Proof 9.15.1
Let E be the set of points in X where the sequence does not converge. Then, by assumption, $\mu(E) = 0$ and

$$f_n I_{E^C} \;\to\; f I_{E^C}, \quad \text{and} \quad \mid f_n I_{E^C} \mid \leq g I_{E^C}$$

Hence, $f I_{E^C}$ is measurable and satisfies $\mid f I_{E^C} \mid \leq g I_{E^C}$. Therefore, since g is summable, we have that $f I_{E^C}$ is summable too.

212   BASIC ANALYSIS IV: MEASURE THEORY AND INTEGRATION

We can write out our fundamental inequality as follows

$$-gI_{E^C} \leq f_n\,I_{E^C} \leq gI_{E^C}. \qquad (\alpha)$$

This implies that $h_n = f_n\,I_{E^C} + gI_{E^C}$ is non-negative and hence, we can apply Fatou's lemma to find

$$\int \liminf h_n\,d\mu \leq \liminf \int h_n\,d\mu$$

However, we know

$$
\begin{aligned}
\liminf h_n &= \liminf\left(f_n\,I_{E^C} + gI_{E^C}\right)\\
&= gI_{E^C} + \liminf f_n\,I_{E^C}\\
&= gI_{E^C} + f\,I_{E^C}
\end{aligned}
$$

because f_n converges pointwise to f on E^C. It then follows that

$$
\begin{aligned}
\int \left(gI_{E^C} + f\,I_{E^C}\right) d\mu &\leq \liminf \int \left(f_n\,I_{E^C} + gI_{E^C}\right) d\mu\\
&= \int gI_{E^C}\,d\mu + \liminf \int f_n\,I_{E^C}\,d\mu
\end{aligned}
$$

Since g is summable, we also know

$$\int (g\,I_{E^C} + f\,I_{E^C})\,d\mu = \int (g\,I_{E^C}\,d\mu + \int f\,I_{E^C})\,d\mu$$

Using this identity, we have

$$\int (g\,I_{E^C}\,d\mu + \int f\,I_{E^C})\,d\mu \leq \int g\,I_{E^C}\,d\mu + \liminf \int f_n\,I_{E^C}\,d\mu$$

The finiteness of the integral of the g term then allows cancellation so that we obtain the inequality

$$\int f\,I_{E^C}\,d\mu \leq \liminf \int f_n\,I_{E^C}\,d\mu$$

Since the integrals of f and f_n are all zero on E, we have shown

$$\int f\,d\mu \leq \liminf \int f_n\,d\mu$$

We now show the reverse inequality holds. Using Equation α, we see $z_n = gI_{E^C} - f_nI_{E^C}$ is also non-negative for all n. Applying Fatou's Lemma, we find

$$\int \liminf z_n\,d\mu \leq \liminf \int z_n\,d\mu$$

Then, we note

$$\liminf z_n = \liminf\left(-f_n\,I_{E^C} + gI_{E^C}\right)$$

$$= gI_{E^C} + \liminf \left(-f_n \, I_{E^C} \right)$$

$$= gI_{E^C} - f \, I_{E^C}$$

because f_n converges pointwise to f on E^C. It then follows that

$$\int \left(gI_{E^C} - f \, I_{E^C} \right) d\mu \;\leq\; \liminf \int \left(-f_n \, I_{E^C} + gI_{E^C} \right) d\mu$$

$$= \int gI_{E^C} \, d\mu + \liminf \int \left(-f_n \, I_{E^C} \right) d\mu$$

Now,

$$\liminf \int \left(-f_n \, I_{E^C} \right) d\mu \;=\; \sup_m \inf_{m \geq n} \int \left(-f_n \, I_{E^C} \right) d\mu$$

$$= \sup_m \left(-\sum_{m \geq n} \int f_n \, I_{E^C} \, d\mu \right)$$

$$= -\inf_m \sup_{m \geq n} \int f_n \, I_{E^C} \, d\mu$$

$$= -\limsup \int f_n \, I_{E^C} \, d\mu$$

Thus, we can conclude

$$\int \left(gI_{E^C} - f \, I_{E^C} \right) d\mu \;\leq\; \int gI_{E^C} \, d\mu - \limsup \int f_n \, I_{E^C} \, d\mu$$

Again, since g is summable, we can write

$$\int gI_{E^C} \, d\mu - \int f \, I_{E^C} \, d\mu \;\leq\; \int gI_{E^C} \, d\mu - \limsup \int f_n \, I_{E^C} \, d\mu$$

After canceling the finite value $\int gI_{E^C} \, d\mu$, we have

$$\int f \, I_{E^C} \, d\mu \;\geq\; \limsup \int f_n \, I_{E^C} \, d\mu$$

This then implies, using arguments similar to the ones used in the first case, that

$$\int f \, d\mu \;\geq\; \limsup \int f_n \, d\mu$$

However, limit inferiors are always less than limit superiors and so we have

$$\limsup \int f_n \, d\mu \;\leq\; \int f \, d\mu \leq \liminf \int f_n \, d\mu \leq \limsup \int f_n \, d\mu$$

It follows immediately that $\lim_n \int f_n d\mu = \int f d\mu$.

Finally, we can now see how to define f in a suitable fashion. The function $f I_{E^C}$ is measurable and is 0 on E. Hence, the limit function f has the form

$$f(x) \;=\; \begin{cases} \lim_n f_n(x) & \textit{when the limit exists, i.e. when } x \in E^C \\ 0 & \textit{when the limit does not exist, i.e. when } x \in E. \end{cases}$$

Homework

Exercise 9.15.1 *Assume $f \in L_1(X, \mathcal{S}, \mu)$ with $f(x) > 0$ on X. Further, assume there is a positive number α so that $\alpha < \mu(X) < \infty$. Prove that*

$$\inf \left\{ \int_E f \, d\mu \mid \mu(E) \geq \alpha \right\} > 0$$

Exercise 9.15.2 *Assume $f \in L_1(X, \mathcal{S}, \mu)$. Let $\alpha > 0$. Prove that*

$$\mu(\{x \mid |f(x)| \geq \alpha\})$$

is finite.

Exercise 9.15.3 *Assume $(f_n) \subseteq L_1(X, \mathcal{S}, \mu)$. Let $f : X \to \overline{\Re}$ be a function. Assume $f_n \to f[ptws\ ae]$. Prove*

$$\int |f_n - f| \, d\mu \to 0 \Rightarrow \int |f_n| \, d\mu \to \int |f| \, d\mu$$

Exercise 9.15.4 *Let $(X, \mathcal{S})$ be a measurable space. Let $\mathcal{C}$ be the collection of all charges on $\mathcal{S}$. Prove that $\mathcal{C}$ is a Banach Space under the operations*

$$\left(c\,\mu \right)(E) \;=\; c\,\mu(E), \ \forall\, c \in \Re, \ \forall\, \mu$$

$$\left(\mu + \nu \right)(E) \;=\; \mu(E) + \nu(E), \ \forall\, \mu, \nu$$

with norm $\|\mu\| = |\mu|(X)$

9.16 Alternative Abstract Integration Schemes

It is possible to define abstract integration in other ways. Let $(X, \mathcal{S}, \mu)$ be a measure space with $\mu(X)$ finite. If $f : X \to \Re$ is bounded and $\mathcal{S}$ measurable, let m denote the lower bound of f on X and M, its upper bound. For any $m_0 \leq m$ and $M_0 > M$ we let π be the set of points $m_0 = y_0 < y_1 < \cdots < y_n = M_0$. In our discussion of Riemann integration and functions of bounded variation, we used the variable x because we often *think* of the symbol x as a *domain* variable; here, we use the variable y because it is often used as a *range* variable in many settings. The important thing to remember is we are *partitioning* the range of f now, rather than the its domain. We define the norm of π as usual following Definition 3.1.4. Further, we define *refinements* and *common refinements* of a partition as in Definition 3.1.1.

Define the following measurable subsets of X:

$$E_j \;=\; \{x : y_{j-1} \leq f(x) < y_j\}, \ 1 \leq j < n, \ E_n = \{x : y_{n-1} \leq f(x) \leq y_n\}$$

It is clear the sets E_j are disjoint and $X = \cup_{j=1}^n E_j$. Thus,

$$\mu(X) \;=\; \sum_{j=1}^n \mu(E_j).$$

Define **Lower** and **Upper** sums as follows: the *Lower* sum is

$$L(f, \boldsymbol{\pi}) = \sum_{\boldsymbol{\pi}} y_{j-1} \, \mu(E_j)$$

and the *upper* sum is

$$U(f, \boldsymbol{\pi}) = \sum_{\boldsymbol{\pi}} y_j \, \mu(E_j)$$

It is clear

$$m_0 \, \mu(X) \le m \, \mu(X) \quad \le \quad L(f, \boldsymbol{\pi}) \; \le \; U(f, \boldsymbol{\pi}) \; \le \; M\mu(X) \le M_0\mu(X) \tag{9.2}$$

Then it follows

$$\sup_{\boldsymbol{\pi}} L(f, \boldsymbol{\pi}) \quad < \quad \infty, \quad \inf_{\boldsymbol{\pi}} U(f, \boldsymbol{\pi}) < \infty$$

We can thus define abstract **Lower** and **Upper** Darboux integrability as usual (see Theorem 4.2.3) for our choice of m_0 and M_0.

Theorem 9.16.1 The Upper and Lower Abstract Darboux Integrals are Finite

Let $(X, \mathcal{S}, \mu)$ be a measure space with $\mu(X)$ finite and $f : X \to \Re$ be bounded and $\mathcal{S}$ measurable. Then the **Lower Integral** $L(f, m_0, M_0)$ and **Upper Integral** $U(f, m_0, M_0)$ defined by

$$L(f, m_0, M_0) \quad = \quad \sup_{\boldsymbol{\pi}} L(f, \boldsymbol{\pi}), \quad U(f, m_0, M_0) = \inf_{\boldsymbol{\pi}} U(f, \boldsymbol{\pi})$$

are both finite.

Proof 9.16.1
This is a consequence of Equation 9.2. ∎

Homework

This way of handling lower and upper sums is a lot different from what we did in the theory of Riemann integration. Here we partition the range of f.

Exercise 9.16.1 *Let $f(x) = x^2$ on $[0, 5]$. The range of f is $[0, 25]$. Let $P = \{0.0, 1.0, 2.0, 5.0, 8.0, 12.0, 15.0, 18.022.025.0\}$ be a partition of the range. Find the associated sets E_j for this partition and compute the lower and upper sums $L(f, P)$ and $U(f, P)$.*

Exercise 9.16.2 *Let $f(x) = x^2 + 2x - 12$ on $[-3, 5]$. Prove the range of f is $[-27, 10]$. Let $P = \{-3.0, -1.0, 0.0, 1.0, 2.0, 5.0\}$ be a partition of the range. Find the associated sets E_j for this partition and compute the lower and upper sums $L(f, P)$ and $U(f, P)$.*

Exercise 9.16.3 *To show you how bad this can get, try to figure out this one. Let $f(x) = x \sin(1/x)$ on $[0, 1]$. The range is $[0, 1]$. Let $P = \{0.0, 0.1, 0.2, 0.4, 0.5, 1.0\}$. Find the associated sets E_j for this partition and compute the lower and upper sums $L(f, P)$ and $U(f, P)$. This one is really messy as it is much harder to find the sets E_j.*

We can then define our *new* abstract integral as follows:

Definition 9.16.1 Abstract Darboux Integrability

BASIC ANALYSIS IV: MEASURE THEORY AND INTEGRATION

Let $(X, \mathcal{S}, \mu)$ be a measure space with $\mu(X)$ finite and $f : X \to \Re$ be bounded and $\mathcal{S}$ measurable. We say f is Abstract Darboux Integrable if $L(f, M_0) = U(f, M_0)$. The common value is then called the Abstract Darboux Integral of f and is denoted by the symbol $DAI(f, M_0)$.

We then prove the following results.

Theorem 9.16.2 π' Refines π Implies $L(f, \pi) \leq L(f, \pi')$ and $U(f, \pi) \geq U(f, \pi')$

Let $(X, \mathcal{S}, \mu)$ be a measure space with $\mu(X)$ finite and $f : X \to \Re$ be bounded and $\mathcal{S}$ measurable. If π' refines π on $[m, M_0]$, then $L(f, \pi) \leq L(f, \pi')$ and $U(f, \pi) \geq U(f, \pi')$.

Proof 9.16.2
Mimic the proof of Theorem 4.2.1, mutatis mutandi. ∎

Theorem 9.16.3 $L(f, \pi_1) \leq U(f, \pi_2)$

Let $(X, \mathcal{S}, \mu)$ be a measure space with $\mu(X)$ finite and $f : X \to \Re$ be bounded and $\mathcal{S}$ measurable. Let π_1 and π_2 be any two partitions of $[m, M_0]$. Then $L(f, \pi_1) \leq U(f, \pi_2)$.

Proof 9.16.3
See the proof of Theorem 4.2.2. ∎

This implies:

Theorem 9.16.4 $L(f, m_0, M_0) \leq U(f, m_0, M_0)$

Let $(X, \mathcal{S}, \mu)$ be a measure space with $\mu(X)$ finite and $f : X \to \Re$ be bounded and $\mathcal{S}$ measurable. Then $L(f, m_0, M_0) \leq U(f, m_0, M_0)$.

Proof 9.16.4
This is an easy argument. ∎

It is easy to see

$$0 \ \leq \ U(f, \pi) \ - \ L(f, \pi) \ \leq \ \|\pi\| \mu(X)$$

Since we can make $\|\pi\|$ as small as we want and $\mu(X)$ is finite,

Theorem 9.16.5 $L(f, m_0, M_0) \geq U(f, m_0, M_0)$

Let $(X, \mathcal{S}, \mu)$ be a measure space with $\mu(X)$ finite and $f : X \to \Re$ be bounded and $\mathcal{S}$ measurable. Then $U(f, m_0, M_0) \leq L(f, m_0, M_0)$.

Proof 9.16.5
Given $\epsilon > 0$, there is a partition π so that $U(f, \pi) - L(f, \pi) < \epsilon$ for any partition π' which refines π. From this it immediately follows that $U(f, m_0, M_0) \leq L(f, m_0, M_0) + \epsilon$ which implies the result.
∎

We conclude that the Abstract Darboux Integral of f exists and $DAI(f, M_0)$ has the common value $L(f, M_0) = U(f, M_0)$.

ABSTRACT INTEGRATION   217

It seems we have finite numbers $DAI(f, m_0, M_0)$ for many choices of m_0 and M_0. Clearly, our development of this abstract integral is without much application if these numbers depend on our choice of m_0 and M_0. This is not the case. We can prove:

Theorem 9.16.6 $DAI(f, m_0, M_0)$ **is Independent of the Choice of** m_0 **and** M_0

> *Let $(X, \mathcal{S}, \mu)$ be a measure space with $\mu(X)$ finite and $f : X \to \Re$ be bounded and $\mathcal{S}$ measurable. Then $DAI(f, m_0, M_0)$ is independent of the choice of m_0 and M_0*

Proof 9.16.6
It is enough to do the argument for the upper bound side, M_0, as the arguments for the m_0 case are similar. Here is a sketch of the argument. Let $M < M_0 < M_0'$ be given. Show that each partition of $[m_0, M_0]$ corresponds to a partition of $[m_0, M_0']$ for which the lower and upper sums for the partition of $[m_0, M_0]$ are the same as the ones for the partition of $[m_0, M_0']$. Then, with that done, show this implies the desired result. ∎

With Theorem 9.16.6 established, we can choose any value of m_0 and M_0 useful in our calculations. This can be of great help in some arguments. Since the value of the abstract Darboux integral is now well-defined, we will begin using the standard notation, $\int_X f \, d\mu$ or simply $\int f \, d\mu$ for this value. Also, sometimes, we will continue to refer to this value as $DAI(f)$ where we no longer need to add the argument m_0 or M_0. Note this is the same symbol we use for the other abstract integral we have discussed. We also define the symbol $\int_E f \, d\mu = DAI(f, E)$ as usual for any measurable set E.

You can then prove the following theorems.

Theorem 9.16.7 Abstract Darboux Integral Lower and Upper Bounds

> *Let $(X, \mathcal{S}, \mu)$ be a measure space with $\mu(X)$ finite and $f : X \to \Re$ be bounded and $\mathcal{S}$ measurable. Then*
>
> $$m \, \mu(X) \leq DAI(f) \leq M \, \mu(X)$$

Theorem 9.16.8 $DAI(f) = 0$ **if** $\mu(X) = 0$

> *Let $(X, \mathcal{S}, \mu)$ be a measure space with $\mu(X)$ finite and $f : X \to \Re$ be bounded and $\mathcal{S}$ measurable. Then $DAI(f) = 0$ if $\mu(X) = 0$.*

Theorem 9.16.9 $DAI(f(x) \equiv c)$ **is** $c\mu(X)$

> *Let $(X, \mathcal{S}, \mu)$ be a measure space with $\mu(X)$ finite and $f : X \to \Re$ be bounded, $\mathcal{S}$ measurable and have constant value c. Then $DAI(f) = c\mu(X)$.*

Theorem 9.16.10 Abstract Darboux Integral Measures

> *Let $(X, \mathcal{S}, \mu)$ be a measure space with $\mu(X)$ finite and $f : X \to \Re$ be bounded, $\mathcal{S}$ measurable and non-negative. Then $\lambda : \mathcal{S} \to \Re$ defined by $\lambda(E) = DAI(f, E)$ defines a measure on $\mathcal{S}$.*

Proof 9.16.10
We will provide a sketch of the proof. First, note our argument for a non-negative f can be used to

218 BASIC ANALYSIS IV: MEASURE THEORY AND INTEGRATION

show a finite valued summable f can be used to create a charge. We will not show that argument here, but it should be straightforward for you to do.

- *First, prove $\lambda(A \cup B) = \lambda(A) + \lambda(B)$ when A and B are disjoint and measurable. Let π be a partition consisting of points (y_j) and corresponding sets (E_j) as usual. Then for some set D and the function $f\,I_D$, we have the sets*

$$F_j = (y_{j-1} \leq f(x)I_D(x) < y_j)$$

This is the same as $F_1 = (0 = y_0 \leq f(x) < y_1) \cap D \cup [0, y_1) \cap D^C$ and for j larger than 1, $F_j = (y_{j-1} \leq f(x) < y_j) \cap D$. We know $E_j = (y_{j-1} \leq f(x) < y_j)$ and so $F_j = E_j \cap D$ if $j > 1$ and $F_1 = E_0 \cap D \cup [0, y_1) \cap D^C$. Hence, since the first term vanishes, we find the lower sum is $\sum_{j>1} y_{j-1}\, \mu(E_j \cap D)$.

Hence, if we apply this idea to the set $D = A \cup B$ for A and B disjoint, we find $L(f I_{A \cup B}, \pi) = L(f I_A, \pi) + L(f I_B, \pi)$ which leads to the inequality $\lambda(A \cup B) \leq \lambda(A) + \lambda(B)$. A similar argument using upper sums gives us the other inequality which proves this first result. We will leave the details for you in an exercise below.

- *It then follows immediately that the result above holds for finite unions of disjoint measurable sets via a standard induction argument. Again, the details are left for you.*

- *Now let (A_n) be a countable union of disjoint measurable sets. Let A be the full countable union, V_N be the union of the first N sets and R_N be $X \setminus V_N$. We can then apply what we know about finite unions to find*

$$\int_A f\, d\mu = \int_{V_N} f\, d\mu + \int_{R_N} f\, d\mu$$

which further expands to

$$\int_A f\, d\mu = \sum_{i=1}^{N} \int_{A_i} f\, d\mu + \int_{R_N} f\, d\mu$$

It is then not hard to show $\int_{R_N} f\, d\mu \to 0$, which proves countable additivity.

$\blacksquare$

Homework

Exercise 9.16.4 *Let $(X, \mathcal{S}, \mu)$ be a measure space with $\mu(X)$ finite and $f : X \to \Re$ be bounded and $\mathcal{S}$-measurable. Prove $\lambda^+ = \mathcal{S} \to \Re$ defined by $\lambda^+(E) = DAI(f^+, E)$ defines a measure on $\mathcal{S}$.*

Exercise 9.16.5 *Let $(X, \mathcal{S}, \mu)$ be a measure space with $\mu(X)$ finite and $f : X \to \Re$ be bounded and $\mathcal{S}$-measurable. Prove $\lambda^- - = \mathcal{S} \to \Re$ defined by $\lambda^-(E) = DAI(f^-, E)$ defines a measure on $\mathcal{S}$.*

Exercise 9.16.6 *Let $(X, \mathcal{S}, \mu)$ be a measure space with $\mu(X)$ finite and $f : X \to \Re$ be bounded and $\mathcal{S}$-measurable. Prove $\lambda- = \mathcal{S} \to \Re$ defined by $\lambda(E) = DAI(f, E)$ defines a charge on $\mathcal{S}$.*

9.16.1 Properties of the Darboux Integral

Theorem 9.16.11 The Abstract Darboux Integral is Monotone

ABSTRACT INTEGRATION

> *Let $(X, \mathcal{S}, \mu)$ be a measure space with $\mu(X)$ finite and $f, g : X \to \Re$ be bounded and $\mathcal{S}$ measurable. Then if $f \leq g$, $DAI(f) \leq DAI(g)$.*

Proof 9.16.11
It is easy to see that g determines a charge. Hence, for any partition π with points (y_j) and associated sets (E_j), we have

$$\int g \, d\mu = \sum_j \int_{E_j} g \, d\mu$$

On each E_j, g dominates f, implying $\int g \, d\mu$ dominates the lower sum for this partition. The result then follows. ∎

Theorem 9.16.12 Abstract Darboux Integral is Additive

> *Let $(X, \mathcal{S}, \mu)$ be a measure space with $\mu(X)$ finite and $f, g : X \to \Re$ be bounded and $\mathcal{S}$ measurable. Then $DAI(f + g) = DAI(f) + DAI(g)$.*

Proof 9.16.12
We prove additivity in two steps.

- *We prove for the case that g is a constant c on X. Let π be a partition with points (y_j) and associated sets (E_j). Then, it is easy to see $\{y_{j-1} \leq f(x) < y_j\}$ is the same set as $\{y_{j-1} + c \leq f(x) + c < y_j + c\}$. Hence, the points $y_j + c$ and the sets (E_j) give a partition π' for $f + g$. Thus, the lower sum is*

$$L(f + g, \pi') = \sum_j (y_{j-1} + c)\mu(E_j)$$

which can be broken into two sums giving

$$L(f + g, \pi') = \sum_j y_{j-1} \, \mu(E_j) + \sum_j c \, \mu(E_j)$$

The first sum on the right is $L(f, \pi)$ and the second is $c\mu(X)$ which is also $\int g \, d\mu$. Hence, we have shown additivity for this case.

- *We now handle the case of an arbitrary g. Since $f + g$ defines a charge, for any partition for f with points (y_j) and associated sets (E_j), we have*

$$\int (f + g) = \sum_j \int_{E_j} (f + g) \, d\mu$$

However, on E_j, we know $f(x) + g(x) \geq y_{j-1} + g(x)$. Hence, $\int_{E_j} (f + g) \, d\mu \geq \int_{E_j} (y_{j-1} + g) \, d\mu$. The additivity result holds for one of the functions a constant and so $\int_{E_j} (y_{j-1} + g) \, d\mu = y_{j-1} \, \mu(E_j) + \int_{E_j} g \, d\mu$. We conclude

$$\int (f + g) \geq L(f, \pi) + \int_{E_j} g \, d\mu$$

220 BASIC ANALYSIS IV: MEASURE THEORY AND INTEGRATION

Since the partition π is arbitrary, this shows $\int (f + g) \geq \int f \, d\mu + \int_{E_j} g \, d\mu$. A similar argument using upper sums completes this proof.

◼

Theorem 9.16.13 Abstract Darboux Integral is Scalable

> *Let $(X, \mathcal{S}, \mu)$ be a measure space with $\mu(X)$ finite and $f : X \to \Re$ be bounded and $\mathcal{S}$ measurable. Then for all real numbers c, $DAI(c\,f) = c\,DAI(f)$.*

Proof 9.16.13

The case $c = 0$ is easy. First prove the case $c > 0$ using a partition argument. Then, if $c < 0$, we can write cf as $|c|(-f)$ and write immediately $\int cf d\mu = |c| \int (-f) d\mu$. However, additivity tells us $\int (-f) d\mu = - \int f d\mu$, which completes the argument.

◼

Theorem 9.16.14 Abstract Darboux Integral Absolute Inequality

> *Let $(X, \mathcal{S}, \mu)$ be a measure space with $\mu(X)$ finite and $f : X \to \Re$ be bounded and $\mathcal{S}$ measurable. Then*
>
> $$\left| DAI(f) \right| \leq DAI(|f|)$$

Theorem 9.16.15 Abstract Darboux Integral Zero Implies $f = 0$ a.e.

> *Let $(X, \mathcal{S}, \mu)$ be a measure space with $\mu(X)$ finite and $f : X \to \Re$ be bounded, $\mathcal{S}$ measurable and non-negative. Then $DAI(f) = 0$ implies $f = 0$ a.e.*

Finally, it is easy to see:

Theorem 9.16.16 Abstract Darboux Integral Measures are Absolutely Continuous

> *Let $(X, \mathcal{S}, \mu)$ be a measure space with $\mu(X)$ finite and $f : X \to \Re$ be bounded, $\mathcal{S}$ measurable and non-negative. Then $\lambda : \mathcal{S} \to \Re$ defined by $\lambda(E) = DAI(f, E)$ defines an absolutely continuous measure on $\mathcal{S}$.*

Homework

Here is a long series of theorems for you to prove!

Exercise 9.16.7 *Prove Theorem 9.16.2.*

Exercise 9.16.8 *Prove Theorem 9.16.3.*

Exercise 9.16.9 *Prove Theorem 9.16.4.*

Exercise 9.16.10 *Prove Theorem 9.16.6.*

Exercise 9.16.11 *Prove Theorem 9.16.7.*

Exercise 9.16.12 *Prove Theorem 9.16.8.*

Exercise 9.16.13 *Prove Theorem 9.16.9.*

Exercise 9.16.14 *Prove Theorem 9.16.10.*

Exercise 9.16.15 *Prove Theorem 9.16.11.*

Exercise 9.16.16 *Prove Theorem 9.16.12.*

Exercise 9.16.17 *Prove Theorem 9.16.13.*

Exercise 9.16.18 *Prove Theorem 9.16.14.*

Exercise 9.16.19 *Prove Theorem 9.16.15.*

Exercise 9.16.20 *Prove Theorem 9.16.16.*

Chapter 10

The $\mathcal{L}_p$ Spaces

In mathematics and other fields, we often group objects of interest into sets and study the properties of these sets. In this book, we have been studying a set X with a sigma - algebra of subsets contained within it, the collection of functions which are measurable with respect to the sigma - algebra and recently, the set of functions which are summable. In addition, we have noted that the sets of measurable and summable functions are closed under scalar multiplication and addition as long as we interpret addition in the right way when the functions are extended real-valued.

We can do more along these lines. We will now study the sets of summable functions as vector spaces with a suitable norm. We begin with a review.

Definition 10.0.1 The Norm on a Vector Space

> *Let X be a non empty vector space over $\Re$. We say $\rho : X \to \Re$ is a norm on X if*
>
> *(N1): $\rho(x)$ is non-negative for all x in X,*
>
> *(N2): $\rho(x) = 0 \Leftrightarrow x = 0$,*
>
> *(N3): $\rho(\alpha x) = |\alpha|\rho(x)$, for all α in $\Re$ and for all x in X,*
>
> *(N4): $\rho(x + y) \leq \rho(x) + \rho(y)$, for all x and y in X.*
>
> *If ρ satisfies only N1, N3 and N4, we say ρ is a semi-norm or pseudo-norm. We will usually denote a norm of x by the symbol $\|x\|$.*
>
> *The pair $(X, \| \cdot \|)$ is called a Normed Linear Space or NLS.*

If a set X has no linear structure, we can still have a notion of the distance between objects in the set, if the set is endowed with a metric. This is defined below.

Definition 10.0.2 The Metric on a Set

> Let X be a non empty set. We say $d : X \times X \to \Re$ is a metric if
>
> *(M1)*: $d(x, y)$ is non-negative for all x and y in X,
>
> *(M2)*: $d(x, y) = 0 \Leftrightarrow x = y$,
>
> *(M3)*: $d(x, y) = d(y, x)$, for all for all x and y in X,
>
> *(M4)*: $d(x, y) \leq d(x, z) + d(y, z)$, for all x, y and z in X.
>
> If d satisfies only M1, M2 and M4, we say d is a semi-metric or pseudo-metric. The pair (X, d) is called a metric space. Note that in a metric space, there is no notion of scaling or adding objects because there is no linear structure.

Comment 10.0.1 *It is a standard result from a linear analysis course that the norm in an NLS $(X, \| \cdot \|)$ induces a metric on X by defining*

$$d(x, y) = \|x - y\|, \ \forall \, x, \, y \in X$$

Given a sequence (x_n) in an NLS $(X, \| \cdot \|)$, we can define what we mean by the convergence of this sequence to another object x in X.

Definition 10.0.3 Norm Convergence

> Let $(X, \| \, \|)$ be a non empty NLS. Let (x_n) be a sequence in X. We say the sequence (x_n) converges to x in X if
>
> $$\forall \epsilon > 0, \, \exists \, N \ni n > N \Rightarrow \|x_n - x\| < \epsilon$$

Now let $(X, \mathcal{S}, \mu)$ be a nonempty measurable space. We are now ready to discuss the space $L_1(X, \mathcal{S}, \mu)$. By Theorem 9.14.2, we know that this space is a vector space with suitably defined addition. We can now define a semi-norm for this space.

Theorem 10.0.1 The L_1 Semi-norm

> Let $(X, \mathcal{S}, \mu)$ be a nonempty measurable space. Define $\|x\|_1$ on $L_1(X, \mathcal{S}, \mu)$ by
>
> $$\|f\|_1 = \int |f| \, d\mu, \forall \, f \in L_1(X, \mathcal{S}, \mu)$$
>
> Then, $\|x\|_1$ is a semi-norm. Moreover, property N3 is almost satisfied: instead of N3, we have
>
> $$\|f\|_1 = 0 \Leftrightarrow f = a.e.$$

Proof 10.0.1

(N1): $\|f\|_1$ is clearly non-negative.
(N2): This proof is an easy calculation.

$$\begin{aligned} \|\alpha f\|_1 &= \int |\alpha \, f| \, d\mu = \int |\alpha| \, |f| \, d\mu \\ &= |\alpha| \int |f| \, d\mu = |\alpha| \, \|f\|_1 \end{aligned}$$

THE $\mathcal{L}_P$ SPACES

(N4): To prove this, we start with the triangle inequality for real numbers. We know that if f and g are summable, then the sum of $f + g$ is defined to be $h = (f + g)I_{E_{fg}^C}$. Let A be the set of points where this sum is ∞ and B be the set where the sum is $-\infty$. Then $\mu(E_{fg} \cup A \cup B) = 0$ and on $(E_{fg} \cup A \cup B)^C$, h is finite. For convenience of exposition, we will simply write h as $f + g$ from now on. So $f + g$ is finite off a set of measure 0. At the points where $f + g$ is finite, we can apply the standard triangle inequality to $f(x) + g(x)$. We have

$$|f(x) + g(x)| \leq |f(x)| + |g(x)|, a.e.$$

This implies

$$\int |f + g| \, d\mu \leq \int |f| \, d\mu + \int |g| \, d\mu$$

At the risk of repeating ourselves too much, let's go through the integral on the left-hand side again. We actually have

$$\int |f + g| \, I_{E_{fg}^C \cap A^C \cap B^C} \, d\mu \;=\; \int h \, I_{A^C \cap B^C} \, d\mu$$

$$=\; \int h \, d\mu$$

since $\mu(A^C \cap B^C) = 0$. Now the above inequality estimates clearly tell us

$$\|f + g\|_1 \leq \|f\|_1 + \|g\|_1$$

Finally, we look at what is happening in condition N2. Since $|f|$ is in $M^+(X, \mathcal{S}, \mu)$, by Lemma 9.9.1, we know

$$|f| \;=\; 0 \, a.e. \;\Leftrightarrow\; \int |f| \, d\mu \;=\; 0$$

Hence, $\|f\|_1 = 0$ if and only if $f = 0$ a.e. ∎

Although $\|x\|_1$ is only a semi-norm, there is a way to think of this class of functions as a normed linear space. Compare what we are doing here to the completion of the space of continuous functions using the L_1 norm as discussed in (Peterson (9) 2020) which generated the space $\mathbb{L}_{|\mathbb{K}}$. Let's define two functions f and g in $L_1(X, \mathcal{S}, \mu)$ to be equivalent or to be precise, μ - equivalent if $f = g$ except of a set of μ measure 0. We use the notation $f \sim g$ to indicate this equivalence. It is easy to see that $\sim$ defines an equivalence relation on $L_1(X, \mathcal{S}, \mu)$. We will let $[\boldsymbol{f}]$ denote the equivalence class defined by f:

$$[\boldsymbol{f}] \;=\; \{g \in L_1(X, \mathcal{S}, \mu) \mid g \sim f\}$$

Any g in $[\boldsymbol{f}]$ is called a representative of the equivalence class $[\boldsymbol{f}]$. A straightforward argument shows that two equivalence classes $[\boldsymbol{f_1}]$ and $[\boldsymbol{f_2}]$ are either equal as sets or disjoint. The collection of all distinct equivalence classes of $L_1(X, \mathcal{S}, \mu)$ under a.e. equivalence will be denoted by $\mathcal{L}_1(X, \mathcal{S}, \mu)$.

Theorem 10.0.2 $\mathcal{L}_1$ is a Normed Linear Space

$\mathcal{L}_1(X, \mathcal{S}, \mu)$ *is a vector space over $\Re$ with scalar multiplication and object addition defined as*

$$
\begin{aligned}
\alpha \, [\boldsymbol{f}] &= [\boldsymbol{\alpha} \, \boldsymbol{f}], \ \forall \, [\boldsymbol{f}] \\
[\boldsymbol{f}] + [\boldsymbol{g}] &= [\boldsymbol{f} + \boldsymbol{g}], \ \forall [\boldsymbol{f}] \ and \ [\boldsymbol{g}]
\end{aligned}
$$

Further, $\|[\boldsymbol{f}]\|_1$ defined by

$$
\|[\boldsymbol{f}]\|_1 \;=\; \int |g| \, d\mu
$$

for any representative g of $[\boldsymbol{f}]$, is a norm on $\mathcal{L}_1(X, \mathcal{S}, \mu)$.

Proof 10.0.2

The definition of scalar multiplication is clear. However, as usual, we can spend some time with addition. We know $f + g$ is defined on E_{fg}^C and that E_{fg} has measure 0. Hence, if $u \in [\boldsymbol{f}]$ and $v \in [\boldsymbol{g}]$, then $u = f$ and $v = g$ except on sets A and B of measure 0. Also, as usual, the sum $u + v$ is defined on E_{uv}^C. Hence,

$$
u + v = f + g, \ x \in E_{uv}^C \cap E_{fg}^C \cap A^C \cap B^C
$$

which is the complement of a set of measure 0. Hence, $u + v \in [\boldsymbol{f} + \boldsymbol{g}]$. Thus, $[\boldsymbol{f}] + [\boldsymbol{g}] \subseteq [\boldsymbol{f} + \boldsymbol{g}]$. Conversely, let $h \in [\boldsymbol{f} + \boldsymbol{g}]$. Now

$$
(f + g) \, I_{E_{fg}C} = f \, I_{E_{fg}C} + g \, I_{E_{fg}C}
$$

Hence, if we let

$$
u = f \, I_{E_{fg}C} \ and \ v = g \, I_{E_{fg}C}
$$

we see $h \sim (u + v)$, with $u \in [\boldsymbol{f}]$ and $v \in [\boldsymbol{g}]$. We conclude $[\boldsymbol{f} + \boldsymbol{g}] \subseteq [\boldsymbol{f}] + [\boldsymbol{g}]$. Hence, the addition of equivalence classes makes sense.

We now turn our attention to the possible norm $\|[\boldsymbol{f}]\|_1$. First, we must show that our definition of norm is independent of the choice of representative chosen from $[\boldsymbol{f}]$. If $g \sim f$, then $g = f$ except on a set A of measure 0. Thus, we know the integral of f and g match by Lemma 9.9.1. Here are the details:

$$
\begin{aligned}
\int |g| \, d\mu &= \int_A |g| \, d\mu + \int_{A^C} |g| \, d\mu \\
&= 0 + \int_{A^C} |f| \, d\mu \\
&= \int_A |f| \, d\mu + \int_{A^C} |f| \, d\mu \\
&= \int |f| \, d\mu
\end{aligned}
$$

We conclude the value of $\|[\boldsymbol{f}]\|_1$ is independent of the choice of representative from $[\boldsymbol{f}]$. Now we prove this is a norm.
(N1): $\|[\boldsymbol{f}]\|_1 = \int |g| d\mu \geq 0$.
(N2): If $\|[\boldsymbol{f}]\|_1 = 0$, then for any representative g of $[\boldsymbol{f}]$, we have $\int |g| d\mu = 0$. By Lemma 9.9.1, this implies that $g = 0$ a.e. and hence, $g \in [\boldsymbol{0}]$ (we abuse notation here by simply writing the zero function $h(x) = 0, \ \forall x$ as 0). But since $g \in [\boldsymbol{f}]$ also, this means $[\boldsymbol{f}] \cap [\boldsymbol{0}]$ is nonempty. This immediately

THE $\mathcal{L}_P$ SPACES 227

implies that $[\boldsymbol{f}] = [\boldsymbol{0}]$. Conversely, if $[\boldsymbol{f}] = [\boldsymbol{0}]$, the result is clear.
(N3): Let α be a real number. Then, if g is any representative of $[\boldsymbol{f}]$, we have α g is a representative of $[\alpha \, \boldsymbol{f}]$. We find

$$\begin{aligned} \|[\alpha \, \boldsymbol{f}]\|_1 &= \int |\alpha \, g| \, d\mu = |\alpha| \int |g| \, d\mu \\ &= |\alpha| \, \|[\boldsymbol{f}]\| \end{aligned}$$

(N4): The triangle inequality follows from the triangle inequality that holds for the representatives.
∎

Homework

Exercise 10.0.1 *We have defined two functions f and g in $L_1(X, \mathcal{S}, \mu)$ to be μ - equivalent if $f = g$ except of a set of μ measure 0 and use the notation $f \sim g$ to indicate this equivalence. Prove $\sim$ defines an equivalence relation on $L_1(X, \mathcal{S}, \mu)$.*

Exercise 10.0.2 *Go back and review how we did this for the Riemann Integral functions and compare what we do there to what we do here. We defined two functions f and g in $RI([a, b])$ to be equivalent if $\int_a^b |f(x) - g(x)| dx = 0$ and use the notation $f \sim g$ to indicate this equivalence. Prove $\sim$ defines an equivalence relation on $RI([a, b])$. Note this says $\int_a^b |f(x) - g(x)| dx = 0$ which implies $f = g$ a.e.*

Exercise 10.0.3 *Now, define two functions f and g in $RI([a, b])$ to be equivalent if $f = g$ a.e and use the notation $f \sim g$ to indicate this equivalence. Prove $\sim$ defines an equivalence relation on $RI([a, b])$. Note this says $\int_a^b |f(x) - g(x)| dx = 0$ if $f \sim g$.*

Exercise 10.0.4 *Let $(X, \mathcal{S}, \mu)$ be a nonempty measurable space and let f and g be in $L_1(X, \mathcal{S}, \mu)$. Prove that $\int |2f - 8g| d\mu \leq 2 \int |f| d\mu + 8 \int g d\mu$ by carefully mimicking all the arguments in the proof above.*

Exercise 10.0.5 *Let $(X, \mathcal{S}, \mu)$ be a nonempty measurable space and let f and g be in $L_1(X, \mathcal{S}, \mu)$. Prove that $\int |12f + 7g| d\mu \leq 12 \int |f| d\mu + 7 \int g d\mu$ by carefully mimicking all the arguments in the proof above.*

10.1 The General L_p Spaces

We can construct additional spaces of summable functions. Let p be a real number satisfying $1 \leq p < \infty$. Then the function $\phi(u) = u^p$ is a continuous function on $[0, \infty)$ that satisfies $\lim_n \phi(n) = \infty$. Thus, by Lemma 8.7.3, if f is an extended real-valued function on X, then the composition $\phi \circ |f|$ or $|f|^p$ is also measurable. Hence, we know the integral $\int |f|^p \, d\mu$ exists as an extended real-valued number. The class of measurable functions that satisfy $\int |f|^p \, d\mu < \infty$ is another interesting class of functions.

We begin with some definitions.

Definition 10.1.1 The Space of p-Summable Functions

228   BASIC ANALYSIS IV: MEASURE THEORY AND INTEGRATION

> *Let $(X, \mathcal{S}, \mu)$ be a nonempty measurable space. Let p be a real number satisfying $1 \leq p < \infty$. Then, $|f|^p$ is a measurable function. We let*
>
> $$L_p(X, \mathcal{S}, \mu) = \{ f \in M(X, \mathcal{S}, \mu) \mid \int |f|^p \, d\mu < \infty \}$$

For later use, we will also define what are called *conjugate index pairs*.

Definition 10.1.2 Conjugate Index Pairs

> *Let p be a real number satisfying $1 \leq p \leq \infty$. If $1 < p$ is finite, the index conjugate to p is the real number q satisfying*
>
> $$\frac{1}{p} + \frac{1}{q} = 1$$
>
> *while if $p = 1$, the index conjugate to p is $q = \infty$.*

We will be able to show that $L_p(X, \mathcal{S}, \mu)$ is a vector space under the usual scalar multiplication and addition operations once we prove some auxiliary results. These are the Hölder's and Minkowski's Inequalities. First, there is a standard lemma we will call the *Real Number Conjugate Indices Inequality*.

Lemma 10.1.1 Real Number Conjugate Indices Inequality

> *Let $1 < p < \infty$ and q be the corresponding conjugate index. Then if α and β are positive numbers,*
>
> $$A B \leq \frac{A^p}{p} + \frac{B^q}{q}$$

Proof 10.1.1
This proof is standard in any linear analysis book and so we will not repeat it here.   ∎

Theorem 10.1.2 Hölder's Inequality: $1 < p < \infty$

> *Let $1 < p < \infty$ and q be the index conjugate to p. Let $f \in L_p(X, \mathcal{S}, \mu)$ and $g \in L_q(X, \mathcal{S}, \mu)$. Then $f g \in L_1(X, \mathcal{S}, \mu)$ and*
>
> $$\int |fg| \, d\mu \leq \left(\int |f|^p \, d\mu \right)^{1/p} \left(\int |g|^q \, d\mu \right)^{1/q}$$

Proof 10.1.2
The result is clearly true if $f = g = 0$ a.e. Also, if $\int |f|^p d\mu = 0$, then $|f|^p = 0$ a.e. which tells us $f = 0$ a.e. and the result follows again. We handle the case where $\int |g|^q d\mu = 0$ in a similar fashion. Thus, we will assume both $I^p = \int |f|^p d\mu > 0$ and $J^q = \int |g|^q d\mu > 0$.

Let E_f and E_g be the sets where f and g are not finite. By our assumption, we know the measure of these sets is 0. Hence, for all x in $E_f^C \cap E_g^C$, the values $f(x)$ and $g(x)$ are finite. We apply Lemma 10.1.1 to conclude

$$\frac{|f(x)|}{I} \frac{|g(x)|}{J} \leq (1/p) \frac{|f(x)|^p}{I^p} + (1/q) \frac{|g(x)|^q}{J^q}$$

holds on $E_f^C \cap E_g^C$. Off of this set, we have that the left-hand side is ∞ and so is the right-hand side. Hence, even on $E_f \cup E_g$, the inequality is satisfied. Thus, since the function on the right-hand side is summable, we must have the left-hand side is a summable function too by Theorem 9.14.1. Hence, $f\,g \in L_1(X, \mathcal{S}, \mu)$. We then have

$$\int \frac{|f(x)|}{I} \frac{|g(x)|}{J} \, d\mu \;\le\; \int (1/p) \frac{|f(x)|^p}{I^p} \, d\mu + \int (1/q) \frac{|g(x)|^q}{J^q} \, d\mu$$

$$= \frac{1}{pI^p} \int |f(x)|^p \, d\mu + \frac{1}{qJ^q} \int |g(x)|^q \, d\mu$$

$$= \frac{1}{p} + \frac{1}{q} = 1$$

Thus,

$$\int |f\,g| \, d\mu \le I\,J \;=\; \left(\int |f|^p \, d\mu \right)^{1/p} \left(\int |g|^q \, d\mu \right)^{1/p}$$

■

The special case of $p = q = 2$ is of great interest. The resulting Hölder's Inequality is often called the Cauchy - Schwartz Inequality. We see:

Theorem 10.1.3 Cauchy - Bunyakovskiĭ - Schwartz Inequality

Let $f, g \in L_2(X, \mathcal{S}, \mu)$. Then $f\,g \in L_1(X, \mathcal{S}, \mu)$ and

$$\int |fg| \, d\mu \le \left(\int |f|^2 \, d\mu \right)^{1/2} \left(\int |g|^2 \, d\mu \right)^{1/2}$$

Homework

Exercise 10.1.1 *Make sure you know how to prove Theorem 10.1.1. You can find this in (Peterson (8) 2020).*

Exercise 10.1.2 *Let $p = 3$ and q be the index conjugate to 3, 3/2. Let $f \in L_3(X, \mathcal{S}, \mu)$ and $g \in L_{3/2}(X, \mathcal{S}, \mu)$. Prove, showing all the relevant details, that $f\,g \in L_1(X, \mathcal{S}, \mu)$ and $\int |fg| \, d\mu \le \left(\int |f|^3 \, d\mu \right)^{1/3} \left(\int |g|^{3/2} \, d\mu \right)^{2/3}.$*

Exercise 10.1.3 *Let $p = 4$ and q be the index conjugate to 4, 4/3. Let $f \in L_4(X, \mathcal{S}, \mu)$ and $g \in L_{4/3}(X, \mathcal{S}, \mu)$. Prove, showing all the relevant details, that $f\,g \in L_1(X, \mathcal{S}, \mu)$ and $\int |fg| \, d\mu \le \left(\int |f|^4 \, d\mu \right)^{1/4} \left(\int |g|^{4/3} \, d\mu \right)^{3/4}.$*

Next, we prove the important result called **Minkowski's Inequality**.

Theorem 10.1.4 Minkowski's Inequality

> Let $1 \le p < \infty$ and let $f, g \in L_p(X, \mathcal{S}, \mu)$. Then $f + g$ is in $L_p(X, \mathcal{S}, \mu)$ and
>
> $$\left(\int |f + g|^p \, d\mu \right)^{1/p} \le \left(\int |f|^p \, d\mu \right)^{1/p} + \left(\int |g|^p \, d\mu \right)^{1/p}$$

Proof 10.1.4

If $p = 1$, this is property N4 of the semi-norm $\| \cdot \|_1$. Thus, we can assume $1 < p < \infty$. Since f and g are measurable, we define the sum of $f + g$ as $h = (f + g) \, I_A$ where $A = E_{fg}^C$ with $\mu(E_{fg}) = 0$. Then as discussed h is measurable. We see on A,

$$|f(x) + g(x)| \le |f(x)| + |g(x)| \le 2 \max\{|f(x)|, |g(x)|\}$$

even when function values are ∞. Hence,

$$|f(x) + g(x)|^p \le 2^p \left(\max\{|f(x)|, |g(x)|\} \right)^p \le 2^p \left(|f(x)|^p + |g(x)|^p \right)$$

Then, since the right-hand side is summable, so is the left-hand side. We conclude $f + g$ is in $L_p(X, \mathcal{S}, \mu)$. Further,

$$|f(x) + g(x)|^p = |f + g| \, |f + g|^{p-1} \le |f| \, |f + g|^{p-1} + |g| \, |f + g|^{p-1}$$

We have the identity

$$|f(x) + g(x)|^p \le |f| \, |f + g|^{p-1} + |g| \, |f + g|^{p-1}. \tag{$*$}$$

Now since p and q are conjugate indices, we know

$$(1/p) + (1/q) = 1 \quad \Rightarrow \quad p + q = pq$$
$$\Rightarrow \quad p = q(p - 1)$$

Thus, the function

$$\left(\left| f + g \right|^{p-1} \right)^q = |f + g|^p$$

and so this function is summable implying $|f + g|^{p-1} \in L_q(X, \mathcal{S}, \mu)$. Now apply Hölder's Inequality to the two parts of the right-hand side of Equation $*$. We find

$$\int |f| \, |f + g|^{p-1} \, d\mu \le \left(\int |f|^p \, d\mu \right)^{1/p} \left(\int \left(|f + g|^{p-1} \right)^q \, d\mu \right)^{1/q}$$

and

$$\int |g| \, |f + g|^{p-1} \, d\mu \le \left(\int |g|^p \, d\mu \right)^{1/p} \left(\int \left(|f + g|^{p-1} \right)^q \, d\mu \right)^{1/q}$$

But we have learned we can rewrite the second terms of the above inequalities to get

$$\int |f| \, |f + g|^{p-1} \, d\mu \le \left(\int |f|^p \, d\mu \right)^{1/p} \left(\int |f + g|^p \, d\mu \right)^{1/q}$$

THE $\mathcal{L}_P$ SPACES 231

and

$$\int |g|\, |f+g|^{p-1}\, d\mu \;\leq\; \left(\int |g|^p\, d\mu\right)^{1/p} \left(\int |f+g|^p\, d\mu\right)^{1/q}$$

Thus, combining, we have

$$\int |f+g|^p\, d\mu \;\leq\; \left(\left(\int |f|^p\, d\mu\right)^{1/p} + \left(\int |g|^p\, d\mu\right)^{1/p}\right) \left(\int |f+g|^p\, d\mu\right)^{1/q}$$

We can rewrite this as

$$\left(\int |f+g|^p\, d\mu\right)^{1-1/q} \;\leq\; \left(\int |f|^p\, d\mu\right)^{1/p} + \left(\int |g|^p\, d\mu\right)^{1/p}$$

Since $1 - 1/q = 1/p$, we have established the desired result. ∎

Homework

Exercise 10.1.4 *Let $p = 3$ and let f and g be in $L_3(X, \mathcal{S}, \mu)$. Prove, showing all the relevant details, that $\left(\int |f+g|^3\, d\mu\right)^{1/3} \leq \left(\int |f|^3\, d\mu\right)^{1/3} + \left(\int |g|^3\, d\mu\right)^{1/3}$.*

Exercise 10.1.5 *Let $p = 4$ and let f and g be in $L_4(X, \mathcal{S}, \mu)$. Prove, showing all the relevant details, that $\left(\int |f+g|^4\, d\mu\right)^{1/4} \leq \left(\int |f|^4\, d\mu\right)^{1/4} + \left(\int |g|^4\, d\mu\right)^{1/4}$.*

Hölder's and Minkowski's Inequalities allow us to prove that the L_p spaces are normed linear spaces.

Theorem 10.1.5 L_p is a Vector Space

> *Let $(X, \mathcal{S}, \mu)$ be a measure space and let $1 \leq p < \infty$. Then, if scalar multiplication and object addition are defined pointwise as usual, $L_p(X, \mathcal{S}, \mu)$ is a vector space.*

Proof 10.1.5
The only thing we must check is that if f and g are in $L_p(X, \mathcal{S}, \mu)$, then so is $f + g$. This follows from Minkowski's inequality. ∎

Since $L_p(X, \mathcal{S}, \mu)$ is a vector space, the next step is to find a norm for the space.

Theorem 10.1.6 The L_p Semi-Norm

> *Let $(X, \mathcal{S}, \mu)$ be a measure space and let $1 \leq p < \infty$. Define $\|\cdot\|_p$ on $L_p(X, \mathcal{S}, \mu)$ by*
>
> $$\|f\|_p \;=\; \left(\int |f|^p\, d\mu\right)^{1/p}$$
>
> *Then, $\|\cdot\|_p$ is a semi-norm.*

Proof 10.1.6
Properties N1 and N3 of a norm are straightforward to prove. To see that the triangle inequality

232 BASIC ANALYSIS IV: MEASURE THEORY AND INTEGRATION

holds, simply note that Minkowski's Inequality can be rewritten as

$$\|f + g\|_p \;\leq\; \|f\|_p + \|g\|_p$$

for arbitrary f and g in $L_p(X, \mathcal{S}, \mu)$. ∎

If we use the same notion of equivalence a.e. as did earlier, we can define the collection of all distinct equivalence classes of $L_p(X, \mathcal{S}, \mu)$ under a.e. equivalence. This will be denoted by $\mathcal{L}_p(X, \mathcal{S}, \mu)$. We can prove that this space is a normed linear space using the norm $\|[\cdot]\|_p$.

Theorem 10.1.7 $\mathcal{L}_p$ is a Normed Linear Space

> *Let $1 \leq p < \infty$. Then $\mathcal{L}_p(X, \mathcal{S}, \mu)$ is a vector space over $\Re$ with scalar multiplication and object addition defined as*
>
> $$\alpha\,[\boldsymbol{f}] \;=\; [\boldsymbol{\alpha}\,\boldsymbol{f}]\,\forall\,[\boldsymbol{f}]$$
> $$[\boldsymbol{f}] + [\boldsymbol{g}] \;=\; [\boldsymbol{f} + \boldsymbol{g}], \forall[\boldsymbol{f}]\ and\ [\boldsymbol{g}]$$
>
> *Further, $\|[\boldsymbol{f}]\|_p$ defined by*
>
> $$\|[\boldsymbol{f}]\|_p \;=\; \left(\int |g|^p\, d\mu \right)^{1/p}$$
>
> *for any representative g of $[\boldsymbol{f}]$ is a norm on $\mathcal{L}_p(X, \mathcal{S}, \mu)$.*

Proof 10.1.7
The proof of this is quite similar to that of Theorem 10.0.2 and so we will not repeat it. ∎

Homework

You will prove Theorem 10.1.7 in the exercises.

Exercise 10.1.6 *Let $p = 3$. Let $f \in \mathcal{L}_3(X, \mathcal{S}, \mu)$. Prove $\alpha f \in \mathcal{L}_3(X, \mathcal{S}, \mu)$ for all real α.*

Exercise 10.1.7 *Let $p = 4$. Let f and g be in $\mathcal{L}_4(X, \mathcal{S}, \mu)$. Prove $f + g \in \mathcal{L}_4(X, \mathcal{S}, \mu)$.*

Exercise 10.1.8 *Let $p = 5$. Let $f \in \mathcal{L}_5(X, \mathcal{S}, \mu)$ and define $\|[\boldsymbol{f}]\|_p$ by $\|[\boldsymbol{f}]\|_5 = \left(\int |g|^5\, d\mu \right)^{1/5}$, for any representative g of $[\boldsymbol{f}]$. Show all the details of the proof that $\|[\cdot]\|_5$ is a norm on $\mathcal{L}_5(X, \mathcal{S}, \mu)$.*

Exercise 10.1.9 *Let $1 \leq p < \infty$. Let $f \in \mathcal{L}_p(X, \mathcal{S}, \mu)$ and define $\|[\boldsymbol{f}]\|_p$ by $\|[\boldsymbol{f}]\|_p = \left(\int |g|^p d\mu \right)^{1/p}$, for any representative g of $[\boldsymbol{f}]$. Show all the details of the proof that $\|[\cdot]\|_p$ is a norm on $\mathcal{L}_p(X, \mathcal{S}, \mu)$.*

We will now show that $\mathcal{L}_p(X, \mathcal{S}, \mu)$ is a complete NLS. Note we can do this without using the completion process for a metric space as we did in (Peterson (9) 2020). First, recall what a Cauchy sequence means.

Definition 10.1.3 Cauchy Sequence in Norm

THE $\mathcal{L}_P$ SPACES 233

> Let $(X, \|\cdot\|)$ be an NLS. We say the sequence (f_n) of X is a Cauchy sequence, if given $\epsilon > 0$, there is a positive integer N so that
>
> $$\|f_n - f_m\| \;<\; \epsilon, \; \forall\, n,\, m > N$$

This leads to the definition of a complete NLS or **Banach space**.

Definition 10.1.4 Complete NLS

> Let $(X, \|\cdot\|)$ be an NLS. We say the X is a complete NLS if every Cauchy sequence in X converges to some object in X.

It is a standard proof to show that any sequence in an NLS that converges must be a Cauchy sequence. Let's prove that in the context of the $\mathcal{L}_p(X, \mathcal{S}, \mu)$ space to get some practice.

Theorem 10.1.8 Sequences That Converge in $\mathcal{L}_p$ are Cauchy

> Let $([\boldsymbol{f_n}])$ be a sequence in $\mathcal{L}_p(X, \mathcal{S}, \mu)$ which converges to $[\boldsymbol{f}]$ in $\mathcal{L}_p(X, \mathcal{S}, \mu)$ in the $\|\cdot\|_p$ norm. Then, $([\boldsymbol{f_n}])$ is a Cauchy sequence.

Proof 10.1.8
Let $\epsilon > 0$ be given. Then, there is a positive integer N so that if $n > N$, then

$$\|[\boldsymbol{f_n} - \boldsymbol{f}]\|_p \;<\; \epsilon/2$$

Thus, if n and m are larger than N, by property N4,

$$\|[\boldsymbol{f_n} - \boldsymbol{f_m}]\|_p \;=\; \|[(\boldsymbol{f_n} - \boldsymbol{f}) + (\boldsymbol{f} - \boldsymbol{f_m})]\|_p \le \|[\boldsymbol{f_n} - \boldsymbol{f}]\|_p + \|[\boldsymbol{f_m} - \boldsymbol{f}]\|_p < \epsilon$$

This shows the sequence in a Cauchy sequence. ∎

We will now show the $\mathcal{L}_p(X, \mathcal{S}, \mu)$ is a Banach space.

Theorem 10.1.9 $\mathcal{L}_p$ is a Banach Space

> Let $1 \le p < \infty$. Then $\mathcal{L}_p(X, \mathcal{S}, \mu)$ is complete with respect to the norm $\|\cdot\|_p$.

Proof 10.1.9
Let $[\boldsymbol{f_n}]$ be a Cauchy sequence. These are the steps of the proof.

*(**Step 1**): We find a subsequence $([\boldsymbol{g_k}])$ so that for all k,*

$$\int |g_{k+1} - g_k|^p \, d\mu \;<\; (1/2^k)^p \tag{α}$$

*(**Step 2**): Define the function g by*

$$g(x) \;=\; |g_1(x)| + \sum_{k=1}^{\infty} |g_{k+1}(x) - g_k(x)|. \tag{β}$$

We show that g satisfies

$$\|g\|_p \;\leq\; \|g_1\|_p + 1. \qquad (\gamma)$$

This implies that g, defined by Equation β, converges and is finite a.e.

(Step 3): Then, we show

$$f(x) \;=\; g_1(x) + \sum_{k=1}^{\infty} \Big(g_{k+1}(x) - g_k(x) \Big)$$

is defined a.e. and is in $L_p(X, \mathcal{S}, \mu)$. This is our candidate for the convergence of the Cauchy sequence.

(Step 4): We show g_k converge to f in $\|\cdot\|_p$.

(Step 5): We show $[\boldsymbol{f_n}]$ converges to $[\boldsymbol{f}]$ in $\|\cdot\|_p$. This last step will complete the proof of completeness.

Now to the proof of these steps.

(Proof Step 1): For $\epsilon = (1/2)$, since $[\boldsymbol{f_n}]$ is a Cauchy sequence, there is a positive integer N_1 so that $n, m > N_1$ implies

$$\int |f_n - f_m|^p \, d\mu \;<\; (1/2)$$

Note we use representative $f_n \in [\boldsymbol{f_n}]$ for simplicity of exposition since the norms are independent of choice of representatives. Define $g_1 = f_{N_1+1}$.

Next, for $\epsilon = (1/2)^2$, there is a positive integer N_2, which we can always choose so that $N_2 > N_1$, so that $n, m > N_2$ implies

$$\int |f_n - f_m|^p \, d\mu \;<\; \left(1/(2^2) \right)^p$$

Let $g_2 = f_{N_2+1}$. Then, again by our choice of indices, we have

$$\int |g_2 - g_1|^p \, d\mu \;<\; (1/2)^p$$

The next step is similar. For $\epsilon = (1/2)^3$, there is a positive integer N_3, which we can always choose so that $N_3 > N_2$, so that $n, m > N_3$ implies

$$\int |f_n - f_m|^p \, d\mu \;<\; \left(1/(2^3) \right)^p$$

Let $g_3 = f_{N_3+1}$. Then, we have

$$\int |g_3 - g_2|^p \, d\mu \;<\; \left(1/(2^2) \right)^p$$

An induction argument thus shows that there is a subsequence $[\boldsymbol{g_k}]$ that satisfies

$$\int |g_{k+1} - g_k|^p \, d\mu \;<\; \left(1/(2^k) \right)^p$$

THE $\mathcal{L}_P$ SPACES

for all $k \geq 1$. This establishes Equation α.

(Proof Step 2): *Define the non-negative sequence (h_n) by*

$$h_n(x) = |g_1(x)| + \sum_{k=1}^{n} |g_{k+1}(x) - g_k(x)|$$

In this definition, there is the usual messiness of where all the differences are defined. Let's clear that up. Each pair (g_k, g_{k+1}) has a potential set E_k of measure zero where the subtraction is not defined. Thus, we need to throw away the set $E = \cup_k E_k$ which also has measure 0. Thus, it is clear that all of the h_n are defined on E^C. Now they may take on the value ∞, but that is acceptable. We see $h_n^p \uparrow g^p$ on E^C. Apply Fatou's Lemma to (h_n). We find

$$\int \left(\liminf h_n^p \, I_{E^C} \right) d\mu \leq \liminf \int h_n^p \, I_{E^C} \, d\mu$$

But, $\liminf h_n^p = g^p$ and so

$$\int g^p \, I_{E^C} \, d\mu \leq \liminf \int h_n^p \, I_{E^C} \, d\mu$$

The pth root function is continuous and so

$$\liminf \left(\int h_n^p \, I_{E^C} \, d\mu \right)^{1/p} = \left(\liminf \int h_n^p \, I_{E^C} \, d\mu \right)^{1/p}$$

Then, since the pth root function is increasing, we have

$$\left(\int g^p \, I_{E^C} \, d\mu \right)^{1/p} \leq \liminf \left(\int h_n^p \, I_{E^C} \, d\mu \right)^{1/p}$$

Next, applying Minkowski's Inequality to a finite sum, we obtain

$$\left(\int h_n^p \, I_{E^C} \right)^{1/p} = \left(\int \left(|g_1| + \sum_{k=1}^{n} |g_{k+1} - g_k| \right)^p I_{E^C} \right)^{1/p}$$

$$\leq \| g_1 I_{E^C} \|_p + \sum_{k=1}^{n} \| (g_{k+1} - g_k) I_{E^C} \|_p$$

Since the finite sum is monotonic increasing, we have immediately that the series

$$\sum_{k=1}^{\infty} \| (g_{k+1} - g_k) I_{E^C} \|_p$$

is a well-defined extended real-valued number. Thus, we have

$$\left(\int h_n^p \, I_{E^C} \right)^{1/p} \leq \| g_1 I_{E^C} \|_p + \sum_{k=1}^{\infty} \| (g_{k+1} - g_k) I_{E^C} \|_p$$

BASIC ANALYSIS IV: MEASURE THEORY AND INTEGRATION

By Equation α, *we also know that*

$$\sum_{k=1}^{\infty} \|(g_{k+1} - g_k)I_{E^C}\|_p \leq \sum_{k=1}^{\infty} 1/(2)^k = 1$$

Hence, we can actually say

$$\left(\int g\, I_{E^C} \right)^{1/p} \leq \|g_1 I_{E^C}\|_p + 1$$

We conclude $g\, I_{E^C}$ *is in* $L_p(X, \mathcal{S}, \mu)$. *Next, note if* $F = \{x \mid g(x)I_{E^C}(x) = \infty\}$, *we know* F *has measure* 0. *Hence,* $g\, I_{E^C \cap F^C}$ *is finite. This completes Step 2.*

(Proof Step 3): *Now define the function* f *by*

$$f(x) = \begin{cases} g_1(x) + \sum_{k=1}^{\infty} \left(g_{k+1}(x) - g_k(x) \right), & x \in E^C \cap F^C \\ 0 & x \in E \cup F. \end{cases}$$

Note, for $x \in E^C \cap F^C$,

$$
\begin{aligned}
|g_k| &= |g_1 + (g_2 - g_1) + (g_3 - g_2) + \ldots + (g_k - g_{k-1})| \\
&\leq |g_1| + \sum_{i=1}^{k} |g_{k+1} - g_k| = h_k
\end{aligned}
$$

However, we already seen that on this set $h_k \uparrow g$. *Hence, we can say*

$$|g_k| \leq g$$

This tells us that the partial sum expansion of g_k *converges absolutely on* $E^C \cap F^C$ *and thus,* g_k *converges to* g. *But* $g = f$ *on this set, so we have shown that* g_k *converges to* f *a.e. We can now apply the Lebesgue Dominated Convergence Theorem to say*

$$\lim_n \int g_n \, d\mu = \int f \, d\mu$$

Since $|g_k| \leq g$ *for all* k, *it follows* $|f|^p \leq |g|^p$. *Since* g *is p-summable, we have established that* f *is in* $L_p(X, \mathcal{S}, \mu)$.

(Proof Step 4): *Now we show* g_k *converges to* f *in* $\| \cdot \|_p$. *To see this, let* $z_k = f - g_k$ *on* $E^C \cap F^C$. *From the definition of* f, *we can write this as* $\sum_{j=k}^{\infty} (g_{j+1} - g_j)$. *The rest of the argument is very similar to the one used in Step 2. Consider the partial sums of this convergent series*

$$z_k^n = \sum_{j=k}^{n} |g_{j+1} - gj|$$

Minkowski's Inequality then gives for all n,

$$\|z_k^n\|_p \leq \sum_{j=k}^{n} \|g_{j+1} - gj\|_p$$

THE $\mathcal{L}_P$ SPACES

Using Equation α, it follows that the right-hand side is bounded above by $\sum_{j=k}^{n} 1/2^j$ which sums to $1/2^{n-1}$. Now apply Fatou's Lemma to find

$$\int \liminf |z_k^n|^p \leq \liminf \int |z_k^n|^p$$

or

$$\int |z_k|^p \leq \liminf \int |z_k^n|^p$$

The continuity and increasing nature of the pth root then give us

$$\left(\int |z_k|^p\right)^{1/p} \leq \liminf \left(\int |z_k^n|^p\right)^{1/p} \leq \liminf (1/2^{n-1}) = 0$$

Thus, $\|f - g_k\| \to 0$.

*(**Proof Step 5**): Finally, given $\epsilon > 0$, since $[f_n]$ is a Cauchy sequence, there is an N so that*

$$\|f_n - f_m\|_p < \epsilon/2, \ \forall \, n, m \, > \, N$$

Since $([g_k])$ is a subsequence of $([f_n])$, there is a K_1 so that if $k > K_1$, we have

$$\|f_m - g_k\|_p < \epsilon/2, \ \forall \, m \, > \, N, \, k \, > \, K_1$$

Also, since $g_k \to f$ in p-norm, there is a K_2 so that

$$\|g_k - f\|_p < \epsilon/2, \ \forall \, k \, > \, K_2$$

We conclude for any given $k > \max(K_1, K_2)$, we have

$$\|f_m - f\|_p \leq \|f_m - g_k\|_p + \|g_k - f\|_p \, < \, \epsilon$$

if $m > N$. Thus, $[f_n] \to [f]$ in p-norm. ∎

Homework

These exercises ask you to go through the details of the proof of Theorem 10.1.9 for each step for a specific p. This will help you get better at handling the particular kinds of manipulations we use for Cauchy sequence arguments and equivalence classes.

Exercise 10.1.10 *Prove Step 1 for $p = 3$.*

Exercise 10.1.11 *Prove Step 2 for $p = 4$.*

Exercise 10.1.12 *Prove Step 3 for $p = 2$.*

Exercise 10.1.13 *Prove Step 4 for $p = 5$.*

Exercise 10.1.14 *Prove Step 5 for $p = 2$*

The proof of the theorem above has buried in it a powerful result. We state this below.

Theorem 10.1.10 Sequences That Converge in p-Norm Possess Subsequences Converging Pointwise a.e.

238   BASIC ANALYSIS IV: MEASURE THEORY AND INTEGRATION

> Let $1 \leq p < \infty$. Let $([\boldsymbol{f_n}])$ be a sequence in $\mathcal{L}_p(X, \mathcal{S}, \mu)$ which converges in norm to $[\boldsymbol{f}]$ in $\mathcal{L}_p(X, \mathcal{S}, \mu)$. Then, there is a subsequence $([\boldsymbol{f_n^1}])$ of $([\boldsymbol{f_n}])$ with f_n^1 converging pointwise a.e. to f.

Proof 10.1.10

Since the sequence here converges, it is a Cauchy sequence. The subsequence we seek is the sequence (g_n) as defined in the proof of Theorem 10.1.9. But it is a bit much to just say that. Let's go through the outline of the proof so you can see how you repurpose it for this argument.

Since $[\boldsymbol{f_n}]$ converges, it is also a Cauchy sequence. We also know $[\boldsymbol{f_n}]$ converges to $[\boldsymbol{f}]$ in norm.

(Step 1): *We find a subsequence $([\boldsymbol{g_k}])$ so that for all k, $\int |g_{k+1} - g_k|^p \, d\mu < (1/2^k)^p$.*

(Step 2): *Define the function g by $g(x) = |g_1(x)| + \sum_{k=1}^{\infty} |g_{k+1}(x) - g_k(x)|$. We show that g satisfies $\|g\|_p \leq \|g_1\|_p + 1$. This implies that g is finite a.e.*

(Step 3): *Then, we show*

$$F(x) \;=\; g_1(x) \;+\; \sum_{k=1}^{\infty} \left(g_{k+1}(x) - g_k(x) \right)$$

is defined a.e., $F \in L_p(X, \mathcal{S}, \mu)$ and g_k converges to F a.e.
(Step 4): *We show g_k converge to F in $\| \cdot \|_p$.*

(Step 5): *We show $[\boldsymbol{f_n}]$ converges to $[\boldsymbol{F}]$ in $\| \cdot \|_p$. But limits are unique and so $[\boldsymbol{F}] = [\boldsymbol{f}]$. Hence $f = F \mu$ a.e. and so g_k converges to f pointwise a.e.* ∎

Homework

Exercise 10.1.15 *Let $f \in C([a, b])$ and assume we know all about Borel measure μ.*

- *Prove there is a sequence (p_n) of Bernstein polynomials on $[a, b]$ that converge to f uniformly on $[a, b]$.*

- *Prove this means p_n converges to f pointwise also.*

- *Prove this sequence of polynomials converges to f is $\| \cdot \|_2$.*

- *Prove there is another subsequence of (p_n) which converges pointwise μ a.e. to f.*

Exercise 10.1.16 *Let $f \in RI([a, b])$ and assume we know all about Borel measure μ. From our discussions about the completion of $\| \cdot \|_2$ in (Peterson (9) 2020) we know the continuous functions on $[a, b]$ are dense in $RI([a, b])$.*

- *Prove there is a sequence of continuous functions (f_n) on $[a, b]$ which converge to f in $\| \cdot \|_2$.*

- *Prove there is subsequence of (f_n) which converges pointwise μ a.e. to f.*

10.2   The World of Counting Measure

Let's see what the previous material means when we use counting measure, μ_C, on the set of natural numbers $\mathbb{N}$. In this case, the sigma - algebra is the power set of $\mathbb{N}$. Note if $f : \mathbb{N} \to \overline{\mathfrak{R}}$, then f is identified with a sequence of extended real-valued numbers, (a_n) so that $f(n) = a_n$. It is therefore possible for $f(n) \infty$ or $f(n) = -\infty$ for some n. Let

$$\phi_N(n) = \begin{cases} |f(n)|, & 1 \le n \le N \\ 0, & n > N \end{cases}$$

Then, $\phi_N \uparrow f$ and so by the Monotone Convergence Theorem,

$$\int |f| \, d\mu_C = \lim_N \int \phi_N(n) \, d\mu_C$$

Now the simple functions ϕ_N are not in their standard representation. Let $\{c_1, \ldots, c_M\}$ be the distinct elements of $\{|a_1|, \ldots, |a_N|\}$. Then we can write

$$\phi_N = \sum_{i=1}^{M} c_i \, I_{E_i}$$

where E_I is the pre-image of each distinct element c_i. The sets E_i are clearly disjoint by construction. It is a straightforward matter to see that

$$\int \phi_N \, d\mu_C = \sum_{i=1}^{M} c_i \, \mu_C E_i = \sum_{i=1}^{N} |a_i|$$

Thus, we have

$$\int |f| \, d\mu_C = \lim_N \sum_{i=1}^{N} |f(i)|$$

Since all the terms $|f(i)|$ are non-negative, we see the sequence of partial sums converges to some extended real-valued number (possibly ∞). For counting measure, the only set of measure 0 is $\emptyset$, so measurable functions cannot differ on a set of measure 0 in this case. We see for $1 \le p < \infty$,

$$L_p(\mathbb{N}, \mathcal{P}(\mathbb{N}), \mu_C) = \mathcal{L}_p(\mathbb{N}, \mathcal{P}(\mathbb{N}), \mu_C)$$

Further,

$$L_p(\mathbb{N}, \mathcal{P}(\mathbb{N}), \mu_C) = \{ \text{ sequences } (a_n) \mid \sum_{i=1}^{\infty} |a_i|^p \text{ converges } \}$$

We typically use the notation

$$\ell_p = L_p(\mathbb{N}, \mathcal{P}(\mathbb{N}), \mu_C) = \{ \text{ sequences } (a_n) \mid \sum_{i=1}^{\infty} |a_i|^p \text{ converges } \}$$

and we call this a sequence space. Note in all cases, summability implies the sequence involved must be finite everywhere. Note this is the same space we used to call ℓ^p. We use a different notation now to make it clearer we are thinking of this in the context of measures.

In this context, Hölder's Inequality becomes:

Theorem 10.2.1 Hölder's Inequality: Sequence Spaces: $1 < p < \infty$

240   *BASIC ANALYSIS IV: MEASURE THEORY AND INTEGRATION*

> *Let $1 < p < \infty$ and q be the index conjugate to p. Let $(a_n) \in \ell_p$ and $(b_n) \in ell_q$. Then $(a_n \, b_n) \in \ell_1$ and*
>
> $$\sum_n |a_n \, b_n| \leq \left(\sum_n |a_n|^p \right)^{1/p} \left(\sum_n |b_n|^q \right)^{1/q}$$

and Minkowski's Inequality becomes:

Theorem 10.2.2 Minkowski's Inequality: Sequence Spaces: $1 \leq p < \infty$

> *Let $1 \leq p < \infty$ and let $(a_n), (b_n) \in \ell_p$. Then $(a_n + b_n)$ is in ℓ_p and*
>
> $$\left(\sum_n |a_n + b_n|^p \right)^{1/p} \leq \left(\sum_n |a_n|^p \right)^{1/p} + \left(\sum_n |b_n|^p \right)^{1/p}$$

Finally, the special case of $p = q = 2$ should be mentioned. The sequence space version of the resulting Hölder's Inequality Cauchy - Schwartz Inequality has this form:

Theorem 10.2.3 Cauchy - Bunyakovskiĭ - Schwartz Inequality: ℓ_2 **Sequence Spaces**

> *Let $(a_n), (b_n) \in \ell_2$. Then $(a_n \, b_n) \in \ell_1$ and*
>
> $$\sum_n |a_n \, b_n| \leq \left(\sum_n |a_n|^2 \right)^{1/2} \left(\sum_n |b_n|^2 \right)^{1/2}$$

Homework

In previous volumes, we have proven Hölder's and Minkowski's Inequalities but these proofs, while essentially the same, use a much more abstract point of view. You should go through these proofs again to help you see how it still works.

Exercise 10.2.1 *Prove in detail Hölder's Inequality for $p = 2$ and $q = 2$ for the sequence space ℓ_p. Make sure you go through all the explanations for converting integrations into appropriate summations.*

Exercise 10.2.2 *Prove in detail Minkowski's Inequality for $p = 2$ and $q = 2$ for the sequence space ℓ_p. Make sure you go through all the explanations for converting integrations into appropriate summations.*

10.3   Equivalence Classes of Essentially Bounded Functions

There is one more space to define. This will be the analogue of the space of bounded functions we use in the definition of the Riemann Integral.

Definition 10.3.1 Essentially Bounded Functions

THE $\mathcal{L}_P$ SPACES

241

> Let $(X, \mathcal{S}, \mu)$ be a measure space and let f be measurable. If E is a set of measure 0, let
>
> $$\xi(E) = \sup_{x \in E^C} |f(x)|$$
>
> and
>
> $$\rho_\infty(f) = \inf \{ \xi(E) \mid E \in \mathcal{S}, \mu(E) = 0 \}$$
>
> If $\rho_\infty(f)$ is finite, we say f is an essentially bounded function.

There are then two more spaces to consider:

Definition 10.3.2 The Space of Essentially Bounded Functions

> Let $(X, \mathcal{S}, \mu)$ be a measure space. Then we define
>
> $$L_\infty(X, \mathcal{S}, \mu) = \{ f : X \to \overline{\Re} \mid f \in M(X, \mathcal{S}), \rho_\infty(f) < \infty \}$$
>
> and defining equivalence classes using a.e. equivalence,
>
> $$\mathcal{L}_\infty(X, \mathcal{S}, \mu) = \{ [f] \mid \rho_\infty(f) < \infty \}$$

There is an equivalent way of characterizing an essentially bounded function. This requires another definition.

Theorem 10.3.1 Alternate Characterization of Essentially Bounded Functions

> Let $(X, \mathcal{S}, \mu)$ be a measure space and f be a measurable function. Define $q_\infty(f)$ by
>
> $$q_\infty(f) = \inf \{ a \mid \mu(\{x \mid |f(x)| > a\}) = 0 \}$$
>
> Then, $\rho_\infty(f) = q_\infty(f)$.

Proof 10.3.1
Let $E_a = \{x \mid |f(x)| > a\}$. If a is a number so that $\mu(E_a) = 0$, then for any other measurable set A with measure 0, we have

$$A^C = A^C \cap E_a \cup A^C \cap E_a^C$$

Thus,

$$\sup_{A^C} |f| \geq \sup_{A^C \cap E_a} |f| \geq a$$

because if $x \in A^C \cap E_a$, then $|f(x)| > a$. Since we can do this for such a, it follows that

$$\sup_{A^C} |f| \geq q_\infty(f)$$

Further, since the measurable set A with measure zero is arbitrary, we must have

$$\rho_\infty(f) \geq q_\infty(f)$$

Next, we prove the reverse inequality. If $\mu(E_a) = 0$, then by the definition of $\rho_\infty(f)$, we have

$$\rho_\infty(f) \leq \sup_{E_a^C} |f| = \sup_{|f(x)| \leq a} |f(x)| \leq a$$

But this is true for all such a. Thus, $\rho_\infty(f)$ is a lower bound for the set $\{a | \mu(E_a) = 0\}$ and we can say

$$\rho_\infty(f) \leq q_\infty(f)$$

∎

Homework

Exercise 10.3.1 Let $f(x) = x^2$ on $[-2, 3]$ and assume you know all about Borel measure μ. Explain why $q_\infty(f) = \|f\|_\infty = 9$ here.

Exercise 10.3.2 Let $f(x) = I_{\mathbb{Q} \cap [0,1]}(x) - 2I_{\mathbb{I} \cap [0,1]}(x)$ and assume you know all about Borel measure μ. Explain why $q_\infty(f) = \|f\|_\infty = 2$ here.

Exercise 10.3.3 Let $f(x) = I_S(x)$ on $\Re$ and assume you know all about Borel measure μ. The set S is any subset of $\Re$. Explain why $q_\infty(f) = \|f\|_\infty = 1$ here.

We need to know that if two functions are equivalent with respect to the measure μ, then their ρ_∞ values agree.

Lemma 10.3.2 Essentially Bounded Functions That are Equivalent Have the Same Essential Bound

Let (X, S, μ) be a measure space and f and g be equivalent measurable functions such that $\rho(f)$ is finite. Then $\rho(g) = \rho(f)$.

Proof 10.3.2
Let E be the set of points where f and g are not equal. Then $\mu(E) = 0$. Now,

$$0 \leq \mu\bigg((|g(x)| > a) \cap E \bigg) \leq \mu(E) = 0$$

Thus,

$$\mu\bigg((|g(x)| > a) \bigg) = \mu\bigg((|g(x)| > a) \cap E \bigg) + \mu\bigg((|g(x)| > a) \cap E^C \bigg)$$

$$= \mu\bigg((|g(x)| > a) \cap E^C \bigg)$$

But on E^C, f and g match, so we have

$$\mu\bigg((|g(x)| > a) \bigg) = \mu\bigg((|f(x)| > a) \cap E^C \bigg) = \mu\bigg((|f(x)| > a) \bigg)$$

by the same sort of argument we used on $\mu\bigg((|g(x)| > a) \bigg)$. Hence, if $\mu\bigg((|f(x)| > a) \bigg) = 0$, then $\mu\bigg((|g(x)| > a) \bigg) = 0$ as well. This immediately implies $q_\infty(g) = q_\infty(f)$. The result then follows because $q_\infty = \rho_\infty$. ∎

Finally, we can show that essentially bounded functions are bounded above by their essential bound a.e.

THE $\mathcal{L}_P$ SPACES

Lemma 10.3.3 Essentially Bounded Functions Bounded Above by Their Essential Bound a.e.

Let $(X, \mathcal{S}, \mu)$ be a measure space and f be a measurable functions such that $\rho(f)$ is finite. Then $|f(x)| \leq \rho(f)$ a.e.

Proof 10.3.3
Let $E = (|f(x) > \rho_\infty(f))$. It is easy to see that

$$E = \bigcup_{k=1}^{\infty} \left(|f(x)| > \rho_\infty(f) + 1/k \right)$$

If you look at how q_∞ is defined, if $\mu(|f(x)| > \rho_\infty(f) + 1/k) > 0$, that would force $q_\infty(f) = \rho_\infty(f) \geq \rho_\infty(f) + 1/k$ which is not possible. Hence, $\mu(|f(x)| > \rho_\infty(f) + 1/k) = 0$ for all k. This means E has measure 0 also. It is then clear from the definition of the set E that $|f(x)| \leq \rho_\infty(f)$ on E^C. ∎

We can now prove that $\mathcal{L}_\infty(X, \mathcal{S}, \mu)$ is a vector space with norm $\|[\boldsymbol{f}]\|_\infty = \rho_\infty(f)$.

Theorem 10.3.4 The L_∞ Semi-Norm

Let $(X, \mathcal{S}, \mu)$ be a measure space. Define $\| \cdot \|_\infty$ on $L_\infty(X, \mathcal{S}, \mu)$ by

$$\|f\|_\infty = \rho_\infty(g)$$

where g is any representative of $[\boldsymbol{f}]$. Then, $\| \cdot \|_\infty$ is a semi-norm.

Proof 10.3.4
We show $\rho_\infty(\cdot)$ satisfies all the properties of a norm except N2 and hence it is a semi-norm.
(N1): It is clear the N1 is satisfied because $\rho_\infty(\cdot)$ is always non-negative.
(N2): Let 0_X be the function defined to be 0 for all x and let $E_a = \{x \mid |0_X(x)| > a\}$. It is clear $E_a = \emptyset$ for all $a > 0$. Thus, since $\rho_\infty = q_\infty$,

$$q_\infty(0_X) = \inf \{ a \mid \mu(E_a) = 0 \} = 0$$

However, if $q_\infty(f) = 0$, let $F_n = (|f_n(x)| > 1/n)$. Then, by definition of $q_\infty(f)$, it follows that $\mu(F_n) = 0$ and $|f(x)| \leq 1/n$ on the complement F_n^C. Let $F = \cup F_n$. Then, $\mu(F) = 0$ and

$$F^C = \bigcap_n F_n^C = \bigcap_n \left(|f(x)| \leq 1/n \right) = \left(f(x) = 0 \right)$$

Thus, f is 0 on F^C and nonzero on F which has measure 0. All that we can say then is that $f = 0$ a.e. and hence, $\| \cdot \|_\infty$ does not satisfy N2.
(N3): If α is 0, the result is clear. If α is a positive number, then

$$
\begin{aligned}
q_\infty(\alpha f) &= \inf \{ a \mid \mu(\{x \mid |\alpha f(x)| > a\}) = 0 \} \\
&= \inf \{ a \mid \mu(\{x \mid |f(x)| > a/\alpha\}) = 0 \}
\end{aligned}
$$

Let $\beta = a/\alpha$ and we have

$$
\begin{aligned}
q_\infty(\alpha f) &= \inf \{ \alpha \beta \mid \mu(\{x \mid |f(x)| > \beta\}) = 0 \} \\
&= \alpha \inf \{ \beta \mid \mu(\{x \mid |f(x)| > \beta\}) = 0 \} \\
&= \alpha \, q_\infty(f)
\end{aligned}
$$

244   BASIC ANALYSIS IV: MEASURE THEORY AND INTEGRATION

If α is negative, simply write αf as $|\alpha| \, (-f)$ and apply the result for a positive α.

(N4): Now let f and g be in $L_\infty(X, \mathcal{S}, \mu)$ with the sum $f + g$ defined in the usual way on E_{fg}^C with $\mu(E_{fg}) = 0$. Note on E_{fg} itself, $f(x) + g(x) = 0$, so the sum is bounded above by $\rho_\infty(f) + \rho_\infty(g)$ there. Now by Lemma 10.3.3, there are sets F and G of measure 0 so that

$$
\begin{aligned}
|f(x)| &\leq \rho_\infty(f), \, \forall x \in F^C, \\
|g(x)| &\leq \rho_\infty(g), \, \forall x \in G^C
\end{aligned}
$$

Thus,

$$
|f(x) + g(x)| \leq \rho_\infty(f) + \rho_\infty(g), \, \forall x \in F^C \cap G^C
$$

Thus, the measure of the set of points where $|f(x)+g(x)| > \rho_\infty(f)+\rho_\infty(g)$ is zero as $\mu(F \cup G) = 0$. By definition of q_∞, it then follows that

$$
q_\infty(f + g) \leq \rho_\infty(f) + \rho_\infty(g)
$$

which implies the result.   ■

Theorem 10.3.5 $\mathcal{L}_\infty$ is a Normed Linear Space

> *Then $\mathcal{L}_\infty(X, \mathcal{S}, \mu)$ is a vector space over $\Re$ with scalar multiplication and object addition defined as*
>
> $$
> \begin{aligned}
> \alpha \, [\boldsymbol{f}] &= [\boldsymbol{\alpha} \, \boldsymbol{f}] \, \forall \, [\boldsymbol{f}] \\
> [\boldsymbol{f}] + [\boldsymbol{g}] &= [\boldsymbol{f} + \boldsymbol{g}], \forall [\boldsymbol{f}] \text{ and } [\boldsymbol{g}]
> \end{aligned}
> $$
>
> *Further, $\|[\boldsymbol{f}]\|_\infty$ defined by $\|[\boldsymbol{f}]\|_\infty = \rho_\infty(g)$, for any representative g of $[\boldsymbol{f}]$ is a norm on $\mathcal{L}_\infty(X, \mathcal{S}, \mu)$.*

Proof 10.3.5

The argument that the scalar multiplication and addition of equivalence classes is the same as the one we used in the proof of Theorem 10.1.5 and so we will not repeat it here. From Lemma 10.3.2 we know that any two functions which are equivalent a.e. will have the same value for ρ_∞ and so $\|[\boldsymbol{f}]\|_\infty$ is independent of the choice of representative from $[\boldsymbol{f}]$. The proofs that properties N1, N3 and N4 hold follow immediately from the fact that they hold for representatives of equivalence classes. It remains to show that if $\|[\boldsymbol{f}]\|_\infty = 0$, then $[\boldsymbol{f}] = [\boldsymbol{0_X}]$ where 0_X is the zero function on X. However, we have already established in the proof of Theorem 10.3.4 that such an f is 0 a.e. This tells us $f \in [\boldsymbol{0_X}]$; thus, $[\boldsymbol{f}] = [\boldsymbol{0_X}]$.   ■

Homework

Exercise 10.3.4 *In the proof of Lemma 10.3.3, we let $E = (|f(x)| > \rho_\infty(f))$. Prove*

$$
E = \bigcup_{k=1}^{\infty} \left(|f(x)| > \rho_\infty(f) + 1/k \right)
$$

Exercise 10.3.5 *Let $f \in \mathcal{L}_\infty(X, \mathcal{S}, \mu)$. Prove $\alpha f \in \mathcal{L}_\infty(X, \mathcal{S}, \mu)$ for all real α.*

Exercise 10.3.6 *Let f and g be in $\mathcal{L}_\infty(X, \mathcal{S}, \mu)$. Prove $f + g \in \mathcal{L}_\infty(X, \mathcal{S}, \mu)$.*

THE $\mathcal{L}_P$ SPACES

245

Exercise 10.3.7 *Define* $\|[f]\|_\infty$ *by* $\|[f]\|_\infty = \rho(g)$ *where g is any representative of $[f]$. Prove this definition is independent of the choice of representative.*

We can now show that $\mathcal{L}_\infty(X, \mathcal{S}, \mu)$ is complete!

Theorem 10.3.6 $\mathcal{L}_\infty$ is a Banach Space

> *Then $\mathcal{L}_\infty(X, \mathcal{S}, \mu)$ is complete with respect to the norm $\|\cdot\|_\infty$.*

Proof 10.3.6

Let $([f_n])$ be a Cauchy sequence of objects in $\mathcal{L}_\infty(X, \mathcal{S}, \mu)$. Now everything is independent of the choice of representative of an equivalence class, so for convenience, we will use as our representatives the functions f_n themselves. Then, by Lemma 10.3.3, there are sets E_n of measure 0 so that

$$|f_n(x)| \leq \rho_\infty(f_n), \, \forall\, x \in E_n^C$$

Also, there are sets F_{nm} of measure 0 so that

$$|f_n(x) - f_m(x)| \leq \rho_\infty(f_n - f_m), \, \forall\, x \in F_{nm}^C$$

Hence, both of the equations above hold on

$$U \;=\; \bigcap_{m=1}^{\infty} \bigcap_{n=1}^{\infty} \left(E_n^C \cap F_{nm}^C \right)$$

We then use De Morgan's Laws to rewrite U as follows:

$$U \;=\; \bigcap_{m=1}^{\infty} \bigcap_{n=1}^{\infty} \left(E_n \cup F_{nm} \right)^C$$

$$=\; \bigcap_{m=1}^{\infty} \left(\bigcup_{n=1}^{\infty} \left(E_n \cup F_{nm} \right) \right)^C$$

$$=\; \left(\bigcup_{m=1}^{\infty} \bigcup_{n=1}^{\infty} \left(E_n \cup F_{nm} \right) \right)^C$$

Clearly, the measure of U is 0 and

$$|f_n(x) - f_m(x)| \;\leq\; \rho_\infty(f_n - f_m), \, \forall\, x \in U^C. \tag{α}$$

Now since $([f_n])$ is a Cauchy sequence with respect to $\|\cdot\|_\infty$, given $\epsilon > 0$, there is a positive integer N so that

$$|f_n(x) - f_m(x)| \;\leq\; \rho_\infty(f_n - f_m) < \epsilon/4, \, \forall\, x \in U^C, \, \forall\, n, m > N. \tag{β}$$

Equation β implies that at each x in U^C, the sequence $(f_n(x))$ is a Cauchy sequence of real numbers. By the completeness of $\Re$, it then follows that $\lim_n f_n(x)$ exists on U^C. Define the function $\hat{f} : X \to \Re$ by

$$\hat{f}(x) \;=\; \begin{cases} \lim_n f_n(x), & x \in U^C, \\ 0 & x \in U. \end{cases}$$

246   BASIC ANALYSIS IV: MEASURE THEORY AND INTEGRATION

From Equation β, we have that

$$\lim_n |f_n(x) - f_m(x)| \leq \epsilon/4, \forall x \in U^C, \forall m > N$$

As usual, since the absolute value function is continuous, we can let the limit operation pass into the absolute value function to obtain

$$|\hat{f}(x) - f_m(x)| \leq \epsilon/4, \forall x \in U^C, \forall m > N. \tag{γ}$$

From the backwards triangle inequality, we then find

$$|\hat{f}(x)| \leq \epsilon/4 + |f_m(x)| < \epsilon/4 + \rho_\infty(f_m), \forall x \in U^C, \forall m > N$$

Now fix $M > N + 1$. Then

$$|\hat{f}(x)| < \epsilon/2 + \rho_\infty(f_M), \forall x \in U^C$$

Since the measure of the set $(|\hat{f}(x)| > \epsilon/4 + \rho_\infty(f_M))$ is 0, from the definition of $q_\infty(\hat{f})$, it then follows that

$$q_\infty(\hat{f}) \leq \epsilon/4 + \rho_\infty(f_M)$$

We have thus shown $\hat{f}$ is essentially bounded and since f equals $\hat{f}$ a.e., we have f is in $\mathcal{L}_\infty(X, \mathcal{S}, \mu)$. It remains to show that $[\boldsymbol{f_n}]$ converges to $[\boldsymbol{f}]$ in norm. Note that Equation γ implies that (f_n) converges uniformly on U^C. Further, the measure of the set $(|f_n(x) - f(x)| > \epsilon/4)$ is 0. Thus, we can conclude

$$q_\infty(f - f_m) \leq \epsilon/4 < \epsilon, \forall m > N$$

This shows the desired convergence in norm.
Thus, we have shown $\mathcal{L}_\infty(X, \mathcal{S}, \mu)$ is complete. ∎

From the proofs above, we see Minkowski's Inequality holds for the case $p = \infty$ because $\|\cdot\|_\infty$ is a norm. Finally, we can complete the last case of Hölder's Inequality: the case of the conjugate indices $p = 1$ and $q = \infty$. We obtain:

Theorem 10.3.7 Hölder's Inequality: $p = 1$

> Let $p = 1$ and $q = \infty$ be the index conjugate to 1. Let $[\boldsymbol{f}] \in \mathcal{L}_1(X, \mathcal{S}, \mu)$ and $[\boldsymbol{g}] \in \mathcal{L}_\infty(X, \mathcal{S}, \mu)$. Then $[\boldsymbol{f} \, \boldsymbol{g}] \in \mathcal{L}_1(X, \mathcal{S}, \mu)$ and
>
> $$\int |fg| \, d\mu \leq \|[\boldsymbol{f}]\|_1 \, \|[\boldsymbol{g}]\|_\infty$$

Proof 10.3.7
It is enough to prove this result for the representatives of the equivalence classes $f \in [\boldsymbol{f}]$ and $g \in [\boldsymbol{g}]$. We know the product fg is measurable. It remains to show that fg is summable. Since g is essentially bounded, by Lemma 10.3.3, there is a set E of measure 0 so that

$$|g(x)| \leq \rho_\infty(g), \forall x \in E^C$$

Thus, $|f(x) \, g(x)| \leq |f(x)| \, \rho_\infty(g)$ a.e. and since the right-hand side is summable, by Theorem 9.14.1, we see fg is also summable and

$$\int |f \, g| \, d\mu \leq \int |f| \, \rho_\infty(g) \, d\mu = \rho_\infty(g) \int |f| \, d\mu$$

Homework

Exercise 10.3.8 *Define $\ell_\infty = \mathcal{L}_\infty(\mathbb{N}, \mathcal{P}(\mathbb{N}), \mu_C)$ as usual.*

- *Show $\ell_\infty = \{(a_n) | \sup_n |a_n| < \infty\}$.*

- *Show any $f \in \mathcal{L}_\infty(\mathbb{N}, \mathcal{P}(\mathbb{N}), \mu_C)$ can be identified with a bounded sequence (a_n).*

- *Show that $\|f\|_\infty = \|(a_n)$
  $_\infty$ for this identification of f with a bounded sequence.*

Exercise 10.3.9 *Prove in detail Hölder's Inequality for the case $p = 1$ and $q = \infty$.*

Exercise 10.3.10 *Prove Minkowski's Inequality for the case $p = \infty$.*

10.4 The Hilbert Space L_2

The space $\mathcal{L}_2(X, \mathcal{S}, \mu)$ is a Normed linear space with norm $\|[\cdot]\|_2$. This space is also an inner product space which is complete. Such a space is called a Hilbert space.

Definition 10.4.1 Inner Product Space

> *Let X be a non empty vector space over $\Re$. Let $\omega\, X \times X \to \Re$ be a mapping which satisfies*
>
> $$
> \begin{aligned}
> (IP1\!:)\, \omega(x + y, z) &= \omega(x, z) + \omega(y, x), \ \forall\, x, y, z \in X \\
> (IP2\!:)\, \omega(\alpha\, x, y) &= \alpha\, \omega(x, y), \ \forall\, \alpha \in \Re, \forall\, x, y \in X \\
> (IP3\!:)\, \omega(x, y) &= \omega(y, x), \ \forall\, x, y \in X \\
> (IP4\!:)\, \omega(x, x) &\geq 0, \forall\, x, \in X, \ \text{and}\ \omega(x, x) = 0 \Leftrightarrow x = 0
> \end{aligned}
> $$
>
> *Such a mapping is called a real inner product on the real vector space X. It is easy to define a similar mapping on complex vector spaces, but we will not do that here. We typically use the symbol $< \cdot, \cdot >$ to denote the value $\omega(\cdot, \cdot)$.*

There is much more we could say on this subject, but instead we will focus on how we can define an inner product on $\mathcal{L}_2(X, \mathcal{S}, \mu)$.

Theorem 10.4.1 The Inner Product on the Space of Square Summable Equivalence Classes

> *For brevity, let $\mathcal{L}_2$ denote $\mathcal{L}_2(X, \mathcal{S}, \mu)$. The mapping $< \cdot, \cdot >$ on $\mathcal{L}_2 \times \mathcal{L}_2$ defined by*
>
> $$
> < [f], [g] > = \int u\, v\, d\mu, \ \forall\, u \in [f],\, v \in [g]
> $$
>
> *is an inner product on $\mathcal{L}_2$. Moreover, it induces the norm $\|[\cdot]\|_2$ by*
>
> $$
> \begin{aligned}
> \|[f]\|_2 &= \sqrt{\int |f|^2\, d\mu} \\
> &= \sqrt{< [f], [f] >}
> \end{aligned}
> $$

Proof 10.4.1

The proof of these assertions is immediate as we have already shown $\| \cdot \|_2$ is a norm and the verification of properties IP1 to IP4 is straightforward. ∎

Finally, from our general $\mathcal{L}_p$ results, we know $\mathcal{L}_2$ is complete. However, for the record, we state this as a theorem.

Theorem 10.4.2 The Space of Square Summable Equivalence Classes is a Hilbert Space

> *For brevity, let $\mathcal{L}_2$ denote $\mathcal{L}_2(X, \mathcal{S}, \mu)$. Then $\mathcal{L}_2$ is complete with respect to the norm induced by the inner product $< [\cdot], [\cdot] >$. The inner product space $(\mathcal{L}_2, < \cdot, \cdot >)$ is often denoted by the symbol $\mathcal{H}$.*

Proof 10.4.2

This has already been done. ∎

Homework

We denote $\mathcal{L}_2$ by $\mathcal{L}_2(X, \mathcal{S}, \mu)$ and define the mapping $< \cdot, \cdot >$ on $\mathcal{L}_2 \times \mathcal{L}_2$ by

$$< [\boldsymbol{f}], [\boldsymbol{g}] > \;\; = \;\; \int u\,v\,d\mu, \, \forall\, u \,\in\, [\boldsymbol{f}],\, v \,\in\, [\boldsymbol{g}]$$

You will show this is an inner product in the exercises.

Exercise 10.4.1 *Verify property IP1.*

Exercise 10.4.2 *Verify property IP3.*

Exercise 10.4.3 *Verify property IP3.*

Exercise 10.4.4 *Verify property IP4.*

Now here are some more general exercises. They should make you scratch your head a bit!

Exercise 10.4.5 *Let $(X, \mathcal{S}, \mu)$ be a measure space. Let f be in $\mathcal{L}_p(X, \mathcal{S}\mu)$ for $1 \leq p < \infty$. Let $E = \{x \mid |f(x)| \neq 0\}$. Prove E is σ - finite.*

Hint 10.4.1 *Divide the indicated set into $(1 \leq |f(x)|)$ and $\cup_n (1/n \leq |f(x)| < 1)$.*

Exercise 10.4.6 *Let $(X, \mathcal{S}, \mu)$ be a finite measure space. If f is measurable, let*

$$E_n = \{x \mid n - 1 \leq |f(x)| < n\}$$

Prove f is in $\mathcal{L}_1(X, \mathcal{S}\mu)$ if and only if $\sum_{n=1}^{\infty} n\mu(E_n) < \infty$. More generally, prove f is in $\mathcal{L}_p(X, \mathcal{S}\mu)$, $1 \leq p < \infty$, if and only if $\sum_{n=1}^{\infty} n^p \mu(E_n) < \infty$.

Hint 10.4.2 *Note E_1 has finite measure because X does.*

Part V

Constructing Measures

Chapter 11

Constructing Measures

Although you now know quite a bit about measures, measurable functions, associated integration and the like, you still do not have many concrete and truly interesting measures to work with. In this chapter, you will learn how to construct interesting measures using some simple procedures. A very good reference for this material is (A. Bruckner and J. Bruckner and B. Thomson (1) 1997). Another good source is (Taylor (14) 1985). We begin with a definition.

11.1 Measures from Outer Measure

Definition 11.1.1 Outer Measure

Let X be a non empty set and let μ^* be an extended real-valued mapping defined on all subsets of X that satisfies

(i): $\mu^*(\emptyset) = 0$.

(ii): If A and B are subsets of X with $A \subseteq B$, then $\mu^*(A) \leq \mu^*(B)$.

(iii): If (A_n) is a sequence of subsets of X, then $\mu^*(\cup_n A_n) \leq \sum_n \mu^*(A_n)$.

Such a mapping is an outer measure on X and condition (iii) is called the countable subadditivity (CSA) condition if the sets are disjoint.

Comment 11.1.1 *Since $\emptyset \subseteq A$ for all A in X, condition (ii) tells us $\mu^*(\emptyset) \leq \mu^*(A)$. Hence, by condition (i), we have $\mu^*(A) \geq 0$ always. Thus, the outer measure is non-negative.*

The outer measure is defined on all the subsets of X. In Chapter 9, we defined the notion of a measure on a σ-algebra of subsets of X. Look back at Definition 9.1.1 again. Recall, the mapping $\mu : S \to \overline{\Re}$ is a measure on S if

(i): $\mu(\emptyset) = 0$,

(ii): $\mu(E) \geq 0$, for all $E \in S$,

(iii): μ is countably additive on S; i.e. if $(E_n) \subseteq S$ is a countable collection of disjoint sets, then $\mu(\cup_n E_n) = \sum_n \mu(E_n)$.

The third condition says the mapping μ is countably additive and hence, we label this condition as condition (CA). The collection of all subsets of X is the largest σ-algebra of subsets of X, so to convert the outer measure μ^* into a measure, we have to convert the countable subadditivity condition

251

252   BASIC ANALYSIS IV: MEASURE THEORY AND INTEGRATION

to countable additivity. This is not that easy to do! Now if T and E are any subsets of X, then we know

$$T = \left(T \cap E\right) \bigcup \left(T \cap E^C\right)$$

The outer measure μ^* is subadditive on finite disjoint unions and so we always have

$$\mu^*(T) \leq \mu^*\left(T \cap E\right) + \mu^*\left(T \cap E^C\right)$$

To have equality, we need to have

$$\mu^*(T) \geq \mu^*\left(T \cap E\right) + \mu^*\left(T \cap E^C\right)$$

also. So, as a first set towards the countable additivity condition we need, why don't we look at all subsets E of X that satisfy the condition

$$\mu^*(T) \geq \mu^*\left(T \cap E\right) + \mu^*\left(T \cap E^C\right), \forall T \subseteq X$$

We don't know how many such sets E there are at this point. But we certainly want finite additivity to hold. Therefore, it seems like a good place to start. This condition is called the Caratheodory Condition.

Definition 11.1.2 The Caratheodory Condition

Let μ^* be an outer measure on the non empty set X. A subset E of X satisfies the Caratheodory Condition if for all subsets T,

$$\mu^*(T) \geq \mu^*\left(T \cap E\right) + \mu^*\left(T \cap E^C\right)$$

Such a set E is called μ^* measurable. The collection of all μ^* measurable subsets of X will be denoted by $\mathcal{M}$.

We will first prove that the collection of μ^* measurable sets is an algebra of sets.

Definition 11.1.3 Algebra of Sets

Let X be a non empty set and let $\mathcal{A}$ be a nonempty family of subsets of X. We say $\mathcal{A}$ is an algebra of sets if the following conditions are true:

(i): $\emptyset$ is in $\mathcal{A}$.

(ii): If A and B are in $\mathcal{A}$, so is $A \cup B$.

(iii): If A is in $\mathcal{A}$, so is $A^C = X \setminus A$.

Homework

Exercise 11.1.1 Let $X = (0, 1]$. Let $\mathcal{A}$ consist of the empty set and all finite unions of half- open intervals of the form $(a, b]$ from X. Prove $\mathcal{A}$ is an algebra of sets of $(0, 1]$.

BUILDING MEASURES 253

Exercise 11.1.2 *Let* $X = (0, 1]$. *Let* $\mathcal{A}$ *consist of the empty set and all finite unions of half- open intervals of the form* $[a, b)$ *from* X. *Prove* $\mathcal{A}$ *is an algebra of sets of* $[0, 1)$.

Exercise 11.1.3 *Let* $X = (0, 1] \times (0, 1]$. *Let* $\mathcal{A}$ *consist of the empty set and all finite unions of half-open rectangles of the form* $(a, b] \times (c, d]$ *from* X. *Prove* $\mathcal{A}$ *is an algebra of sets of* $(0, 1] \times (0, 1]$.

Exercise 11.1.4 *Let* $X = [0, 1) \times [0, 1)$. *Let* $\mathcal{A}$ *consist of the empty set and all finite unions of half-open rectangles of the form* $[a, b) \times [c, d)$ *from* X. *Prove* $\mathcal{A}$ *is an algebra of sets of* $[0, 1) \times [0, 1)$.

Theorem 11.1.1 The μ^* Measurable Sets Form an Algebra

Let X *be a non empty set,* μ^* *an outer measure on* X *and* $\mathcal{M}$ *be the collection of* μ^* *measurable subsets of* X. *Then* $\mathcal{M}$ *is a algebra.*

Proof 11.1.1

For the empty set,

$$
\mu^* \left(T \cap \emptyset \right) + \mu^* \left(T \cap \emptyset^C \right) = \mu^* \left(\emptyset \right) + \mu^* \left(T \cap X \right)
$$
$$
= 0 + \mu^* \left(T \right)
$$

Hence $\emptyset$ *satisfies the Caratheodory condition and so* $\emptyset \in \mathcal{M}$.

Next, if $A \in \mathcal{M}$, *we note the Caratheodory condition is symmetric with respect to complementation and so* $A^C \in \mathcal{M}$ *also.*

To show $\mathcal{M}$ *is closed under countable unions, we will start with the union of just two sets and then proceed by induction. Let* E_1 *and* E_2 *be in* $\mathcal{M}$. *Let* T *be in* X. *Then, since* E_1 *and* E_2 *both satisfy Caratheodory's condition, we know*

$$
\mu^*(T) = \mu^*(T \cap E_1) + \mu^*(T \cap E_1^C) \tag{a}
$$

and

$$
\mu^*(T) = \mu^*(T \cap E_2) + \mu^*(T \cap E_2^C). \tag{b}
$$

In Equation **b**, *let "T" be "$T \cap E_1^C$". This gives*

$$
\mu^*(T \cap E_1^C) = \mu^*(T \cap E_1^C \cap E_2) + \mu^*(T \cap E_1^C \cap E_2^C). \tag{c}
$$

We also know that

$$
T \cap E_1 = T \cap (E_1 \cup E_2) \cap E_1, \; T \cap E_1^C \cap E_2 = T \cap (E_1 \cup E_2) \cap E_1^C. \tag{d}
$$

Now replace the term "$\mu^*(T \cap E_1^C)$" in Equation **a** *by the one in Equation* **c**. *This gives*

$$
\mu^*(T) = \mu^*(T \cap E_1) + \mu^*(T \cap E_1^C \cap E_2) + \mu^*(T \cap E_1^C \cap E_2^C)
$$

Next, replace the sets in the first two terms on the right side in the equation above by what is shown in Equation **d**. *We obtain*

$$
\mu^*(T) = \mu^*(T \cap (E_1 \cup E_2) \cap E_1) + \mu^*(T \cap (E_1 \cup E_2) \cap E_1^C) + \mu^*(T \cap E_1^C \cap E_2^C)
$$

254   BASIC ANALYSIS IV: MEASURE THEORY AND INTEGRATION

But E_1 is in $\mathcal{M}$ and so

$$\mu^*(T \cap (E_1 \cup E_2)) = \mu^*(T \cap (E_1 \cup E_2) \cap E_1) + \mu^*(T \cap (E_1 \cup E_2) \cap E_1^C)$$

Using this identity, we then have

$$\begin{aligned} \mu^*(T) &= \mu^*(T \cap (E_1 \cup E_2)) + \mu^*(T \cap E_1^C \cap E_2^C) \\ &= \mu^*(T \cap (E_1 \cup E_2)) + \mu^*(T \cap (E_1 \cup E_2)^C) \end{aligned}$$

using De Morgan's laws. Since the set T is arbitrary, we have shown $E_1 \cup E_2$ is also in $\mathcal{M}$.

Since, E_1 and E_2 are in $\mathcal{M}$, we now know $E_1^C \cup E_2^C$ is in $\mathcal{M}$ too. But this set is the same as $E_1 \cap E_2$. Thus, $\mathcal{M}$ is closed under intersection.

It then follows that $E_1 \setminus E_2 = E_1 \cap E_2^C$ is in $\mathcal{M}$. So $\mathcal{M}$ is also closed under set differences. Hence, $\mathcal{M}$ is an algebra. ■

Homework

Exercise 11.1.5 *Let $\mathcal{M}$ be the algebra of μ^* sets. If A and B are disjoint sets in $\mathcal{M}$, prove $\mu^*(A \cup B) = \mu^*(A) + \mu^*(B)$. To do this, use the fact that $\mu*(T) = \mu^*(T \cap (A \cup B)) + \mu^*(T \cap (A \cup B)^C)$ for all T. Now use $T = A \cup B$.*

Exercise 11.1.6 *Let $\mathcal{M}$ be the algebra of μ^* sets. If A, B and C are disjoint sets in $\mathcal{M}$, prove $\mu^*(A \cup B \cup C) = \mu^*(A) + \mu^*(B) + \mu^*(C)$.*

Exercise 11.1.7 *Let $\mathcal{M}$ be the algebra of μ^* sets. If $A_i, 1 \leq i \leq N$ are disjoint sets in $\mathcal{M}$, prove $\mu^*(\cup_{i=1}^{N}) = \sum_{i=1}^{N} \mu^*(A_i)$.*

Exercise 11.1.8 *Let $\mathcal{M}$ be the algebra of μ^* sets. If A is in $\mathcal{M}$ and $\mu^*(X)$ is finite, prove $\mu^*(A^C) = \mu^a st(X) - \mu^*(A)$.*

11.2   The Properties of the Outer Measure

Theorem 11.2.1 μ^* Measurable Sets Properties

Let X be a non empty set, μ^* an outer measure on X and $\mathcal{M}$ be the collection of μ^* measurable subsets of X. Then, if (E_n) is a countable disjoint sequence from $\mathcal{M}$, $\cup_n E_n$ is in $\mathcal{M}$ and

$$\mu^*(T \cap \cup_{i=1}^{\infty} E_i) = \sum_{i=1}^{\infty} \mu^* \left(T \cap E_i \right)$$

for all T in X.

Proof 11.2.1
Let "T" be "$T \cap (E_1 \cup E_2)$" in the Caratheodory condition of E_2. Then, we have

$$\mu^*(T \cap (E_1 \cup E_2)) = \mu^*(T \cap (E_1 \cup E_2) \cap E_2) + \mu^*(T \cap (E_1 \cup E_2) \cap E_2^C)$$

This simplifies to

$$\mu^*(T \cap (E_1 \cup E_2)) = \mu^*(T \cap E_2) + \mu^*(T \cap E_1 \cap E_2^C)$$

BUILDING MEASURES   255

But E_1 and E_2 are disjoint. Hence, E_1 is contained in E_2^C. Hence, we can further simplify to

$$\mu^*(T \cap (E_1 \cup E_2)) \;=\; \mu^*(T \cap E_2) + \mu^*(T \cap E_1)$$

Let's do another step. Since E_3 is in $\mathcal{M}$, we have

$$\mu^*(T \cap (E_1 \cup E_2 \cup E_3)) \;=\; \mu^*(T \cap (E_1 \cup E_2 \cup E_3) \cap E_3)$$
$$+ \mu^*(T \cap (E_1 \cup E_2 \cup E_3) \cap E_3^C)$$

This can be rewritten as

$$\mu^*(T \cap (E_1 \cup E_2 \cup E_3)) \;=\; \mu^*(T \cap E_3) + \mu^*(T \cap E_1 \cap E_3^C \cup T \cap E_2 \cap E_3^C)$$
$$=\; \mu^*(T \cap E_3) + \mu^*(T \cap E_1 \cup T \cap E_2)$$
$$=\; \mu^*(T \cap E_3) + \mu^*(T \cap (E_1 \cup E_2))$$

because $E_1 \subseteq E_3^C$ and $E_2 \subseteq E_3^C$ since all the E_n are disjoint. Then, we can apply the first step to conclude

$$\mu^*(T \cap (E_1 \cup E_2 \cup E_3)) \;=\; \mu^*(T \cap E_3) + \mu^*(T \cap E_2) + \mu^*(T \cap E_1)$$

We have therefore shown

$$\mu^*(T \cap (\cup_{i=1}^{3} E_i)) \;=\; \sum_{i=1}^{3} \mu^*(T \cap E_i)$$

It is now clear, we can continue this argument by induction to show

$$\mu^*(T \cap (\cup_{i=1}^{n} E_i)) \;=\; \sum_{i=1}^{n} \mu^*(T \cap E_i) \qquad \textbf{(a)}$$

for any positive integer n. Further, since $\mathcal{M}$ is an algebra, induction also shows $\cup_{i=1}^{n} E_i$ is in $\mathcal{M}$ for any such n. It then follows that for any T in X,

$$\mu^*(T) \;=\; \mu^*(T \cap (\cup_{i=1}^{n} E_i)) + \mu^*(T \cap (\cup_{i=1}^{n} E_i)^C)$$

*Using Equation **a**, we then have*

$$\mu^*(T) \;=\; \sum_{i=1}^{n} \mu^*(T \cap E_i) + \mu^*(T \cap (\cup_{i=1}^{n} E_i)^C). \qquad \textbf{(b)}$$

Next, note for all n

$$T \cap (\cup_{i=1}^{n} E_i)^C \;\supseteq\; T \cap (\cup_{i=1}^{\infty} E_i)^C$$

and hence

$$\mu^*\!\left(T \cap (\cup_{i=1}^{\infty} E_i)^C\right) \;\leq\; \mu^*\!\left(T \cap (\cup_{i=1}^{n} E_i)^C\right)$$

*Using this in Equation **b**, we find*

$$\mu^*(T) \;\geq\; \sum_{i=1}^{n} \mu^*(T \cap E_i) + \mu^*(T \cap (\cup_{i=1}^{\infty} E_i)^C). \qquad \textbf{(c)}$$

256 BASIC ANALYSIS IV: MEASURE THEORY AND INTEGRATION

Since this holds for all n, letting $n \to \infty$, we obtain

$$\mu^*(T) \;\geq\; \sum_{i=1}^{\infty} \mu^*(T \cap E_i) \;+\; \mu^*(T \cap (\cup_{i=1}^{\infty} E_i)^C). \qquad (d)$$

Finally, since

$$\bigcup_{i=1}^{\infty}(T \cap E_i) \;=\; T \bigcap (\cup_{i=1}^{\infty} E_i)$$

by the countable subadditivity of μ^*, it follows that

$$\mu^*\left(T \bigcap (\cup_{i=1}^{\infty} E_i)\right) \;=\; \mu^*\left(\bigcup_{i=1}^{\infty}(T \cap E_i)\right) \leq \sum_{i=1}^{\infty} \mu^*\left(T \cap E_i\right)$$

Using this in Equation c, we have

$$\mu^*(T) \;\geq\; \mu^*\left(T \bigcap (\cup_{i=1}^{\infty} E_i)\right) \;+\; \mu^*\left(T \cap (\cup_{i=1}^{\infty} E_i)^C\right). \qquad (e)$$

Since this holds for all subsets T, this tells us $\cup_n E_n$ is in $\mathcal{M}$. This proves that $\mathcal{M}$ is a σ-algebra.

Countable subadditivity of μ^* then gives us

$$\mu^*(T) \;\leq\; \mu^*\left(T \bigcap (\cup_{i=1}^{\infty} E_i)\right) \;+\; \mu^*\left(T \cap (\cup_{i=1}^{\infty} E_i)^C\right)$$

Using countable subadditivity again,

$$\mu^*(T) \;\leq\; \sum_{i=1}^{\infty} \mu^*\left(T \cap E_i\right) \;+\; \mu^*\left(T \cap (\cup_{i=1}^{\infty} E_i)^C\right). \qquad (f)$$

Combining Equation d and Equation f, we find

$$\mu^*(T) \;=\; \sum_{i=1}^{\infty} \mu^*\left(T \cap E_i\right) \;+\; \mu^*\left(T \cap (\cup_{i=1}^{\infty} E_i)^C\right)$$

Thus, letting our choice of T be $T \cap (\cup_n E_n)$, we find

$$\mu^*(T \cap \cup_{i=1}^{\infty} E_i) \;=\; \sum_{i=1}^{\infty} \mu^*\left(T \cap E_i\right). \qquad (g)$$

■

Homework

Exercise 11.2.1 *If A and B are any subsets of X, not necessarily disjoint, prove $A \cup B = A \cup (B \setminus A)$ and hence we can rewrite $A \cup B$ as a disjoint union.*

Exercise 11.2.2 *If A, B and C are any subsets of X, not necessarily disjoint, prove $A \cup B \cup C = A \cup (B \setminus A) \cup (C \setminus (A \cup B))$. Hence we can rewrite $A \cup B \cup C$ as a disjoint union.*

Exercise 11.2.3 *If $A_i, 1 \leq i \leq N$ are subsets of X, not necessarily disjoint, prove $\cup_{i=1}^{N} A_i =$*

$\cup_{i=1}^N E_i$ where the sets E_i are disjoint and are defined recursively by $E_1 = A_1$, $E_2 = A_2 \setminus A_1$, $E_3 = A_3 \setminus (A_1 \cup A_2)$ and so on. Hence we can rewrite $\cup_{i=1}^N A_i$ as a disjoint union. Note this is an induction proof.

11.3   Measures Induced by Outer Measures

Theorem 11.3.1 The Measure Induced by an Outer Measure

Let X be a non empty set, μ^* an outer measure on X and $\mathcal{M}$ be the collection of μ^* measurable subsets of X. Then, $\mathcal{M}$ is a σ-algebra and μ^* restricted to $\mathcal{M}$ is a measure we will denote by μ.

Proof 11.3.1
Recall that $\mathcal{M}$ is a σ-algebra if

(i) $\emptyset$, $X \in \mathcal{M}$.

(ii) If $A \in \mathcal{M}$, so is A^C.

(iii) If $\{A_n\}_{n=1}^\infty \in \mathcal{M}$, then $\cup_{n=1}^\infty A_n \in \mathcal{M}$.

Since we know $\mathcal{M}$ is an algebra of sets, all that remains is to show it is closed under countable unions. We have already shown all the properties of a σ-algebra except closure under arbitrary countable unions. The previous theorem, however, does give us closure under countable disjoint unions. So, let (A_n) be a countable collection of sets in $\mathcal{M}$. Letting

$$
\begin{aligned}
E_1 &= A_1 \\
E_2 &= A_2 \setminus A_1 \\
\vdots &= \vdots \\
E_n &= A_n \setminus \left(\cup_{i=1}^{n-1} A_i \right) \\
\vdots &= \vdots
\end{aligned}
$$

we see each E_n is in $\mathcal{M}$ by Theorem 11.1.1. Further, they are pairwise disjoint and so by Theorem 11.2.1, we can conclude $\cup_n E_n$ is in $\mathcal{M}$. But it is easy to see that $\cup_n E_n = \cup_n A_n$. Thus, $\mathcal{M}$ is a σ-algebra.

To show μ^* restricted to $\mathcal{M}$, μ, is a measure, we must show

(i): $\mu(\emptyset) = 0$,

(ii): $\mu(E) \geq 0$, for all $E \in \mathcal{S}$,

(iii): μ is countably additive on $\mathcal{S}$; i.e. if $(E_n) \subseteq \mathcal{S}$ is a countable collection of disjoint sets, then $\mu(\cup_n E_n) = \sum_n \mu(E_n)$.

*Since $\mu^*(\emptyset) = 0$, condition (i) follows immediately. Also, we know $\mu^*(E) \geq 0$ for all subsets E, and so condition (ii) is valid. It remains to show countable additivity. Let (B_n) be a countable disjoint family in $\mathcal{M}$. We can apply Equation **g** to conclude, using $T = X$, that*

$$
\mu^*(\cup_{i=1}^\infty B_i) = \sum_{i=1}^\infty \mu^*(B_i)
$$

258   *BASIC ANALYSIS IV: MEASURE THEORY AND INTEGRATION*

Finally, since $\mu^* = \mu$ on these sets, we have shown μ is countably additive and so is a measure. ■

It is also true that the measure constructed from an outer measure in this fashion is a complete measure.

Theorem 11.3.2 The Measure Induced by an Outer Measure is Complete

> *If E is a subset of X satisfying $\mu^*(E) = 0$, then $E \in \mathcal{M}$. Also, if $F \subseteq E$, then $F \in \mathcal{M}$ as well, with $\mu^*(F) = 0$. Note, this tells us that if $\mu(E) = 0$, then subsets of E are also in $\mathcal{M}$ with $\mu(F) = 0$; i.e. μ is a complete measure.*

Proof 11.3.2
We know $\mu^*(T \cap E) \leq \mu^*(E)$ for all T; hence, $\mu^*(T \cap E) = 0$ here. Thus, for any T,

$$\mu^*(T \cap E) + \mu^*(T \cap E^C) = \mu^*(T \cap E^C) \leq \mu^*(T)$$

This tells us E satisfies the Caratheodory condition and so is in $\mathcal{M}$. Thus, we have $\mu(E) = 0$. Now, let $F \subseteq E$. Then, $\mu^*(F) = 0$ also; hence, by the argument above, we can conclude $F \in \mathcal{M}$ with $\mu(F) = 0$. ■

Homework

Exercise 11.3.1 *If all the singleton sets $\{x\}$ are in $\mathcal{M}$ and X was infinite, prove $\mu^*(\cup_{1=1}^{\infty}\{x_i\}) = \sum_{i=1}^{\infty} \mu^*(\{x_i\})$.*

Exercise 11.3.2 *Let $X = \{x_1, \ldots, x_5\}$ and define μ^* on the subsets of X by $\mu^*(x_i) = \alpha_i > 0$ and extending μ^* to subsets of more than one element by summation. Is μ^* an outer measure on X and what are the μ^* measurable sets?*

Exercise 11.3.3 *Let $X = \{x_1, \ldots, x_i\}$ be a countable sets and define μ^* on the subsets of X by $\mu^*(x_i) = \alpha_i > 0$ with $\sum_{i=1}^{\infty} \alpha_i < \infty$. Extend μ^* to subsets of more than one element by summation. Is μ^* an outer measure on X and what are the μ^* measurable sets?*

11.4   Measures from Metric Outer Measures

Another outer measure is called the **metric outer measure**.

Definition 11.4.1 Metric Outer Measure

> *Let (X, d) be a non empty metric space and for two subsets A and B of X, define the distance between A and B by*
>
> $$D(A, B) = \inf\{ d(a, b) \mid a \in A, \ b \in B \}$$
>
> *If μ^* is an outer measure on X which satisfies*
>
> $$\mu^*(A \cup B) = \mu^*(A) + \mu^*(B)$$
>
> *whenever $D(A, B) > 0$, we say μ^* is a metric outer measure.*

The σ algebra of open subsets of X is called the Borel σ algebra $\mathcal{B}$. We can use the construction process in Section 11.1 to construct a σ algebra of subsets, $\mathcal{M}$, which satisfies the Caratheodory

BUILDING MEASURES

259

condition for this metric outer measure μ^*. This gives us a measure on $\mathcal{M}$. We would like to be able to say that open sets in the metric space are μ^* measurable. Thus, we want to prove $\mathcal{B} \subseteq \mathcal{M}$. This is what we do in the next theorem. It is becoming a bit cumbersome to keep saying μ^* measurable for the sets in $\mathcal{M}$. We will make the following convention for later use: a set in $\mathcal{M}$ will be called **OMI** measurable, where **OMI** stands for outer measure induced.

Theorem 11.4.1 Open Sets in a Metric Space are OMI Measurable

> Let (X, d) be a non empty metric space and μ^* a metric outer measure on X. Then open sets are OMI measurable.

Proof 11.4.1

Let E be open in X. To show E is μ^* measurable we must show

$$\mu^*(T) \;\geq\; \mu^*(T \cap E) + \mu^*(T \cap E^C)$$

for all subsets T in X. Since this is true for all subsets with $\mu^*(T) = \infty$, it suffices to prove the inequality is valid for all subsets with $\mu^*(T)$ finite. Also, we already know $\emptyset$ and X are μ^* measurable, so we can further restrict our attention to nonempty strict subsets E of X. We will prove this in a series of steps:

Step (i): *Let E_n be defined for each positive integer n by*

$$E_n \;=\; \{\, x \mid D(x, E^C) > \frac{1}{n} \}$$

It is clear $E_n \subseteq E$ and that $E_n \subseteq E_{n+1}$.

Note, if $y \in E_n$ and $x \in E^c$, we have $d(y, x) > 1/n$ and so

$$\inf_{y \in E_n, \, x \in E^c} d(y, x) \;\geq\; \frac{1}{n}$$

and so $D(E_n, E^C) \geq 1/n$. This immediately tells us

$$D(T \cap E_n, T \cap E^C) \geq 1/n$$

also for all T.

Since μ^* is a metric outer measure, we then have

$$\mu^*\Big((T \cap E_n) \cup (T \cap E^C) \Big) \;=\; \mu^*(T \cap E_n) + \mu^*(T \cap E^C)$$

However, we also know E_n is a subset of E and so

$$(T \cap E_n) \cup (T \cap E^C) \;\subseteq\; (T \cap E) \cup (T \cap E^C) = T$$

We conclude then

$$\mu^*\Big((T \cap E_n) \cup (T \cap E^C) \Big) \;\leq\; \mu^*(T)$$

Hence, for all T, we have

$$\mu^* \left(T \cap E_n \right) + \mu^* \left(T \cap E^C \right) \leq \mu^*(T). \tag{$*$}$$

Step (ii)*: If $\lim_n \mu^*(T \cap E_n) = \mu^*(T)$, then letting n go to infinity in Equation $*$, we would find*

$$\mu^* \left(T \cap E \right) + \mu^* \left(T \cap E^C \right) \leq \mu^*(T)$$

This means E satisfies the Caratheodory condition and so is μ^* measurable.

To show this limit acts in this way, we will construct a new sequence of sets (W_n) that are disjoint from one another with $E = \cup_n W_n$ so that the new sets W_n have useful properties. Since E is open, every point p in E is an interior point. Thus, there is a positive r so that $B(p; r) \subseteq E$. So, if $z \in E^C$, we must have and $d(p, z) \geq r$. It follows that $D(p, E^C) \geq r > r/2$. We therefore know that $p \in E_n$ for some n. Since our choice of p is arbitrary, we have shown

$$E \subseteq \cup_n E_n$$

It was already clear that $\cup_n E_n \subseteq E$; we conclude $E = \cup_n E_n$. We then define the needed disjoint collection (W_n) as follows

$$
\begin{aligned}
W_1 &= E_1 \\
W_2 &= E_2 \setminus E_1 \\
W_2 &= E_3 \setminus E_2 \\
&\vdots \quad \vdots \quad \vdots \\
W_n &= E_n \setminus E_{n-1}
\end{aligned}
$$

It helps to draw a picture here for yourself in terms of the annuli $E_n \setminus E_{n-1}$. We can see that for any n, we can write

$$T \cap E = (T \cap E_n) \bigcup \cup_{k=n+1}^{\infty} (T \cap W_k)$$

as the terms $T \cap W_k$ give the contributions of each annuli or strip outside of the core E_n. Hence,

$$\mu^*(T \cap E) \leq \mu^*(T \cap E_n) + \sum_{k=n+1}^{\infty} (T \cap W_k) \tag{$**$}$$

because μ^* is subadditive. At this point, the series sum $\sum_{k=n+1}^{\infty} (T \cap W_k)$ could be ∞; we haven't determined if it is finite yet.

For any $k > 1$, if $x \in W_k$, then $x \in E_k \setminus E_{k-1}$ and so

$$\frac{1}{k} \leq D(x, E^C) \leq \frac{1}{k-1}$$

Next, if $x \in W_k$ and $y \in W_{k+p}$ for any $p \geq 2$, we can use the triangle inequality with an arbitrary $z \in E^C$ to conclude

$$d(x, z) \leq d(x, y) + d(y, z)$$

But, this says

$$d(x,y) \;\geq\; d(x,z) \;-\; d(y,z)$$
$$\geq\; D(x, E^C) \;-\; d(y,z) \;>\; \frac{1}{k} \;-\; d(y,z)$$

We have shown the fundamental inequality

$$d(x,y) \;>\; \frac{1}{k} \;-\; d(y,z), \;\forall\, x \in W_k, \forall\, y \in W_{k+p} \qquad (\pmb{\alpha})$$

holds for $p \geq 2$. The definition of the set E_{k+p} then implies for these p,

$$\frac{1}{k+p} \;<\; D(y, E^C) \;\leq\; \frac{1}{k+p-1} \qquad (\pmb{\beta})$$

Now consider how $D(y, E^C)$ is defined. Since this is an infimum, by the Infimum Tolerance Lemma, given a positive ϵ, there is a $z_\epsilon \in E^C$ so that

$$D(y, E^C) \;\leq\; d(y, z_\epsilon) \;<\; D(y, E^C) + \epsilon$$

Hence, using Equation β, we have

$$-d(y, z_\epsilon) \;>\; -D(y, E^C) - \epsilon$$
$$>\; -\frac{1}{k+p-1} - \epsilon$$

Also, using Equation α, we find

$$d(x,y) \;>\; \frac{1}{k} \;-\; d(y, z_\epsilon)$$
$$>\; \frac{1}{k} \;-\; -\frac{1}{k+p-1} \;-\; \epsilon$$
$$=\; \frac{p-1}{k(k+p-1)} \;-\; \epsilon$$

Since $\epsilon > 0$ is arbitrary, we conclude

$$d(x,y) \;\geq\; \frac{p-1}{k(k+p-1)} \;>\; 0$$

for all $x \in W_k$ and $y \in W_{k+p}$ with $p \geq 2$. Hence,

$$D(W_k, W_{k+p}) \;\geq\; \frac{p-1}{k(k+p-1)} \;>\; 0$$

It follows that

$$D(W_1, W_3) \;>\; 0$$

and, in general, we find this is true for the successive odd integers

$$D(W_{2k+1}, W_{2k+3}) \;>\; 0$$

262   BASIC ANALYSIS IV: MEASURE THEORY AND INTEGRATION

Since μ^* is a metric outer measure, this allows us to say

$$\sum_{k=0}^{n} \mu^*(T \cap W_{2k+1}) \;=\; \mu^*\left(\cup_{k=0}^{n} T \cap W_{2k+1}\right)$$

$$\leq \;\; \mu^*\left(\cup_{k=0}^{\infty} T \cap W_{2k+1}\right) \;\leq\; \mu^*(T)$$

A similar argument shows that successive even integers satisfy

$$D(W_{2k}, W_{2k+2}) \;>\; 0$$

Again, as μ^* is a metric outer measure, this allows us to say

$$\sum_{k=0}^{n} \mu^*(T \cap W_{2k}) \;=\; \mu^*\left(\cup_{k=0}^{n} T \cap W_{2k}\right)$$

Therefore, we have

$$\sum_{k=0}^{n} \mu^*(T \cap W_{2k}) \;\leq\; \mu^*\left(\cup_{k=0}^{\infty} T \cap W_{2k}\right)$$

$$\leq \mu^*(T)$$

We conclude

$$\sum_{k=0}^{n} \mu^*(T \cap W_k) \;=\; \sum_{k \text{ even}} \mu^*(T \cap W_k) + \sum_{k \text{ odd}} \mu^*(T \cap W_k)$$

$$\leq \;\; 2\,\mu^*(T)$$

for all n. This implies the sum $\sum_k \mu^*(T \cap W_k)$ converges to a finite number.

Since the series converges, we now know given $\epsilon > 0$, there is an N so that

$$\sum_{k=n}^{\infty} \mu^*(T \cap W_k) \;<\; \epsilon$$

for all $n > N$. Now go back to Equation $**$. We have for any $n > N$,

$$\mu^*(T \cap E) \;\leq\; \mu^*(T \cap E_n) + \epsilon$$

This tells us $\mu^*(T \cap E_n) \to \mu^*(T \cap E)$. By our earlier remark, this completes the proof. ■

We can even prove more.

Theorem 11.4.2 Open Sets in a Metric Space are μ^* Measurable if and only if μ^* is a Metric Outer Measure

> *Let X be a non empty metric space. Then open sets are μ^* measurable if and only if μ^* is a metric outer measure.*

Proof 11.4.2
If we assume μ^* is a metric outer measure, then opens sets are μ^* measurable by Theorem 11.4.1.

BUILDING MEASURES

On the other hand if we know that all the open sets are μ^* measurable, this implies all Borel sets are μ^* measurable as well. Let A and B be any two sets with $D(A,B) = r > 0$. For each $x \in A$, let

$$G(x) \;=\; \{\, u \mid d(x,u) < r/2 \,\}$$

and

$$G \;=\; \bigcup_{x \in A} G(x)$$

Then G is open, $A \subseteq G$ and $G \cap B = \emptyset$. Since G is measurable, it satisfies the Caratheodory condition using test set $T = A \cup B$; thus,

$$\mu^*\Big(A \cup B\Big) \;=\; \mu^*\Big((A \cup B) \cap G\Big) + \mu^*\Big((A \cup B) \cap G^C\Big)$$

But $(A \cup B) \cap G$ is simplified to A because $A \subseteq B$ and B is disjoint from G. Further since A is disjoint from G^C and $B \subseteq G^C$, we have $(A \cup B) \cap G^C = B$. We conclude

$$\mu^*(A \cup B) \;=\; \mu^*(A) + \mu^*(B)$$

This shows μ^* is a metric outer measure. ∎

Homework

Exercise 11.4.1 *Let $A = [0,1]$, define f by $f(x) = dist(x,A)$ for $x \in \Re$. Prove f is continuous on $\Re$.*

Exercise 11.4.2 *Let A be any subset in $\Re$ and define $f(x) = dist(x,A)$ for all $x \in \Re$. Prove f is continuous on $\Re$.*

Exercise 11.4.3 *Let $F = [0,2]$ and $G = (-1,3)$. Define $f(x) = \frac{d(x,G^C)}{d(x,G^C)+d(x,F)}$ for all $x \in \Re$. Prove f is continuous on $\Re$.*

Exercise 11.4.4 *Let $F = [0,2]$ and $G = (-1,3)$. Compute $D(F,G)$.*

Exercise 11.4.5 *Let $F = [0,2] \times (-3,5]$ and $G = (1,2] \times [-1,1)$. Compute $d(F,G)$.*

The proofs above are really technical, so let's do a few exercises to make sure you have all the messy details down.

Exercise 11.4.6 *The sets E_n are important and so is the decomposition into the annular like sets W_n. For a given open set E in $\Re$, draw the sets E_1 through E_5 and also sketch the sets W_1 through W_5. Of course an open set can be quite complicated so this is just a convenient way to help with how the argument is organized.*

Exercise 11.4.7 *The sets E_n are important and so is the decomposition into the annular like sets W_n. For a given open set E in $\Re^2$, draw the sets E_1 through E_5 and also sketch the sets W_1 through W_5. Of course an open set can be quite complicated so this is just a convenient way to help with how the argument is organized. The pictures are harder in $\Re^2$, of course, but persevere!*

Exercise 11.4.8 *For the open set in $\Re$ case, verify for your example the inequality $D(W_k, W_{k+p}) \geq \frac{p-1}{k(k+p-1)} > 0$ for the indices you have drawn.*

Exercise 11.4.9 *For the open set in $\Re62$ case, verify for your example the inequality $D(W_k, W_{k+p}) \geq \frac{p-1}{k(k+p-1)} > 0$ for the indices you have drawn.*

264   *BASIC ANALYSIS IV: MEASURE THEORY AND INTEGRATION*

Exercise 11.4.10 *If μ^* is a metric outer measure, are singleton sets $\{x\}$ μ^* measurable?*

11.5   Constructing Outer Measure

We still have to find ways to construct outer measures. We want the resulting OMI measure we induce to have certain properties useful to us. Let's discuss how to do this now.

Definition 11.5.1 Premeasures and Covering Families

Let X be a nonempty set. Let $\mathscr{T}$ be a family of subsets of X that contains the empty set. This family is called a covering family for X. Let τ be a mapping on $\mathscr{T}$ so that $\tau(\emptyset) = 0$. The mapping τ is called a premeasure.

It is hard to believe, but even with virtually no restrictions on τ and $\mathscr{T}$, we can build an outer measure.

Theorem 11.5.1 Constructing Outer Measures via Premeasures

Let X be a nonempty set. Let $\mathscr{T}$ be a covering family of subsets of X and $\tau : \mathscr{T} \to [0, \infty]$ be a premeasure. For any A in X, define

$$\mu^*(A) \;\; = \;\; \inf\left\{ \sum_n \tau(T_n) \,|\, T_n \in \mathscr{T}, \, A \subseteq \cup_n T_n \right\}$$

where the sequence of sets (T_n) from $\mathscr{T}$ is finite or countably infinite. Such a sequence is called a cover. In the case where there are no sets from $\mathscr{T}$ that cover A, we define the infimum over the resulting empty set to be ∞. Then μ^* is an outer measure on X.

Proof 11.5.1
To verify the mapping μ^* is an outer measure on X, we must show:

(i): $\mu^*(\emptyset) = 0$.

(ii): If A and B are subsets of X with $A \subseteq B$, then $\mu^*(A) \leq \mu^*(B)$.

(iii): If (A_n) is a sequence of disjoint subsets of X, then $\mu^*(\cup_n A_n) \leq \sum_n \mu^*(A_n)$.

It is straightforward to see conditions (i) and (ii) are true. It suffices to prove condition (iii) is valid. Let (A_n) be a countable collection, finite or infinite, of subsets of X. If there is an index n with $\tau(A_n)$ infinite, then since $\mu^*(\cup_n A_n) \leq \infty$ anyway, it is clear

$$\mu^*(\cup_{i=1}^{\infty} A_i) \;\; \leq \;\; \sum_{i=1}^{\infty} \mu^*(A_i) \;\; = \;\; \infty$$

On the other hand, if $\mu^*(A_n)$ is finite for all n, given any $\epsilon > 0$, we can use the Infimum Tolerance Lemma to find a sequence of families $(T_{n\,k})$ in $\mathscr{T}$ so that

$$\sum_{k=1}^{\infty} \tau(T_{n\,k}) \;\; < \;\; \mu^*(A_n) + \frac{\epsilon}{2^n}$$

We also know that

$$\bigcup_{n=1}^{\infty} A_n \;\; \subseteq \;\; \bigcup_{n=1}^{\infty} \bigcup_{k=1}^{\infty} T_{n\,k}$$

BUILDING MEASURES

Hence, the collection $\cup_n \cup_k T_{n\,k}$ is a covering family for $\cup_n A_n$ and so by the definition of μ^*, we have

$$\mu^*\left(\bigcup_{n=1}^{\infty} A_n\right) \leq \sum_{n=1}^{\infty} \sum_{k=1}^{\infty} \mu^*\left(T_{n\,k}\right)$$

$$\leq \sum_{n=1}^{\infty} \{\mu^*(A_n) + \frac{\epsilon}{2^n}\}$$

$$\leq \sum_{n=1}^{\infty} \mu^*(A_n) + \epsilon$$

Since ϵ is arbitrary, we see μ^* is countable subadditive and so is an outer measure. ■

Homework

Exercise 11.5.1 *Let $X = \{1.\,2,\,3\}$ and $\mathcal{T}$ consist of $\emptyset$, X and all doubleton subsets $\{x,\,y\}$ of X. Let τ satisfy:*

(i): $\tau(\emptyset) = 0$.

(ii): $\tau\left(\{x,\,y\}\right) = 1$ *for all* $x \neq y$ *in* X.

(iii): $\tau(X) = 2$.

Use the method of Theorem 11.5.1 to construct the outer measure μ^* defined by $\mu^*(\emptyset) = 0$, $\mu^*(X) = 2$ and $\mu^*(A) = 1$ for any other subset A of X.

Exercise 11.5.2 *Let $X = \{1.\,2,\,3\}$ and $\mathcal{T}$ consist of $\emptyset$, X and all doubleton subsets $\{x,\,y\}$ of X. Let τ satisfy:*

(i): $\tau(\emptyset) = 0$.

(ii): $\tau\left(\{x,\,y\}\right) = 1$ *for all* $x \neq y$ *in* X.

(iii): $\tau(X) = 3$.

Prove the method of Theorem 11.5.1 to construct an outer measure like the previous problem. What changes because we altered $\tau(X)$ from 2 to 3?

Exercise 11.5.3 *Let X be the natural numbers $\mathbb{N}$ and let τ consist of $\emptyset$, $\mathbb{N}$ and all singleton sets. Define $\tau(\emptyset) = 0$ and $\tau(\{x\}) = 1$ for all x in $\mathbb{N}$. Let $\tau(\mathbb{N}) = 2$. Construct the outer measure μ^* and determine the family of measurable sets (i.e. the sets that satisfy the Caratheodory Condition).*

Exercise 11.5.4 *Let X be the natural numbers $\mathbb{N}$ and let τ consist of $\emptyset$, $\mathbb{N}$ and all singleton sets. Define $\tau(\emptyset) = 0$ and $\tau(\{x\}) = 1$ for all x in $\mathbb{N}$. Now let $\tau(\mathbb{N}) = \infty$. Construct the outer measure μ^* and determine the family of measurable sets (i.e. the sets that satisfy the Caratheodory Condition).*

Exercise 11.5.5 *Let X be the natural numbers $\mathbb{N}$ and let τ consist of $\emptyset$, $\mathbb{N}$ and all singleton sets. Define $\tau(\emptyset) = 0$ and $\tau(\{x\}) = 1$ for all x in $\mathbb{N}$. Let $\tau(\mathbb{N}) = 2$ and set $\tau(\{x\}) = 2^{-(x-1)}$. Construct the outer measure μ^* and determine the family of measurable sets (i.e. the sets that satisfy the Caratheodory Condition).*

Exercise 11.5.6 *Let X be the natural numbers $\mathbb{N}$ and let τ consist of $\emptyset$, $\mathbb{N}$ and all singleton sets. Define $\tau(\emptyset) = 0$ and $\tau(\{x\}) = 1$ for all x in $\mathbb{N}$. Let $\tau(\mathbb{N}) = \infty$ and set $\tau(\{x\}) = 2^{-(x-1)}$.*

266   BASIC ANALYSIS IV: MEASURE THEORY AND INTEGRATION

Construct the outer measure μ^* and determine the family of measurable sets (i.e. the sets that satisfy the Caratheodory Condition). You should see $\mathbb{N}$ is measurable but $\tau(\mathbb{N}) \neq \mu(\mathbb{N})$, where μ denotes the measure constructed in the process of Part (a).

Exercise 11.5.7 *Let X be the natural numbers $\mathbb{N}$ and let τ consist of $\emptyset$, $\mathbb{N}$ and all singleton sets. Define $\tau(\emptyset) = 0$ and $\tau(\{x\}) = 1$ for all x in $\mathbb{N}$. Let $\tau(\mathbb{N}) = 1$ and set $\tau(\{x\}) = 2^{-(x-1)}$. Construct the outer measure μ^* and determine the family of measurable sets (i.e. the sets that satisfy the Caratheodory Condition).*

There is so little known about τ and $\mathscr{T}$, that it is not clear at all that:

(i): $\mathscr{T} \subseteq \mathcal{M}$, where $\mathcal{M}$ is the σ-algebra of sets that satisfy the Caratheodory condition for the outer measure μ^* generated by τ. If this is true, we will call $\mathcal{M}$ an OMI-F σ-algebra, where the "F" denotes the fact that the covering family is in $\mathcal{M}$.

(ii): If $A \in \mathscr{T}$, then $\tau(A) = \mu(A)$ where μ is the measure obtained by restricting μ^* to $\mathcal{M}$. If this is true, we will call the constructed σ-algebra, an OMI-FE σ-algebra, where the "E" indicates the fact the μ restricted to $\mathscr{T}$ recovers τ.

If τ represents some primitive notion of size of special sets, like length of intervals on the real line, we normally want both condition (i) and (ii) above to be valid. We can obtain these results if we add a few more properties to τ and $\mathscr{T}$. First, $\mathscr{T}$ needs to be an algebra (which we have already defined) and τ needs to be additive on the algebra.

Definition 11.5.2 Additive Set Function

> *Let $\mathcal{A}$ of subsets of the set X be an algebra. Let ν be an extended real-valued function defined on $\mathcal{A}$ which satisfies:*
>
> *(i): $\nu(\emptyset) = 0$.*
>
> *(ii): If A and B in $\mathcal{A}$ are disjoint, then $\nu(A \cup B) = \nu(A) + \nu(B)$.*
>
> *Then ν is called an additive set function on $\mathcal{A}$.*

We also need a property of outer measures called regularity.

Definition 11.5.3 Regular Outer Measures

> *Let X be a nonempty set, μ^* be an outer measure on X and $\mathcal{M}$ be the set of all μ^* measurable sets of X. The outer measure μ^* is called regular if for all E in X there is a μ^* measurable $F \in \mathcal{M}$ so that $E \subseteq F$ with $\mu^*(E) = \mu(F)$, where μ is the measure induced by μ^* on $\mathcal{M}$. The set F is often called a measurable cover for E.*

We begin with a technical lemma.

Lemma 11.5.2 Condition for Outer Measure to be Regular

> *Let X be a nonempty set, $\mathscr{T}$ a covering family and τ a premeasure. Then if the σ-algebra, $\mathcal{M}$, generated by τ using $\mathscr{T}$ contains $\mathscr{T}$, μ^* is regular.*

Proof 11.5.2
Let A be a subset in X. We need to show there is measurable set B containing A so that $\mu^*(A) = \mu(B)$. If the $\mu^*(A) = \infty$, then we can choose X as the needed set. Otherwise, we have $\mu^*(A)$ is finite. Applying the Infimum Tolerance Lemma, for each m, there is a family of sets (E_n^m) so that

$$\text{BUILDING MEASURES} \qquad 267$$

$A \subseteq \cup_n E_n^m$ and

$$\sum_n \tau(E_n^m) \;<\; \mu^*(A) + \frac{1}{m}$$

Let

$$E_m \;=\; \bigcup_n E_n^m$$

$$H \;=\; \bigcap_m E_m$$

these sets are measurable by assumption. Also, $A \subseteq H$ and $H \subseteq E_m$. Hence, $\mu^*(A) \le \mu(H)$. We now show the reverse inequality. For each m, we have

$$\mu^*(E_m) \;\le\; \sum_n \mu^*(E_n^m) \le \sum_n \tau(E_n^m)$$

$$\le\; \mu^*(A) + \frac{1}{m}$$

Further, since $H \subseteq E_m$ for each m, we find

$$\mu(H) \;\le\; \mu^*(E_m) \le \mu^*(A) + \frac{1}{m}$$

This is true for all m; hence, it follows that $\mu(H) \le \mu^*(A)$. Combining inequalities, we have $\mu(H) = \mu^*(A)$ and so H is a measurable cover. Thus, μ^* is regular. ∎

Homework

Exercise 11.5.8 *Let $\mathcal{A}$ be the algebra of subsets of $(0,1]$ which consists of the empty set and all finite unions of half open intervals of the form $(a,b]$ from $(0,1]$. In a previous exercise, we showed this is an algebra. Let f be an arbitrary function on $[0,1]$. Define ν_f on $\mathcal{A}$ by*

$$\nu_f\Big((a,b]\Big) \;=\; f(b) - f(a)$$

Extend ν_f to be additive on finite disjoint intervals as follows: if $(A_i) = ((a_i, b_i])$ is a finite collection of disjoint intervals of $(0,1]$, we define

$$\nu_f\Big(\cup_{i=1}^n (a_i, b_i]\Big) \;=\; \sum_{i=1}^n f(b_i) - f(a_i)$$

1. *Prove that ν_f is additive on $\mathcal{A}$.*

    Hint 11.5.1 *It is enough to show that the value of $\nu_f(A)$ is independent of the way in which we write A as a finite disjoint union.*

2. *Prove ν_f is non-negative if and only if f is non-decreasing.*

Exercise 11.5.9 *If λ is an additive set function on an algebra of subsets $\mathcal{A}$, prove that λ cannot take on both the value ∞ and $-\infty$.*

Hint 11.5.2 *If there is a set A in the algebra with $\lambda(A) = \infty$ and there is a set B in the algebra with $\lambda(B) = -\infty$, then we can find disjoint sets A' and B' in $\mathcal{A}$ so that $\lambda(A') = \infty$ and $\lambda(B') = -\infty$.*

268   *BASIC ANALYSIS IV: MEASURE THEORY AND INTEGRATION*

But this is not permitted as the value of $\lambda(A' \cup B')$ must be a well-defined extended real value not the undefined value $\infty - \infty$.

Theorem 11.5.3 Conditions for OMI-F Measures

Let X be a nonempty set, $\mathscr{T}$ a covering family which is an algebra and τ an additive set function on $\mathscr{T}$. Then the σ-algebra, $\mathcal{M}$, generated by τ using $\mathscr{T}$ contains $\mathscr{T}$ and μ^* is regular.

Proof 11.5.3

By Lemma 11.5.2, it is enough to show each member of $\mathscr{T}$ is measurable. So, let A be in $\mathscr{T}$. As usual, it suffices to show that

$$\mu^*(T) \;\geq\; \mu^*(T \cap A) + \mu^*(T \cap A^C)$$

for all sets T of finite outer measure. This will show A satisfies the Caratheodory condition and hence, is measurable. Let $\epsilon > 0$ be given. By the Infimum Tolerance Lemma, there is a family (A_n) from $\mathscr{T}$ so that $T \subseteq \cup_n A_n$ and

$$\sum_n \tau(A_n) \;<\; \mu^*(T) + \epsilon$$

Since τ is additive on $\mathscr{T}$, we know

$$\tau(A_n) \;=\; \tau(A \cap A_n) + \tau(A^C \cap A_n)$$

Also, we have

$$A \bigcap T \;\subseteq\; \bigcup_n (A \cap A_n), \;\; and \;\; A^C \bigcap T \subseteq \bigcup_n (A^C \cap A_n)$$

Hence,

$$\mu^*(A \cap T) \;\leq\; \sum_n \mu^*(A \cap A_n), \; \mu^*(A^C \cap T) \leq \sum_n \mu^*(A^C \cap A_n). \qquad (\alpha)$$

$$\begin{aligned}
\mu^*(T) + \epsilon \;>\; &\sum_n \tau(A_n) = \sum_n \tau(A_n \cap A) + \sum_n \tau(A_n \cap A^C) \\
\geq\; &\sum_n \mu^*(A_n \cap A) + \sum_n \mu^*(A_n \cap A^C) \\
\geq\; &\mu^*(A \cap T) + \mu^*(A^C \cap T)
\end{aligned}$$

by Equation $\boldsymbol{\alpha}$. Thus, A satisfies the Caratheodory condition and is measurable.   ∎

In order for condition (ii) to hold, we need to add one more additional property to τ: it needs to be a pseudo-measure.

Definition 11.5.4 Pseudo-Measure

BUILDING MEASURES

269

> *Let the mapping $\tau : \mathcal{A} \to [0, \infty]$ be additive on the algebra $\mathcal{A}$. Assume whenever (A_i) is a countable collection of disjoint sets in $\mathcal{A}$ whose union is also in $\mathcal{A}$ (note this is not always true because $\mathcal{A}$ is not a σ-algebra), then it is true that*
>
> $$\tau(\cup_i A_i) \;=\; \sum_i \tau(A_i)$$
>
> *Such a mapping τ is called a pseudo-measure on $\mathcal{A}$.*

Theorem 11.5.4 Conditions for OMI-FE Measures

> *Let X be a nonempty set, $\mathcal{T}$ a covering family which is an algebra and τ an additive set function on $\mathcal{T}$ which is a pseudo-measure. Then the σ-algebra, $\mathcal{M}$, generated by τ using $\mathcal{T}$ contains $\mathcal{T}$, μ^* is regular and $\mu(T) = \tau(T)$ for all T in $\mathcal{T}$.*

Proof 11.5.4

A typical proof is shown in (A. Bruckner and J. Bruckner and B. Thomson (1) 1997) and you should look at it. We have done enough with this for now though, and we will leave it to you to do the requisite reading. ∎

Comment 11.5.1 *The results above tell us that we can construct measures so that $\mathcal{T}$ is contained in $\mathcal{M}$ and the measure recovers τ as long as the premeasure is a pseudo-measure and the covering family is an algebra. This means the covering family must be closed under complementation. Hence, if we a covering family such as the collection of all open intervals (which we do when we construct Lebesgue measure later) these theorems do not apply.*

Homework

We will now work out some examples using exercises. We need you to be engaged! Note the covering families here do not simply contain open intervals!

Exercise 11.5.10 *Let $\mathcal{U}$ be the family of subsets of $\Re$ of the form $(a, b]$, $(-\infty, b]$, (a, ∞) and $(-\infty, \infty)$ and the empty set.*

- *Prove $\mathcal{F}$, the collection of all finite unions of sets from $\mathcal{U}$, is an algebra of subsets of $\Re$.*

- *Let τ be the usual length of an interval and extend τ to $\mathcal{F}$ additively. This extended τ is a premeasure on $\mathcal{F}$ which can then be used to define an outer measure as usual $\mu^*(\tau)$. There is then an associated σ-algebra of μ_τ^* measurable sets of $\Re$, $\mathcal{M}_\tau$ and μ_τ^*, restricted to $\mathcal{M}_\tau$, is a measure μ_τ.*

- *You will now prove $\mathcal{F}$ is contained in $\mathcal{M}_\tau$.*

  1. *Let's consider the set I from $\mathcal{U}$. Let T be any subset of $\Re$ and let $\epsilon > 0$ be given. Explain why it is enough to consider sets T with $\mu^*(F)$ finite.*

  2. *Prove there is then a cover (A_n) of sets from the algebra $\mathcal{F}$ so that*

     $$\sum_n \tau(A_n) \;\leq\; \mu_\tau^*(T) \;+\; \epsilon$$

  3. *Prove $I \cap T \subseteq \cup_n (A_n \cap I)$ and $I^C \cap T \subseteq \cup_n (A_n \cap I^C)$.*

  4. *So because $\mathcal{F}$ is an algebra, prove this means $(A_n \cap I)$ covers $I \cap T$ and $(A_n \cap I^C)$ covers $I^C \cap T$.*

270   BASIC ANALYSIS IV: MEASURE THEORY AND INTEGRATION

5. *Prove it then follows that*

$$\mu_\tau^*(T \cap I) \ \leq \ \sum_n \tau(A_n \cap I),$$

$$\mu_\tau^*(T \cap I^C) \ \leq \ \sum_n \tau(A_n \cap I^C)$$

6. *Combining, note you can prove*

$$\mu_\tau^*(T \cap I) + \mu_\tau^*(T \cap I^C) \ \leq \ \sum_n \left(\tau(A_n \cap I) + \tau(A_n \cap I^C) \right)$$

7. *But τ is additive on $\mathcal{F}$, and hence you can show*

$$\sum_n \left(\tau(A_n \cap I) + \tau(A_n \cap I^C) \right) \ = \ \sum_n \tau(A_n)$$

8. *Thus, you can prove*

$$\mu_\tau^*(T \cap I) + \mu_\tau^*(T \cap I^C) \ \leq \ \mu_\tau^*(T) + \epsilon$$

9. *Since $\epsilon > 0$ is arbitrary, prove this shows I satisfies the Caratheodory condition. This shows that I is OMI measurable and so $\mathcal{F} \subseteq \mathcal{M}_\tau$.*

Exercise 11.5.11 *Let $\mathcal{U}$ be the family of subsets of $\Re$ of the form $(a, b]$, $(-\infty, b]$, (a, ∞) and $(-\infty, \infty)$ and the empty set.*

- *Prove $\mathcal{F}$, the collection of all finite unions of sets from $\mathcal{U}$ is an algebra of subsets of $\Re$.*

- *Let g be the monotone increasing function on $\Re$ defined by $g(x) = x^3$. Note g is right continuous which means*

$$\lim_{h \to 0^+} g(x + h) \ exists \ , \ \forall \, x, \quad \lim_{x \to -\infty} g(x) \ exists, \quad \lim_{x \to \infty} g(x) \ exists$$

where the last two limits are $-\infty$ and ∞ respectively. Define the mapping τ_g on $\mathcal{U}$ by

$$\tau_g\left((a, b] \right) \ = \ g(b) - g(a),$$

$$\tau_g\left((-\infty, b) \right) \ = \ g(b) - \lim_{x \to -\infty} g(x),$$

$$\tau_g\left((a, \infty) \right) \ = \ \lim_{x \to \infty} g(x) - g(a),$$

$$\tau_g\left((-\infty, \infty) \right) \ = \ \lim_{x \to \infty} g(x) - \lim_{x \to -\infty} g(x)$$

Extend τ_g to $\mathcal{F}$ additively as usual. This extended τ_g is a premeasure on $\mathcal{F}$. τ_g can then be used to define an outer measure as usual $\mu^*(g)$. There is then an associated σ-algebra of μ_g^* measurable sets of $\Re$, $\mathcal{M}_g$, and μ_g^* restricted to $\mathcal{M}_g$ is a measure, μ_g. We will now prove $\mathcal{F}$ is contained in $\mathcal{M}_g$.

1. *Consider the set I from $\mathcal{U}$. Let T be any subset of $\Re$ and let $\epsilon > 0$ be given.*

2. *Explain why it is enough to consider sets T with $\mu^*(F)$ finite.*

BUILDING MEASURES

3. Prove there is a cover (A_n) of sets from the algebra $\mathcal{F}$ so that

$$\sum_n \tau_g(A_n) \ \le \ \mu_g^*(T) + \epsilon$$

4. Now $I \cap T \subseteq \cup_n (A_n \cap I)$ and $I^C \cap T \subseteq \cup_n (A_n \cap I^C)$. So

$$\mu_g^*(T \cap I) \ \le \ \sum_n \tau_g(A_n \cap I),$$

$$\mu_g^*(T \cap I^C) \ \le \ \sum_n \tau_g(A_n \cap I^C)$$

Combining, prove

$$\mu_g^*(T \cap I) + \mu_g^*(T \cap I^C) \ \le \ \sum_n \left(\tau_g(A_n \cap I) + \tau_g(A_n \cap I^C) \right)$$

5. But τ_g is additive on $\mathcal{F}$, and hence prove

$$\sum_n \left(\tau_g(A_n \cap I) + \tau_g(A_n \cap I^C) \right) \ = \ \sum_n \tau_g(A_n)$$

6. Thus, you can prove

$$\mu_g^*(T \cap I) + \mu_g^*(T \cap I^C) \ \le \ \mu_g^*(T) + \epsilon$$

7. Since $\epsilon > 0$ is arbitrary, we have shown I satisfies the Caratheodory condition. This shows that I is OMI measurable and so $\mathcal{F} \subseteq \mathcal{M}_g$.

Exercise 11.5.12 *Do the same thing as in the last exercise, but use $g(x) = x$.*

Exercise 11.5.13 *Do the same thing as in the last exercise, but use $g(x) = x^5$.*

Exercise 11.5.14 *What changes, if anything, in the last two exercises if we use $g(x) = \tanh(x)$?*

Chapter 12

Lebesgue Measure

We will now construct Lebesgue measure on $\Re^k$. We will begin by defining the mapping μ^* on the subsets of $\Re^k$ which will turn out to be an outer measure. The σ-algebra of subsets that satisfy the Caratheodory condition will be called the σ-algebra of Lebesgue measurable subsets. We will denote this σ-algebra by $\mathcal{M}$ as usual. We will usually be able to tell from context what σ-algebra of subsets we are working with in a given study area or problem. The primary references here are again (A. Bruckner and J. Bruckner and B. Thomson (1) 1997). and (Taylor (14) 1985). We like the development of Lebesgue measure in (Taylor (14) 1985) better than that of (A. Bruckner and J. Bruckner and B. Thomson (1) 1997) and so our coverage reflects that. In all cases, we have added more detail to the proofs of propositions to help you build your analysis skills by looking hard at many interesting and varied proof techniques.

12.1 Lebesgue Outer Measure

We will be working in $\Re^k$ for any positive integer k. We have to work our way through a fair bit of definitional material; so be patient while we set the stage. We let $x = (x_1, x_2, \ldots, x_k)$ denote a point in the Euclidean space $\Re^k$. An open interval in $\Re^k$ will be denoted by I and it is determined by the cross - product of k intervals of the form (a_i, b_i) where each a_i and b_i is a finite real number. Hence, the interval I has the form

$$I = \Pi_{i=1}^k (a_i, b_i)$$

The interval (a_i, b_i) is called the ith edge of I and the number $\ell_i = b_i - a_i$ is the length of the ith edge. The **content** of the open interval I is the product of the edge lengths and is denoted by $|I|$; i.e.

$$|I| = \Pi_{i=1}^k \left(b_i - a_i \right)$$

We need additional terminology. The center of I is the point

$$p = \left(\frac{a_1 + b_1}{2}, \frac{a_2 + b_2}{2}, \ldots, \frac{a_k + b_k}{2} \right)$$

if the interval J has the same center as the interval I, we say the intervals are concentric.

If I and J are intervals, for convenience of notation, let ℓ_J and ℓ_I denote the vector of edge lengths of J and I, respectively. In general, there is no relationship between ℓ_J and ℓ_I. However, there is a special case of interest. We note that if J is concentric with I and each edge in ℓ_J is a fixed multiple of the corresponding edge length in ℓ_I, we can say $\ell_J = \lambda \, \ell_I$ for some constant λ. In this case, we

273

274   BASIC ANALYSIS IV: MEASURE THEORY AND INTEGRATION

write $J = \lambda I$. It then follows that $|J| = \lambda^k |I|$.

We are now ready to define outer measure on $\Re^k$. Following Definition 11.5.1, we define a suitable covering family $\mathscr{T}$ and premeasure τ. Then, the mapping μ^* defined in Theorem 11.5.1 will be an outer measure. For ease of exposition, let's define this here.

Definition 12.1.1 Lebesgue Outer Measure

> Let $\mathscr{T}$ be the collection of all open intervals in $\Re^k$ and define the premeasure τ by $\tau(I) = |I|$ for all I in $\mathscr{T}$. For any A in X, define
>
> $$\mu^*(A) \;=\; \inf \left\{ \sum_n |I_n| \,\big|\, I_n \in \mathscr{T}, \, A \subseteq \cup_n I_n \right\}$$
>
> We will call a collection (I_n) whose union contains A a Lebesgue Cover of A.

Then, μ^* is an outer measure on $\Re^k$ and as such induces a measure through the usual Caratheodory condition route. It remains to find its properties. **The covering family here is not an algebra**, so **we cannot use** Theorem 11.5.3 and Theorem 11.5.4 to conclude:

(i): $\mathscr{T} \subseteq \mathcal{M}$; i.e. $\mathcal{M}$ is an OMI-F σ-algebra.

(ii): If $A \in \mathscr{T}$, then $|A| = \mu(A)$; i.e. $\mathcal{M}$ is an OMI-FE σ-algebra.

However, we will be able to alter our original proofs to get these results with just a little work.

Homework

Let's figure out some basic stuff in the exercises below.

Exercise 12.1.1 *Prove if I is an open interval in $\Re^k$, $\mu^*(I) \leq |I|$; this is true because I covers itself.*

Exercise 12.1.2 *Prove if $\{x\}$ is a singleton set, $\mu^*(\{x\}) = 0$. Do this by choosing any open interval I that has x as its center. Then, I is a cover of $\{x\}$ and so $\mu^*(\{x\}) \leq |I|$. Now use the concentric intervals $1/2^n \, I$ to finish the proof.*

Exercise 12.1.3 *Prove if E is a finite set, then $\mu^*(E) = 0$. Note these arguments are essentially the same as the ones we use when we talked about a set being content zero when we first discussed Riemann integration in (Peterson (8) 2020).*

Exercise 12.1.4 *If E is countable, label its points by (a_n). Let $\epsilon > 0$ be given. Then by the Infimum Tolerance Lemma, there are intervals I_n having a_n as a center so that $|I_n| < \epsilon/2^n$. Then the intervals (I_n) cover E and by definition,*

$$\mu^*(E) \leq \sum_n |I_n| \leq \sum_n \epsilon/2^n \;=\; \epsilon$$

Use this to prove $\mu^*(E) = 0$.

We want to see if $\mu^*(I) = |I|$. This is not clear since our covering family is not an algebra. We now need a technical lemma.

Lemma 12.1.1 Sums over Finite Lebesgue Covers of $\overline{I}$ Dominate $|I|$

> Let I be any interval of $\Re^k$ and let $(I_1, \ldots, I_N)$ be any finite Lebesgue cover of $\overline{I}$. Then
> $$\sum_{n=1}^{N} |I_n| \geq |I|$$

The proof is based on an algorithm that cycles through the covering sets I_i one by one and picks out certain relevant subintervals. We can motivate this by looking at an interval I in $\Re^2$ whose closure is covered by 3 overlapping intervals I_1, I_2 and I_3. This is shown in Figure 12.1. We do not attempt to indicate the closure of I in this figure nor the fact that the intervals I_1 and so forth are open. We simply draw boxes and you can easily remove or add edges in your mind to open an interval or close it.

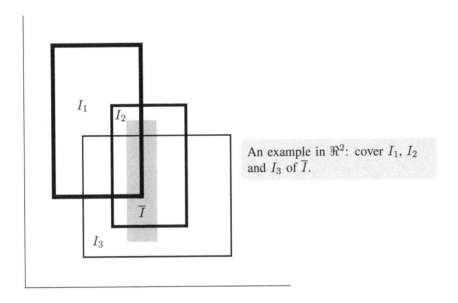

An example in $\Re^2$: cover I_1, I_2 and I_3 of $\overline{I}$.

Figure 12.1: Motivational Lebesgue cover.

These four intervals all have endpoints on both the x and y axes. If we draw all the possible constant x and constant y lines corresponding to these endpoints, we subdivide the original four intervals into many smaller intervals as shown in Figure 12.2.

In particular, if we looked at interval I_1, it is divided into 16 subintervals (J_1, i), for $1 \leq i \leq 16$ as shown in Figure 12.3.

These rectangles are all disjoint and

$$\overline{I}_1 = \bigcup_{i=1}^{16} \overline{J}_{1,i}$$

although we won't show it in a figure, I_2 and I_3 are also sliced up into smaller intervals; using the same left to right and then downward labeling scheme that we used for I_1, we have

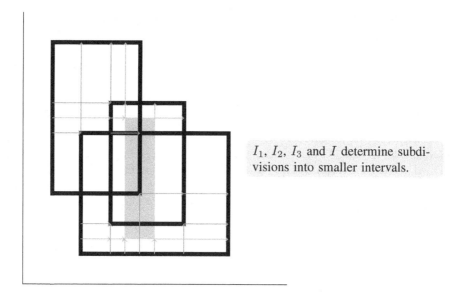

I_1, I_2, I_3 and I determine subdivisions into smaller intervals.

Figure 12.2: Subdivided Lebesgue cover.

- I_2 is divided by 4 horizontal and 4 vertical lines into 16 disjoint subintervals, $J_{2,1}$ to $J_{2,16}$. Further,

$$\overline{I}_2 = \bigcup_{i=1}^{16} \overline{J}_{2,i}$$

- I_3 is divided by 4 horizontal and 6 vertical lines into 24 disjoint subintervals, $J_{3,1}$ to $J_{3,24}$. We thus know

$$\overline{I}_3 = \bigcup_{i=1}^{24} \overline{J}_{3,i}$$

Finally, I is also subdivided into subintervals: it is divided by 4 horizontal and 2 vertical lines into 8 disjoint subintervals, J_1 to J_8 and

$$\overline{I} = \bigcup_{i=1}^{8} \overline{J}_i$$

We also know

$$|I| = \sum_{i=1}^{8} |J_i|, \quad |I_1| = \sum_{i=1}^{16} |J_{1,i}|,$$
$$|I_2| = \sum_{i=1}^{16} |J_{2,i}|, \quad |I_3| = \sum_{i=1}^{24} |J_{3,i}|$$

Now look at Figure 12.2 and you see immediately that the intervals J_{kj} and J_{pq} are either the same or are disjoint. For example, the subintervals match when interval I_2 and I_3 overlap. We can conclude each J_i is disjoint from a J_{kj} or it equals J_{kj} for some choice of k and j. Here is the algorithm we

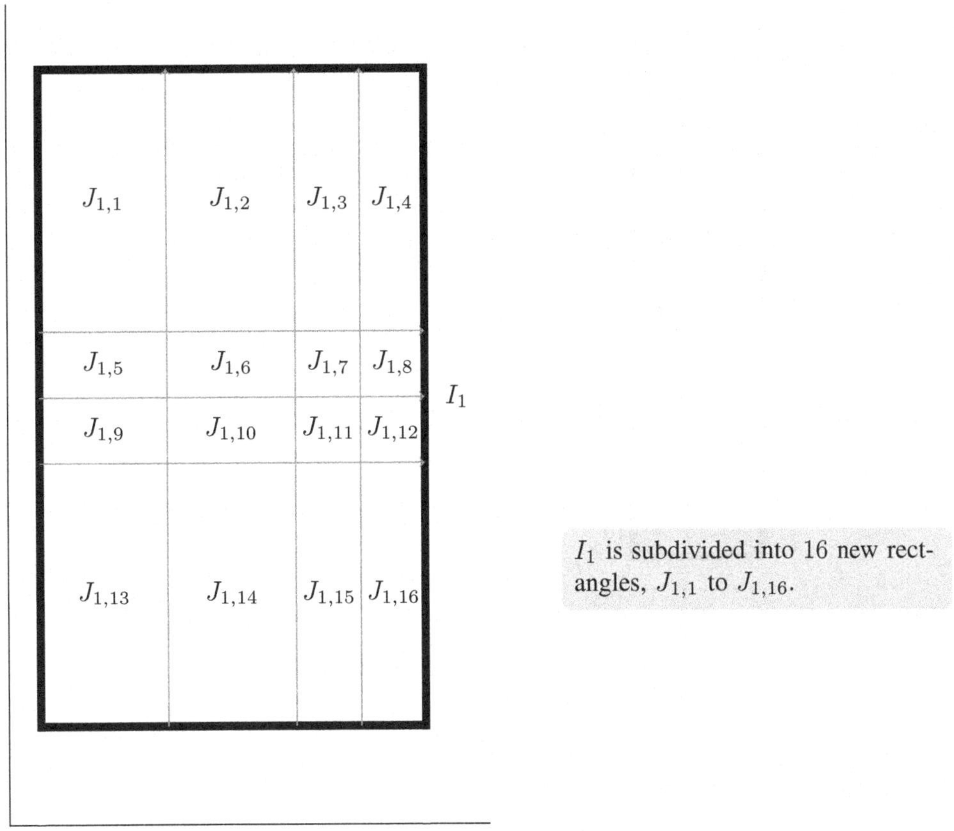

LEBESGUE MEASURE

I_1 is subdivided into 16 new rectangles, $J_{1,1}$ to $J_{1,16}$.

Figure 12.3: Subdivided I_1.

want to use:

Step 1: We know $I \subseteq I_1 \cup I_2 \cup I_3$ and $J_1 = J_{n_1,q_1}$ where n_1 is the smallest index from 1, 2 or 3 which works. For this fixed n_1, consider the collection

$$S_{n_1} = \{J_{n_1,1}, \ldots, J_{n_1,p(n_1)}\}$$

where we are using the symbol $p(n_1)$ to denote the number of subintervals for I_{n_1}. Thus, $p(1) = p(2) = 16$ and $p(3) = 24$ in our example. In our example, we find $n_1 = 1$ and

$$\begin{aligned} J_1 &= J_{1,12} \\ S_1 &= \{J_{1,1}, \ldots, J_{1,16}\} \end{aligned}$$

Look at Figure 12.4 to see what we have done so far.

By referring to Figure 12.2, you can see $J_1 = J_{1,12}$ and $J_3 = J_{1,16}$. Now, let

$$T_{n_1} \equiv T_1 = \{i \mid \exists k \ni J_i = J_{n_1,k}\}$$

Here $T_1 = \{1, 3\}$. Also, let

$$U_{n_1} \equiv U_1 = \{k \mid \exists i \ni J_{n_1,k} = J_i\}$$

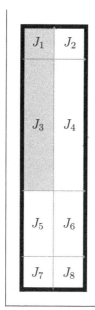

Figure 12.4: The part of I covered by I_1.

We see $U_1 = \{12, 16\}$.

Step 2: Now look at the indices

$$\begin{aligned} V_{n_1} \equiv V_1 &= \{1,2,3,\ldots,8\} \setminus T_1 \\ &= \{2,4,5,6,7,8\} \end{aligned}$$

The smallest index in this set is 2. Next, find the smallest index n_2 so that

$$J_2 = J_{n_2,k}$$

for some index k. From Figure 12.2, we see both I_2 and I_3 intersect $I \setminus I_1$. The smallest index n_2 is thus $n_2 = 2$. The index k that works is 7 and so $J_2 = J_{2,7}$. In figure 12.5, we have now shaded the part of I not in I_1 that lies in I_2.

We can see that $J_2 = J_{2,7}$, $J_4 = J_{2,11}$, $J_5 = J_{2,14}$ and $J_6 = J_{2,15}$. Let

$$T_{n_2} \equiv T_2 = \{i \in V_1 \mid \exists k \ni J_i = J_{n_2,k}\}$$

Here $T_2 = \{2, 4, 5, 6\}$. Also, let

$$U_{n_2} \equiv U_2 = \{k \mid \exists i \ni J_{n_2,k} = J_i\}$$

We see $U_1 = \{7, 11, 14, 15\}$.

Step 3: Now look at the indices

$$\begin{aligned} V_{n_2} \equiv V_2 &= \{1,2,3,\ldots,8\} \setminus (T_1 \cup T_2) \\ &= \{7,8\} \end{aligned}$$

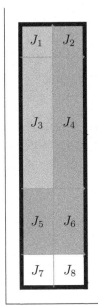

Figure 12.5: The part of I covered by I_1 and I_2.

The smallest index in this set is 7. Next, find the smallest index n_3 so that

$$J_7 = J_{n_3,k}$$

for some index k. From Figure 12.2, we see both I_2 and I_3 intersect $I \setminus (I_1 \cup I_2)$. The smallest index n_3 must be 3 and so $n_3 = 3$. The index k that works now is 15 and we have $J_7 = J_{3,15}$. In figure 12.6, we have now shaded the part of I not in $I_1 \cup I_2$ that lies in I_3.

In fact, we have $J_7 = J_{3,15}$ and $J_8 = J_{3,16}$. Thus, we set

$$\begin{aligned} T_{n_3} \equiv T_3 &= \{i \in V_2 \mid \exists k \ni J_i = J_{n_3,k}\} \\ &= \{7, 8\} \end{aligned}$$

Also, we let

$$U_{n_3} \equiv U_3 = \{k \mid \exists i \ni J_{n_3,k} = J_i\}$$

We see $U_1 = \{15, 16\}$.

We have now expressed each J_i as some $J_{n_1,k}$ through $J_{n_3,k}$. We are now ready to finish our argument.

Step 4: We have

$$\begin{aligned} \{1, \ldots, 8\} &= T_{n_1} \cup T_{n_2} \cup T_{n_3} \\ &= T_1 \cup T_2 \cup T_3 \end{aligned}$$

Thus,

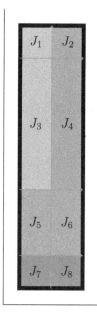

Figure 12.6: The part of I covered by I_1, I_2 and I_3.

$$\sum_{k \in U_{n_3} = U_3} |J_{n_3,k}| \leq \sum_{k=1}^{p(n_3)} |J_{n_3,k}| = \sum_{k=1}^{24} |J_{3,k}| \leq |I_3|,$$

$$\sum_{k \in U_{n_2} = U_2} |J_{n_2,k}| \leq \sum_{k=1}^{p(n_2)} |J_{n_2,k}| = \sum_{k=1}^{16} |J_{2,k}| \leq |I_2|,$$

$$\sum_{k \in U_{n_1} = U_1} |J_{n_1,k}| \leq \sum_{k=1}^{p(n_1)} |J_{n_1,k}| = \sum_{k=1}^{16} |J_{1,k}| \leq |I_{n_1}|$$

Thus,

$$|I| = \sum_{i=1}^{8} |J_i| = \sum_{p=1}^{3} \sum_{k \in U(n_p)} |J_{n_p,k}|$$

$$\leq \sum_{p=1}^{3} |I_{n_p}|$$

This proves that

$$|I| \leq \sum_{i=1}^{3} |I_i|$$

This is our desired proposition for a particular example set in $\Re^2$ using three intervals. We are now ready to adapt this algorithm to prove the general result.

LEBESGUE MEASURE   281

Proof 12.1.1

We are given intervals I_1 to I_N in $\Re^k$ whose union covers $\overline{I}$. Each interval I_i is the product

$$(\alpha_{i1}, \beta_{i1}) \times \cdots \times (\alpha_{ik}, \beta_{ik})$$

and I is the product

$$(\alpha_1, \beta_1) \times \cdots \times (\alpha_k, \beta_k)$$

On the x_j axis, the N intervals and the interval I determine a collection of points

$$\left\{ \begin{array}{l} (\alpha_{1j}, \beta_{1j}), \ x_j \ edge \ from \ interval \ I_1; \\[6pt] (\alpha_{2j}, \beta_{2j}), \ x_j \ edge \ from \ interval \ I_2; \\[6pt] \qquad\qquad\qquad \vdots \\[6pt] (\alpha_{Nj}, \beta_{Nj}), \ x_j \ edge \ from \ interval \ I_N; \\[6pt] (\alpha_j, \beta_j), \ x_j \ edge \ from \ interval \ I \end{array} \right\}$$

We do not care if these points are ordered. These x_j axis points, for $1 \le j \le k$, "slice" the intervals I_1 through I_N and I into smaller intervals just as we did in the example for $\Re^2$ shown in Figure 12.2. We have

$$\begin{array}{rcl} I & \longrightarrow & J_1, \ \ldots, \ J_p \\[4pt] I_1 & \longrightarrow & J_{11}, \ \ldots, \ J_{1,p(1)} \\[4pt] \vdots & & \\[4pt] I_N & \longrightarrow & J_{N1}, \ \ldots, \ J_{N,p(1)} \end{array}$$

Step 1: *Look at J_1. There is a smallest index n_1 so that $J - 1 = J_{n_1,\ell}$ for some ℓ. Let*

$$\begin{array}{rcl} T_{n_1} & = & \{\, i\{1,\ldots,p\} \,|\, \exists \ell \ni J_i = J_{n_1,\ell}\}, \\[4pt] U_{n_1} & = & \{\, \ell \,|\, \exists i \ni J_i = J_{n_1,\ell}\} \end{array}$$

This uses up T_{n_1} of the indices $\{1, \ldots, p\}$. You can see this process in Figure 12.4.

Step 2: *Let*

$$V_1 \ = \ \{1, \ \ldots, \ p\} \setminus T_{n_1}$$

and let q be the smallest index from the set V_1. For this q, find the smallest index $n_2 \ne n_1$ so that $J_q = J_{n_2,\ell}$ for some ℓ. This is the process we are showing in Figure 12.5. We define

$$\begin{array}{rcl} T_{n_2} & = & \{\, i \in V_1 \,|\, \exists \ell \ni J_i = J_{n_2,\ell}\} \\[4pt] U_{n_2} & = & \{\, \ell \,|\, \exists i \in V_1 \ni J_i = J_{n_2,\ell}\} \end{array}$$

This uses up more of the smaller subintervals I_1 to I_p.

Additional Steps : *Let*

$$V_2 \ = \ \{1, \ \ldots, \ p\} \setminus (T_{n_1} \cup T_{n_2})$$

We see V_2 is a smaller subset of the original $\{1, \ \ldots, \ p\}$ than V_1. We continue this construction

282 BASIC ANALYSIS IV: MEASURE THEORY AND INTEGRATION

process until we have used up all the indices in $\{1, \ldots, p\}$. This takes say Q steps and we know $Q \leq p$.

Final Step: *After the process terminates, we have*

$$
\begin{aligned}
|I| &= \sum_{i=1}^{p} |J_i| \\
&= \sum_{p=1}^{Q} \sum_{\ell \in U(n_p)} |J_{n_p, \ell}| \\
&\leq \sum_{p=1}^{Q} |I_{n_p}| \leq \sum_{i=1}^{N} |I_i|
\end{aligned}
$$

This completes the proof. ■

Homework

Now we ask you to redo our motivational argument with a different choice of $\overline{I}$ and three different intervals I_1, I_2 and I_3 that provide a cover. This is a very instructive exercise. Orient $\overline{I}$ in different ways to see how the decompositions work out.

Exercise 12.1.5 $I = (1,3) \times (2,4]$ *and* $I_1 = (-1,2) \times (1,3)$, $I_2 = (0.5, 2.5) \times (-1, 6)$ *and* $I_3 = (1.5, 7) \times (1.5, 4.5)$.

Exercise 12.1.6 $I = (1,7) \times [1,2]$ *and* $I_1 = (-1,4) \times (-1,5)$, $I_2 = (3,6) \times (-0.5, 8)$ *and* $I_3 = (4,8) \times (-3,3)$.

We can now finally prove that $\mu^*(I) = |I|$. Note that we have to work this hard because our original covering family was not an algebra! The final arguments are presented in the next two lemmas.

Lemma 12.1.2 $\mu^*(\overline{I}) = |I|$: **the Outer Measure of the Closure of Interval Equals Content of Interval**

> *Let I be an open interval in $\Re^k$. Then $\mu^*(\overline{I}) = |I|$.*

Proof 12.1.2
Let (I_n) be any Lebesgue cover of $\overline{I}$. Since $\overline{I}$ is compact, this cover has a finite subcover, $I_{n_1}, \ldots, I_{n_N}$. Applying Lemma 12.1.1, we see

$$
|I| \leq \sum_{i=1}^{N} |I_{n_i}| \leq \sum_{i} |I_i|
$$

Since (I_n) is an arbitrary cover of $\overline{I}$, we then have $|I|$ is a lower bound for the set

$$
\left\{ \sum_{n} |I_n| \,\Big|\, (I_n) \text{ is a cover of } \overline{I} \right\}
$$

It follows that

$$
|I| \leq \mu^*(\overline{I})
$$

To prove the reverse inequality holds, let U be an open interval concentric with I so that $\overline{I} \subseteq U$. Then U is a cover of $\overline{I}$ and so $\mu^*(\overline{I}) \leq |U|$. Hence, for any concentric interval, $\lambda I, 1 < \lambda < 2$, we have $\mu^*(\overline{I}) \leq \lambda^k |I|$. Since this holds for all $\lambda > 1$, we can let $\lambda \to 1$ to obtain $\mu^*(\overline{I}) \leq |I|$. ∎

Lemma 12.1.3 $\mu^*(I) = |I|$

> *If I is an open interval of $\Re^k$, then $\mu^*(I) = |I|$.*

Proof 12.1.3
We know I is a cover of itself, so it is immediate that $\mu^*(I) \leq |I|$. To prove the reverse inequality, let λI be concentric with I for any $0 < \lambda < 1$. Then, $\overline{\lambda I} \subseteq I$ and since μ^* is an outer measure, it is monotonic and so

$$\mu^*(\overline{\lambda I}) \leq \mu^*(I)$$

But $\mu^*(\overline{\lambda I}) = \lambda^k |I|$. We thus have $\lambda^k |I| \leq \mu^*(I)$ for all $\lambda \in (0, 1)$. Letting $\lambda \to 1$, we obtain the desired inequality. ∎

Homework

Exercise 12.1.7 *Prove $\mu^*(\overline{\lambda I}) = \lambda^k |I|$.*

Exercise 12.1.8 *If $I = (2, 4]$, what is $\mu^*([2, 4])$?*

Exercise 12.1.9 *If $I = (2, 4] \times [-1, 6)$, what is $\mu^*(\overline{I})$?*

Exercise 12.1.10 *If $I = (2, 4] \times [-1, 6) \times (-10, 25]$, what is $\mu^*(\overline{I})$?*

12.2 Lebesgue Outer Measure is a Metric Outer Measure

We have now shown that if $I \in \mathscr{T}$, then $|I| = \mu^*(I)$. However, we still do not know that the intervals I from $\mathscr{T}$ are μ^* measurable. We will do this by showing that Lebesgue outer measure is a metric outer measure. Then, it will follow from Theorem 11.4.1 that the open sets in $\Re^k$ are μ^* measurable, i.e. are in $\mathcal{M}$. Of course, this implies $\mathscr{T} \subseteq \mathcal{M}$ as well. Then, since an interval I is measurable, we have $|I| = \mu(I)$. Let's prove μ^* is a metric outer measure. We begin with a technical definition.

Definition 12.2.1 Rewriting Lebesgue Outer Measure Using Edge Length Restricted Covers: the M_δ Form of μ^*

> *For any set E is $\Re^k$ and any $\delta > 0$, let (I_n) be a cover of E with each I_n an interval in $\Re^k$ with each edge of an I_n having a length less than δ. Then*
>
> $$M_\delta(E) = \inf \{ \sum_n |I_n| \, | \, (I_n) \}$$

Next, we need a technical lemma concerning finite Lebesgue covers.

Lemma 12.2.1 Approximate Finite Lebesgue Covers of $\overline{I}$

> Let I be an open interval and let $\overline{I}$ denote its closure. Let ϵ and δ be given positive numbers. Then there exists a finite Lebesgue Covering of $\overline{I}$, $I_1, \ldots, I_N$ so that each edge of I_i has length less than δ and
>
> $$|I_1| + \cdots + |I_N| \ < \ |I| + \epsilon$$

Proof 12.2.1

Let

$$I \ = \ \Pi_{i=1}^k (a_i, b_i)$$

and divide each component interval (a_i, b_i) into n_i uniform pieces so that $(b_i - a_i)/2 < \delta/2$. This determines n_i open intervals of the form (a_{ij}, b_{ij}) for $1 \leq j \leq n_i$ with $b_{ij} - a_{ij} < \delta/2$.

Let $N = n_1 n_2 \cdot n_k$ and let $J = (j_1, \ldots, j_k)$ denote the k - tuple of indices chosen so that $1 \leq j_i \leq n_i$. There are N of these indices. Let $\boldsymbol{j}$ indicate any such k - tuple. Then $\boldsymbol{j}$ determines an interval $I_{\boldsymbol{j}}$ where

$$I_{\boldsymbol{j}} \ = \ \Pi_{i=1}^k (a_{ij}, b_{ij}), \quad \text{with } (b_{ij} - a_{ij}) < \delta/2$$

Hence, $|I_{\boldsymbol{j}}| < (\delta/2)^k$. It is also clear that

$$\sum |I_{\boldsymbol{j}}| \ = \ |I|$$

Now choose concentric open intervals $\lambda I_{\boldsymbol{j}}$ for any λ with $1 < \lambda < 2$. Then since $\lambda > 1$, $(\lambda I_{\boldsymbol{j}})$ over all k - tuples $\boldsymbol{j}$ is a Lebesgue cover of $\overline{I}$, we have

$$|\lambda I_{\boldsymbol{j}}| \ = \ \lambda^k |I_{\boldsymbol{j}}|$$

and so

$$\sum |\lambda I_{\boldsymbol{j}}| \ = \ \lambda^k \sum |I_{\boldsymbol{j}}|$$
$$= \ \lambda^k |I|$$

Since $\lambda^k \to 1$, for our given $\epsilon > 0$, there is a $\eta > 0$ so that if $1 < \lambda < 1 + \eta < 2$, we have

$$\lambda^k - 1 \ < \ \frac{\epsilon}{|I| + 1}$$

In particular, if we pick $\lambda = (1 + \eta)/2$, then

$$|\lambda I_{\boldsymbol{j}}| \ < \ \left(1 + \frac{\epsilon}{|I| + 1} \right) |I| < |I| + \epsilon$$

Thus, the finite collection $((1 + \eta)/2) I_{\boldsymbol{j}})$ is the one we seek as each edge has length $((1 + \eta)/2)\, \delta/2$ which is less than δ. ∎

Lemma 12.2.2 $M_\delta \ = \ \mu^*$

> *For any subset E of $\Re^k$, we have $M_\delta(E) \ = \ \mu^*(E)$.*

Proof 12.2.2

Let's pick a given $\delta > 0$. The way M_δ is defined then tells us immediately that $\mu^*(E) \leq M_\delta(E)$ for any $\delta > 0$ and subset E. It remains to prove the reverse inequality. If $\mu^*(E)$ was infinite, we would have $\mu^*(E) \geq M_\delta(E)$; hence, it is enough to handle the case where $\mu^*(E)$ is finite. By the Infimum Tolerance Lemma for a given $\epsilon > 0$, there is a Lebesgue cover (I_n) of E so that

$$\sum_n |I_n| \;\; < \;\; \mu^*(E) + \frac{\epsilon}{2}$$

By Lemma 12.2.1, there is a finite Lebesgue cover of each $(\overline{I_n})$ which we will denote by (J_{nj}), $1 \leq j \leq p(n)$ so that each interval J_{nj} has edge length less than δ and satisfies

$$\sum_{j=1}^{p(n)} |J_{nj}| \;\; < \;\; |I_n| + \frac{\epsilon}{2^{n+1}}$$

The combined family of intervals (J_{nj}) for all n and $1 \leq j \leq p(n)$ is clearly a Lebesgue cover of E also. Thus, by definition of μ^*, we have

$$\sum_{n=1}^{\infty} \sum_{j=1}^{p(n)} |J_{nj}| \;\; < \;\; \sum_{n=1}^{\infty} |I_n| + \sum_{n=1}^{\infty} \frac{\epsilon}{2^{n+1}}$$
$$< \;\; \mu^*(E) + \epsilon$$

Now each edge length of the interval I_{nj} is less than δ and so

$$M_\delta \;\; \leq \;\; \sum_{n=1}^{\infty} \sum_{j=1}^{p(n)} |J_{nj}|$$

by definition. We see we have established

$$M_\delta \;\; \leq \;\; \mu^*(E) + \epsilon$$

for an arbitrary ϵ; hence, $M_\delta \leq \mu^*(E)$. ∎

We now have enough "ammunition" to prove Lebesgue outer measure is a metric outer measure; i.e. LOM is a MOM!

Theorem 12.2.3 Lebesgue Outer Measure is a Metric Outer Measure

The Lebesgue Outer Measure, μ^* is a metric outer measure; i.e., if A and B are two sets in $\Re^k$ with $D(A, B) > 0$, then $\mu^*(A \cup B) = \mu^*(A) + \mu^*(B)$.

Proof 12.2.3

We always know that $\mu^*(A \cup B) \leq \mu^*(A) + \mu^*(B)$ for any A and B. Hence, for two sets A and B with $D(A, B) = \delta > 0$, it is enough to show $\mu^*(A) + \mu^*(B) \leq \mu^*(A \cup B)$. Let $\epsilon > 0$ be chosen. Since $M_\delta = \mu^*$, there is a cover of $A \cup B$ so that the edge length of each I_n is less than δ/k and

$$M_\delta(A \cup B) \;\; = \;\; \mu^*(A \cup B) \leq \sum_n |I_n| < \mu^*(A \cup B) + \epsilon$$

by an application of the Infimum Tolerance Lemma.

286 — BASIC ANALYSIS IV: MEASURE THEORY AND INTEGRATION

If x and y in $A \cup B$ are both in a given I_n, then

$$d(x,y) \;=\; \sqrt{\sum_{i=1}^{k} (x_i - y_i)^2} \;<\; \sqrt{\sum_{i=1}^{k} (\frac{\delta}{k})^2} \;=\; \sqrt{k^2 \frac{\delta^2}{k^2}} \;=\; \delta \quad .$$

However, $D(A,B) = \delta$ by assumption. Thus, a given I_n cannot contain points of both A and B. We can therefore separate the family (I_n) into two collections indexed by U and V, respectively. If $n \in U$, then $I_n \cap A$ is non empty and if $n \in V$, $I_n \cap B$ is non empty. We see $\{I_n\}_{n \in U}$ is a cover for A and $\{I_n\}_{n \in V}$ is a cover for B. Thus, $\mu^*(A) \le \sum_{n \in U} |I_n|$ and $\mu^*(B) \le \sum_{n \in V} |I_n|$. It then follows that

$$\mu^*(A \cup B) + \epsilon \;\ge\; \sum_{n} |I_n| = \sum_{n \in U} |I_n| + \sum_{n \in V} |I_n|$$
$$\ge\; \mu^*(A) + \mu^*(B)$$

Since ϵ is arbitrary, we have shown $\mu^*(A) + \mu^*(B) \le \mu^*(A \cup B)$. This completes the proof that Lebesgue outer measure is a metric outer measure. ∎

Homework

Exercise 12.2.1 *Let $E = (0,2] \times [-1,4]$. Find a finite $\{I_i\}$ cover of E where the length of each edge of I_i has length smaller than $\delta = 0.1$.*

Exercise 12.2.2 *Let $E = (0,2] \times [-1,4] \cup [3,5] \times (-2,1]$. Find a finite $\{I_i\}$ cover of E where the length of each edge of I_i has length smaller than $\delta = 0.1$.*

Exercise 12.2.3 *Let $E = (0,2] \times [-1,4]$. Let $\epsilon = 0.1$. Find a finite $\{I_i\}$ cover of E where the length of each edge of I_i has length smaller than $\delta = 0.05$ and $|I_1| + \cdots + |I_N| < |I| + \epsilon$.*

Exercise 12.2.4 *Let $E = (0,2] \times [-1,4] \cup [3,5] \times (-2,1]$. Let $\epsilon = 0.2$. Find a finite $\{I_i\}$ cover of E where the length of each edge of I_i has length smaller than $\delta = 0.03$ and $|I_1| + \cdots + |I_N| < |I| + \epsilon$.*

Exercise 12.2.5 *Compute $\mu^* E$ where $E = [2,3] \times \{p\}$ for a fixed p in $\Re$.*

Exercise 12.2.6 *Compute $\mu^* E$ where $E = [2,3] \times (4,8) \times \{p\}$ for a fixed p in $\Re$.*

Exercise 12.2.7 *Compute $\mu^* E$ where $E = [2,3] \times \{q\} \times \{p\}$ for a fixed p and q in $\Re$.*

This theorem is the final piece we need to fully establish the two conditions

(i): $\mathscr{T} \subseteq \mathcal{M}$; i.e. $\mathcal{M}$ is an OMI-F σ-algebra.

(ii): If $I \in \mathscr{T}$, then $|I| = \mu(I)$; i.e. $\mathcal{M}$ is an OMI-FE σ-algebra.

Comment 12.2.1 *We see immediately that since Lebesgue outer measure is a metric outer measure, the σ-algebra of μ^* measurable subsets contains all the open sets of $\Re^k$. In particular, any open interval I is measurable. As mentioned previously, we thus know the Borel σ-algebra of subsets is contained in $\mathcal{M}$.*

By Theorem 11.3.2, we know Lebesgue measure μ is complete.

12.3 Lebesgue Measure is Regular

We can also prove Lebesgue measure μ is regular.

Theorem 12.3.1 Lebesgue Measure is Regular

For any set E in $\Re^k$,

$$\mu^*(E) = \inf \{\mu(U) \mid U, \, E \subseteq U, \, U \text{ is open }\}$$
$$\mu^*(E) = \inf \{\mu(F) \mid E, \, E \subseteq F, \, F \text{ is Lebesgue measurable }\}$$

Thus, Lebesgue measure is regular.

Proof 12.3.1
Since U is open, U is Lebesgue measurable and so $\mu^*(U) = \mu(U)$. It follows immediately that $\mu^*(E) \leq \mu(U)$ for such U. Hence,

$$\mu^*(E) \leq \inf \{\mu(U) \mid U, \, E \subseteq U, \, U \text{ is open }\}$$

On the other hand, if $\epsilon > 0$ is given, the Infimum Tolerance Lemma tells us there is a Lebesgue cover of E, (I_n), so that

$$\mu^*(E) \leq \sum_n |I_n| < \mu^*(E) + \epsilon$$

However, this open cover generates an open set $G = \cup_n I_n$ containing E with $\mu(G) \leq \sum_n |I_n|$ because $\mu(I_n) = |I_n|$. We conclude, using the definition of μ^* that

$$\mu(G) \leq \sum_n |I_n| < \mu^*(E) + \epsilon$$

Hence, we must have

$$\inf \{\mu(U) \mid U, \, E \subseteq U, \, U \text{ is open }\} \leq \mu^*(E) + \epsilon$$

Since ϵ is arbitrary, the result follows.

Since each open U is measurable, we then know

$$\mu^*(E) = \inf \{\mu(U) \mid U, \, E \subseteq U, \, U \text{ is open }\}$$
$$\geq \inf \{\mu(F) \mid E, \, E \subseteq F, \, F \in \mathcal{M}\}$$

by the first argument. To obtain the reverse inequality, note that since $\mu^*(F) = \mu(F)$ for all measurable F, monotonicity of μ^* says $\mu^*(E) \leq \mu^*(F)$ for all measurable F. We conclude

$$\mu^*(E) \leq \inf \{\mu(F) \mid E, \, E \subseteq F, \, F \in \mathcal{M}\}$$

Now recall the definition of a regular measure from Definition 11.5.3. Using the Infimum Tolerance Lemma again, there are measurable sets (F_n) so that $E \subseteq F_n$ for all n and

$$\mu^*(E) \leq \mu(F_n) < \mu^*(E) + \frac{1}{n}$$

288   *BASIC ANALYSIS IV: MEASURE THEORY AND INTEGRATION*

Then, $\cap_n F_n$ is also measurable and so by our equivalent form of μ^*, we have $\mu^*(E) \leq \mu(\cap_n F_n)$. However, $\cap_n F_n \subseteq F_n$ always and hence,

$$\mu^*(E) \;\; \leq \;\; \mu(\cap_n F_n) \leq \mu(F_n) \; < \; \mu^*(E) + \frac{1}{n}$$

We conclude for all n, $\mu^*(E) \leq \mu(\cap_n F_n) \; < \; \mu^*(E) + \frac{1}{n}$. Letting n go to infinity, we find $\mu^*(E) = \mu(\cap_n F_n)$ which shows μ is regular. ∎

Homework

Exercise 12.3.1 *If $F \subset E$ where $E = [2,3] \times \{p\}$ for a fixed p in $\Re$, are E and F in $\mathcal{M}$, the set of μ^* sets?*

Exercise 12.3.2 *If $F \subset E$ where $E = [2,3] \times (1,8) \times \{p\}$ for a fixed p in $\Re$, are E and F in $\mathcal{M}$, the set of μ^* sets?*

Exercise 12.3.3 *If $F \subset E$ where $E = [2,3] \times \{q\} \times \{p\}$ for a fixed p and q in $\Re$, are E and F in $\mathcal{M}$, the set of μ^* sets?*

Exercise 12.3.4 *If E is any set in $\Re^2$ and F is an open set containing E, is it true $m^*(E) \leq \mu(F)$?*

Exercise 12.3.5 *If E is any set in $\Re^3$ and F is an open set containing E, is it true $m^*(E) \leq \mu(F)$?*

Exercise 12.3.6 *If E is a convex set in $\Re^2$ is $\overline{E}$ Lebesgue measurable?*

Exercise 12.3.7 *If E is a compact set in $\Re^6$, is it Lebesgue measurable?*

Exercise 12.3.8 *If the real-valued function f is continuous on $\Omega \subset \Re^3$, let $D = f(\Omega)$. Is D Lebesgue measurable in $\Re$?*

12.4 Approximation Results

We now present some approximation results for Lebesgue measurable sets and some applications to the $\mathcal{L}_p$ spaces. We begin with the following result.

12.4.1 Approximating Measurable Sets

Theorem 12.4.1 Measurability and Approximation Conditions

> *A set E in $\Re^k$ is Lebesgue measurable if and only if for all $\epsilon > 0$, there is a pair of sets (F, G) so that F is closed, G is open with $F \subseteq E \subseteq G$ and $\mu(G \setminus F) < \epsilon$. Moreover, if E is bounded, we can choose the open set G so that $\overline{G}$ is compact.*

Proof 12.4.1
First, we prove that if the condition holds for all $\epsilon > 0$, then E must be measurable. For each positive integer n, there are therefore closed sets F_n and open sets G_n so that

$$F_n \; \subseteq \; E \; \subseteq \; G_n$$

with $\mu(G_n \setminus F_n) < \frac{1}{n}$. Let $F = \cup_n F_n$ and $G = \cap_n G_n$. Then, G and F are measurable and

$$G \setminus F \; \subseteq \; G_n \setminus F_n$$

with $\mu(G \setminus F) \leq \mu(G_n \setminus F_n) < \frac{1}{n}$. Hence, as n goes to ∞, we see $\mu(G \setminus F) = 0$. Since Lebesgue measure is complete and $E \setminus F \subseteq G \setminus F$, we also know $E \setminus F$ is measurable with measure 0. Finally, since $E = F \cup (E \setminus F)$ and each piece is measurable, E is measurable.

To prove the other direction is a bit more complicated. We now start with the assumption that E is measurable. First, let's assume E is a bounded set. Since E is bounded, there is a bounded open interval I so that $\mu(E) \leq |I|$ with $E \subseteq I$. Further, for any ϵ that is positive, by Theorem 12.3.1, there is an open set H so that $E \subseteq H$ and $\mu(H) < \mu(E) + \frac{\epsilon}{2}$. But then the set $G = H \cap I$ is also open and since it contains E, we must have

$$\mu(E) \leq \mu(G) \leq \mu(H) < \mu(E) + \frac{\epsilon}{2}.$$

It is also clear $\overline{G}$ is compact.

Now choose a compact set A so that $E \subseteq A$ (we can do this because E is bounded). Next, since the set $A \setminus E$ is measurable, by Theorem 12.3.1 we can find an open set B so that $A \setminus E \subseteq B$ and

$$\mu(B) < \mu(A \setminus E) + \frac{\epsilon}{2}.$$

Define the closed set F by $F = A \setminus B$. Then, an easy calculation shows $F \subseteq E$. We thus have shown

$$\begin{aligned}
\mu(F) &= \mu(A) - \mu(A \cap B) \geq \mu(A) - \mu(B) \\
&> \mu(A) - \mu(A \setminus E) - \frac{\epsilon}{2} \\
&= \mu(E) - \frac{\epsilon}{2}
\end{aligned}$$

We conclude $\mu(G \setminus F) = \mu(G) - \mu(F) < \epsilon$. This shows the result holds if E is bounded.

To show the result is valid if E is not bounded, let $S_n = \{x : ||x|| \leq n\}$ be the closed ball of radius n in $\Re^n$. Define the sets E_n as follows:

$$\begin{aligned}
E_1 &= E \cap S_1 \\
E_2 &= E \cap (S_2 \setminus S_1) \\
E_3 &= E \cap (S_3 \setminus S_2) \\
&\vdots \\
E_n &= E \cap (S_n \setminus S_{n-1}) \\
&\vdots
\end{aligned}$$

Then $E = \cup_n E_n$ and each E_n is bounded and measurable. Hence, for each n, the result for a bounded measurable set applies. Thus, there are closed sets F_n and open sets G_n with $F_n \subseteq E_n \subseteq G_n$ and

$$\mu(G_n \setminus F_n) < \frac{\epsilon}{2^n}.$$

Let $F = \cup F_n$ and $G = \cup G_n$. We see

$$G \setminus F \subseteq \cup_n (G_n \setminus F_n)$$

and

$$\mu(G \setminus F) \leq \sum_n \mu(G_n \setminus F_n) < \sum_n \frac{\epsilon}{2^n} < \epsilon.$$

290   *BASIC ANALYSIS IV: MEASURE THEORY AND INTEGRATION*

Since $F \subseteq E \subseteq G$, all that is left to prove is that F is closed.

To do this, let x_i be a sequence in F which converges to x. Since the sequence converges (x_i) is bounded, all the x_i live in some S_N. But $F_n \subseteq E \cap (S_n \subseteq S_n)$ for all $n \geq 2$ and hence

$$
\begin{aligned}
F_{N+1} &\subseteq S_{N+1} \setminus S_N \\
F_{N+2} &\subseteq S_{N+2} \setminus S_{N+1} \subseteq S_{N+2} \setminus S_N
\end{aligned}
$$

and so forth. Thus, $F_n \subseteq S_n \setminus S_N \subseteq S_N^C$ for all $n > N$. Hence, the points x_i must all live in $\cup_{n=1}^N F_n$ which is a bounded set. But then the limit point x must also be in this set which tells us x is in F. Thus, F is closed and we have proven the desired result. ∎

We easily prove Theorem 12.4.1 in a more general setting. First, let's introduce a common definition.

Definition 12.4.1 Borel Measure

Let X be a metric space and $\mathcal{S}$ a sigma-algebra of subsets of X. We say the measure μ on $\mathcal{S}$ is a Borel measure if the Borel sets in X are μ measurable.

Theorem 12.4.2 Finite Measure Borel Sets Can be Approximated by Closed Sets

Let X be a metric space, μ a Borel measure on X, $\epsilon > 0$ and B a Borel set with $\mu(B) < \infty$. Then B contains a closed set F with $\mu(B \setminus F) < \epsilon$.

Proof 12.4.2
First, let's assume $\mu(X) < \infty$. Let $\mathcal{F}$ be the defined as follows:

$$
\mathcal{F} = \{E \subseteq X \text{ and } \forall \gamma > 0, \exists \text{ closed set } K \subseteq E \text{ with } \mu(E \setminus K) < \gamma\}
$$

It is easy to see that closed sets with finite measure are in $\mathcal{F}$. Let $\epsilon > 0$ be chosen. Now suppose sets E_1 to E_N are in $\mathcal{F}$ for some positive integer N. Then, there are closed sets K_1 to K_N so that $K_i \subseteq E_i$ and $\mu(E_i \setminus K_i) < \frac{\epsilon}{2^i}$. Then

$$
\begin{aligned}
\mu(\cap_{i=1}^N E_i \setminus \cap_{i=1}^N K_i) &\leq \mu(\cap_{i=1}^N (E_i \setminus K_i)) \\
&< \sum_{i=1}^N \frac{\epsilon}{2^i} = 1 - \frac{1}{2^N} < \epsilon
\end{aligned}
$$

Hence, the closed set $\cap_{i=1}^N K_i$ is contained in the set $\cap_{i=1}^N E_i$ and satisfies the γ condition. Thus, $\mathcal{F}$ is closed under finite intersections. From this argument, it is also clear that $\mathcal{F}$ is closed under countable intersections using minor changes in the reasoning as a countable intersection of closed sets is closed. Now, let's look at a finite union. We again suppose sets E_1 to E_N are in $\mathcal{F}$ for some positive integer N. Then, there are closed sets K_1 to K_N so that $K_i \subseteq E_i$ and $\mu(E_i \setminus K_i) < \frac{\epsilon}{2^i}$. Then,

$$
\begin{aligned}
\mu(\cup_{i=1}^N E_i \setminus \cup_{i=1}^N K_i) &\leq \mu(\cup_{i=1}^N (E_i \setminus K_i)) \\
&< \sum_{i=1}^N \frac{\epsilon}{2^i} = 1 - \frac{1}{2^N} < \epsilon
\end{aligned}
$$

Since the finite union of closed sets is closed, we see the closed set $\cup_{i=1}^N K_i$ is contained in the set $\cup_{i=1}^N E_i$ and satisfies the γ condition also. Hence, $\mathcal{F}$ is closed under finite unions. To handle

LEBESGUE MEASURE

countable unions, define the sequence of sets

$$F_i \quad = \quad \cup_{i=1}^{\infty} E_i \setminus \cup_{i=1}^{N} K_i$$

Then, $\ldots, F_N \subseteq F_{N-1} \subseteq \ldots \subseteq F_1$ *and* $\mu(F_1) < \infty$ *because* $\mu(X) < \infty$. *Hence, we can invoke Lemma 9.2.2 to conclude*

$$\lim_{N} \mu(\cup_{i=1}^{\infty} E_i \setminus \cup_{i=1}^{N} K_i) \quad = \quad \mu(\cup_{i=1}^{\infty} E_i \setminus \cup_{i=1}^{\infty} K_i)$$

$$\leq \quad \mu(\cup_{i=1}^{\infty} (E_i \setminus K_i))$$

$$< \quad \sum_{i=1}^{N} \frac{\epsilon}{2^i} = 1 - \frac{1}{2^N} < \epsilon$$

Thus, for large enough choice of N, *we must have*

$$\mu(\cup_{i=1}^{\infty} E_i \setminus \cup_{i=1}^{N} K_i) \quad < \quad \epsilon$$

The set $\cup_{i=1}^{N} K_i$ *is a closed subset of* $\cup_{i=1}^{\infty} E_i$ *and satisfies the* γ *condition. Hence,* $\mathcal{F}$ *is closed under countable unions. Since* $\mathcal{F}$ *contains the closed subsets, it now contains the open sets as well. We conclude* $\mathcal{F}$ *contains the Borel sets.*

If $\mu(X) = \infty$, *given a set* B *with* $\mu(B) < \infty$, *note* $\mathcal{S}_B = \{E \cap B : E \in \mathcal{S}\}$ *is a sigma-algebra for* X *and* μ_B *defined as the restriction of* μ *to this sigma-algebra is finite. Apply the argument for the finite measure case to* μ_B *to get the desired result.* ∎

Theorem 12.4.3 Finite Measure Borel Sets Can be Approximated by Open Sets

Let X *be a metric space,* μ *as Borel measure on* X, $\epsilon > 0$ *and* B *a Borel set. If* $\mu(X) < \infty$ *or more generally, if* B *is contained in a countable union of open sets* U_i, *each of finite measure, then* B *is contained in an open set* G *with* $\mu(G \setminus B) < \epsilon$.

Proof 12.4.3

We use Theorem 12.4.2 here. Since the $\mu(X)$ *is finite,* $\mu(B^C)$ *is finite. So there is a closed set* K *in* B^C *so that* $\mu(B^C \setminus K) < \epsilon$. *Let the open set* $G = K^C$. *Then* $\mu(G \setminus B) < \epsilon$ *and we are done.*

In the more general situation, choose a closed set K_i *contained in each* $U_i \setminus B$ *so that*

$$\mu\left((U_i \setminus B) \setminus K_i \right) \quad < \quad \frac{\epsilon}{2^i}$$

Next, note that $B \cap U_i \subseteq U_i \setminus K_i$, *which is an open set. Define* $G = \cup_{i=1}^{\infty} (U_i \setminus K_i)$. *Then* G *is open, contains* B *and* $\mu(G \setminus B) < \epsilon$. ∎

Homework

Exercise 12.4.1 *Let* $A \subset X$ *where* X *is a metric space with metric* d. *Define* $f : \Re^x \to \Re$ *by* $f(x) = d(x, A)$. *Prove* f *is continuous.*

Exercise 12.4.2 *Let* $A \subset \Re^k$ *and define* $f : \Re^x \to \Re$ *by* $f(x) = d(x, A)$. *Prove* f *is continuous.*

Exercise 12.4.3 *Let* f *be real-valued and continuous on* $\Re^k$. *Then* f *is Lebesgue measurable on* $\Re^k$.

Exercise 12.4.4 *Let* $f(x) = I_E(x)$ *where* E *is bounded and Lebesgue measurable in* $\Re$. *Then given* $\epsilon > 0$, *there is a closed set* F *and an open set* G *so that* $F \subset E \subset G$ *where* $\overline{G}$ *is compact and*

$\mu(G \setminus F) < \epsilon$. Define $g : \Re \to \Re$ by

$$g(x) \;=\; \frac{d(x, G^C)}{d(x, G^C) + d(x, F)}$$

Prove

- *Prove g is continuous and $0 \le f(x) \le 1$.*

- *Show $g = 1$ on F and $g = 0$ on G^C and hence g smoothly transitions from 0 on G^C to 1 on F.*

- *Prove both I_E and f are Lebesgue measurable.*

- *Prove $\int (f - I_E) d\mu < \epsilon$*

Hence, given $\epsilon > 0$, we can approximate I_E by a continuous function g so that $\|g - I_E\|_1 < \epsilon$. Note this result depends on the fact that we can trap a measurable set between two Borel sets. And that results follows because Lebesgue measure is regular.

Exercise 12.4.5 *Let $f(x) = 2I_E(x)$ where E is bounded and Lebesgue measurable in $\Re$. Hence, given $\epsilon > 0$, we can approximate f by a continuous function g so that $\|g - f\|_1 < \epsilon$.*

Exercise 12.4.6 *Let $f(x) = 2I_E(x)$ where E is bounded and Lebesgue measurable in $\Re^2$. Hence, given $\epsilon > 0$, we can approximate f by a continuous function g so that $\|f - g\|_1 < \epsilon$.*

Exercise 12.4.7 *Let $f(x) = 2I_E(x) + 3I_F(x)$ where E and F are bounded, disjoint and Lebesgue measurable in $\Re$. Prove given $\epsilon > 0$, we can approximate f by a continuous function g so that $\|f - g\|_1 < \epsilon$.*

Exercise 12.4.8 *Let $f(x) = 2I_E(x) + 3I_F(x)$ where E and F are bounded, disjoint and Lebesgue measurable in $\Re^2$. Prove, given $\epsilon > 0$, we can approximate f by a continuous function so that $\|f - g\|_1 < \epsilon$.*

Exercise 12.4.9 *Let $f(x) = 2I_E(x) + 3I_F(x)$ where E and F are bounded, disjoint and Lebesgue measurable in $\Re^3$. Prove, given $\epsilon > 0$, we can approximate f by a continuous function so that $\|f - g\|_1 < \epsilon$.*

12.4.2 Approximating Measurable Functions

We can use Theorem 12.4.1 in many ways. One way is to construct continuous functions on suitable domains which approximate summable functions. Let's start with the characteristic function of the measurable set E, I_E.

Theorem 12.4.4 Continuous Approximation of a Characteristic Function

Let E in $\Re^k$ be Lebesgue measurable. Then given $\epsilon > 0$, there is a continuous function ϕ_E so that $\|I_E - \phi_E\|_1 < \epsilon$. Moreover, if E is bounded, the support of ϕ_E is compact.

Proof 12.4.4

Using Theorem 12.4.1, given $\epsilon > 0$, we see there is a closed set F and an open set G with $F \subseteq E \subseteq G$ with $\mu(F^C \cap G) < \frac{\epsilon}{2}$. For any subset C in $\Re^k$, it is easy to see the distance function

$$d(x, C) \;=\; \inf\{d(x, y) : y \in C\}$$

is continuous on $\Re^k$, where d denotes the standard Euclidean metric. It follows that the function f defined on $\Re^k$ by

$$f(x) \;=\; \frac{d(x, G^C)}{d(x, G^C) + d(x, F)}$$

is continuous and satisfies ϕ_E is 1 on F and 0 on G^C and $0 \le f(x) \le 1$ always. Since ϕ_E is nonzero on G^C, it follows that ϕ_E has compact support if E is bounded.

Finally, note

$$\int |\phi_E - I_E|\, d\mu \;=\; \int_F |\phi_E - I_E|\, d\mu + \int_{F^C \cap E} |\phi_E - I_E|\, d\mu$$
$$+ \int_{E^C \cap G} |\phi_E - I_E|\, d\mu + \int_{G^C} |\phi_E - I_E|\, d\mu$$

However, since $E \subseteq G$, $G^C \subseteq E^C$ which tells us both I_E and ϕ_E are 0 on G^C. Hence, the last integral vanishes. Also, note $F^C \cap E \subseteq F^C \cap G$ and $E^C \cap G \subseteq F^C \cap G$. Thus,

$$\int_{F^C \cap E} |\phi_E - I_E|\, d\mu \;\; + \;\; \int_{E^C \cap G} |\phi_E - I_E|\, d\mu$$
$$\le \;\; \mu(F^C \cap E) + \mu(E^C \cap E) \;\le\; 2\mu(F^C \cap G) \;<\; \epsilon$$

This allows us to conclude $\int |\phi_E - I_E|\, d\mu < \epsilon$ which is the desired result. ∎

Next, we see we can approximate simple functions arbitrarily close in the L_1 norm.

Theorem 12.4.5 Continuous Approximation of a Simple Function

Let μ denote Lebesgue measure in $\Re^k$. Let $\phi : \Re^k -> \Re$ be a simple function. Then given any $\epsilon > 0$, there is a continuous function g so that $\int |\phi - g|\, d\mu < \epsilon$. Moreover, if the support of ϕ is bounded, the support of g is compact.

Proof 12.4.5

The simple function ϕ has the standard representation $\phi = \sum_{i=1}^{n} a_i I_{E_i}$ where the numbers a_i are distinct and nonzero and the sets E_i are measurable and disjoint. Let $A = \max_{1 \le i \le n} |a_i|$. By Theorem 12.4.4, there are continuous functions g_i so that $\int (I_{E_i} - g_i)\, d\mu < \frac{\epsilon}{A\,n}$. Let $g = \sum_{i=1}^{n} a_i g_i$. Then

$$\int |g - \phi|\, d\mu \;=\; \sum_{i=1}^{n} |a_i| |I_{E_i} - g_i|\, d\mu$$
$$\le \;\; A \sum_{i=1}^{n} \frac{\epsilon}{A\,n} \;<\; \epsilon$$

It is clear that g has compact support if $\cup_{i=1}^{n} E_i$ is bounded. ∎

Now, we can approximate a summable function f with a continuous function as well.

Theorem 12.4.6 Approximation of a Summable Function with a Continuous Function

Let f be summable on $\Re^k$ with respect to Lebesgue measure. Then given $\epsilon > 0$, there is a continuous function of compact support g so that $\int |f - g|\, d\mu < \epsilon$.

294 BASIC ANALYSIS IV: MEASURE THEORY AND INTEGRATION

Proof 12.4.6
First, define the functions f_n by

$$f_n = f\, I_{[-n,n]}$$

Then, for all n, f_n is dominated by the summable function f and we also know $f_n \to f$. By the Dominated Convergence Theorem, we then have $\int f_n\, d\mu \to \int f\, d\mu$. Hence, given $\epsilon > 0$, there is an N so that if $n > N$, then

$$\int |f - f_n|\, d\mu = \int_{[-n,n]^C} |f|\, d\mu < \frac{\epsilon}{4}$$

Choose a particular $p > N$. Then since f_p is summable, we know there are sequences of simple functions, (ϕ_j) and (ψ_j) so that $\phi_j \uparrow f_p^+$ and $\psi_j \uparrow f_p^-$ with $f_p = f_p^+ - f_p^-$. We know also that $\int \phi_j\, d\mu \uparrow \int f_p^+\, d\mu$ and $\int \psi_j\, d\mu \uparrow \int f_p^-\, d\mu$. These simple functions are constructed using the technique given in Theorem 8.7.2. Since f_p has compact support, it follows that for each j these simple functions have compact support also. We see there is an M and a simple function $\zeta = \phi_M - \psi_M$ so that $\int (f_p - \zeta)\, d\mu < \frac{\epsilon}{4}$. Using Theorem 12.4.5, there is a continuous function of compact support g so that $\int (\zeta - g)\, d\mu < \frac{\epsilon}{4}$. Combining, we see

$$\int |f - g|\, d\mu \leq \int |f - f_p|\, d\mu + \int |f_p - \zeta|\, d\mu + \int |\zeta - g| < \epsilon$$

∎

Homework

Exercise 12.4.10 *Let $f(x) = 2I_E(x) + 3I_F(x)$ where E and F are bounded, disjoint and Lebesgue measurable in $[0,1]$ with $E \cup F = [0,1]$. Prove given $\epsilon > 0$, we can approximate f by a continuous function so that $\|f - I_E\|_1 < \epsilon$.*

Exercise 12.4.11 *Let f be Riemann integrable on $[0,1]$ and let $P = \{x_0, x_1, \ldots, x_n\}$ be a partition of $[0,1]$. Then P induces a decomposition of $[0,1]$ as usual, $E_i = [x_i, x_{i+1})$ for $1 \leq i < n$ and $E_n = [x_{n-1}, x_n]$. Let $m_i = \inf_{x \in E_i} f(x)$ and $h = \sum_{i=1}^{N} m_i \mu(E_i)$.*

- *Prove h is Lebesgue integrable with $\int h\, d\mu = \sum_{i=1}^{N} m_i \mu(E_i)$.*

- *Prove given $\epsilon > 0$, there is a continuous function ϕ so that $\|h - \phi\|_1 < \epsilon$.*

12.5 The Summable Functions are Separable

This result allows us to show that $\mathcal{L}_1$ is separable.

Theorem 12.5.1 $\mathcal{L}_1$ is Separable

> *Let $[a,b]$ be a finite interval. The space $\mathcal{L}_1([a,b], \mathcal{M}, \mu)$ where μ is Lebesgue measure on $[a,b]$ and $\mathcal{M}$ is the sigma-algebra of Lebesgue measurable subsets is separable.*

Proof 12.5.1
Let f be a summable function. By Theorem 12.4.6 there is a continuous function g with $\int_a^b |f - g|\, d\mu < \frac{\epsilon}{4}$. Further, there is a polynomial p on $[a,b]$ by the Weierstrass Approximation Theorem 5.1.2 so that $\sup_{x \in [a,b]} |g(x) - p(x)| < \frac{\epsilon}{4(b-a)}$. Further, there is a polynomial q with rational coefficients

so that $\sup_{x \in [a,b]} |p(x) - q(x)| < \frac{\epsilon}{4(b-a)}$. *Combining, we have*

$$\int |f - q| \, d\mu \quad \leq \quad \int |f - g| \, d\mu + \int |g - p| \, d\mu + \int |p - q| \, d\mu$$

$$\leq \quad \frac{\epsilon}{4} + (b-a) \, ||g - p||_\infty + (b-a) \, ||p - q||_\infty$$

$$\leq \quad \frac{\epsilon}{4} + \frac{\epsilon}{4} + \frac{\epsilon}{4} < \epsilon$$

Since the set of all polynomials on $[a, b]$ with rational coefficients is countable, we see we have shown $\mathcal{L}_1$ is separable here. ∎

Comment 12.5.1 *It is straightforward to extend these results to any $1 \leq p < \infty$. We can also extend the separability result to a bounded subset Ω of $\Re^k$ but we will not do that here.*

Comment 12.5.2 *We see Theorem 12.4.6 tells us the continuous functions with compact support are dense in $\mathcal{L}_1$ on $\Re^k$.*

Exercise 12.5.1 *Let f be Riemann integrable on $[a, b]$, Then any sequence of partitions (P_n) with $||P_n|| \to 0$ gives $S(f, P_n, E_n) \to RI(f, a, b)$ where E_n is any choice of inbetween sets in P_n.*

- *Prove $L(f, P_n) \to RI(f, a, b)$ also.*

- *Hence, prove there is a sequence of continuous functions (ϕ_n) with $\int (h_n - \phi_n) d\mu < 1/n$ (the $\int \cdot d\mu$ denotes the Lebesgue integral). Here h_n is the simple function we defined in the previous exercise. Hence, there is a sequence of continuous functions which converge in $|| \cdot ||_1$ in $\mathcal{L}_1([a, b])$ to the Riemann integrable function f.*

- *Prove this implies the continuous functions are dense in $\mathcal{L}_1[a, b]$.*

Exercise 12.5.2 *Do you think the space we get by completing the metric space $RI([a, b])$ using the $|| \cdot ||_2$ distance measure or equivalently the space we get by completing the metric space $C([a, b])$ with the $|| \cdot ||_2$ metric and $\mathcal{L}_2([a, b])$, as we construct it here using Lebesgue measure, are the same even though we constructed them differently? Look back at our discussions of these metric completions in (Peterson (9) 2020) to refresh your mind on these matters. How would you prove this assertion?*

12.6   The Existence of Non-Lebesgue Measurable Sets

We now show that there are subsets of $\Re^n$ which are not Lebesgue measurable. We begin by showing Lebesgue measure is translation invariant: this means if E is measurable, so it $t + E$ for any t in $\Re^n$ and $\mu(E) = \mu(E + t)$.

Theorem 12.6.1 Lebesgue Measure is Translation Invariant

> *Let E in $\Re^n$ be Lebesgue measurable and let t in $\Re^n$ be arbitrary. Then, $\mu^*(E) = \mu^*(t + E)$, $t + E$ is also measurable and hence, $\mu(E) = \mu(T + E)$.*

Proof 12.6.1
We will provide a sketch. The proof is left as an exercise.

Step 1: *First, to show $\mu^*(E) = \mu^*(t + E)$ for all t, we use a standard Lebesgue Cover argument.*

Step 2:

We show if E is measurable, so is $t + E$. This is done by showing the set $t + E$ satisfies the Caratheodory condition.

BASIC ANALYSIS IV: MEASURE THEORY AND INTEGRATION

- *Prove $T \cap (t + B) = (T - t) \cap B + t$ for all t and sets B and T.*

- *Prove $(t + B)^C = t + B^C$.*

- *Finally, prove $t + E$ satisfies the Caratheodory condition which essentially finishes the proof.*

■

The translation invariance of Lebesgue measure is important in the construction of a non-Lebesgue measurable set. We establish this via two preliminary results which we put into Lemmas and then the final theorem.

Lemma 12.6.2 Non-Measurable Set Lemma 1

Let θ be in $(0, 1)$. Let E be a measurable set in $\Re^n$ with $\mu(E) > 0$. Then, there is an open interval I so that $\mu(E \cap I) > \theta\mu(I)$.

Proof 12.6.2

First do the case $\mu(E)$ is finite. Start with a Lebesgue cover of E using the supremum tolerance lemma for positive epsilon δ. If you think about this carefully, by choosing δ right this can be phrased as

$$\mu(E) > \frac{1}{1 + \delta} \sum_n |I_n|.$$

Now pick δ so that $\frac{1}{1+\delta} = \theta$. Then we know $\mu(E) > \theta \sum_n |I_n|$. The result then follows.

If the measure of E is not finite and for all intervals J we had $\mu(E \cap J) = 0$, we would find $\mu(E) = 0$ as well which is not possible. So there must be one interval J with $\mu(E \cap J)$ positive. Now use the first part to get the result. ■

Lemma 12.6.3 Non-Measurable Set Lemma 2

Given E in $\Re^n$, we define the set of all differences $x - y$ for x and y in E to the difference set of E which we denote by E_d. If E is measurable in $\Re^n$ and if $\mu(E) > 0$, the E_d contains a neighborhood of 0.

Proof 12.6.3

This one is a bit trickier. First choose λ so that

$$1 - \frac{1}{2^{n+1}} < \lambda < 1.$$

Then by Lemma 12.6.2, we can find an open interval I so that $\lambda\mu(I) < \mu(E \cap I)$. Let δ be the smallest edge length of the interval I and let J be the open interval whose i^{th} edge is $(-\frac{\delta}{2}, \frac{\delta}{2})$.

- *Show if x is from J, then $I + x$ contains the center of I.*

- *For any x from J, look at I and $I + x$ on their i^{th} edge. For convenience, let the point x from J satisfy $0 \leq x_i < \frac{d_i}{2}$. Then, show the common part of I and $I + x$ here has length larger than $\frac{d_i}{2}$. Thus, $\mu(I \cap (I + x)) > \frac{1}{2^n} \mu(I)$.*

- *Then show for x in J, $\mu(I \cap (I + x)) < 2\lambda\mu(I)$.*

LEBESGUE MEASURE

- *For any x from J, let $A = E \cap I$ and $B = E \cap I + x$. Both $\mu(A)$ and $\mu(B)$ are larger than $\lambda \mu(I)$. If A and B were disjoint, this would imply $\mu(A \cup B) > 2\lambda\mu(I)$. However, we have shown $\mu(A \cup B) < 2\lambda\mu(I)$ which is a contradiction. Hence A and B cannot be disjoint.*

The above tells us $E \cap I$ and $(E \cap I) + x$ have a common point for each x in J. It then follows we can write x as $u - v$ for some u and v from E implying x is in E_d. This completes the proof. ∎

We can now show there is a set in $\Re^n$ which is not Lebesgue measurable. From this, we can then prove every measurable set of positive measure contains a non-measurable set.

Theorem 12.6.4 Non-Lebesgue Measurable Set

> *There is a non-Lebesgue measurable set in $\Re^n$. Further, any set of positive measure contains a non-measurable set.*

Proof 12.6.4
Let z be a fixed irrational number and let the set $M = Q$. It is easy to show M^n is countable and closed under addition and subtraction and dense in $\Re^n$.
Then, to prove the result:

- *Prove the relation $\sim$ on $\Re^n \times \Re^n$ defined by $x \sim y$ if $x - y$ is in M^n is an equivalence relation.*

- *Let $\mu_*(E)$ be defined by*

$$\mu_*(E) = \sup\{\mu(F) : F \subseteq E, \ F \ measurable\}.$$

  Prove if E is measurable, then $\mu(E) = \mu_*(E) = \mu^*(E)$.

- *Let $\mathscr{M}$ be the collection of equivalence classes determined by $\sim$ and use the Axiom of Choice (look this up!) to pick a unique point from each equivalence class to form the set S. Recall, we know that $\mu_*(S) = \sup\{\mu(E) : E \subseteq S, E \ measurable\}$. If $\mu(E) > 0$, use Lemma 12.6.3 to note E_d contains a neighborhood of 0. Since M is dense, $E_d \cap M$ contains a point x which is not 0. Prove this point $x = u - v$ with u, v from E implying $u \sim v$ with $x \neq 0$ which is not possible by the way S was constructed. This implies $\mu_*(E) = 0$. Thus, by the previous result, if S were measurable, $\mu(S) = 0$.*

- *Let (x_n) be an enumeration of M. Then, show $\Re^n = \cup_n (S + x_n)$ and prove that S measurable implies $\infty = 0$ giving a contradiction. We must therefore conclude that S cannot be a measurable set.*

- *Show if E is measurable with positive measure, then E contains a non-measurable set.*

∎

Homework

Exercise 12.6.1 *Prove Lemma 12.6.2 in detail.*

Exercise 12.6.2 *Prove Lemma 12.6.3 in detail.*

Exercise 12.6.3 *Prove Theorem 12.6.4 in detail.*

12.7 Metric Spaces of Finite Measure Sets

We can do many things with measurable sets. This material is a connection of sorts to the standard discussions of linear analysis.

Theorem 12.7.1 The Metric Space of Finite Measurable Sets

Let M be the collection of all Lebesgue measurable sets in $\Re^n$ with finite measure. Recall, the symmetric difference of sets A and B is

$$A\Delta B = \left(A \cap B^C\right) \cup \left(A^C \cap B\right)$$

Then

1. The relation $\sim$ on $M \times M$ defined by

$$A \sim B \;\; if \mu(A\Delta B) = 0$$

is an equivalence relation.

2. Let $\mathcal{M}$ denote the set of all equivalence classes with respect to this relation and let $[A]$ be a typical equivalence class with representative A. Then $\mathcal{M}$ is a metric space with metric D defined by

$$D([A],[B]) := \mu(A\Delta B).$$

Proof 12.7.1

A sketch of the proof is given below:

- First prove if A and B are in $\mathcal{M}$,

$$|\mu(A) - \mu(B)| \leq \mu(A\Delta B).$$

- Then, prove if $A \sim E$ and $B \sim F$, then $\mu(A\Delta B) = \mu(E\Delta F)$.
  Hint: Show
$$|\mu(A\Delta B) - \mu(E\Delta F)| \leq \mu(A\Delta E) + \mu(B\Delta F).$$

- Prove D is a metric which requires you prove D is well-defined by showing its value is independent of the choice of equivalence class representatives.

$\blacksquare$

Theorem 12.7.2 The Metric Space of All Finite Measure Sets is Complete

The metric space $(\mathcal{M}, D)$ is complete.

Proof 12.7.2

We provide a detailed sketch. Let (E_n) be a Cauchy sequence.

1. Prove there is a subsequence (E_{n_m}) of (E_n) so that

$$\mu(E_{n_m}\Delta E_n) < \frac{1}{2^m} \;\; for \; all \;\; n > n_m,$$

$$\mu(E_{n_k}\Delta E_{n_m}) < \frac{1}{2^m} \;\; for \; all \;\; n_k > n_m$$

2. For expositional convenience, let $G_m = E_{n_m}$ in what follows. Let $H = \limsup G_m$ and $G = \liminf G_m$.

- *Prove $\mu(H \setminus G_m) \to 0$ and $\mu(G_m \setminus G) \to 0$.*

- *Prove $\mu(H \Delta G_m) \to 0$. This argument uses the disjoint decomposition idea from Lemma 9.3.3.*
  Step 1: Since
  $$\mu(H^C \cap G_j) = \lim_m \cap_{n \geq m}(G_n^C \cap G_j)$$

  overestimate the right hand side by $\mu(G_n^C \cap G_j)$ which is also bounded above by $\mu(G_n \Delta G_j)$ which is as small as we want. This shows $\mu(H^C \cap G_j) \to 0$.
  Step 2: First note
  $$H \cap G_j^C = \cap_m \cup_{n \geq m} G_n \cap G_j^C$$

  which shows $H \cap G_j^C$ is a decreasing limit of sets. Now look at $\cup_{n \geq m} G_n \cap G_j^C$ which is contained in G_j^C always. However, since the measure of G_j is finite, the measure of its complement is not, so this upper bound is not very useful. For convenience, let $B_m = \cup_{n \geq m} G_n$. To understand how to use the properties of our sequence G_n, we rewrite the union B_m like this:

  $$
  \begin{aligned}
  U_m &= G_m \\
  U_{m+1} &= G_{m+1} \setminus (G_m) \\
  U_{m+2} &= G_{m+2} \setminus (G_m \cup G_{m+1}) \\
  &\vdots \\
  U_{m+n} &= G_{m+n} \setminus (G_m \cup G_{m+1} \cup \cdots \cup G_{m+n-1}) \\
  &\vdots
  \end{aligned}
  $$

  From the above, we see for all $m > j$

  $$\mu(B_m \cap G_j) = \sum_{n=0}^{\infty} \mu(U_{m+n} \cap G_j^C)$$

  However, we can overestimate a bit using

  $$
  \begin{aligned}
  \mu(U_m \cap G_j^C) &< \frac{1}{2^m} \\
  \mu(U_{m+1} \cap G_j^C) &= \mu(G_{m+1} \cap G_m^C \cap G_j^C) \\
  &\leq \mu(G_{m+1} \cap G_m^C) \\
  &< \frac{1}{2^m}. \\
  \mu(U_{m+2} \cap G_j^C) &= \mu(G_{m+2} \cap G_m^C \cap G_{m+1}^C \cap G_j^C) \\
  &\leq \mu(G_{m+2} \cap G_{m+1}^C) \\
  &< \frac{1}{2^{m+1}}. \\
  &\vdots \\
  \mu(U_{m+n} \cap G_j^C) &= \mu(G_{m+n} \cap G_m^C \cap \cdots \cap G_{m+n-1}^C \cap G_j^C) \\
  &\leq \mu(G_{m+n} \cap G_{m+n-1}^C) \\
  &< \frac{1}{2^{m+n-1}}
  \end{aligned}
  $$

BASIC ANALYSIS IV: MEASURE THEORY AND INTEGRATION

Thus,

$$
\begin{aligned}
\mu(B_m \cap G_j) \quad &< \quad \frac{1}{2^m} + \sum_{n=1}^{\infty} \frac{1}{2^{m+n-1}} \\
&= \quad \frac{1}{2^m} + \frac{1}{2^{m-1}}
\end{aligned}
$$

We conclude

$$
\mu(H \setminus G_j) \quad = \quad \lim_m \mu(B_m \cap G_j^C) \leq \lim_n \left(\frac{1}{2^m} + \frac{1}{2^{m-1}} \right) = 0
$$

This shows the desired result.

3. *Prove* $\mu(H \Delta E_m) \to 0$ *also which proves completeness.*

Comment 12.7.1 *We can also prove* $\mu(G \Delta H) = 0$ *so that* $G \sim H$.

Homework

Exercise 12.7.1 *Prove if A and B are in $\mathscr{M}$, $|\mu(A) - \mu(B)| \leq \mu(A \Delta B)$ from Theorem 12.7.1.*

Exercise 12.7.2 *Prove if $A \sim E$ and $B \sim F$, then $\mu(A \Delta B) = \mu(E \Delta F)$ from Theorem 12.7.2.*

Exercise 12.7.3 *Prove D is a metric which requires you prove D is well-defined by showing its value is independent of the choice of equivalence class representatives.*

Exercise 12.7.4 *In Theorem 12.7.2 prove $\mu(H \setminus G_m) \to 0$ and $\mu(G_m \setminus G) \to 0$.*

Exercise 12.7.5 *In Theorem 12.7.2 prove $\mu(H \Delta E_m) \to 0$.*

Chapter 13

Cantor Set Experiments

We now begin a series of investigations into the construction of an important subset of $[0, 1]$ called the Cantor Set. We follow a great series of homework exercises outlined, without solutions, in a really hard but extraordinarily useful classical analysis text by Stromberg, (Stromberg (13) 1981).

13.1 The Generalized Cantor Set

Let (a_n) for $n \geq 0$ be a fixed sequence of real numbers which satisfy

$$a_0 = 1, \ 0 < 2a_n < a_{n-1} \tag{13.1}$$

Define the sequence (d_n) by

$$d_n = a_{n-1} - 2a_n$$

Note each $d_n > 0$. We can use the sequence (a_n) to define a collection of intervals $J_{n,k}$ and $I_{n,k}$ as follows.

(0) $J_{0,1} = [0, 1]$ which has length a_0.

(1) $J_{1,1} = [0, a_1]$ and $J_{1,2} = [1 - a_1, 1]$. You can see each of these intervals has length a_1. We let $W_{1,1} = J_{1,1} \cup J_{1,2}$ and $I_{1,1} = J_{0,1} - W_{1,1}$ where the *minus* symbol used here represents the set difference. This step creates an open interval of $[0, 1]$ which has length $d_1 > 0$. Let $P_1 = J_{1,1} \cup J_{1,2}$. This is a closed set.

(2) Set $J_{2,1} = [0, a_2]$, $J_{2,2} = [a_1 - a_2, a_1]$, $J_{2,3} = [1 - a_1, 1 + a_2 - a_1]$, and $J_{2,4} = [1 - a_2, 1]$. These 4 closed subintervals have length a_2. It is not so mysterious how we set up the $J_{2,k}$ intervals. Step (1) created a closed interval $[0, a_1]$, an open interval $(a_1, 1 - a_1)$ and another closed interval $[1 - a_1, 1]$. The first closed subinterval is what we have called $J_{1,1}$. Divide it into three parts; the first part will be a closed interval that starts at the beginning of $J_{1,1}$ and has length a_2 and the third part will be a closed interval of length a_2 that ends at the last point of $J_{1,1}$. When these two closed intervals are subtracted from $J_{1,1}$, an open interval will remain. The length of $J_{1,1}$ is a_1. So the open interval must have length $a_1 - 2a_2 = d_2$. A little thought tells us that the first interval must be $[0, a_2]$ (which we have named $J_{2,1}$) and the third interval must be $[a_1 - a_2, a_1]$ (which we have named $J_{2,2}$). To get the intervals $J_{2,3}$ and $J_{2,4}$, we divide $J_{1,2}$ into the same type of three subintervals as we did for $J_{1,1}$. The first and third must have length a_2 which will give an open interval in the inside of length d_2. This will give $J_{2,3} = [1 - a_1, 1 - a_1 + a_2]$ and $j_{2,4} = [1 - a_2, 1]$.

301

Then let $W_{2,1} = J_{2,1} \cup J_{2,2}$, and $W_{2,2} = J_{2,3} \cup J_{2,4}$. Then create new intervals by letting $I2,1 = J_{1,1} - W_{2,1}$ and $I2,2 = J_{1,2} - W_{2,2}$. We have now created 4 open subintervals of length d_2. Let $P_2 = J_{2,1} \cup J_{2,2} \cup J_{2,3} \cup J_{2,4}$. We can write this more succinctly as $P_2 = \cup\{J_{2,k} | 1 <= k <= 2^2\}$. Again, notice that P_2 is a closed set that consists of 4 closed subintervals of length a_2.

Let's look even more closely at the details. A careful examination of the process above with pen and paper in hand gives the following table that characterizes the left-hand endpoint of each of the intervals $J_{2,k}$.

$J_{2,1}$	0
$J_{2,2}$	$a_2 + d_2$
$J_{2,3}$	$2a_2 + d_2 + d_1$
$J_{2,4}$	$3a_2 + 2d_2 + d_1$

Since we know the left-hand endpoint and the length is always a_2, this fully characterizes the subintervals $J_{2,k}$. Also, as a check, the last endpoint $3a_2 + 2d_2 + d_1$ plus one more a_2 should add up to 1. We find

$$
\begin{aligned}
4a_2 + 2d_2 + d_1 &= 4a_2 + 2(a_1 - 2a_2) + (a_0 - 2a_1) \\
&= a_0 = 1
\end{aligned}
$$

(3) Step (2) has created 4 closed subintervals $J_{2,k}$ of length a_2 and 2 new open intervals $I_{2,i}$ of length d_2. There is also the first open interval $I_{1,1}$ of length d_1 which was abstracted from $[0, 1]$. Now we repeat the process described in Step (2) on each closed subinterval $J_{2,k}$. We do not need to use the auxiliary sets $W_{3,i}$ now as we can go straight into the subdivision algorithm. We divide each of these intervals into 3 pieces. The first and third will be of length a_3. This leaves an open interval of length d_3 between them. We label the new closed subintervals so created by $J_{3,k}$ where k now ranges from 1 to 8. The new intervals have left-hand endpoints

$J_{3,1}$	0
$J_{3,2}$	$a_3 + d_3$
$J_{3,3}$	$2a_3 + d_3 + d_2$
$J_{3,4}$	$3a_3 + 2d_3 + d_2$
$J_{3,5}$	$4a_3 + 2d_3 + d_2 + d_1$
$J_{3,6}$	$5a_3 + 3d_3 + d_2 + d_1$
$J_{3,7}$	$6a_3 + 3d_3 + 2d_2 + d_1$
$J_{3,8}$	$7a_3 + 4d_3 + 2d_2 + d_1$

Each of these subintervals have length a_3 and a simple calculation shows $(7a_3 + 4d_3 + 2d_2 + d_1) + a_3 = 1$ as desired. There are now 4 more open intervals $I_{3,i}$ giving a total of 6 open subintervals arranged as follows:

	Parent	Length
$I_{1,1}$	$J_{0,1}$	d_1
$I_{2,1}$	$J_{1,1}$	d_2
$I_{2,2}$	$J_{1,2}$	d_2
$I_{3,1}$	$J_{2,1}$	d_3
$I_{3,2}$	$J_{2,2}$	d_3
$I_{3,3}$	$J_{2,3}$	d_3
$I_{3,4}$	$J_{2,4}$	d_4

CANTOR SETS

We define $P_3 = \cup\{J_{3,k} | 1 <= k <= 2^3\}$ and note that $P_1 \cap P_2 \cap P_3 = P_3$.

We can, of course, continue this process recursively. Thus, after Step n, we have constructed 2^n closed subintervals $J_{n,k}$ each of length a_n. The union of these subintervals is labeled P_n and is therefore defined by $P_n = \cup\{J_{n,k} | 1 <= k <= 2^n\}$. The left-hand endpoints of $J_{n,k}$ can be written in a compact and illuminating form, but we will delay working that out until later. Now, we can easily see the form of the left-hand endpoints for the first few intervals:

$J_{n,1}$	0
$J_{n,2}$	$a_n + d_n$
$J_{n,3}$	$2a_n + d_n + d_{n-1}$
$J_{n,4}$	$3a_n + 2d_n + d_{n-1}$

Definition 13.1.1 The Generalized Cantor Set

> Let (a_n), $n \geq 0$ satisfy Equation 13.1. Let (a_n) for $n \geq 0$ be a fixed sequence of real numbers which satisfy $a_0 = 1$, $0 < 2a_n < a_{n-1}$. We call such a sequence a **Cantor Set Generating Sequence** and we define the generalized Cantor Set C_a generated by (a_n) to be the set $P = \cap_{n=1}^{\infty} P_n$, where the sets P_n are defined recursively via the discussion in this section.

Comment 13.1.1 *The Cantor set generated by the sequence $(1/3^n)$, $n \geq 0$ is very famous and is called the* Middle Thirds *set because we are always removing the middle third of each interval in the construction process. We will denote the Middle Third Cantor set by* C.

Comment 13.1.2 *For example, if $a_n = (2/3)(1/5^n) + (1/3)(1/2^n)$, it is a generalized Cantor Generating Sequence and $\lim_n 2^n a_n = 1/3$.*

Homework

The details of working out the Cantor construction and its consequences are given as a project in (Peterson (8) 2020) in the second semester.

Exercise 13.1.1 *Write out the explicit endpoints of all these intervals up to and including Step 4. Illustrate this process with clearly drawn tables and graphs.*

Exercise 13.1.2 *Write out explicitly P_1, P_2, P_3 and P_4. Illustrate this process with clearly drawn tables and graphs.*

Exercise 13.1.3 *Do the above two steps for the choice $a_n = 3^{-n}$ for $n >= 0$. Illustrate this process with clearly drawn tables and graphs.*

Exercise 13.1.4 *Do the above two steps for the choice $a_n = 5^{-n}$ for $n >= 0$. Illustrate this process with clearly drawn tables and graphs.*

Exercise 13.1.5 *As mentioned, the above construction process can clearly be handled via induction. Prove the following:*

(a) $P_{n-1} - P_n = \cup\{I_{n,k} \mid 1 <= k <= 2^{n-1}\}$

(b) Let $P = \cap_{n=0}^{\infty} P_n$. Then $P_0 - P = \cup_{n=1}^{\infty} \left(P_{n-1} - P_n \right)$

Exercise 13.1.6 *Prove any Cantor set is in the Borel σ-algebra $\mathcal{B}$.*

Exercise 13.1.7 *Explain why the Cantor sets are measurable and their measure is the same as their outer measure.*

13.2 Representing the Generalized Cantor Set

We are now in a position to prove additional properties about the Cantor Set C_a for a Cantor generating sequence (a_n). Associate with (a_n) the sequence (r_n) whose entries are defined by $r_n = a_{n-1} - a_n$. Let S denote the set of all sequences of real numbers whose values are either 0 or 1; i.e. $S = \{x = (x_n) \mid x_n = 0 \text{ or } x_n = 1\}$. Now define the mapping $f_a : S \to C_a$ by

$$f_a(x) \;=\; \sum_{n=1}^{\infty} x_n r_n \tag{13.2}$$

Theorem 13.2.1 Representing the Cantor Set

> 1. f_a is well-defined.
>
> 2. $f_a(x)$ is an element of C_a.
>
> 3. f_a is $1-1$ from S to C_a.
>
> 4. f_a is onto C_a.

Proof 13.2.1
You will prove this Theorem by establishing a series of results in the exercises below. ∎

Homework

Exercise 13.2.1 *For any Cantor generating sequence (a_n), we have $\lim_n a_n = 0$.*

Exercise 13.2.2 *Show*

$$\sum_{j=n+1}^{\infty} r_j \;=\; \lim_m \sum_{j=n+1}^{m} r_j \;=\; \lim_m (a_n - a_m) \;=\; a_n$$

Exercise 13.2.3 $r_n > \sum_{j=n+1}^{\infty} r_j.$

Exercise 13.2.4 *For $n >= 1$ and any finite sequence $(x_1, x_2, ..., x_n)$ of 0's and 1's, define the closed interval*

$$J(x_1, ..., x_n) \;=\; [\sum_{j=1}^{n} x_j r_j, a_n + \sum_{j=1}^{n} x_j r_j]$$

Show

1. *Show $J(0) = [0, a_1] = J_{1,1}$.*

2. *Show $J(1) = [r_1, a_1 + r_1] = [1 - a_1, 1] = J_{1,2}$.*

3. *Now use induction on n to show that the intervals $J(x_1, ..., x_n)$ are exactly the 2^n intervals $J_{n,k}$ for $1 <= k <= 2^n$ that we described in the previous section.*

   Hint 13.2.1 *Assume true for $n - 1$. Then we can assume that there is a unique $(x_1, ..., x_{n-1})$ choice so that $J_{n-1,k} = J(x_1, ..., x_{n-1})$.*

CANTOR SETS   305

Recall how the J's are constructed. At Step $n-1$, the interval $J_{n-1,k}$ is used to create 2 more intervals on level n by removing a piece. The 2 intervals that are left both have length a_n and we would denote them by $J_{n,2k-1}$ and $J_{n,2k}$. Now use the definition of the closed intervals $J(x_1, ..., x_n)$ to show that (remember our $x_1, ..., x_{n-1}$ are fixed)

$$J(x_1, ..., x_{n-1}, 0) \;=\; J_{n,2k-1}$$
$$J(x_1, ..., x_{n-1}, 1) \;=\; J_{n,2k}$$

This will complete the induction. $\square$

Exercise 13.2.5 *Let x be in S. Show that $f_a(x)$ is in $J(x_1, ..., x_n)$ for each n.*
Sketch of Argument: We know that each $J(x_1, ..., x_n) = J_{n,k}$ for some k. Let this k be written $k(x, n)$ to help us remember that it depends on the x and the n. Also remember that $1 <= k(x, n) <= 2^n$. So $f_a(x)$ is in $J_{n,k(x,n)}$ which is contained in P_n Hence, $f_a(x)$ is in P_n for all n which shows $f_a(x)$ is in C_a. This shows f_a maps S into C_a.

Exercise 13.2.6 *Now let x and y be distinct in S. Choose an index j so that x_j is different from y_j. Show this implies that $f_a(x)$ and $f_a(y)$ then belong to different closed intervals on the jth level. This implies $f_a(x)$ is not the same as $f_a(y)$ and so f is $1 - 1$ on S.*

Exercise 13.2.7 *Show f_a is surjective. To do this, let z be in C_a. Since z is in P_1, either z is in $J(0)$ or z is in $J(1)$. Choose x_1 so that z is in $J(x_1)$. Then assuming $x_1, ..., x_{n-1}$ have been chosen, we have z is in $J(x_1, ..., x_{n-1})$. Now z is in*

$$P_n \cap J(x_1, ..., x_{n-1}) = J(x_1, ..., x_{n-1}, 0) \cup J(x_1, ..., x_{n-1}, 1)$$

This tells us how to choose x_n.

Hence, by induction, we can find a sequence (x_n) in S so that z is in intersection over n of $J(x_1, ..., x_n)$. But by our earlier arguments, $f_a(x)$ is in the same intersection!

Finally, each of these closed intervals has length a_n which we know goes to 0 in the limit on n. So z and $f_a(x)$ are both in a decreasing sequence of sets whose lengths go to 0. Hence z and $f_a(x)$ must be the same (this uses what is called the Cantor Intersection Theorem).

We can also prove a result about the internal structure of the generalized Cantor set: it cannot contain any open intervals.

Exercise 13.2.8 *Prove C_a contains no open intervals.*

Exercise 13.2.9 *In addition, we have the following result: The limit of $2^n a_n$ always exists and is in $[0, 1]$.*

13.3   The Cantor Functions

We now prove additional interesting results that arise from the use of generalized Cantor sets via a series of exercises that you complete. As usual, let (a_n) be a Cantor Set generating sequence. Using the function f_a defined in the previous section, let's define the mapping ϕ_a by

$$\phi((x_n)) \;=\; \sum_{j=1}^{\infty} x_j \, (1/2^j)$$

306   *BASIC ANALYSIS IV: MEASURE THEORY AND INTEGRATION*

Hence, $\phi : S \to [0,1]$. and $\phi \circ f_a : S \to [0,1]$. Let the mapping $\Psi_a = \phi \circ f_a^{-1}$. Note $\Psi_a : C_a \to [0,1]$. We call Ψ_a the Cantor Function associated with the Cantor set (a_n).

Homework

Exercise 13.3.1 *ϕ maps S one to one and onto $[0,1]$ with a suitable restriction on the base 2 representation of a number in $[0,1]$.*

Exercise 13.3.2 *$x < y$ in C_a implies $\Psi_a(x) \leq \Psi_a(y)$.*

Exercise 13.3.3 *$\Psi_a(x) = \Psi_a(y)$ if and only if (x,y) is one of the intervals removed in the Cantor set construction process, i.e.*

$$(x,y) \quad = \quad \left(\sum_{j-1}^{n-1} x_j r_j + a_n, \sum_{j=1}^{n-1} x_j r_j + r_n \right)$$

Exercise 13.3.4 *In the case where $\Psi_a(x) = \Psi_a(y)$ extend the mapping Ψ_a to $[0,1] - C_a$ by*

$$\Psi_a(t) = \Psi_a(x) = \Psi_a(y), \ x < t < y$$

Finally, define $\Psi_a(0) = 0$ and $\Psi_a(1) = 1$. Prove $\Psi_a : [0,1] \to [0,1]$ is a non-decreasing continuous map of $[0,1]$ onto $[0,1]$ and is constant on each component interval *of $[0,1] - C_a$ where component interval means the $I_{n,k}$ sets we constructed in the Cantor set construction process.*

Comment 13.3.1 *For C, the Cantor set constructed from the sequence $(1/3^n)$, we denote Ψ_a simply as Ψ and call it the* **Lebesgue Singular Function**.

Exercise 13.3.5 *Define the mapping $g : [0,1] \to [0,1]$ by $g(x) = (\Psi(x) + x)/2$. Prove g is strictly increasing and continuous from $[0,1]$ onto $[0,1]$.*

Exercise 13.3.6 *Prove that*

$$g(\sum_{j=1}^{\infty} x_j \, r_j) = \sum_{j=1}^{\infty} x_j \, r_j'$$

where $r_j{}' = (1/2^j + r_j)/2$.

Exercise 13.3.7 *Prove $C' = g(C)$ is also a generalized Cantor set.*

Comment 13.3.2 *Note that the sequence $a_j' = (1/2)(1/2^j + a_j)$ is also a Cantor generating sequence that gives the desired $r_j{}'$ for the previous exercise.*

Exercise 13.3.8 *Compute the Lebesgue measure of the Cantor set generated by (a_n) and also the Lebesgue measure of the Cantor set $C' = g(C)$.*

Hint 13.3.1 *We know P_n is measurable as it is a finite union of intervals. We know the intervals are disjoint and $\mu^*(P_n) = 2^n a_n$. Hence $\mu^*(C_a) = \lim_n 2^n a_n$ which we know is some number in $[0,1]$. The Cantor middle third set has measure zero but the cantor set generated by $a_n = 1/(5^n) + (1/3)(1/2^n)$ satisfies $\lim_n 2^n a_n = 1/3$ and so has measure $1/3$.*

Exercise 13.3.9 *Prove the Lebesgue measure of $g(C) = 1/2$ and compute the Lebesgue measure of $(g(C))^C$.*

CANTOR SETS 307

Exercise 13.3.10 *Define the mapping* $g_a : [0,1] \to [0,1]$ *by* $g_a(x) = (\Psi_a(x) + x)/2$. *Prove* g_a *is strictly increasing and continuous from* $[0,1]$ *onto* $[0,1]$.

Exercise 13.3.11 *Compute Lebesgue measure of the Cantor set* $C'_a = g_a(C_a)$ *and compute the Lebesgue measure of* $(g_a(C))^C$.

Comment 13.3.3 *Thus if we call the Cantor set generated by* $a_n = 1/(5^n) + (1/3)(1/2^n)$, D, *we have*

$$1 = \mu([0,1]) \;=\; \mu(D) + \mu(D^C)$$

implying $\mu(D^C) = 2/3$. **Note D is a subset of** $[0,1]$ **with measure** $1/3$, **which does not contain any open intervals**! *We can do this sort of argument for* $g_a(D)$ *also.*

Next, we show this function g is of great importance in developing a better understanding of measures.

13.4 Interesting Consequences

These constructions allow us to understand some subtle things.

There are Lebesgue measurable sets that are not Borel sets: As usual, let C be the middle thirds Cantor set with Lebesgue measure 0. Further, if Ψ is the mapping we defined earlier associated with C, the mapping $g : [0,1] \to [0,1]$ defined by $g(x) = (\Psi(x) + x)/2$ is quite nice. It is $1 - 1$, onto, strictly increasing and continuous. We also showed in the exercises that $g(C)$ is another Cantor set with measure $1/2$. We denote Borel measure by μ_B as usual and to be perfectly clear, we let μ_L be Lebesgue measure. We know

$$\mu_B(C) \;=\; \mu_L(C) = 0$$
$$\mu_B(g(C)) \;=\; \mu_L(g(C)) = 1/2$$

Theorem 12.6.4 then says any Lebesgue measurable set with positive Lebesgue measure must contain a subset which is not in the Lebesgue sigma-algebra. Since $\mu_L(g(C)) = 1/2$, there is a set $F \subseteq g(C)$ which is not in L. Thus, $g^{-1}(F) \subseteq C$ which has Lebesgue measure 0. Lebesgue measure is a measure which has the property that every subset of a set of measure 0 must be in the Lebesgue sigma-algebra. Then, using the monotonicity of μ_L, we have $\mu_L(g^{-1}(F))$ is also 0. From the above remarks, we can infer something remarkable.

Let the mapping h be defined to be g^{-1}. Then h is also continuous and hence it is measurable with respect to the Borel sigma-algebra. Note since $B \subseteq L$, this tells us immediately that h is also measurable with respect to the Lebesgue sigma-algebra. Thus, $h^{-1}(U)$ is in the Borel sigma-algebra for all Borel sets U. But we know $h^{-1} = g$, so this tells us $g(U)$ is in the Borel sigma -algebra if U is a Borel set. Hence, if we chose $U = g^{-1}(F)$, then $g(U) = F$ would have to be a Borel set if U is a Borel set. However, we know that F is not in L and so it is also not a Borel set. We can only conclude that $g^{-1}(F)$ cannot be a Borel set. However, $g^{-1}(F)$ is in the Lebesgue sigma-algebra. Thus, there are Lebesgue measurable sets which are not Borel! Thus, the Borel sigma-algebra is strictly contained in the Lebesgue sigma-algebra!

There are functions equal a.e. with respect to Borel measure with only one Borel measurable: We can use this example to construct another remarkable thing. Using all the notations from above, note the indicator function of C^C, the complement of C, is defined by

$$I_{C^C}(x) \;=\; \begin{cases} 1 & x \in C^C \\ 0 & x \in C \end{cases}$$

We see $f = I_{C^C}$ is Borel measurable. Next, define a new mapping like this:

$$\phi(x) = \begin{cases} 1 & x \in \boldsymbol{C}^C \\ 2 & x \in \boldsymbol{C} \setminus g^{-1}(F) \\ 3 & x \in g^{-1}(F) \end{cases}$$

Note that $\phi = f$ a.e. with respect to Borel measure. However, ϕ is **not** Borel measurable because $\phi^{-1}(3)$ is the set $g^{-1}(F)$ which is not a Borel set. We conclude that in this case, even though the two functions were equal a.e. with respect to Borel measure, only one was Borel measurable! The reason this happens is that even though $\boldsymbol{C}$ has Borel measure 0, there are subsets of $\boldsymbol{C}$ which are not Borel sets! We discussed this example earlier in Chapter 9 as well.

Also, note the g is Riemann Integrable as it is monotone and continuous; indeed uniformly continuous on $[0, 1]$.

Chapter 14

Lebesgue - Stieltjes Measure

We now show how to construct a very important class of measures using right continuous monotonic functions on $\Re$. We know that if λ on $\Re$ is a measure, then for any point b

$$\lim_{x \to b} \lambda\Big((x, b] \Big) = \lambda\Big((a, b] \Big).$$

In fact, if we define the function h by $h(x) = \lambda([a, x])$, then we see that h must be continuous from the right at each a; i.e. $\lim_{x \to a^+} h(x) = h(a^+) = h(a)$. On the other hand, using a monotonic sequence of sets that decrease, we see

$$\lim_{x \to b^-} \lambda\Big((a, x] \Big) = h(b^-).$$

Hence, it is not required that h must be continuous from the left at each b. Now if g is any monotone increasing function on $[a, b]$, g is of bounded variation and so it always has right- and left-hand limits at any point. From the discussion above, we see we can define outer measures in a more general fashion. For Lebesgue outer measure, we use the premeasure $\tau((a, b]) = b - a$ which corresponds to the continuous, strictly monotone increasing choice $g(x) = x$. We can also define a premeasure τ_g by $\tau_g((a, b]) = g(b) - g(a)$. This will generate a measure in the usual way. Call this measure λ_g. Then if the set $[a, b]$ is measurable, we know we can write

$$\lambda_g\Big([a, b] \Big) = \lambda_g\Big(\{a\} \Big) + \lambda_g\Big((a, b] \Big).$$

But since

$$\lambda_g\Big(\{a\} \Big) = \lim_{\epsilon \to 0^+} \lambda_g\Big((a - \epsilon, a + \epsilon) \Big)$$
$$= g(a^+) - g(a^-),$$

we see for all $x > a$,

$$\lambda_g\Big([a, x] \Big) = g(a^+) - g(a^-) + g(x) - g(a).$$

Letting $x \to a^+$, we find

$$\lambda_g\Big(\{a\} \Big) = \lim_{x \to a^+} \Big(g(a^+) - g(a^-) + g(x) - g(a) \Big)$$

$$= g(a^+) - g(a^-) + g(a^+) - g(a) = \lambda_g\Big(\{a\}\Big) + g(a^+) - g(a)$$

Simplifying, we have

$$g(a^+) = \lim_{x \to a^+} g(x) = g(a).$$

which tells us the monotonic non-decreasing function g we use to construct a new measure must be right continuous if we want the measurable sets to contain the closed sets too.

Homework

Exercise 14.0.1 *Let g be defined by*

$$g(x) = \begin{cases} x, & x < 0 \\ 2x, & 0 \le x < 1 \\ 3, & 1 = x \\ 5x^2 + 1, & 1 < x \le 2 \\ 22, & x > 2 \end{cases}$$

- *The premeasure $\tau_g((a,b]) = g(b) - g(a)$ generates the measure λ_g.*

- *Does this g satisfy the requirement to generate the measure λ_g we would like?*

- *How could you redefine g to make λ_g behave as we would like?*

Exercise 14.0.2 *Let g be defined by*

$$g(x) = \begin{cases} 0.5x, & x < -1 \\ 2, & x = -1 \\ x + 4, & -1 < x < 1 \\ 6, & 1 = x \\ x^2 + 7, & 1 < x < 2 \\ 12, & x = 2 \\ 2x^3, & 2 < x \le 3 \\ x^4, & x > 3 \end{cases}$$

- *The premeasure $\tau_g((a,b]) = g(b) - g(a)$ generates the measure λ_g.*

- *Does this g satisfy the requirement to generate the measure λ_g we would like?*

- *How could you redefine g to make λ_g behave as we would like?*

Exercise 14.0.3 *Let g_n, for $n \ge 2$, be defined by*

$$g_n(x) = \begin{cases} x, & x > 1 \\ 1, & x = 1 \\ 1/(k+1), & 1/(k+1) \le x < 1/k,\ 1 \le k < n \\ 0, & 0 \le x < 1/n \\ x, & x < 0 \end{cases}$$

- *The premeasure $\tau_{g_n}((a,b]) = g_n(b) - g_n(a)$ generates the measure λ_{g_n}.*

- *Does g_n satisfy the requirement to generate the measure λ_{g_n} we would like?*

- *How could you redefine g_n to make λ_{g_n} behave as we would like?*

LEBESGUE - STIELTJES MEASURE

Exercise 14.0.4 *Let g be the pointwise limit of the sequence of functions (g_n) for $n \geq 2$ defined by*

$$
g_n(x) \quad = \quad \begin{cases} x, & x > 1 \\ 1, & x = 1 \\ 1/(k+1), & 1/(k+1) \leq x < 1/k,\ 1 \leq k < n \\ 0, & 0 \leq x < 1/n \\ x, & x < 0 \end{cases}
$$

- *The premeasure $\tau_g((a,b]) = g_n(b) - g_n(a)$ generates the measure λ_g.*

- *Does g satisfy the requirement to generate the measure λ_g we would like?*

- *How could you redefine g to make λ_g behave as we would like?*

14.1   Lebesgue - Stieltjes Outer Measure and Measure

From the discussion above, to construct new measures, we need to choose any g which is non-decreasing function on $\Re$ and continuous from the right. Moreover, the unbounded limits are well-defined $\lim_{x \to -\infty} g(x)$ and $\lim_{x \to \infty} g(x)$. These last two limits could be $-\infty$ and ∞ respectively. Then, define the mapping τ_g on $\mathcal{U}$ by

$$
\begin{aligned}
\tau_g(\emptyset) &= 0, \\
\tau_g\Big((a,b]\Big) &= g(b) - g(a), \\
\tau_g\Big((-\infty,b]\Big) &= g(b) - \lim_{x \to -\infty} g(x), \\
\tau_g\Big((a,\infty)\Big) &= \lim_{x \to \infty} g(x) - g(a), \\
\tau_g\Big((-\infty,\infty)\Big) &= \lim_{x \to \infty} g(x) - \lim_{x \to -\infty} g(x)
\end{aligned}
$$

This defines τ_g on the collection of sets $\mathcal{U}$ consisting of the empty set, intervals of the form $(a,b]$ for finite numbers a and b and unbounded intervals of the form $(-\infty,b]$ and (a,∞).

Let $\mathcal{A}$ be the algebra generated by finite unions of sets from $\mathcal{U}$. Note $\mathcal{A}$ contains $\Re$.

Let's extend the mapping τ_g to be additive on $\mathcal{A}$. If $E_1, E_2, \ldots, E_n$ is a finite collection of disjoint sets in $\mathcal{A}$, we extend the definition of τ_g to this finite disjoint union as follows:

$$
\tau_g\left(\bigcup_{i-1}^{n} E_i\right) \quad = \quad \sum_{i=1}^{n} \tau_g(E_i) \tag{14.1}
$$

Lemma 14.1.1 Extending τ_g to Additive is Well-Defined

The extension of τ_g from $\mathcal{U}$ to the algebra $\mathcal{A}$ is well-defined; hence, τ_g is additive on $\mathcal{A}$.

BASIC ANALYSIS IV: MEASURE THEORY AND INTEGRATION

Proof 14.1.1

For $(a, b] \in \mathcal{A}$, write

$$(a, b] \;=\; \bigcup_{i=1}^{n} (a_i, b_i]$$

for any positive integer n with $a_1 = a$, $b_n = b$ and the in-between points satisfy $a_{i+1} = b_i$ for all i. Of course, there are many such decompositions of $(a, b]$ we could choose. Also, these are the only decompositions we can have. If we use the unbounded sets, we cannot recapture $(a, b]$ using a finite number of unions! Then, using Equation 14.1, we have

$$
\begin{aligned}
\tau_g((a, b]) \;&=\; \sum_{i=1}^{n} \tau_g((a_i, b_i]) \\
&=\; \sum_{i=1}^{n} g(b_i) \;-\; g(a_i)
\end{aligned}
$$

But since $a_{i+1} = b_i$, this sum collapses to

$$\tau_g((a, b]) \;=\; g(b) \;-\; g(a)$$

This was the original definition of τ_g on the element $(a, b]$ in $\mathcal{U}$. We conclude the value of τ_g on elements of the form $(a, b]$ is independent of the choice of decomposition of it into a finite union of sets from $\mathcal{U}$.

For an unbounded interval of the form (a, ∞), any finite disjoint decomposition can have only one interval of the form (b, ∞) giving $(a, \infty) = (a, b] \cup (b, \infty)$, with the piece $(a, b]$ written as any finite disjoint union $(a, b] = \cup_{i=1}^{n} (a_i, b_i]$ as before. The same arguments as used above then show τ_g is well-defined on this type of element of $\mathcal{U}$ also. We handle the sets $(-\infty, b]$ is a similar fashion.

Next, if we look at any arbitrary A in $\mathcal{A}$, then A can be written as a finite union of members $A_1, \ldots, A_p$ of $\mathcal{U}$. Each of these elements A_i can then be written using a finite disjoint decomposition into intervals $(a_{ij}, b_{ij}]$, $1 \leq j \leq p(i)$ as we have done above. Thus,

$$A \;=\; \cup_{i=1}^{m} \cup_{j=1}^{p(i)} (a_{ij}, b_{ij}]$$

where we abuse notation, for convenience, by noting it is possible $a_{11} = -\infty$ and $b_{m\,p(m)} = \infty$. We simply interpret a set of the form $(a, \infty]$ as (a, ∞). We then combine these intervals and relabel as necessary to write A as a finite disjoint union

$$A \;=\; \cup_{i=1}^{N} (a_i, b_i]$$

with $b_i \leq a_{i+1}$ and again it is possible that $a_1 = -\infty$ and $b_N = \infty$. We therefore know that

$$\tau_g(A) \;=\; \cup_{i=1}^{N} \tau_g((a_i, b_i])$$

Now assume A has been decomposed into another finite disjoint union, $A = \cup_{j=1}^{M} B_j$, each $B_j \in \mathcal{A}$. Let

$$C_j \;=\; \{ i \,|\, (a_i, b_i] \subseteq B_j \}$$

Note a given interval $(a_i, b_i]$ cannot be in two different sets B_j and B_k because they are assumed disjoint. Hence, we have

$$B_j \;=\; \cup_{i \in C_j} (a_i, b_i]$$

and

$$\tau_g(B_j) \;=\; \sum_{i \in C_j} \tau_g((a_i, b_i])$$

Thus,

$$\sum_{j=1}^{M} \tau_g(B_j) \;=\; \sum_{j=1}^{M} \sum_{i \in C_j} \tau_g((a_i, b_i])$$

$$=\; \sum_{i=1}^{N} \tau_g((a_i, b_i])$$

This shows that our extension for τ_g is independent of the choice of finite decomposition and so the extension of τ_g is a well-defined additive map on $\mathcal{A}$. ∎

We can now apply Theorem 11.5.3 to conclude that since the covering family $\mathscr{A}$ is an algebra and τ_g is additive on $\mathscr{A}$, the σ-algebra, $\mathcal{M}_g$, generated by τ_g contains $\mathscr{A}$ and the induced measure, μ_g, is regular. Next, we want to know that $\mu_g(A) = \tau_g(A)$ for all A in $\mathcal{A}$. To do this, we will prove the extension τ_g is actually a pseudo-measure. Thus, we will be able to invoke Theorem 11.5.4 to get the desired result.

Lemma 14.1.2 τ_g is a Pseudo-Measure on $\mathcal{A}$

The mapping τ_g is a pseudo-measure on $\mathcal{A}$.

Proof 14.1.2
We need to show that if (T_n) is a sequence of disjoint sets from $\mathcal{A}$ whose union $\cup_n T_n$ is also in $\mathcal{A}$, then

$$\tau_g(\cup_n T_n) \;=\; \sum_n \tau_g(T_n)$$

First, notice that if there was an index n_0 so that $\tau_g(T_{n_0}) = \infty$, then letting $B = \cup_n T_n \setminus T_{n_0}$, we can write $\cup_n T_n$ as the finite disjoint union $B \cup T_{n_0}$ and hence

$$\tau_g(\cup_n T_n) \;=\; \tau_g(B) + \tau_g(T_{n_0}) = \infty$$

Since the right-hand side sums to ∞ in this case also, we see there is equality for the two expressions. Therefore, we can restrict our attention to the case where all the individual T_n sets have finite $\tau_g(T_n)$ values. This means no elements of the form $(-\infty, b]$ or (a, ∞) can be part of any decomposition of the sets T_n. Hence, we can assume each T_n can be written as a finite union of intervals of the form $(a, b]$. It follows then that it suffices to prove the result for a single interval of the form $(a, b]$.

Since τ_g is additive on finite unions, if $C \subseteq D$, we have

$$\tau_g(D) \;=\; \tau_g(C) + \tau_g(D \setminus C) \geq \tau_g(C)$$

Now assume we can write the interval (a, b) as follows:

$$(a, b] \;=\; \cup_{n=1}^{\infty} (a_i, b_i]$$

with the sets $(a_i, b_i]$ disjoint. For any n, we have

$$(a, b] \;=\; \cup_{k=1}^{n} (a_k, b_k] \;\cup\; \cup_{k=n+1}^{\infty} (a_k, b_k]$$

Therefore

$$\tau_g\Big((a, b] \Big) \;=\; \tau_g\Big(\cup_{k=1}^{n} (a_k, b_k] \Big) \;+\; \tau_g\Big(\cup_{k=n+1}^{\infty} (a_k, b_k] \Big)$$

The finite additivity on disjoint intervals then gives us

$$\tau_g\Big(\cup_{k=1}^{n} (a_k, b_k] \Big) \;=\; \sum_{k=1}^{n} \tau_g\Big((a_k, b_k] \Big)$$
$$=\; g(b_1) \,-\, g(a_1) \,+\, g(b_2) \,-\, g(a_2) \,+\, \ldots \,+\, g(b_n) \,-\, g(a_n)$$

We know g is non-decreasing, thus $g(b_1) - g(a_2) \le 0$, $g(b_2) - g(a_3) \le 0$, and so forth until we reach $g(b_{n-1}) - g(a_n) \le 0$. Dropping these terms, we find

$$\tau_g\Big(\cup_{k=1}^{n} (a_k, b_k] \Big) \;\le\; g(b_n) \,-\, g(a_1) \;\le\; g(b) \,-\, g(a)$$

Thus, these partial sums are bounded above and so the series of non-negative terms $\sum_n \tau_g((a_k, b_k])$ converges. This tells us that

$$\tau_g\Big(\cup_{k=1}^{\infty} (a_k, b_k] \Big) \;\le\; \tau_g\Big((a, b] \Big)$$

To obtain the reverse inequality, let $\epsilon > 0$ be given. Then, since the series above converges, there must be a positive integer N so that if $n \ge N$,

$$\sum_{k=n+1}^{\infty} \tau_g\Big((a_k, b_k] \Big) \;<\; \epsilon$$

We conclude that

$$\tau_g\Big((a, b] \Big) \;=\; \sum_{k=1}^{n} \tau_g\Big((a_k, b_k] \Big) \;+\; \tau_g\Big(\cup_{k=n+1}^{\infty} (a_k, b_k] \Big)$$
$$\ge\; \sum_{k=1}^{n} \tau_g\Big((a_k, b_k] \Big) \;+\; \tau_g\Big(\cup_{k=n+1}^{K} (a_k, b_k] \Big)$$
$$=\; \sum_{k=1}^{n} \tau_g\Big((a_k, b_k] \Big) \;+\; \sum_{k=n+1}^{K} \tau_g\Big((a_k, b_k] \Big)$$

We know that

$$\lim_{K} \sum_{k=n+1}^{K} \tau_g\Big((a_k, b_k] \Big) \;=\; 0$$

LEBESGUE - STIELTJES MEASURE

Thus, letting $K \to \infty$, we find for all $n > N$, that

$$\tau_g\Big((a,b] \Big) \geq \sum_{k=1}^{n} \tau_g\Big((a_k, b_k] \Big)$$

However, the sequence of partial sums above converges. We have then the inequality

$$\tau_g\Big((a,b] \Big) \geq \sum_{k=1}^{\infty} \tau_g\Big((a_k, b_k] \Big)$$

Combining the two inequalities, we have that our extension τ_g is a pseudo-measure. ∎

Comment 14.1.1 *It is worthwhile to summarize what we have accomplished at this point. We know now that the premeasure τ_g defined by the non-decreasing and right continuous map g on the algebra of sets, $\mathcal{A}$, generated by the collection $\mathcal{U}$ consisting of the empty set, finite intervals like $(a,b]$ and unbounded intervals of the form $(-\infty, b]$ and (a, ∞) when defined to be additive on $\mathcal{A}$ generates an interesting outer measure μ_g^* which is called Lebesgue - Stieltjes Outer Measure. We have also proven that the extension τ_g becomes a pseudo-measure on $\mathcal{A}$. Thus, we see*

- *(i): The sets A in $\mathcal{A}$ are in the σ-algebra of sets that satisfy the Caratheodory condition using μ_g^* which we denote by $\mathcal{M}_g$. We denote the resulting measure by μ_g. This is because τ_g is additive on the algebra $\mathcal{A}$ by Theorem 11.5.3.*

- *(ii): We know μ_g is regular by Theorem 11.5.3 and complete by Theorem 11.3.2.*

- *(iii): We know that $\mu_g(A) = \tau_g(A)$ for all A in $\mathcal{A}$ since τ_g is a pseudo-measure by Theorem 11.5.4.*

- *(iv): Since any open set can be written as a countable disjoint union of open intervals, this means any open set is in $\mathcal{M}_g$ because $\mathcal{M}_g$ contains open intervals as they are in $\mathcal{A}$ and the σ-algebra $\mathcal{M}_g$ is closed under countable disjoint unions. This also tells us that the Borel σ-algebra is contained in $\mathcal{M}_g$.*

- *(v): Since open sets are μ_g^* measurable, by Theorem 11.4.2, it follows that μ_g^* is a metric outer measure.*

Comment 14.1.2 *The measures μ_g induced by the outer measures μ_g^* are called Lebesgue - Stieltjes measures. Since open sets are measurable here, these measures are also called Borel measures.*

Comment 14.1.3 *So for a given non-decreasing right continuous g, we can construct a Lebesgue - Stieltjes measure satisfying*

$$\mu_g\Big((a,b] \Big) = g(b) - g(a)$$

So what about the open interval (a,b)? We know that

$$(a,b) = \bigcup_n (a, b - \frac{1}{n}]$$

Then

$$\mu_g\Big((a,b) \Big) = \lim_n g\left(b - \frac{1}{n} \right) - g(a)$$
$$= g(b^-) - g(a)$$

316 BASIC ANALYSIS IV: MEASURE THEORY AND INTEGRATION

What about the singleton $\{b\}$? We know

$$\{b\} \;=\; \bigcap_n \left(b - \frac{1}{n}, b \right]$$

and so

$$\mu_g\left(\{b\}\right) \;=\; \lim_n g(b) - g\left(b - \frac{1}{n} \right)$$
$$=\; g(b) - g(b^-)$$

Note this tells us that the Lebesgue - Stieltjes measure of a singleton need not be 0. However, at any point b where g is continuous, this measure will be zero. Since our g can have at most a countable number of discontinuities, we see there are only a countable number of singleton sets whose measure is nonzero.

Homework

Exercise 14.1.1 *Let g be defined by*

$$g(x) \;=\; \begin{cases} x, & x < 0 \\ 2x, & 0 \le x < 1 \\ 3, & 1 = x \\ 5x^2 + 1, & 1 < x \le 2 \\ 22, & x > 2 \end{cases}$$

- *Redefine g to make it right continuous and then use it as described in this section to generate the measure μ_g.*

- *Find all the singleton points which have positive measure.*

Exercise 14.1.2 *Let g be defined by*

$$g(x) \;=\; \begin{cases} 0.5x, & x < -1 \\ 2, & x = -1 \\ x + 4, & -1 < x < 1 \\ 6, & 1 = x \\ x^2 + 7, & 1 < x < 2 \\ 12, & x = 2 \\ 2x^3, & 2 < x \le 3 \\ x^4, & x > 3 \end{cases}$$

- *Redefine g to make it right continuous and then use it as described in this section to generate the measure μ_g.*

- *Find all the singleton points which have positive measure.*

Exercise 14.1.3 *Let g_n, for $n \ge 2$, be defined by*

$$g_n(x) \;=\; \begin{cases} x, & x > 1 \\ 1, & x = 1 \\ 1/(k+1), & 1/(k+1) \le x < 1/k, \; 1 \le k < n \\ 0, & 0 \le x < 1/n \\ x, & x < 0 \end{cases}$$

- *Redefine g to make it right continuous and then use it as described in this section to generate the measure μ_g.*

- *Find all the singleton points which have positive measure.*

Exercise 14.1.4 *Let g be the pointwise limit of the sequence of functions (g_n) for $n \geq 2$ defined by*

$$
g_n(x) \;=\;
\begin{cases}
x, & x > 1 \\
1, & x = 1 \\
1/(k+1), & 1/(k+1) \leq x < 1/k,\ 1 \leq k < n \\
0, & 0 \leq x < 1/n \\
x, & x < 0
\end{cases}
$$

- *Redefine g to make it right continuous and then use it as described in this section to generate the measure μ_g.*

- *Find all the singleton points which have positive measure.*

Exercise 14.1.5 *Let g be the staircase function $g(x) = n$ when $n \leq x < n+1$ for all integers n.*

- *Is g right continuous? If not, define it to make it so and then use it, as described in this section, to generate the measure μ_g.*

- *Find all the singleton points which have positive measure.*

14.2 Approximation Results

We can also prove approximation lemmas for Lebesgue-Stieltjes measures also. A typical one is given below.

Theorem 14.2.1 Approximating Sets with a Lebesgue - Stieltjes Outer Measure

> *Let X be a metric space, μ_g^* be a Lebesgue - Stieltjes outer measure, $\mathcal{M}_g$ the induced sigma-algebra of measurable sets and μ_g the induced measure. Then given any E contained in X, there is a G_δ set G (i.e. a countable intersection of open sets) and an F_σ set (i.e. a countable union of closed sets) F so that $F \subseteq E \subseteq G$ and $\mu_g(F) = \mu_g^*(E) = \mu_g(G)$.*

Proof 14.2.1
If $\mu_g^*(E) = \infty$, we can simply choose $G = X$ and be done. Hence, we can assume $\mu_g^*(E) < \infty$. Therefore, we know μ_g is regular and complete and so there is a set F with $\mu_g^*(E) = \mu_g(T)$ and $E_m = \cup_n E_n^m$, $F = \cap_m E_m$ where for each m, the family of sets (E_n^m) is in the algebra $\mathcal{A}$ used to construct the outer measure, satisfies $A \subseteq \cup_n E_n^m$ and

$$
\sum_n \tau_g(E_n^m) \;<\; \mu^*(A) + \frac{1}{m}
$$

It is straightforward to see we can rewrite each E_n^m as a countable union of sets of the form $(a_n^m, b_n^m]$ and thus we have $E_n^m = \cup_n (a_n^m, b_n^m]$. For any $r > 0$, we see $E_n^m = \cup_n (a_n^m, b_n^m + r)$, a countable union of open sets of finite μ_g measure. Hence, applying Theorem 12.4.3, for any j there is an open set G_j so that $\mu_g(G_j \setminus F) < \frac{1}{j}$. Hence, if $G = \cap_j G_j$, a G_δ set, we have $F \subseteq G$ and

$$
\mu_g(\cap_j G_j \setminus F) \;\leq\; \mu_g\left(\bigcap_{j=1}^{N} G_j \setminus F \right)
$$

$$= \mu_g\left(\bigcap_{j=1}^{N}(G_j \setminus F)\right)$$

$$\leq \mu_g(G_N \setminus F) < \frac{1}{N}$$

As $N \to \infty$, we see $\mu_g(G \setminus F) = 0$ and $\mu_g^*(E) = \mu_g(F) = \mu_g(G)$.

Now if $\mu_g(E)$ is bounded, we have $\mu_g^*(E) = \mu_g(T)$ where $T = \cap_m \cup_n E_n^m$ as before. Thus, T is a Borel set for which $\mu_g(T) < \infty$. We can therefore apply Theorem 12.4.2, since μ_g is a Borel measure and $\Re$ is a metric space, to obtain a closed set $F_n \subseteq T$ so that $\mu_g(T \setminus F_n) < \frac{1}{n}$ for each positive integer n. Let $F = \cup_n F_n$, an F_σ set. Then, for all n, we have

$$\mu_g(T \setminus F) = \mu_g(\cap_n T \setminus F_n)$$

$$\leq \mu_g(T \setminus F_n) < \frac{1}{n}$$

This immediately implies that $\mu_g(T \setminus F) = 0$. Thus, in this case, $\mu_g^*(E) = \mu_g(T) = \mu_g(F)$.

In the case that $\mu_g^*(E)$ is infinite, we let $E_j = E \cap (j, j+1]$ for all integers n. Then $E = \cup_j E_j$ and the sets E_j are disjoint. Hence, there are Borel sets T_j with bounded μ_g^* values satisfying $\mu_g^*(E_j) = \mu_f(T_j)$ where $T_j = \cap_m \cup_n E_n^{j,m})$ where the component sets $E_n^{j,m}$ are defined as before. The argument in the first part now applies. There are F_σ sets F_j so that $\mu_g^*(E_j) = \mu_g(T_j) = \mu_g(F_j)$. Let $F = \cup_j F_j$. Then

$$\mu_g(E \setminus F) = \mu_g(\cup_j E_j \setminus F) = \mu_g(\cup_j E_j \setminus F)$$

$$\leq \mu_g(\cup_j E_j \setminus F_j) = \sum_j \mu_g(E_j \setminus F_j) = 0$$

This proves the result. ∎

14.3   Properties of Lebesgue - Stieltjes Measures

It is clear Lebesgue measure on $\Re$, as developed in Chapter 12 should coincide with the Lebesgue - Stieltjes measure generated by the function $g(x) = x$. In Theorem 14.3.1 below, we see that since Lebesgue measure satisfies Conditions (1), (2) and (3), the function g constructed in the proof is exactly this function, $g(x) = x$. Hence, the Lebesgue - Stieltjes measure for $g(x) = x$ is just Lebesgue measure.

It is also clear that if g is of bounded variation and right continuous, g can be written as a difference of two non-decreasing functions u and v which are right continuous. The function u determines a Lebesgue - Stieltjes measure μ_u; the function v, the Lebesgue - Stieltjes measure μ_v and finally g defines the charge $\mu_g = \mu_u - \mu_v$. Then we see $\int f d\mu_g = \int f d\mu_u - \int f d\mu_v$.

We now summarize our discussions and give converse result.

Theorem 14.3.1 Characterizations of Lebesgue - Stieltjes Measures

LEBESGUE - STIELTJES MEASURE   319

> *Let g be non-decreasing and right continuous on $\Re$. Let μ_g^* be the outer measure associated with g as discussed above. Let $\mathcal{M}_g$ be the associated set of μ_g^* measurable functions and μ_g be the generated measure. Then:*
>
> 1. *μ_g^* is a metric outer measure and so all Borel sets are μ_g^* measurable.*
>
> 2. *If A is a bounded Borel set, then $\mu_g(A)$ is finite.*
>
> 3. *Each set $A \subseteq \Re$ has a measurable cover U which is of type G_δ; i.e. U is a countable intersection of open sets. This means $A \subseteq U$ and $\mu_g^*(A) = \mu_g(U)$.*
>
> 4. *For every half open interval $(a, b]$, $\mu_g((a, b]) = g(b) - g(a)$.*
>
> *Conversely, if μ^* is an outer measure on $\Re$ with resulting measure space $(\Re, \mathcal{M}, \mu)$, then if conditions (1), (2) and (3) above are satisfied by μ^* and μ, then there is a non-decreasing right continuous function g on $\Re$ so that $\mu_g^*(A) = \mu^*(A)$ for all $A \subseteq \Re$. In particular, $\mu_g(A) = \mu(A)$ for all $A \in \mathcal{M}$.*

Proof 14.3.1

The first half of the theorem has been proved in our extensive previous comments and using Theorem 14.2.1. It remains to prove the converse. Define g on $\Re$ by

$$g(x) = \begin{cases} \mu((0, x]), & \text{if } x > 0 \\ 0, & \text{if } x = 0 \\ -\mu((x, 0]) & \text{if } x < 0. \end{cases}$$

It is clear that g is non-decreasing on $\Re$. Is g right continuous? Let (h_n) be any sequence of positive numbers so that $h_n \to 0$. Assume $x > 0$. Then, we know

$$(0, x] = \bigcap_{n=1}^{\infty} (0, x + h_n]$$

By Condition (2) on the measure μ, we see $\mu((0, x + h_1])$ is finite. Thus, by Lemma 9.2.2, we have

$$\mu((0, x]) = \lim_{n \to \infty} \mu((0, x + h_n])$$

or using our definition of g, $g(x) = \lim_n g(x + h_n)$. We conclude g is right continuous for positive x. The arguments for $x < 0$ and $x = 0$ are similar.

Next, we show $\mu_g^* = \mu^*$. By definition of g, $\mu_g((a, b]) = \mu((a, b])$ for all half-open intervals $(a, b]$. Now both μ and μ_g are countably additive. Since every open interval is a countable disjoint union of half-open intervals, it follows that $\mu_g(I) = \mu(I)$ for any open interval I. Since any open set G is a countable union of open intervals, we then see $\mu_g(G) = \mu(G)$ for all open sets G. Further, if H is a countable intersection of open sets, we can write $H = \cap_n G_n$ where each G_n is open. Let

$$\begin{aligned} H_1 &= G_1 \\ H_2 &= G_1 \cap G_2 \\ &\vdots \\ H_n &= \cap_{j=1}^n G_j \end{aligned}$$

Then (H_n) is a decreasing sequence of open sets because finite intersections of open sets are still

320   BASIC ANALYSIS IV: MEASURE THEORY AND INTEGRATION

open. If we assume H is bounded, then using Lemma 9.2.2 again, we have $\mu(H) = \lim_n \mu(H_n)$. But we already have shown that $\mu_g(H_n) = \mu(H_n)$ since H_n is an open set. We conclude therefore that $\mu_g(H) = \lim_n \mu_g(H_n) = \lim_n \mu(H_n) = \mu(H)$. Thus, μ and μ_g match on bounded sets of type G_δ.

Now let A be any bounded subset of $\Re$. Since μ satisfies Condition (3), there is a G_δ set H_1 so that $\mu^*(A) = \mu(H_1)$ with $A \subseteq H$. Since g is non-decreasing and right continuous, we know there is also a G_δ set H_2 so that $A \subseteq H_2$ and $\mu_g^*(A) = \mu_g(H_2)$. Let $H = H_1 \cap H_2$. Then $A \subseteq H$. Since Borel sets are in $\mathcal{M}_g$ and $\mathcal{M}$, we also know $\mu^*(H) = \mu(H)$ and $\mu_g^*(H) = \mu_g(H)$. It then follows that

$$\mu_g^*(H) \geq \mu_g^*(A) = \mu_g(H_2) \geq \mu_g(H) = \mu_g^*(H)$$
$$\mu^*(H) \geq \mu^*(A) = \mu(H_1) \geq \mu(H) = \mu^*(H)$$

We see $\mu_g^*(A) = \mu_g(H)$ and $\mu^*(A) = \mu(H)$. But H is of type G_δ and so the values $\mu(H)$ and $\mu_g(H)$ must match. We conclude

$$\mu_g^*(A) = \mu(H) = \mu^*(A)$$

for all bounded subsets A. To handle unbounded sets A, note $A_n = A \cap [-n, n]$ is bounded and $A = \cup_n A_n$. It is clear A_n increases monotonically to A. For each A_n, there is a G_δ set H_n so that $\mu_g^*(A_n) = \mu_g(H_n) = \mu(H_n) = \mu^*(A_n)$.

Claim 1 *If ν^* is a regular outer measure, then for any sequence of sets (A_n),*

$$\nu^*(\liminf A_n) \leq \liminf \nu^*(A_n)$$

Proof *Since ν^* is regular, there is a measurable set F so that $\nu^*(\liminf A_n) = \nu(F)$ with $\liminf (A_n) \subseteq F$. Further, there are measurable sets F_n with $A_n \subseteq F_n$ with $\nu^*(A_n) = \nu(F_n)$. Hence,*

$$\nu(F) = \nu^*(\liminf A_n) \leq \nu^*(\liminf F_n)$$

Recall

$$\liminf F_n = \bigcup_m \bigcap_{n \geq m} F_n$$

Hence, we see that for all N, we have $\nu(\cup_{m=1}^N \cap_{n \geq m} F_n) \leq \sup_m \inf_{n \geq m} \nu(F_n)$. This implies

$$\lim_N \nu\left(\cup_{m=1}^N \cap_{n \geq m} F_n \right) \leq \liminf \nu(F_n)$$

To finish we note that since the set convergence is monotonic upward to $\liminf F_n$, we have $\nu(\liminf F_n) \leq \liminf \nu(F_n)$. Combining inequalities, we find

$$\nu(F) = \nu^*(\liminf A_n) \leq \nu^*(\liminf F_n)$$
$$\leq \liminf \nu(F_n) = \liminf \nu^*(A_n)$$

which completes the proof. $\qquad\qquad\square$

Claim 2: *If (A_n) is a monotonically increasing sequence with limit A and ν^* is a regular outer measure, then $\nu^*(A_n) \to \nu^*(A)$.*

Proof *Since the sequence is monotonic increasing, it is clear the limit exists and is bounded above by*

LEBESGUE - STIELTJES MEASURE

$\nu^*(A)$. *On the other hand, we know from Claim 1 that* $\nu^*(\liminf A_n) = \nu^*(A) \leq \liminf \nu^*(A_n)$. *But since the limit exists, the* $\liminf$ *matches the limit and we are done.* $\qquad\square$

Using Claim 2 applied to μ_g^* *and* μ^* *on the monotonically increasing sets* A_n, *we have* $\mu_g^*(A_n) \to \mu_g^*(A)$ *and* $\mu^*(A_n) \to \mu^*(A)$. *Since* $\mu_g^*(A_n) = \mu^*(A_n)$, *we have* $\mu_g^*(A) = \mu^*(A)$. *This shows the* $\mu_g^* = \mu^*$ *on* $\Re$. $\qquad\blacksquare$

Homework

Exercise 14.3.1 *Prove if* g *is bounded variation and right continuous,* g *can be rewritten as the difference* $u - v$ *where both* u *and* v *are nondecreasing functions which are right continuous.*

Exercise 14.3.2 *Let* $g(x) = x^2$. *Find a decomposition of* g *into* $u - v$ *where both* u *and* v *are nondecreasing functions which are continuous. Can you characterize* μ_u *and* μ_v?

Exercise 14.3.3 *If* $g(x) = x^3$ *when* $x < 0$ *and* $g(x) = x^4 + 1$ *when* $x \geq 0$, g *determines the measure* μ_g. *What is* $\mu_g([-2, 5])$?

Exercise 14.3.4 *If* $g(x) = x + 1$ *when* $x < 0$ *and* $g(x) = x^2 + 2$ *when* $x \geq 0$, g *determines the measure* μ_g. *What is* $\mu_g([-12, 5])$ *and* $\mu_g([-6, 7])$?

Exercise 14.3.5 *Let* h *be our Cantor function*

$$h(x) \;=\; (x + \Psi(x))/2$$

- *Explain why* τ_h *defines a Lebesgue - Stieltjes measure and since the resulting sigma-algebra contains the Borel sets, it is a Borel-Stieltjes measure.*

- *Determine if* τ_h *is absolutely continuous with respect to the Borel measure on* $\Re$.

Exercise 14.3.6 *Let* h_a *be our Cantor function*

$$h_a(x) \;=\; (x + \Psi_a(x))/2$$

- *Explain why* τ_{h_a} *defines a Lebesgue - Stieltjes measure and since the resulting sigma-algebra contains the Borel sets, it is a Borel-Stieltjes measure.*

- *Determine if* τ_{h_a} *is absolutely continuous with respect to the Borel measure on* $\Re$.

Part VI

Abstract Measure Theory Two

Chapter 15

Modes of Convergence

There are many ways a sequence of functions in a measure space can converge. In this chapter, we will explore some of them and the relationships between them. There are several types of convergence here:

(i): Convergence pointwise.

(ii): Convergence uniformly.

(iii): Convergence almost uniformly.

(iv): Convergence in measure.

(v): Convergence in $\mathcal{L}_p$ norm for $1 \leq p < \infty$.

(vi): Convergence in $\mathcal{L}_\infty$ norm.

We will explore each in turn. We have already discussed the p norm convergence in Chapter 10 so there is no need to go over those ideas again. However, some of the other types of convergence in the list above are probably not familiar to you. Pointwise and pointwise a.e. convergence have certainly been mentioned before, but let's make a formal definition so it is easy to compare it to other types of convergence later.

Definition 15.0.1 Convergence Pointwise and Pointwise a.e.

> *Let $(X, \mathcal{S})$ be a measurable space. Let (f_n) be a sequence of extended real-valued measurable functions: i.e. $(f_n) \subseteq M(X, \mathcal{S})$. Let $f : X \to \overline{\Re}$ be a function. Then, we say f_n converges pointwise to f on X if $\lim_n f_n(x) = f(x)$ for all x in X. Note that this type of convergence does not involve a measure, although it does use the standard metric, $\| \ \|$ on $\Re$. We can write this as*
> $$f_n \to f \ [ptws]$$
> *If there is a measure μ on $\mathcal{S}$, we can also say the sequence converges almost everywhere if $\mu(\{x \mid f_n(x) \nrightarrow f(x)\}) = 0$. We would write this as*
> $$f_n \to f \ [ptws \ a.e.]$$

Next, you have probably already seen uniform convergence in the context of advanced calculus. We can define it nicely in a measure space also.

Definition 15.0.2 Convergence Uniformly

> Let $(X, \mathcal{S})$ be a measurable space. Let (f_n) be a sequence of real-valued measurable functions: i.e. $(f_n) \subseteq M(X, \mathcal{S})$. Let $f : X \to \overline{\Re}$ be a function. Then, we say f_n converges uniformly to f on X if for any $\epsilon > 0$, there is a positive integer N (depending on the choice of ϵ) so that if $n > N$, then $| f_n(x) - f(x) | < \epsilon$ for all x in X. We can write this as
>
> $$f_n \to f \ [unif]$$

However, if we are in a measure space, we can relax the idea of uniform convergence of the whole space by taking advantage of the underlying measure.

Definition 15.0.3 Almost Uniform Convergence

> Let $(X, \mathcal{S}\mu)$ be a measure space. Let $(f_n) \subseteq M(X, \mathcal{S}, \mu)$ be a sequence of functions which are finite a.e. Let $f : X \to \overline{\Re}$ be a function. We say f_n converges almost uniformly to f on X if for any $\epsilon > 0$, there is a measurable set E such that $\mu(E^C) < \epsilon$ and (f_n) converges uniformly to f on E. We write this as
>
> $$f_n \to f \ [a.u.]$$

Theorem 15.0.1 Almost Uniform Convergence Implies Pointwise Convergence and so Limit Function is Measurable

> Let $(X, \mathcal{S}\mu)$ be a measure space. Let $(f_n) \subseteq M(X, \mathcal{S}, \mu)$ be a sequence of functions which are finite a.e. Let $f : X \to \overline{\Re}$ be a function. if $f_n \to f \ [a.u.]$, then $f_n \to f$ pointwise a.e. and so f is measurable.

Proof 15.0.1

Let $\epsilon_n = 1/n$. Then there is a sequence of measurable sets E_n with $\mu(E_n) < 1/n$ and $f_n \to f$ uniformly on E_n^C. Let $E = \cap_{n=1}^{\infty} E_n$. Then $\mu(E) = 0$ and $f_n \to f$ at each $t \in E^C$. ∎

Finally, we can talk about a brand new idea: convergence using only measure itself.

Definition 15.0.4 Convergence in Measure

> Let $(X, \mathcal{S}\mu)$ be a measure space. Let $(f_n) \subseteq M(X, \mathcal{S}, \mu)$ be a sequence of functions which are finite a.e. Let $f : X \to \overline{\Re}$ be a function. Let E be a measurable set. We say f_n converges in measure to f on E if for any pair (ϵ, δ) of positive numbers, there exists a positive integer N (depending on ϵ and δ) so that if $n > N$, then
>
> $$\mu(\{x \mid \ | f_n(x) - f(x) | \geq \delta\}) < \epsilon$$
>
> We write this as
>
> $$f_n \to f \ [meas \ on \ E]$$
>
> If E is all of X, we would just write
>
> $$f_n \to f \ [meas]$$

CONVERGENCE MODES

Comment 15.0.1 *We do not know yet that the limit function f we see in convergence in measure is measurable. We will prove this later.*

Homework

Exercise 15.0.1 *Prove the sequence $x_n(t) = t^n$ on $[0,1]$ converges pointwise to the function $x(t) = 0$ on $[0,1)$ and $x(1) = 1$.*

Exercise 15.0.2 *Prove the sequence $x_n(t) = t + \sin(n^3 t^2 + 5)/n^2$ converges uniformly to $x(t) = t$ on $\Re$.*

Exercise 15.0.3 *Let the sequence x_n on $[0,1]$ be defined by*

$$x_n(t) = \begin{cases} 0, & 0 \le t < 1/2 \\ n(t - 1/2), & 1/2 \le t < 1/2 + 1/n \\ 1, & 1/2 + 1/2 \le t \le 1 \end{cases}$$

Prove (x_n) converges to $[x]$ where

$$x(t) = \begin{cases} 0, & 0 \le t < 1/2 \\ 1, & 1/2 \le t \le 1 \end{cases}$$

in $\| \cdot \|_1$.

Exercise 15.0.4 *Let the sequence x_n on $[0,1]$ be defined by*

$$x_n(t) = \begin{cases} 0, & 0 \le t < 1/2 \\ n(t - 1/2), & 1/2 \le t < 1/2 + 1/n \\ 1, & 1/2 + 1/2 \le t \le 1 \end{cases}$$

Prove (x_n) converges to $[x]$ where

$$x(t) = \begin{cases} 0, & 0 \le t < 1/2 \\ 1, & 1/2 \le t \le 1 \end{cases}$$

in $\| \cdot \|_2$.

Exercise 15.0.5 *Let f be in $\mathcal{L}_1([a,b])$. Given $\epsilon > 0$, there is a continuous function f_c so that $\int |f - f_c| d\mu < \epsilon$ and there is a sequence of polynomials (p_n) so that $p_n \overset{unif}{\longrightarrow} f_c$ on $[a,b]$. Let $E_\epsilon = \{t : |f(t) = f_c(t)| \ge \epsilon/2\}$. Then on E_ϵ^C, $|f(t) - p_n(t)| < |f(t) = p_n(t)| + |f_c(t) - f(t)|$. Using this idea, prove (p_n) converges a.u. to f.*

Exercise 15.0.6 *Let the sequence (x_n) be defined by*

$$x_n(t) = \begin{cases} n, & 0 \le t \le 1/n \\ 0, & 1/2 < t \le 1 \end{cases}$$

- *Prove this sequence converges pointwise to*

$$x(t) = \begin{cases} \infty, & t = 0 \\ 0, & 0 < t \le 1 \end{cases}$$

- *Prove the limit function is Lebesgue measurable.*

- *Prove the sequence converges in measure to x.*

328　　*BASIC ANALYSIS IV: MEASURE THEORY AND INTEGRATION*

- *Prove this sequence converges a.u. also.*

Exercise 15.0.7 *Let $(X, \mathcal{S}\mu)$ be a measure space. Let $(f_n) \subseteq M(X, \mathcal{S}, \mu)$ be a sequence of functions which are finite a.e. and converges a.u to a function $f : X \to \overline{\Re}$. Prove the convergence is a.e. and so the limit function must be measurable. So we don't have to assume the limit function is measurable in the definition of a.u. convergence. This is the proof of Theorem 15.0.1 we sketched.*

15.1　Extracting Subsequences

In some cases, when a sequence of functions converges in one way, it is possible to prove that there is at least one subsequence that converges in a different manner. We will now make this idea precise.

Definition 15.1.1 Cauchy Sequences in Measure

> *Let $(X, \mathcal{S}, \mu)$ be a measure space and (f_n) be a sequence of extended real-valued measurable functions. We say (f_n) is Cauchy in Measure if for all $\alpha > 0$ and $\epsilon > 0$, there is a positive integer N so that*
>
> $$\mu\Big(|f_n(x) - f_m(x)| \geq \alpha\Big) < \epsilon, \ \forall n, m > N$$

We can prove a kind of completeness result next.

Theorem 15.1.1 Cauchy in Measure Implies a Convergent Subsequence

> *Let $(X, \mathcal{S}, \mu)$ be a measure space and (f_n) be a sequence of extended real-valued measurable functions which is Cauchy in Measure. Then there is a subsequence (f_n^1) and an extended real-valued measurable function f such that $f_n^1 \to f$ [a.e.], $f_n^1 \to f$ [a.u.] and $f_n^1 \to f$ [meas].*

Proof 15.1.1
For each pair of indices n and m, there is a measurable set E_{nm} on which the definition of the difference $f_n - f_m$ is not defined. Hence, the set

$$E = \bigcup_n \bigcup_m E_{nm}$$

is measurable and on E^C, all differences are well-defined. We do not know the sets E_{nm} have measure 0 here as the members of the sequence do not have to be summable or essentially bounded.

Now, let's get started with the proof.

(Step 1): Let $\alpha_1 = 1/2$ and $\epsilon_1 = 1/2$ also. Then, (f_n) Cauchy in Measure implies

$$\exists N_1 \ni n, m > N_1 \ \Rightarrow \ \mu\Big(|f_n(x) - f_m(x)| \geq 1/2\Big) < 1/2$$

Let $g_1 = f_{N_1+1}$.

(Step 2): Let $\alpha_2 = 1/2^2$ and $\epsilon_1 = 1/2^2$ also. Then, (f_n) Cauchy in Measure again implies there is an $N_2 > N_1$ so that

$$n, m > N_2 \;\Rightarrow\; \mu\left(|f_n(x) - f_m(x)| \geq 1/4\right) < 1/4$$

Let $g_2 = f_{N_2+1}$. It is then clear by our construction that

$$\mu\left(|g_2(x) - g_1(x)| \geq 1/2\right) < 1/2$$

(Step 3): Let $\alpha_3 = 1/2^3$ and $\epsilon_1 = 1/2^3$ also. Then, (f_n) Cauchy in Measure again implies there is an $N_3 > N_2$ so that

$$n, m > N_3 \;\Rightarrow\; \mu\left(|f_n(x) - f_m(x)| \geq 1/8\right) < 1/8$$

Let $g_3 = f_{N_3+1}$. It follows by construction that

$$\mu\left(|g_3(x) - g_2(x)| \geq 1/4\right) < 1/4$$

Continuing this process by induction, we find a subsequence (g_n) of the original sequence (f_n) so that for all $k \geq 1$,

$$\mu\left(|g_{k+1}(x) - g_k(x)| \geq 1/2^k\right) < 1/2^k$$

Define the sets

$$E_j = \left(|g_{j+1}(x) - g_j(x)| \geq 1/2^j\right)$$

and

$$F_k = \bigcup_{j=k}^{\infty} E_j$$

Note if $x \in F_k^C$,

$$|g_{j+1}(x) - g_j(x)| \;<\; 1/2^j$$

for any index $j \geq k$. Each set F_k is then measurable and they form an increasing sequence. Let's get a bound on $\mu(F_k)$. First, if A and B are measurable sets, then

$$\mu\left(A \cup B\right) = \mu\left(A \cup B^C\right) : + \mu\left(A \cap B\right) + \mu\left(A^C \cup B\right)$$

But adding in $\mu\left(A \cap B\right)$ simply makes the sum larger. We see

$$\mu\left(A \cup B\right) \leq \mu\left(A \cup B^C\right) : + \mu\left(A \cap B\right) + \mu\left(A \cap B\right) + \mu\left(A^C \cup B\right)$$
$$= \mu(A) + \mu(B)$$

330 BASIC ANALYSIS IV: MEASURE THEORY AND INTEGRATION

This result then extends easily to finite unions. Thus, if (A_n) is a sequence of measurable sets, then by the sub additive result above,

$$\mu\left(\bigcup_{i=1}^{n} A_i\right) \leq \sum_{i=1}^{n} \mu(A_i)$$

Hence, the sets $\cup_{i=1}^{n} A_i$ form an increasing sequence and we clearly have

$$\mu\left(\bigcup_{i=1}^{\infty} A_i\right) = \lim_{n} \mu\left(\bigcup_{i=1}^{n} A_i\right) \leq \sum_{i=1}^{\infty} \mu(A_i)$$

We can apply this idea to the increasing sequence (F_k) to obtain

$$\mu(F_k) \leq \sum_{j=k}^{\infty} \mu(E_j) < \sum_{j=k}^{\infty} 1/2^j = 1/2^{k-1}$$

Now, for any $i > j$, we have

$$|g_i(x) - g_j(x)| \leq \sum_{\ell=j}^{i-1} |g_{\ell+1} - g_\ell|$$

Choosing the indices i and j so that $i > j \geq k$, we then find if $x \notin F_k$, that

$$|g_{\ell+1}(x) - g_\ell(x)| < 1/2^\ell$$

Hence, for these indices,

$$|g_i(x) - g_j(x)| \leq \sum_{\ell=j}^{i-1} |g_{\ell+1} - g_\ell|$$

$$< \sum_{\ell=j}^{i-1} 1/2^\ell = \sum_{\ell=j}^{\infty} 1/2^\ell = 1/2^{j-1}$$

We conclude that if $x \in F_k^C$ and $i > j \geq k$ we have

$$|g_i(x) - g_j(x)| \leq 1/2j - 1 \tag{$*$}$$

Now let $F = \cap_k F_k$. The F is measurable and $\mu(F) = \lim_k \mu(F_k) = 0$. Let x be in F^C. By De Morgan's Laws, $x \in \cup_k F_K^C$ which implies x is in some F_k^C. Call this k^*. Then given $\epsilon > 0$, choose J so that $1/2^{J-1} < \epsilon$. Then, by Equation $*$, if $i > j \geq J \geq k^*$,

$$|g_i(x) - g_j(x)| \leq 1/2^{j-1} < 1/2^{J-1} < \epsilon$$

Thus, the sequence $g_k(x)$ is a Cauchy sequence of real numbers for each x in F^C. Hence, $\lim_k g_k(x)$ exists for such x. Defining f by

$$f(x) = \begin{cases} \lim_k g_k(x), & x \in F^C \\ 0, & x \in F, \end{cases}$$

CONVERGENCE MODES

we see f is measurable and it is the pointwise limit a.e. of the subsequence (g_k). This completes the proof of the first claim. To see that (g_k) converges in measure to f, look again at Equation $*$:

$$|g_i(x) - g_j(x)| \leq 1/2j - 1, \ \forall\, i > j \geq k, \ \forall\, x \in F_k^C$$

Now let $i \to \infty$ and use the continuity of the absolute value function to obtain

$$|f(x) - g_j(x)| \leq 1/2j - 1, \ \forall\, j \geq k, \ \forall\, x \in F_k^C \qquad (**)$$

Equation $**$ says that (g_k) converges to f uniformly on F_k^C. Further, recall $\mu(F_k) < 1/2^{k-1}$. Note given any $\delta > 0$, there is an integer k^* so that $1/2^{k^*-1} < \delta$ and g_k converges uniformly on $F_{k^*}^C$. We therefore conclude that (g_k) converges almost uniformly to f as well.

To show the last claim, given an arbitrary $\alpha > 0$ and $\epsilon > 0$, choose a positive integer k^* so that

$$\mu(F_k^*) < 1/2^{k^*-1} < \min(\alpha, \epsilon)$$

Then, by Equation $**$, we have

$$\left(|f(x) - g_j(x)| \geq \alpha\right) \subseteq \left(|f(x) - g_j(x)| > 1/2^{k^*-1}\right)$$

Then, again by Equation $**$, we have

$$\subseteq \left(|f(x) - g_j(x)| > 1/2^{k^*-1}\right)$$
$$\subseteq F_{k^*}$$

Combining, we have

$$\mu\left(|f(x) - g_j(x)| \geq \alpha\right) \leq \mu\left(F_{k^*}\right) < 1/2^{k^*-1} < \epsilon$$

This shows that (g_k) converges to f in measure. ∎

The result above allows us to prove that Cauchy in Measure implies there is a function which the Cauchy sequence actually converges to.

Theorem 15.1.2 Cauchy in Measure Implies Completeness

Let $(X, \mathcal{S}, \mu)$ be a measure space and (f_n) be a sequence of extended real-valued measurable functions which is Cauchy in Measure. Then there is an extended real-valued measurable function f such that $f_n \to f$ [meas] and the limit function f is determined uniquely a.e.

Proof 15.1.2

By Theorem 15.1.1, there is a subsequence (f_n^1) and a real-valued function measurable function f so that $f_n^1 \to f$ [meas]. Let $\alpha > 0$ be given. If $|f(x) - f_n(x)| \geq \alpha$, then given any f_n^1 in the subsequence, we have

$$\alpha \leq |f(x) - f_n(x)| \leq |f(x) - f_n^1(x)| + |f_n(x) - f_n^1(x)|$$

Note, just as in the previous proof, there is a measurable set E where all additions and subtractions of functions are well-defined. Now, let $\beta = |f(x) - f_n^1(x)|$ and $\gamma = |f_n(x) - f_n^1(x)|$. The equation

332   BASIC ANALYSIS IV: MEASURE THEORY AND INTEGRATION

above thus says

$$\beta + \gamma \leq \alpha$$

Since β and γ are non-negative and both are less than or equal to α, we can think about this inequality in a different way. If there was equality

$$\beta^* + \gamma^* = \alpha$$

with both β^* and γ^* not zero, then we could let $t = \beta^*/\alpha$ and we could say $\beta^* = t\,\alpha$ and $\gamma^* = (1-t)\,\alpha$ as $\gamma^* = \alpha - \beta^*$. Now imagine β and γ being larger α. Then, β and γ would have to be bigger than or equal to the values $\beta^* = t\,\alpha$ and $\gamma^* = (1-t)\,\alpha$ for some t in $(0,1)$. Similar arguments work for the cases of $\beta = 0$ and $\gamma = 0$ which will correspond to the cases of $t = 0$ and $t = 1$. Hence, we can say that if $|f(x) - f_n(x)| \geq \alpha$, then there is some $t \in [0,1]$ so that

$$
\begin{aligned}
|f(x) - f_n^1(x)| &\geq t\,\alpha, \\
|f_n(x) - f_n^1(x)| &\geq (1-t)\,\alpha
\end{aligned}
$$

*The following reasoning is a bit involved, so bear with us. First, if x is a value where $|f(x) - f_n(x)| \geq \alpha$, we must have that $|f(x) - f_n^1(x)| \geq t\,\alpha$ (call this Condition I) **and** $|f_n(x) - f_n^1(x)| \geq (1-t)\,\alpha$ (call this Condition II).*

If $0 \leq t \leq 1/2$, then since an x which satisfies Condition I must also satisfy Condition II, we see for these values of t, we have

$$
\begin{aligned}
\{x \mid |f(x) - f_n^1(x)| \geq t\,\alpha\} &\subseteq \{x \mid |f_n(x) - f_n^1(x)| \geq (1-t)\,\alpha\} \\
&\subseteq \{x \mid |f_n(x) - f_n^1(x)| \geq 1/2\,\alpha\}
\end{aligned}
$$

Hence, for $0 \leq t \leq 1/2$, we conclude

$$
\begin{aligned}
\{x \mid |f(x) - f_n^1(x)| \geq t\,\alpha\} \bigcup \{x \mid |f_n(x) - f_n^1(x)| \geq (1-t)\,\alpha\} \\
\subseteq \{x \mid |f_n(x) - f_n^1(x)| \geq 1/2\,\alpha\}
\end{aligned}
$$

A similar argument shows that if $1/2 \leq t \leq 1$, any x satisfying Condition II must satisfy Condition I. Hence, for these t,

$$
\begin{aligned}
\{x \mid |f(x) - f_n^1(x)| \geq t\,\alpha\} \bigcup \{x \mid |f_n(x) - f_n^1(x)| \geq (1-t)\,\alpha\} \\
\subseteq \{x \mid |f(x) - f_n^1(x)| \geq (1-t)\,\alpha\} \\
\subseteq \{x \mid |f(x) - f_n^1(x)| \geq 1/2\,\alpha\}
\end{aligned}
$$

Combining these results, we find

$$
\bigcup_{0 \leq t \leq 1} \left(\{x \mid |f(x) - f_n^1(x)| \geq t\,\alpha\} \bigcup \{x \mid |f_n(x) - f_n^1(x)| \geq (1-t)\,\alpha\} \right)
$$
$$
\subseteq \{x \mid |f_n(x) - f_n^1(x)| \geq 1/2\,\alpha\} \bigcup \{x \mid |f(x) - f_n^1(x)| \geq 1/2\,\alpha\}
$$

Finally, from the triangle inequality,

$$|f(x) - f_n(x)| \leq |f(x) - f_n^1(x)| + |f_n^1(x) - f_n(x)|$$

CONVERGENCE MODES

and so, we have

$$\{x \mid |f(x) - f_n(x)| \geq \alpha\}$$

$$\subseteq \bigcup_{0 \leq t \leq 1} \left(\{x \mid |f(x) - f_n^1(x)| \geq t\,\alpha\} \bigcup \{x \mid |f_n(x) - f_n^1(x)| \geq (1-t)\,\alpha\} \right)$$

$$\subseteq \{x \mid |f_n(x) - f_n^1(x)| \geq 1/2\,\alpha\} \bigcup \{x \mid |f(x) - f_n^1(x)| \geq 1/2\,\alpha\}$$

Next, pick an arbitrary $\epsilon > 0$. Since $f_n^1 \to f$ [meas], there is a positive integer N_1 so that

$$\mu\left(|f(x) - f_n^1(x)| \geq \alpha/2 \right) \;<\; \epsilon/2, \, \forall\, n^1 > N_1$$

where n^1 denotes the index of the function f_n^1. Further, since (f_n) is Cauchy in measure, there is a positive integer N_2 so that

$$\mu\left(|f_n(x) - f_n^1(x)| \geq \alpha/2 \right) \;<\; \epsilon/2, \, \forall\, n,\, n^1 > N_2$$

So if n^1 is larger than $N = \max(N_1, N_2)$, we have

$$\mu\left(|f(x) - f_n(x)| \geq \alpha/2 \right) \;<\; \epsilon, \, \forall\, n > N$$

This shows $f_n \to f$ [meas] as desired.

To show the uniqueness a.e. of f, assume there is another function g so that $f_n \to g$ [meas]. Then, by arguments similar to ones we have already used, we find

$$\{x \mid |f(x) - g(x)| \geq \alpha\} \;\subseteq\; \{x \mid |f_n(x) - f(x)| \geq 1/2\,\alpha\}$$

Then, mutatis mutandi, we obtain

$$\mu\left(\{x \mid |f(x) - g(x)| \geq \alpha\} \right) \;\leq\; \mu\left(\{x \mid |f_n(x) - f(x)| \geq 1/2\,\alpha\} \right)$$

$$+ \;\; \mu\left(\{x \mid |f_n(x) - g(x)| \geq 1/2\,\alpha\} \right)$$

$$<\; \epsilon$$

Since $\epsilon > 0$ is arbitrary, we see for any $\alpha > 0$,

$$\mu\left(\{x \mid |f(x) - g(x)| \geq \alpha\} \right) \;=\; 0$$

However, we know

$$\mu\left(\{x \mid |f(x) - g(x)| > 0\} \right) \;=\; \bigcup_n \left(\{x \mid |f(x) - g(x)| \geq 1/n\} \right)$$

which immediately tells us that

$$\mu\left(\{x \mid |f(x) - g(x)| > 0\} \right) \;=\; 0$$

334   *BASIC ANALYSIS IV: MEASURE THEORY AND INTEGRATION*

This says $f = g$ a.e. and we are done. ■

Comment 15.1.1 *Note this last theorem implies that if a sequence converges in measure the limit function must be measurable. You will prove this in the exercises!*

Homework

Exercise 15.1.1 *Prove $f_n \to f$ in measure implies (f_n) is Cauchy in Measure.*

Exercise 15.1.2 *Prove $f_n \to f$ in $\| \cdot \|_p$ implies (f_n) is Cauchy in Measure.*

Exercise 15.1.3 *Prove $f_n \to f$ a.u. implies (f_n) is Cauchy in Measure.*

Exercise 15.1.4 *Let $(X, \mathcal{S}\mu)$ be a measure space. Let $(f_n) \subseteq M(X, \mathcal{S}, \mu)$ be a sequence of functions which are finite a.e. and converges a.u to $f : X \to \overline{\mathfrak{R}}$. prove f is measurable.*

Exercise 15.1.5 *Let $(X, \mathcal{S}\mu)$ be a measure space. Let $(f_n) \subseteq M(X, \mathcal{S}, \mu)$ be a sequence of functions which are finite a.e. and converges in measure to measurable $f : X \to \overline{\mathfrak{R}}$. Let $(g_n) \subseteq M(X, \mathcal{S}, \mu)$ be a sequence of functions which are finite a.e. and converges a.u to measurable $g : X \to \overline{\mathfrak{R}}$. Prove $(f_n + g_n)$ converges in measure to $f + g$.*

Theorem 15.1.3 p-Norm Convergence Implies Convergence in Measure

Assume $1 \leq p < \infty$. Let (f_n) be a sequence in $\mathcal{L}_p(X, \mathcal{S}, \mu)$ and let $f \in \mathcal{L}_p(X, \mathcal{S}, \mu)$ so that $f_n \to f \ [p - norm]$. Then $f_n \to f \ [meas]$ which is Cauchy in Measure.

Proof 15.1.3
Let $\alpha > 0$ be given and let

$$E_n(\alpha) \;\; = \;\; \{x \mid |f_n(x) - f(x)| \geq \alpha\}$$

Then, given $\epsilon > 0$, there is a positive integer N so that

$$\int |f_n - f|^p \, d\mu \;\; < \;\; \alpha^p \, \epsilon, \; \forall \, n > N$$

Thus,

$$\int_{E_n(\alpha)} |f_n - f|^p \, d\mu \leq \int |f_n - f|^p \, d\mu \;\; < \;\; \alpha^p \, \epsilon, \; \forall \, n > N$$

But on $E_n(\alpha)$, the integrand in the first term is bigger than or equal to α^p. We obtain

$$\alpha^p \, \mu(E_n(\alpha)) \;\; < \;\; \alpha^p \, \epsilon, \; \forall \, n > N$$

Canceling the α^p term, we have $\mu(E_n(\alpha)) < \epsilon$, for all $n > N$. This implies $f_n \to f \ [meas]$. ■

Comment 15.1.2 *Let's assess what we have learned so far. We have shown*

(i):

$$f_n \to f \ [p - norm] \;\; \Rightarrow \;\; f_n \to f \ [meas]$$

CONVERGENCE MODES

by Theorem 15.1.3.

(ii):

$$f_n \to f \,[meas] \quad \Rightarrow \quad (f_n) \; Cauchy \; in \; Measure$$

Then,

$$(f_n) \; Cauchy \; in \; Measure \quad \Rightarrow \quad \exists \, (f_n^1) \subseteq (f_n) \ni f_n^1 \to f \,[a.e.]$$

by Theorem 15.1.1. Note, we proved the existence of such a subsequence already in the proof of the completeness of $\mathcal{L}_p$ as discussed in Theorem 10.1.10.

(iii): Finally, we can also apply Theorem 15.1.1 to infer

$$f_n \to f \,[meas] \quad \Rightarrow \quad \exists \, (f_n^1) \subseteq (f_n) \ni f_n^1 \to f \,[a.u.]$$

(iv): Also, using Theorem 15.1.2, we can show if a sequence converges in measure to a function f, the limit function is measurable.

Theorem 15.1.4 Almost Uniform Convergence Implies Convergence in Measure

> *Let $(X, \mathcal{S})$ be a measurable space. Let (f_n) be a sequence of real-valued measurable functions: i.e. $(f_n) \subseteq M(X, \mathcal{S})$. Let $f : X \to \overline{\Re}$ be measurable. Then*
>
> $$f_n \to f \,[a.u.] \quad \Rightarrow \quad f_n \to f \,[meas]$$

Proof 15.1.4

If f_n converges to f a.u., given arbitrary $\epsilon > 0$, there is a measurable set E_ϵ so that $\mu(E_\epsilon) < \epsilon$ and f_n converges uniformly on E_ϵ^C. Now let $\alpha > 0$ be chosen. Then, there is a positive integer N_α so that

$$|f_n(x) - f(x)| \quad < \quad \epsilon, \; \forall \, n > N_\alpha, \; \forall \, x \in E_\epsilon^C$$

Hence, if $n > N_\alpha$ and x satisfies $|f_n(x) - f(x)| \geq \alpha$, we must have that $x \in E_\epsilon$. We conclude

$$\left(|f_n(x) - f(x)| \geq \alpha \right) \quad \subseteq \quad E_\epsilon, \; \forall \, n > N_\alpha$$

This implies immediately that

$$\mu\left(|f_n(x) - f(x)| \geq \alpha \right) \quad \leq \quad \mu(E_\epsilon) < \epsilon, \; \forall \, n > N_\alpha$$

This proves $f_n \to f \,[meas]$. ∎

Comment 15.1.3 *We have now shown*

$$f_n \to f \,[p - norm] \quad \Rightarrow \quad f_n \to f \,[meas]$$

by Theorem 15.1.3. This then implies by Theorem 15.1.1

$$\exists \, (f_n^1) \subseteq (f_n) \ni f_n^1 \to f \,[a.u.]$$

336 BASIC ANALYSIS IV: MEASURE THEORY AND INTEGRATION

Homework

Exercise 15.1.6 *Consider the ODE $u'' = f$, $u(0) = 0$ and $u(L) = 0$ for some $L > 0$ and any $f \in L_2([0, L])$. This can be rewritten as $u(t) = \int_0^L k(t, s) f(s) ds$ where the kernel function k is continuous on $[0, L] \times [0, L]$ and so is bounded. We know given $\epsilon > 0$, there is a continuous function f_c with $\|f - f - c\|_2 < \epsilon$ and there is a sequence of polynomials (p_n) with $p_n \to f_c$ uniformly.*

- *Prove (p_n) converges to f in $\| \cdot \|_2$.*

- *Prove (p_n) is Cauchy in $\| \cdot \|_2$.*

- *Prove (u_n) is Cauchy in $\| \cdot \|_\infty$ and hence converges to a continuous function u with $u(t) = \int_0^L k(t, s) f(s) ds$. Hence, in a sense, we can solve this non-classical ODE with simply L_2 data.*

- *Prove there are subsequences (u_n^1) and p_n^1 so that $(u_n^1)'' = p_n^1$ with $\lim_{n \to \infty} (u_n^1)'' = f$ a.e. Note this is like an extension of the classical ODE to this non-classical case.*

Exercise 15.1.7 *Consider the ODE $u'' + 4u = f$, $u(0) = 0$ and $u(10) = 0$ and any $f \in L_2([0, 10])$. This can be rewritten as $u(t) = \int_0^{10} k(t, s) f(s) ds$ where the kernel function k is continuous on $[0, 10] \times [0, 10]$ and so is bounded. We know given $\epsilon > 0$, there is a continuous function f_c with $\|f - f - c\|_2 < \epsilon$ and there is a sequence of polynomials (p_n) with $p_n \to f_c$ uniformly.*

- *Prove (p_n) converges to f in $\| \cdot \|_2$.*

- *Prove (p_n) is Cauchy in $\| \cdot \|_2$.*

- *Prove (u_n) is Cauchy in $\| \cdot \|_\infty$ and hence converges to a continuous function u with $u(t) = \int_0^{10} k(t, s) f(s) ds$.*

- *Prove there are subsequences (u_n^1) and p_n^1 so that $(u_n^1)'' = p_n^1$ with $\lim_{n \to \infty} (u_n^1)'' = f$ in measure*

Exercise 15.1.8 *Consider the bounded linear operator $T : L_2([0, L]) \to L_2([0, L])$ defined by $(T(f))(t) = \int_0^L k(t, s) f(s) ds$ where the kernel function k is continuous on $[0, 10] \times [0, 10]$ and so is bounded. We know given $\epsilon > 0$, there is a continuous function f_c with $\|f - f - c\|_2 < \epsilon$ and there is a sequence of polynomials (p_n) with $p_n \to f_c$ uniformly.*

- *Prove there is a subsequence $T(p_n^1)$ which converges a.e to $\int_0^L k(t, s) f(s) ds$.*

- *Prove there is a subsequence $T(p_n^1)$ which converges in measure to $\int_0^L k(t, s) f(s) ds$.*

- *Prove there is another subsequence $T(p_n^2)$ which converges a.u. to $\int_0^L k(t, s) f(s) ds$.*

15.2 Egoroff's Theorem

A famous theorem tells us how pointwise a.e. convergence can be phrased "almost" like uniform convergence. This is **Egoroff's Theorem**.

Theorem 15.2.1 Egoroff's Theorem

> *Let $(X, \mathcal{S}, \mu)$ be a measure space with $\mu(X) < \infty$. Let f be an extended real-valued function which is measurable. Also, let (f_n) be a sequence of functions in $M(X, \mathcal{S})$ such that $f_n \to f$ [a.e.]. Then, $f_n \to f$ [a.u.] and $f_n \to f$ [meas].*

CONVERGENCE MODES

Proof 15.2.1

From previous arguments, the way we handle converge a.e. is now quite familiar. Also, we know how we deal with the measurable set on which addition of the function f_n is not well-defined. Hence, we may assume without loss of generality that the convergence here is on all of X and that addition is defined on all of X. With that said, let

$$E_{nk} \;=\; \bigcup_{k=n}^{\infty} \Big(|f_k(x) - f(x)| \;\geq\; 1/m \Big)$$

Note that each E_{nk} is measurable and $E_{n+1,k} \subseteq E_{nk}$ so that this is an decreasing sequence of sets in the index n. Given x in X, we have $f_n \to f(x)$. Hence, for $\epsilon = 1/m$, there is a positive integer $N(x,\epsilon)$ so that

$$|f_n(x) - f(x)| \;<\; \epsilon = 1/m, \;\; \forall\, n > N(x,\epsilon)$$

Thus,

$$\Big(|f_n(x) - f(x)| \;\geq\; 1/m \Big) \;=\; \emptyset, \;\; \forall\, n > N(x,\epsilon). \tag{$*$}$$

Now consider $F_m = \bigcap_{n=1}^{\infty} E_{nm}$. If $x \in F_m$, then x is in E_{nm} for all n. In particular, letting $n^* = N(x,\epsilon) + 1$, we have $x \in E_{n^* m}$. Looking at how we defined $E_{n^* m}$, we see this implies that there is a positive integer $k' > n^*$, so that $|f'_k(x) - f(x)| \geq 1/m$. However, by Equation $*$, this set is empty. This contradiction means our original assumption that F_m was non empty is wrong. Hence, $F_m = \emptyset$. Now, since $\mu(X) < \infty$, $\mu(E_{1\,m})$ is finite also. Hence, by Lemma 9.2.2,

$$0 = \mu(F_m) \;=\; \lim_n \mu(E_{n\,m})$$

This implies that given $\delta > 0$, there is a positive integer N_m so that

$$\mu(E_{n\,m}) \;<\; \delta/2^m, \;\; \forall\, m > N_m$$

since $\lim_m \mu(E_{nm}) = 0$. For each integer m, choose a positive integer $n_m > N_m$. We can arrange for these integers to be increasing; i.e., $n_m < n_{m+1}$. Then,

$$\mu\Big(E_{n_m\,m} \Big) \;<\; \delta/2^m$$

and letting

$$E_\delta \;=\; \bigcup_{m=1}^{\infty} E_{n_m\,m}$$

we have

$$\mu(E_\delta) \;\leq\; \sum_{m=1}^{\infty} \delta/2^m < \delta$$

Finally, note

$$E_\delta^C \;=\; \left(\bigcup_{m=1}^{\infty} E_{n_m\,m} \right)^C = \bigcap_{m=1}^{\infty} E_{n_m\,m}^C$$

338   BASIC ANALYSIS IV: MEASURE THEORY AND INTEGRATION

Next, note

$$
\begin{aligned}
E^C_{n_m\, m} &= \left(\bigcup_{k=n_m}^{\infty} \left(|f_k(x) - f(x)| \geq 1/m \right) \right)^C \\
&= \bigcap_{k=n_m}^{\infty} \left(|f_k(x) - f(x)| \geq 1/m \right)^C \\
&= \bigcap_{k=n_m}^{\infty} \left(|f_k(x) - f(x)| < 1/m \right)
\end{aligned}
$$

Thus, since $x \in E^C_\delta$ means x is in $E^C_{n_m\, m}$ for all m, the above says $|f_k(x) - f(x)| < 1/m$ for all $k > n_m$. Therefore, given $\epsilon > 0$, pick a positive integer M so that $1/M < \epsilon$. Then, for all x in E^C_δ, we have

$$
|f_k(x) - f(x)| \ < \ 1/M \ < \ \epsilon, \ \forall\, k \geq n_M
$$

This says f_n converges uniformly to f on E^C_δ with $\mu(E_\delta) < \delta$. Hence, we have shown $f_n \to f$ [a.u.]
     Finally, if $f_n \to f$ [a.u.], by Theorem 15.1.4, we have $f_n \to f$ [meas] also. ∎

Next, let's see what we can do with domination by a p-summable function.

Theorem 15.2.2 Pointwise a.e. Convergence Plus Domination Implies p-Norm Convergence

> *Let $1 \leq p < \infty$ and $(X, \mathcal{S}, \mu)$ be a measure space. Let f be an extended real-valued function which is measurable. Also, let (f_n) be a sequence of functions in $\mathcal{L}_p(X, \mathcal{S})$ such that $f_n \to f$ [a.e.]. Assume there is a dominating function g which is p-summable; i.e. $|f_n(x)| \leq g(x)$ a.e. Then, if $f_n \to f$ [a.e.], f is p-summable and $f_n \to f$ [p − norm].*

Proof 15.2.2

Since $|f_n(x)| \leq g(x)$ a.e., we have immediately that $|f| \leq g$ a.e. since $f_n \to f$ [a.e.]. Thus, $|f|^p \leq g^p$ and we know f is in $\mathcal{L}_p(X, \mathcal{S})$. Since all the functions here are p-summable, the set where all additions is not defined has measure zero. So, we can assume without loss of generality that this set has been incorporated into the set on which convergence fails. Hence, we can say

$$
|f_n(x) - f(x)| \ \leq \ |f_n(x)| + |f(x)| \ \leq \ 2\, g(x), \ a.e.
$$

So,

$$
|f_n(x) - f(x)|^p \ \leq \ 2^p\, |g(x)|^p, \ a.e.
$$

By assumption, g is p-summable, so we have $2^p\, g^p$ is in $\mathcal{L}_1(X, \mathcal{S})$. Applying Lebesgue's Dominated Convergence Theorem, we find

$$
\lim_n \int |f_n(x) - f(x)|^p \, d\mu \ = \ \int \lim_n |f_n(x) - f(x)|^p \, d\mu \ = \ 0
$$

Thus, $f_n \to f$ [p − norm]. ∎

Homework

Exercise 15.2.1 *Consider the ODE $u'' = f$, $u(0) = 0$ and $u(L) = 0$ for some $L > 0$ and any*

CONVERGENCE MODES

$f \in L_2([0, L])$. *This can be rewritten as $u(t) = \int_0^L k(t, s) f(s) ds$ where the kernel function k is continuous on $[0, L] \times [0, L]$ and so is bounded. We know given $\epsilon > 0$, there is a continuous function f_c with $\|f - f - c\|_2 < \epsilon$ and there is a sequence of polynomials (p_n) with $p_n \to f_c$ uniformly. Prove there is a subsequence (u_n^1) which converges a.u. to $\int_0^L k(t, s) f(s) ds$.*

Exercise 15.2.2 *Consider the ODE $u'' + 4u = f$, $u(0) = 0$ and $u(10) = 0$ and any $f \in L_2([0, 10])$. This can be rewritten as $u(t) = \int_0^{10} k(t, s) f(s) ds$ where the kernel function k is continuous on $[0, 10] \times [0, 10]$ and so is bounded. We know given $\epsilon > 0$, there is a continuous function f_c with $\|f - f - c\|_2 < \epsilon$ and there is a sequence of polynomials (p_n) with $p_n \to f_c$ uniformly. Prove there is a subsequence (u_n^1) which converges a.u. to $\int_0^L k(t, s) f(s) ds$.*

Exercise 15.2.3 *Consider the bounded linear operator $T : L_2([0, L]) \to L_2([0, L])$ defined by $(T(f))(t) = \int_0^L k(t, s) f(s) ds$ where the kernel function k is continuous on $[0, 10] \times [0, 10]$ and so is bounded. We know given $\epsilon > 0$, there is a continuous function f_c with $\|f - f - c\|_2 < \epsilon$ and there is a sequence of polynomials (p_n) with $p_n \to f_c$ uniformly. Prove there is a subsequence $T(p_n^1)$ which converges a.u. to $\int_0^L k(t, s) f(s) ds$.*

Exercise 15.2.4 *Consider the bounded linear operator $T : L_2([0, L]) \to L_2([0, L])$ defined by $(T(f))(t) = \int_0^L k(t, s) f(s) ds$ where the kernel function k is continuous on $[0, 10] \times [0, 10]$ and so is bounded. We know given $\epsilon > 0$, there is a continuous function f_c with $\|f - f - c\|_2 < \epsilon$ and there is a sequence of polynomials (p_n) with $p_n \to f_c$ uniformly. Assume (f_n) is a sequence which converges a.e. to a function f. Further assume there is a dominating function $g \in L_2([0, L])$ with $f_n \leq g$ a.e.*

- *Prove $(T(f_n))$ converges to $T(f)$ in $\|\cdot\|_\infty$.*

- *Prove $(T(f_n))$ converges to $T(f)$ in $\|\cdot\|_2$.*

15.3 Vitali's Theorem

This important theorem is one that gives us more technical tools to characterize p-norm convergence for a sequence of functions. We need a certain amount of technical infrastructure to pull this off; so bear with us as we establish a series of lemmas.

Lemma 15.3.1 p-Summable Functions Have p-Norm Arbitrarily Small Off a Set

> *Let $1 \leq p < \infty$ and $(X, \mathcal{S}, \mu)$ be a measure space. Let f be in $\mathcal{L}_p(X, \mathcal{S})$. Then given $\epsilon > 0$, there is a measurable set E_ϵ so that $\mu(E_\epsilon) < \infty$ and if $F \subseteq E_\epsilon^C$ is measurable, then $\|f I_F\|_p < \epsilon$.*

Proof 15.3.1
Let $E_n = (|f_n(x)| \geq 1/n)$. Note $E_n \in \mathcal{S}$ and the sequence (E_n) is increasing and $\cup_n E_n = X$. Let $f_n = f I_{E_n}$. It is straightforward to verify that $f_n \uparrow f$ as $f_n \leq f_{n+1}$ for all n. Further, $|f_n|^p \leq |f|^p$; hence, by the Dominated Convergence Theorem,

$$\lim_n \int |f_n|^p \, d\mu = \int \lim_n |f_n|^p \, d\mu = \int |f|^p \, d\mu < \infty$$

The definition of f_n and E_n then implies

$$\mu(E_n)/n^p \leq \int_{E_n} |f|^p \, d\mu \leq \int |f|^p \, d\mu < \infty$$

340   BASIC ANALYSIS IV: MEASURE THEORY AND INTEGRATION

This tells us $\mu(E_n) < \infty$ for all n.

Now choose $\epsilon > 0$ arbitrarily. Then there is a positive integer N so that

$$\int |f|^p \, d\mu \; - \; \int |f_n|^p \, d\mu \;\; < \;\; \epsilon^p, \, \forall \, n \, > \, N$$

Thus, since $f_n = f I_{E_n}$, we can say

$$\int_{E_n} |f|^p \, d\mu \; + \; \int_{E_n^C} |f|^p \, d\mu \; - \; \int_{E_n} |f|^p \, d\mu \;\; < \;\; \epsilon^p, \, \forall \, n \, > \, N$$

or

$$\int_{E_n^C} |f|^p \, d\mu \;\; < \;\; \epsilon^p, \, \forall \, n \, > \, N$$

So choose $E_\epsilon = E_{N+1}$ and we have

$$\int_{E_\epsilon^C} |f|^p \, d\mu \;\; < \;\; \epsilon^p$$

which implies the desired result. ∎

Lemma 15.3.2 p-Summable Inequality

> *Let $1 \leq p < \infty$ and $(X, \mathcal{S}, \mu)$ be a measure space. Let (f_n) be a sequence of functions in $\mathcal{L}_p(X, \mathcal{S})$. Define β_n on $\mathcal{S}$ by*
>
> $$\beta_n(E) \;\; = \;\; \|f_n \, I_E\|_p, \, \forall \, E$$
>
> *Then,*
>
> $$|\beta_n(E) \, - \, \beta_m(E)| \;\; \leq \;\; \|f_n - f_m\|_p, \, \forall \, E, \, \forall \, n, \, m$$

Proof 15.3.2

By the backwards triangle inequality, for any measurable E,

$$\|f_n \, - \, f_m\|_p \;\; \geq \;\; |\|f_n \, I_E\|_p \, - \, \|f_m \, I_E\|_p| \; = \; |\beta_n(E) \, - \, \beta_m(E)|$$

∎

Lemma 15.3.3 p-Summable Cauchy Sequence Condition I

> *Let $1 \leq p < \infty$ and $(X, \mathcal{S}, \mu)$ be a measure space. Let (f_n) be a Cauchy sequence in $\mathcal{L}_p(X, \mathcal{S})$. Define β_n on $\mathcal{S}$ as done in Lemma 15.3.2. Then, there is a positive integer N and a measurable set E_ϵ of finite measure, so that if F is a measurable subset of E_ϵ, then $\beta_n(E) < \epsilon$ for all $n > N$.*

Proof 15.3.3

Since (f_n) is a Cauchy sequence in p-norm, there is a function f in $\mathcal{L}_p(X, \mathcal{S})$ so that $f_n \to f$ [p −

CONVERGENCE MODES 341

norm]. By Lemma 15.3.1, given $\epsilon > 0$, there is a measurable set E_ϵ with finite measure so that

$$\int_{E_\epsilon^C} |f|^p \, d\mu \;\; < \;\; (\epsilon/2)^p$$

Now given a measurable F contained in E_ϵ^C, recalling the meaning of $\beta_n(F)$ as described in Lemma 15.3.2, we can write

$$\begin{aligned}
\beta_n(F) \;\; &\leq \;\; \|f_n\|_p \leq \|f_n \, I_{E_\epsilon^C}\|_p \\
&\leq \;\; \|(f_n - f) \, I_{E_\epsilon^C}\|_p + \|f \, I_{E_\epsilon^C}\|_p \\
&< \;\; \epsilon/2 + \|(f_n - f) \, I_{E_\epsilon^C}\|_p
\end{aligned}$$

Since $f_n \to f \, [p - norm]$, there is a positive integer N so that if $n > N$,

$$\|(f_n - f) \, I_{E_\epsilon^C}\|_p \;\; < \;\; \epsilon/2$$

This shows $\beta_n(F) < \epsilon$ when $n > N$ as desired. ∎

Lemma 15.3.4 Continuity of the Integral

Let $(X, \mathcal{S}, \mu)$ be a measure space and f be a summable function. Then for all $\epsilon > 0$ there is a $\delta > 0$, so that

$$\left| \int_E f \, d\mu \right| \;\; < \;\; \epsilon, \, \forall \, E \in \mathcal{S}, \, \text{with } \mu(E) < \delta$$

Proof 15.3.4

Define the measure γ on $\mathcal{S}$ by $\gamma(E) = \int_E |f| \, d\mu$. Note, by Comment 9.9.1, we know that γ is absolutely continuous with respect to μ. Now assume the proposition is false. Then, there is an $\epsilon > 0$ so that for all choices of $\delta > 0$, we have a measurable set E_δ for which $\mu(E_\delta) < \delta$ and $|\int_{E_\delta} f \, d\mu| \geq \epsilon$. In particular, for the sequence $\delta_n = 1/2^n$, we have a sequence of sets E_n with $\mu(E_n) < 1/2^n$ and

$$\left| \int_{E_n} f \, d\mu \right| \geq \epsilon$$

Let

$$G_n \;\; = \;\; \bigcup_{k=n}^{\infty} E_k, \; G = \bigcap_{n=1}^{\infty} G_n$$

Then,

$$\mu(G) \;\; \leq \;\; \mu(G_n) \leq \sum_{k=n}^{\infty} \mu(E_k) < \sum_{k=n}^{\infty} 1/2^k \; = \; 1/2^{n-1}$$

This implies $\mu(G) = 0$ and thus, since γ is absolutely continuous with respect to μ, $\gamma(G) = 0$ also. We also know the sequence G_n is decreasing and so $\gamma(G_n) \to \gamma(G) = 0$. Finally, since

$$\gamma(G_n) \;\; \geq \;\; \gamma(E_n) \geq |\int_E f \, d\mu| \geq \epsilon$$

342   BASIC ANALYSIS IV: MEASURE THEORY AND INTEGRATION

we have $\gamma(G) = \lim_n \gamma(G_n) \geq \epsilon$ as well. This is impossible. Hence, our assumption that the proposition is false is wrong. ∎

Lemma 15.3.5 p-Summable Cauchy Sequence Condition II

> *Let $1 \leq p < \infty$ and $(X, \mathcal{S}, \mu)$ be a measure space. Let (f_n) be a Cauchy sequence in $\mathcal{L}_p(X, \mathcal{S})$. Define β_n on $\mathcal{S}$ as done in Lemma 15.3.2. Then, given $\epsilon > 0$, there is a $\delta > 0$ and a positive integer N so that if $n > N$, then*
>
> $$\beta_n(E) \; < \; \epsilon, \, \forall \, E \in \mathcal{S}, \, with \, \mu(E) < \delta$$

Proof 15.3.5

Since $\mathcal{L}_p(X, \mathcal{S}, \mu)$ is complete, there is a p-summable function f so that $f_n \to f \, [p - norm]$. Then, by Lemma 15.3.4, given an $\epsilon > 0$, there is a $\delta > 0$, so that

$$\int_E |f|^p \, d\mu \; < \; (\epsilon/2)^p, \, if \, \mu(E) < \delta$$

Hence, using the convenience mapping $\beta_n(E)$ previously defined in Lemma 15.3.2, we see

$$\begin{aligned} \beta_n(E) \; &= \; \|f_n \, I_E\|_p = \|(f - f_n) \, I_E\|_p + \|f \, I_E\|_p \\ &\leq \; \|(f - f_n) \, I_E\|_p + \epsilon/2 \end{aligned}$$

when $\mu(E) < \delta$. Finally, since $f_n \to f \, [p - norm]$, there is a positive integer N so that if $n > N$, then $\|(f - f_n) \, I_E\|_p < \epsilon/2$. Combining, we have $\beta_n(E) < \epsilon$ if $n > N$ for $\mu(E) < \delta$. ∎

Theorem 15.3.6 Vitali Convergence Theorem

> *Let $1 \leq p < \infty$ and $(X, \mathcal{S}, \mu)$ be a measure space. Let (f_n) be a sequence of functions in $\mathcal{L}_p(X, \mathcal{S})$. Then, $f_n \to f \, [p - norm]$ if and only if the following three conditions hold.*
>
> *(i): $f_n \to f \, [meas]$.*
> *(ii): For all $\epsilon > 0$, there exists N and a set $E_\epsilon \in \mathcal{S}$ with $\mu(E_\epsilon) < \infty$ so that if F is a measurable set in E_ϵ^C, then $\int_F |f_n|^p \, d\mu < \epsilon^p$, for all $n > N$.*
> *(iii): For all $\epsilon > 0$, there is a $\delta > 0$ and an N so that if E is measurable with $\mu(E) < \delta$, then $\int_E |f_n|^p \, d\mu < \epsilon^p$, for all $n > N$.*

Proof 15.3.6

$\Rightarrow$: *If $f_n \to f \, [p - norm]$, then by Theorem 15.1.3, $f_n \to f \, [meas]$ which shows (i) holds. Then, since $f_n \to f \, [p - norm]$, (f_n) a Cauchy sequence. Thus, by Lemma 15.3.1, condition (ii) holds. Finally, since (f_n) is a Cauchy sequence, by Lemma 15.3.5, condition (iii) holds.*

$\Rightarrow$: *Now assume conditions (i), (ii) and (iii) hold. Let $\epsilon > 0$ be given. From condition (ii), we see there is a measurable set E_ϵ of finite measure and a positive integer N_1 so that*

$$\int_{E_\epsilon^C} |f_n|^p \, d\mu \; < \; (\epsilon/4)^p$$

CONVERGENCE MODES

if $n > N_1$. Thus, for indices n and m larger than N_1, we have

$$
\begin{aligned}
\|f_n - f_m\|_p &= \|(f_n - f_m)\,I_{E_\epsilon} + (f_n - f_m)\,I_{E_\epsilon^C}\|_p \\
&\leq \|(f_n - f_m)\,I_{E_\epsilon}\|_p + \|(f_n - f_m)\,I_{E_\epsilon^C}\|_p \\
&\leq \|(f_n - f_m)\,I_{E_\epsilon}\|_p + \|f_n\,I_{E_\epsilon^C}\|_p + \|f_m\,I_{E_\epsilon^C}\|_p \\
&< \|(f_n - f_m)\,I_{E_\epsilon}\|_p + \epsilon/2
\end{aligned}
$$

We conclude

$$
\|f_n - f_m\|_p < \|f_m\,I_{E_\epsilon}\|_p + \epsilon/2, \ \forall\, n,\, m > N_1. \tag{$*$}
$$

Now let $\beta - \mu(E_\epsilon)$ and set $\alpha = \frac{\epsilon}{4\,\beta^{1/p}}$ and define the sets H_{nm} by

$$
H_{nm} = \{x \mid |f_n(x) - f_m(x)| \geq \alpha\}
$$

Apply condition (ii) for our given ϵ now. Thus, there is a $\delta(\epsilon)$ and a positive integer N_2 so that

$$
\int_E |f_n|^p \, d\mu < (\epsilon/8)^p, \ n > N - 2, \ \text{when } \mu(E) < \delta(\epsilon). \tag{$**$}
$$

Since $f_n \to f$ [meas], (f_n) is Cauchy in measure. Hence, there is a positive integer N_3 so that

$$
\mu(H_{nm}) < \delta(\epsilon), \ \forall\, n,\, m > N_3. \tag{$***$}
$$

Finally, using the Minkowski Inequality, we have

$$
\begin{aligned}
\|(f_n - f_m)\,I_{E_\epsilon}\|_p &= \|(f_n - f_m)\,I_{E_\epsilon \setminus H_{nm}} + (f_n - f_m)\,I_{H_{nm}}\|_p \\
&\leq \|(f_n - f_m)\,I_{E_\epsilon \setminus H_{nm}}\|_p + \|f_n\,I_{H_{nm}}\|_p + \|f_m\,I_{H_{nm}}\|_p
\end{aligned}
$$

Now let $N = \max\{N_1, N_2, N_3\}$. Then, if n and m exceed N, we have Equation $*$, Equation $**$ and Equation $***$ all hold. This implies

$$
\begin{aligned}
\|(f_n - f_m)\,I_{E_\epsilon}\|_p &\leq \left(\alpha^p\,\mu(E_\epsilon \setminus H_{nm})\right)^{1/p} + \epsilon/8 + \epsilon/8 \\
&\leq \alpha\left(\mu(E_\epsilon)\right)^{1/p} + \epsilon/4 \\
&= \alpha\,\beta^{1/p} + \epsilon/4 = \epsilon/(4\,\beta^{1/p})\,\beta^{1/p} + \epsilon/4 \\
&= \epsilon/2
\end{aligned}
$$

From Equation $*$, we have for these indices n and m,

$$
\begin{aligned}
\|f_n - f_m\|_p &< \|f_m\,I_{E_\epsilon}\|_p + \epsilon/2 \\
&< \epsilon
\end{aligned}
$$

Thus, (f_n) is a Cauchy sequence in p-norm. Since $\mathcal{L}_p(X, \mathcal{S}, \mu)$ is complete, there is a function g so that $f_n \to g\,[p-norm]$. So by Theorem 15.1.3, $f_n \to g$ [meas]. It is then straightforward to show that $f = g$ a.e. This tells us f and g belong to the same equivalence class of $\mathcal{L}_p(X, \mathcal{S}, \mu)$. ∎

Homework

344 BASIC ANALYSIS IV: MEASURE THEORY AND INTEGRATION

Exercise 15.3.1 *Let g be defined by*

$$g(x) \;=\; \begin{cases} 2x, & 0 \leq x < 1 \\ 5x^2 + 1, & 1 \leq x \leq 2 \end{cases}$$

This g generates the Lebesgue - Stieltjes measure μ_g on $[0, 2]$.

- *Prove for all $\epsilon > 0$ there is a $\delta > 0$ so that $\mu_g(E) < \epsilon$ if $\mu(E) < \delta$.*

- *If (E_n) is a monotonically decreasing sequence of measurable sets all contained in E which has finite measure, prove given $\epsilon > 0$ there is an N so that $\mu_g(E_n) < \epsilon$ if $n > N$ which implies $\mu_g(E_n) \to 0$.*

Exercise 15.3.2 *Let g be defined by*

$$g(x) \;=\; \begin{cases} x + 4, & -1 \leq x < 1 \\ x^2 + 7, & 1 \leq x < 2 \\ 2x^3, & 2 \leq x \leq 3 \end{cases}$$

This g generates the Lebesgue - Stieltjes measure μ_g on $[0, 3]$.

- *Prove for all $\epsilon > 0$ there is a $\delta > 0$ so that $\mu_g(E) < \epsilon$ if $\mu(E) < \delta$.*

- *If (E_n) is a monotonically decreasing sequence of measurable sets all contained in E which has finite measure, prove given $\epsilon > 0$ there is an N so that $\mu_g(E_n) < \epsilon$ if $n > N$ which implies $\mu_g(E_n) \to 0$.*

15.4 Summary

We can summarize the results of this chapter as follows. If we know the measure is finite, we can say quite a bit.

Theorem 15.4.1 Convergence Relationships on Finite Measure Space One

> Let $(X, \mathcal{S}, \mu)$ be a measure space with $\mu(X) < \infty$. Let f and (f_n) be in $M(X, \mathcal{S})$. Then,
>
> $$\left. \begin{array}{l} f_n \to f \; [p-norm] \\ f_n \to f \; [unif] \\ f_n \to f \; [a.u.] \\ f_n \to f \; [a.e.] \end{array} \right\} \;\Rightarrow\; f_n \to f \; [meas]$$

Certain types of convergence give us pointwise convergence.

Theorem 15.4.2 Convergence Relationships on Finite Measure Space Two

> Let $(X, \mathcal{S}, \mu)$ be a measure space with $\mu(X) < \infty$. Let f and (f_n) be in $M(X, \mathcal{S})$. Then,
>
> $$\left. \begin{array}{l} f_n \to f \; [unif] \\ f_n \to f \; [a.u.] \end{array} \right\} \;\Rightarrow\; f_n \to f \; [a.e.]$$

Uniform convergence on a finite measure space implies p norm convergence.

Theorem 15.4.3 Convergence Relationships on Finite Measure Space Three

CONVERGENCE MODES

> Let $(X, \mathcal{S}, \mu)$ be a measure space with $\mu(X) < \infty$. Let f and (f_n) be in $M(X, \mathcal{S})$. Then,
>
> $$f_n \to f \ [unif] \ \Rightarrow \ \begin{cases} f_n \to f \ [a.u.] \\ f_n \to f \ [p-norm] \end{cases}$$

The final result says that pointwise convergence is "almost" uniform convergence.

Theorem 15.4.4 Convergence Relationships on Finite Measure Space Four

> Let $(X, \mathcal{S}, \mu)$ be a measure space with $\mu(X) < \infty$. Let f and (f_n) be in $M(X, \mathcal{S})$. Then,
>
> $$f_n \to f \ [a.e.] \ \Rightarrow \ f_n \to f \ [a.u.]$$

If the measure of X is infinite, we have many one way implications.

Theorem 15.4.5 Convergence Relationships on General Measurable Space

> Let $(X, \mathcal{S}, \mu)$ be a measure space. Let f and (f_n) be in $M(X, \mathcal{S})$. Then,
> (i):
>
> $$\left. \begin{array}{l} f_n \to f \ [p-norm] \\ f_n \to f \ [unif] \\ f_n \to f \ [a.u.] \end{array} \right\} \ \Rightarrow \ f_n \to f \ [meas]$$
>
> (ii):
>
> $$\left. \begin{array}{l} f_n \to f \ [unif] \\ f_n \to f \ [a.u.] \end{array} \right\} \ \Rightarrow \ f_n \to f \ [a.e.]$$
>
> (iii):
>
> $$f_n \to f \ [unif] \ \Rightarrow \ f_n \to f \ [a.u.]$$

Next, if we can dominate the sequence by an $\mathcal{L}_p$ function, we can say even more.

Theorem 15.4.6 Convergence Relationships with p-Domination One

> Let $(X, \mathcal{S}, \mu)$ be a measure space. Let f and (f_n) be in $M(X, \mathcal{S})$. Assume there is a $g \in L_p$ so that $|f_n| \le g$. Then, we know the following implications:
>
> $$\left. \begin{array}{l} f_n \to f \ [p-norm] \\ f_n \to f \ [unif] \\ f_n \to f \ [a.u.] \\ f_n \to f \ [a.e.] \end{array} \right\} \ \Rightarrow \ f_n \to f \ [meas]$$

Theorem 15.4.7 Convergence Relationships with p-Domination Two

BASIC ANALYSIS IV: MEASURE THEORY AND INTEGRATION

Let $(X, \mathcal{S}, \mu)$ be a measure space. Let f and (f_n) be in $M(X, \mathcal{S})$. Assume there is a $g \in L_p$ so that $|f_n| \leq g$. Then, we know the following implications:

$$\left. \begin{array}{l} f_n \to f \ [unif] \\ f_n \to f \ [a.u.] \end{array} \right\} \ \Rightarrow \ f_n \to f \ [a.e.]$$

Theorem 15.4.8 Convergence Relationships with p-Domination Three

Let $(X, \mathcal{S}, \mu)$ be a measure space. Let f and (f_n) be in $M(X, \mathcal{S})$. Assume there is a $g \in L_p$ so that $|f_n| \leq g$. Then, we know the following implications:

$$f_n \to f \ [unif] \ \Rightarrow \ \left\{ \begin{array}{l} f_n \to f \ [a.u.] \\ f_n \to f \ [p-norm] \end{array} \right.$$

Theorem 15.4.9 Convergence Relationships with p-Domination Four

Let $(X, \mathcal{S}, \mu)$ be a measure space. Let f and (f_n) be in $M(X, \mathcal{S})$. Assume there is a $g \in L_p$ so that $|f_n| \leq g$. Then, we know the following implications:

$$f_n \to f \ [a.e.] \ \Rightarrow \ f_n \to f \ [a.u.]$$

Theorem 15.4.10 Convergence Relationships with p-Domination Five

Let $(X, \mathcal{S}, \mu)$ be a measure space. Let f and (f_n) be in $M(X, \mathcal{S})$. Assume there is a $g \in L_p$ so that $|f_n| \leq g$. Then, we know the following implications:

$$\left. \begin{array}{l} f_n \to f \ [a.e.] \\ f_n \to f \ [a.u.] \end{array} \right\} \ \Rightarrow \ f_n \to f \ [p-norm]$$

Theorem 15.4.11 Convergence Relationships with p-Domination Six

Let $(X, \mathcal{S}, \mu)$ be a measure space. Let f and (f_n) be in $M(X, \mathcal{S})$. Assume there is a $g \in L_p$ so that $|f_n| \leq g$. Then, we know the following implications:

$$f_n \to f \ [meas] \ \Rightarrow \ f_n \to f \ [p-norm]$$

There are circumstances where we can be sure we can extract a subsequence that converges in some fashion.

Theorem 15.4.12 Convergent Subsequences Exist

CONVERGENCE MODES

Let $(X, \mathcal{S}, \mu)$ be a measure space. It doesn't matter whether or not $\mu(X)$ is finite. Let f and (f_n) be in $M(X, \mathcal{S})$. Then, we know the following implications:

(i):
$$\left. \begin{array}{c} f_n \to f \ [p-norm] \\ f_n \to f \ [meas] \end{array} \right\} \quad \Rightarrow \quad \exists \ subsequence \ f_n^1 \to f \ [a.u.]$$

(ii):
$$\left. \begin{array}{c} f_n \to f \ [p-norm] \\ f_n \to f \ [meas] \end{array} \right\} \quad \Rightarrow \quad \exists \ subsequence \ f_n^1 \to f \ [a.e.]$$

Further, the same implications hold if we know there is a $g \in L_p$ so that $|f_n| \leq g$.

Homework

Exercise 15.4.1 *Characterize convergence in measure when the measure is counting measure.*

Exercise 15.4.2 *Let $(X, \mathcal{S}\mu)$ be a measure space. Let $(f_n), (g_n) \subseteq M(X, \mathcal{S}, \mu)$ be sequences of functions which are finite a.e. Let $f, g : X \to \overline{\mathfrak{R}}$ be functions. Prove if $f_n \to f[meas \ on \ E]$ and $g_n \to g[meas \ on \ E]$, then $(f_n + g_n) \to (f + g)[meas \ on \ E]$.*

Exercise 15.4.3 *Let $(X, \mathcal{S}\mu)$ be a measure space with $\mu(X) < \infty$. Let $(f_n), (g_n) \subseteq M(X, \mathcal{S}, \mu)$ be sequences of functions which are finite a.e. Let $f, g : X \to \overline{\mathfrak{R}}$ be functions. Prove if $f_n \to f[meas \ on \ E]$ and $g_n \to g[meas \ on \ E]$, then $(f_n \, g_n) \to (f \, g)[meas \ on \ E]$. Hint: first consider the case that $f_n \to 0[meas \ on \ E]$ and $g_n \to 0[meas \ on \ E]$.*

Exercise 15.4.4 *Let $(X, \mathcal{S}, \mu)$ be a measure space. Let $(f_n) \subseteq M(X, \mathcal{S}, \mu)$ be a sequence of functions which are finite a.e. Let $f : X \to \overline{\mathfrak{R}}$ be a function. Prove if $f_n \to f[a.u.]$, then $f_n \to f[ptws \ a.e.]$ and $f_n \to f[meas]$.*

Exercise 15.4.5 *Let $(X, \mathcal{S}, \mu)$ be a finite measure space. For any pair of measurable functions f and g, define*
$$d(f, g) = \int \frac{|f - g|}{1 + |f - g|} \, d\mu$$

(i): *Prove $M(X, \mathcal{S}, \mu)$ is a semi-metric space.*

(ii): *Prove if (f_n) is a sequence of measurable functions and f is another measurable function, then $f_n \to f[meas]$ if and only if $d(f_n, f) \to 0$.*

Hint: You don't need any high-power theorems here. First, let $\phi(t) = t/(1 + t)$ so that $d(f, g) = \int \phi(|f - g|) d\mu$. Then try this:
$(\Rightarrow)$: We assume $f_n \to f[meas]$. Then, given any pair of positive numbers (δ, ϵ), we have there is an N so that if $n > N$, we have

$$\mu(|f_n(x) - f(x)| \geq \delta) < \epsilon/2$$

Let E_δ denote the set above. Now for such $n > N$, note

$$d(f_n, f) = \int_{E_\delta} \phi(|f_n - f|) \, d\mu + \int_{E_\delta^C} \phi(|f_n - f|) \, d\mu$$

348   *BASIC ANALYSIS IV: MEASURE THEORY AND INTEGRATION*

Since ϕ is increasing, we see that on E_δ^C, $\phi(|f_n(x) - f(x)|) < \phi(\delta)$. Thus, you should be able to show that if $n > N$, we have

$$d(f_n, f) < \mu(E_\delta) + \phi(\delta)\,\mu(X) = \epsilon/2 + \phi(\delta)\,\mu(X)$$

Then a suitable choice of δ does the job.

($\Leftarrow$): If we know $d(f_n, f)$ goes to zero, break the integral up the same way into a piece on E_δ and E_δ^C. This tells us right away that given $\epsilon > 0$, there is an N so that $n > N$ implies

$$\phi(\delta)\,\mu(E_\delta) < \epsilon$$

This gives us the result with a little manipulation.

Exercise 15.4.6 *Let $(\Re, \mathcal{M}, \mu)$ denote the measure space consisting of the Lebesgue measurable sets $\mathcal{M}$ and Lebesgue measure μ. Let the sequence (f_n) of measurable functions be defined by*

$$f_n = n\, I_{[1/n, 2/n]}$$

Prove $f_n \to 0$ on all $\Re$, $f_n \to 0$ [meas] but $f_n \not\to 0$ [p - norm] for $1 \leq p < \infty$.

Chapter 16

Decomposition of Measures

We now examine the structure of a charge λ on a σ - algebra $\mathcal{S}$. For convenience, let's recall that a charge is a mapping on $\mathcal{S}$ to $\Re$ which assigns the value 0 to $\emptyset$ and which is countably additive. We need some beginning definitional material before we go further.

16.1 The Jordan Decomposition of a Charge

Definition 16.1.1 Positive and Negative Sets for a Charge

Let λ be a charge on $(X, \mathcal{S})$. We say $P \in \mathcal{S}$ is a positive set with respect to λ if

$$\lambda(E \cap P) \geq 0, \forall E \in \mathcal{S}$$

Further, we say $N \in \mathcal{S}$ is a negative set with respect to λ if

$$\lambda(E \cap N) \leq 0, \forall E \in \mathcal{S}$$

Finally, $M \in \mathcal{S}$ is a null set with respect to λ is

$$\lambda(E \cap M) = 0, \forall E \in \mathcal{S}$$

Definition 16.1.2 The Positive and Negative Parts of a Charge

---

Let λ be a charge on $(X, \mathcal{S})$. Define the mapping λ^+ on $\mathcal{S}$ by

$$\lambda^+(E) = \sup \{ \lambda(A) \,|\, A \in \mathcal{S}, \, A \subseteq E \}$$

Also, define the mapping λ^- on $\mathcal{S}$ by

$$\lambda^-(E) = -\inf \{ \lambda(A) \,|\, A \in \mathcal{S}, \, A \subseteq E \}$$

Theorem 16.1.1 The Jordan Decomposition of a Finite Charge

Let λ be a finite valued charge on $(X, \mathcal{S})$. Then, λ^+ and λ^- are finite measures on $\mathcal{S}$ and $\lambda = \lambda^+ - \lambda^-$. The pair (λ^+, λ^-) is called the Jordan Decomposition of λ.

BASIC ANALYSIS IV: MEASURE THEORY AND INTEGRATION

Proof 16.1.1

Let's look at λ^+ first. Given any measurable E, since $\emptyset$ is contained in E, by the definition of λ^+, we must have $\lambda^+(E) \geq \lambda(\emptyset) = 0$. Hence, λ^+ is non-negative.

Next, if A and B are measurable and disjoint, By definition of λ^+, for $C_1 \subseteq A$ $C_2 \subseteq B$, we must have

$$\begin{aligned} \lambda^+(A \cup B) &\geq \lambda(C_1 \cup C_2) \\ &= \lambda(C_1) + \lambda(C_2) \end{aligned}$$

This says $\lambda^+(A \cup B) - \lambda(C_2)$ is an upper bound for the set of numbers $\{\lambda(C_1)\}$. Hence, by definition of $\lambda^+(A)$, we have

$$\begin{aligned} \lambda^+(A \cup B) &\geq \lambda(C_1 \cup C_2) \\ &= \lambda^+(A) + \lambda(C_2) \end{aligned}$$

A similar argument then shows that $\lambda(C_2)$ is bounded above by $\lambda^+(A \cup B) - \lambda^+(A)$. Thus, we have

$$\begin{aligned} \lambda^+(A \cup B) &\geq \lambda(C_1 \cup C_2) \\ &= \lambda^+(A) + \lambda^+(B) \end{aligned}$$

On the other hand, if $C \subseteq A \cup B$, then we have

$$\begin{aligned} \lambda(C) &= \lambda(C \cap A \cup C \cap B) \\ &\leq \lambda^+(A) + \lambda^+(B) \end{aligned}$$

This immediately implies that

$$\lambda^+(A \cup B) \leq \lambda^+(A) + \lambda^+(B)$$

Thus, it is clear λ^+ is additive on finite disjoint unions.

We now address the question of the finiteness of λ^+. To see λ^+ is finite, assume that it is not. So there is some set E with $\lambda^+(E) = \infty$. Hence, by definition, there is a measurable set A_1 so that $\lambda(A_1) > 1$. Thus, by additivity of λ^+, we have

$$\lambda^+(A_1) + \lambda^+(E \setminus A_1) = \lambda^+(E) = \infty$$

Thus, at least one of $\lambda^+(A_1)$ and $\lambda^+(E \setminus A_1)$ is also ∞. Pick one such set and call it B_1. Thus, $\lambda^+(B_1) = \infty$. Let's do one more step. Since $\lambda^+(B_1) = \infty$, there is a measurable set A_2 inside it so that $\lambda(A_2) > 2$. Then,

$$\lambda^+(A_2) + \lambda^+(B_1 \setminus A_2) = \lambda^+(B_1) = \infty$$

Thus, at least one of $\lambda^+(A_2)$ and $\lambda^+(B_1 \setminus A_2)$ is also ∞. Pick one such set and call it B_2. Thus, we have $\lambda^+(B_2) = \infty$. You should be able to see how we construct the two sequences (A_n) and (B_n). When we are done, we know $A_n \subseteq B_{n-1}$, $\lambda(A_n) > n$ and $\lambda(B_n) = \infty$ for all n.

Now, if for an infinite number of indices n_k, $B_{n_k} = B_{n_k-1} \setminus A_{n_k}$, what happens? It is easiest to see with an example. Suppose $B_5 = B_4 \setminus A_5$ and $B_8 = B_7 \setminus A_8$. By the way we construct these sets, we see A_6 does not intersect A_5. Hence, $A_7 \cap A_5 = \emptyset$ also. Finally, we have $A_8 \cap A_5 = \emptyset$ too. Hence, extrapolating from this simple example, we can infer that the sequence (A_{n_k}) is disjoint. By *the countable additivity of λ, we then have*

$$\lambda\left(\bigcup_k A_{n_k}\right) \;=\; \sum_k \lambda(A_{n_k}) \;>\; \sum_k n_k \;=\; \infty$$

But λ is finite on all members of S. This is therefore a contradiction.

Another possibility is that there is an index N so that if $n > N$, the choice is always that of $B_n = A_n$. In this case, we have

$$E \;\supseteq\; A_{N+1} \supseteq A_{N+2} \cdots$$

Since λ is finite and additive,

$$\lambda(A_{N+j-1} \setminus A_{N+j}) \;=\; \lambda(A_{N+j-1}) - \lambda(A_{N+j})$$

for $j > 2$ since all the λ values are finite. We now follow the construction given in the proof of the second part of Lemma 9.2.2 to finish our argument. Construct the sequence of sets (E_n) by

$$\begin{aligned}
E_1 &= \emptyset \\
E_2 &= A_{N+1} \setminus A_{N+2} \\
E_3 &= A_{N+1} \setminus A_{N+3} \\
&\vdots \quad \vdots \quad \vdots \\
E_n &= A_{N+1} \setminus A_{N+n-1}
\end{aligned}$$

Then (E_n) is an increasing sequence of sets which are disjoint and so $\lambda(\cup_n E_n) = \lim_n \lambda(E_n)$. Since $\lambda(A_{N+1})$ is finite, we then know that $\lambda(E_n) = \lambda(A_{N+1}) - \lambda(A_{N+n})$. Hence, $\lambda(\cup_n E_n) = \lambda(A_{N+1}) - \lim_n \lambda(A_{N+n})$. Next, note by De Morgan's Laws,

$$\begin{aligned}
\lambda\left(\cup_n E_n\right) &= \lambda\left(\bigcup_n A_{N+1} \cap A_{N+n}^C\right) \\
&= \lambda\left(A_{N+1} \bigcap \cup_n A_{N+n}^C\right) \\
&= \lambda\left(A_{N+1} \bigcap \left(\cap_n A_{N+n}\right)^C\right) \\
&= \lambda\left(A_{N+1} \setminus \left(\cap_n A_{N+n}\right)\right)
\end{aligned}$$

Thus, since $\lambda(A_{N+1})$ is finite and $\cap_n A_{N+n} \subseteq A_{N+1}$, it follows that

$$\lambda(\cup_n E_n) \;=\; \lambda(A_{N+1}) - \lambda(\cap_n A_{N+n})$$

Combining these results, we have

$$\lambda(A_{N+1}) - \lim_n \lambda(A_{N+n}) \;=\; \lambda(A_{N+1}) - \lambda(\cap_n A_{N+n})$$

Canceling $\lambda(A_{N+1})$ from both sides, we find

$$\lambda(\cap_n A_{N+n}) \;=\; \lim_n \lambda(A_{N+n}) \;\geq\; \lim_n N+n \;=\; \infty$$

BASIC ANALYSIS IV: MEASURE THEORY AND INTEGRATION

We again find a set $\cap_n A_{N+n}$ with λ value ∞ inside E. However, λ is always finite. Thus, in this case also, we arrive at a contradiction.

We conclude at this point that if $\lambda^+(E) = \infty$, we force λ to become infinite for some subsets. Since that is not possible, we have shown λ^+ is finite. Since $\lambda^- = (-\lambda)^+$, we have established that λ^- is finite also. Next, given the relationship between λ^+ and λ^-, it is enough to prove λ^+ is a measure to complete this proof.

It is enough to prove that λ^+ is countably additive. Let (E_n) be a countable sequence of measurable sets and let E be their union. If $A \subseteq E$, then $A = \cup_n A \cap E_n$ and so

$$\lambda(A) \;=\; \sum_n \lambda(A \cap E_n)$$
$$\leq\; \sum_n \lambda^+(E_n)$$

by the definition of λ^+. Since this holds for all such subsets A, we conclude $\sum_n \lambda^+(E_n)$ is an upper bound for the collection of all such $\lambda(A)$. Hence, by the definition of a supremum, we have $\lambda^+(E) \leq \sum_n \lambda^+(E_n)$.

To show the reverse, note $\lambda^+(E)$ is finite by the arguments in the first part of this proof. Now, pick $\epsilon > 0$. Then, by the Supremum Tolerance Lemma, there is a sequence (A_n) of measurable sets, each $A_n \subseteq E_n$ so that

$$\lambda^+(E_n) \;-\; \epsilon/2^n \;<\; \lambda(A_n) \;\leq\; \lambda^+(E_n)$$

Let $A = \cup_n A_n$. Then $A \subseteq E$ and so we have $\lambda(A) \leq \lambda^+(E)$. Hence,

$$\sum_n \lambda^+(E_n) \;<\; \sum_n \left(\lambda(A_n) \;+\; \epsilon/2^n \right)$$
$$<\; \sum_n \lambda(A_n) \;+\; \epsilon$$

since the second term is a standard geometric series. Next, since (A_n) is a disjoint sequence, the countable additivity of λ gives

$$\sum_n \lambda^+(E_n) \;<\; \lambda\left(\cup_n A_n \right) \;+\; \epsilon$$

But $A = \cup_n A_n$ and since this holds for all $\epsilon > 0$, we can conclude

$$\sum_n \lambda^+(E_n) \;<\; \lambda(A) \;\leq\; \lambda^+(E)$$

Combining these inequalities, we see λ^+ is countably additive and hence is a measure. ∎

Comment 16.1.1 *If we had allowed the charge in Definition 9.1.2 to be extended real-valued; i.e. take on the values of ∞ and $-\infty$, what would happen? First, note by applying the arguments in the first part of the proof above, we can say if $\lambda^+(E) = \infty$, $\lambda(E) = \infty$ and similarly, if $\lambda^-(E) = \infty$, $\lambda(E) = -\infty$. Conversely, note by definition of λ^+, if $\lambda(E) = \infty$, then $\lambda^+(E) = \infty$ also and if $\lambda(E) = -\infty$, then $\lambda^-(E) = \infty$. So if $\lambda^+(E) = \infty$, what about $\lambda^-(E)$? If $\lambda^-(E) = \infty$, that would force $\lambda(E) = -\infty$ contradicting the value it already has. Hence $\lambda^-(E)$ is finite. Next, given any measurable set F, what about $\lambda^-(F)$? There are several cases. First, if $F \subseteq E$, then*

$$\lambda^-(E) \;=\; \lambda^-(F) + \lambda^-(F \setminus E)$$

DECOMPOSING MEASURES 353

Since $\lambda^- \geq 0$, if $\lambda^-(F) = \infty$, we get $\lambda^-(E)$, which is finite, and is also infinite. Hence, this cannot happen. Second, if F and E are disjoint, with $\lambda^-(F) = \infty$, we find

$$\lambda^-(E \cup F) \;=\; \lambda^-(E) \,+\, \lambda^-(F)$$

The right-hand side is ∞ and so since $\lambda^-(E \cup F)$ is infinite, this forces $\lambda(E \cup F) = -\infty$. But since λ is additive on disjoint sets, this leads to the undefined expression

$$\lambda(E \cup F) \;=\; \lambda(E) \,+\, \lambda(F)$$
$$-\infty \;=\; (\infty) \,+\, (-\infty)$$

This is not possible because by assumption, λ takes on a well-defined value in $\overline{\mathfrak{R}}$ for all measurable subsets. Thus, we conclude if there is a measurable set E so that $\lambda^+(E)$ is infinite, then λ^- will be finite everywhere. The converse is also true: if $\lambda^-(E)$ is infinite, then λ^+ will be finite everywhere. Thus, we can conclude if λ is extended real-valued, only one of λ^+ or λ^- can take on ∞ values.

Homework

Since this proof is so technical, some parts have been left out and are perfect as exercises! Working these out will help your understanding of the manipulations we do here.

Exercise 16.1.1 *In the proof we need to show $\lambda(C_2)$ is bounded above by $\lambda^+(A \cup B) - \lambda^+(A)$. Provide these details.*

Exercise 16.1.2 *In the proof, when we establish that λ^+ is bounded using a contradiction argument, we need to construct the two sequences (A_n) and (B_n). When we are done, we know $A_n \subseteq B_{n-1}$, $\lambda(A_n) > n$ and $\lambda(B_n) = \infty$ for all n. Show the details of this construction beyond the case A_2 and B_2 which are discussed.*

Exercise 16.1.3 *We prove that λ^+ is countably additive because the details for the proof that λ^- is countably additive are quite similar. Still, going through those details helps you understand the proof even better. So show λ^- is countably additive.*

Theorem 16.1.2 The Jordan Decomposition of a Charge

> Let λ be an extended real-valued charge on $(X, \mathcal{S})$. Then, λ^+ and λ^- are measures on $\mathcal{S}$ and $\lambda = \lambda^+ - \lambda^-$. The pair (λ^+, λ^-) is called the Jordan Decomposition of λ.

Proof 16.1.2
By Comment 16.1.1, we can assume the charge λ takes on only ∞ on some measurable sets. The first part of the proof of Theorem 16.1.1 is the same. It is clear both λ^+ and λ^- are non-negative and additive on finite unions of disjoint sets. The argument λ^- is countably additive is the same as the one in the proof of Theorem 16.1.1 because we have assumed λ^- is finite. There are two cases to show λ^+ is countably additive. These are:

1. *$\lambda^+(\cup_n E_n) < \infty$. In this case, we can use the argument of Theorem 16.1.1 to show that $\lambda^+(\cup_n E_n) \geq \sum_n \lambda^+(E_n)$ which shows countable additivity in this case.*

2. *$\lambda^+(\cup_n E_n) = \infty$. Here, the argument in the previous theorem shows there is a set $A \subseteq \cup_n E_n$ so that $\lambda(A) = \infty$. Hence, in this case, there is at least one N so that $\lambda(A \cap E_N) = \infty$. However, this implies $\lambda^+(E_n) = \infty$ and so $\sum_n \lambda^+(E_n) = \infty$ also. Hence, the two sides match and again we have established countable additivity.*

354 BASIC ANALYSIS IV: MEASURE THEORY AND INTEGRATION

Homework

The proof of the theorem above relies heavily on the arguments used in Theorem 16.1.1 and so it is easy to just nod your head wisely and take our word for it. But it is always important to do the work for yourself. The next two exercises ask you to do that.

Exercise 16.1.4 *One of the cases to show λ^+ is countably additive here is when $\lambda^+(\cup_n E_n) < \infty$. In this case, show the details of the argument based on the one given in Theorem 16.1.1 to show that $\lambda^+(\cup_n E_n) \geq \sum_n \lambda^+(E_n)$ which shows countable additivity in this case.*

Exercise 16.1.5 *The other case to show λ^+ is countably additive here is when $\lambda^+(\cup_n E_n) = \infty$. Provide the details for this argument.*

16.2 The Hahn Decomposition Associated with a Charge

Now we show that any charge λ has associated positive and negative sets.

Theorem 16.2.1 The Hahn Decomposition Associated With a Charge

> Let λ be a charge on $(X, \mathcal{S})$. Then, there is a positive set P and a negative set N so that $X = P \cup N$ and $P \cap N = \emptyset$. The pair (P, N) is called a Hahn Decomposition associated with the charge λ.

Proof 16.2.1
We may assume that λ^+ is finite; hence, by the Supremum Tolerance Lemma there are measurable sets A_n so that

$$\lambda(A_n) \; > \; \lambda^+(X) - \frac{1}{2^n} \qquad (\alpha)$$

for all n. Let $A = \limsup A_n$. Then, $A^C = \liminf A_n$. Recall

$$\limsup A_n \; = \; \bigcap_{m=1}^{\infty} \bigcup_{n=m}^{\infty} A_n, \quad \liminf A_n = \bigcup_{m=1}^{\infty} \bigcap_{n=m}^{\infty} A_n^C$$

Let $B_m = \cap_{n=m}^{\infty} A_n$. Since $B_m \subseteq A_n$ for $m \geq n$, we see $\lambda^+(B_m) \leq \lambda^+(A_m)$. Hence, we have

$$\liminf \lambda^+(B_m) \; \leq \; \liminf \lambda^+(A_n)$$

But the sequence of sets B_m increases monotonically to $\cup_m B_m$ and so we know by Lemma 9.2.2 that $\lim \lambda^+(B_m) = \lambda^+(\cup_m B_m)$. Since $\liminf \lambda^+(B_m) = \lim \lambda^+(B_m)$ here, we obtain

$$\lambda^+(\cup_m B_m) \; = \; \lambda^+(\liminf \lambda^+(A_n)) \; \leq \; \liminf_m \lambda^+(A_n)$$

Now, from Equation α, it then follows since λ^+ is finite, that

$$\lambda^+(A_n^C) \; = \; \lambda^+(X) - \lambda^+(A_n) \leq \lambda^+(X) - \lambda(A_n) \leq \frac{1}{2^n}$$

DECOMPOSING MEASURES

Thus,

$$0 \leq \lambda^+(A^C) \leq \liminf \lambda^+(A_n^C) = 0$$

and we have shown that $\lambda^+(A^C) = 0$. It remains to show $\lambda^-(A) = 0$. Note

$$0 \leq \lambda^-(A_n) = \lambda^+(A_n) - \lambda(A_n)$$
$$\leq \lambda^+(X) - \lambda(A_n) \leq \frac{1}{2^n}$$

Hence, for all m,

$$0 \leq \lambda^-(A) \leq \lambda^-\left(\bigcup_{n=m}^{\infty} A_n\right)$$
$$\leq \sum_{n=m}^{\infty} \lambda^-(A_n) \leq \sum_{n=m}^{\infty} \frac{1}{2^n}$$
$$< \frac{1}{2^{m-1}}$$

This clearly implies that $\lambda^-(A) = 0$ and we have the desired decomposition. ∎

We can use the Hahn Decomposition to characterize λ^+ and λ^- in a new way.

Lemma 16.2.2 The Hahn Decomposition Characterization of a Charge

> Let (A, B) be a Hahn Decomposition for the charge λ on $(X, \mathcal{S})$. Then, if E is measurable, $\lambda^+(E) = \lambda(E \cap A)$ and $\lambda^-(E) = -\lambda(E \cap B)$.

Proof 16.2.2

Let D be a measurable subset of $E \cap A$. Then $\lambda(D) \geq 0$ by the definition of the positive set A. Since λ is countably additive, we then have

$$\lambda\left(E \cap A\right) = \lambda\left((E \cap A) \cap D\right) + \lambda\left((E \cap A) \cap D^C\right)$$
$$= \lambda\left(D\right) + \lambda\left((E \cap A) \cap D^C\right)$$

But the second set is contained in $E \cap A$ and so its λ measure is non-negative. Hence, we can overestimate the left-hand side as

$$\lambda\left(E \cap A\right) \geq \lambda\left(D\right) \geq 0$$

Since this is true for all subsets D, the definition of λ^+ implies $\lambda^+(E \cap A) \leq \lambda(E \cap A)$. Now,

$$\lambda^+(E) = \lambda^+(E \cap A) + \lambda^+(E \cap B)$$

If F is a measurable subset of $E \cap B$, then $\lambda(F) \leq 0$ and so $\sup\{\lambda(F)\} \leq 0$. This tells us $\lambda^+(E \cap B) = 0$. Thus, we have established that $\lambda^+(E) = \lambda^+(E \cap A)$, and so $\lambda^+(E) \leq \lambda(E \cap A)$. The reverse inequality is easier. Since $E \cap A$ is a measurable subset of E, the definition of λ^+ implies $\lambda(E \cap A) \leq \lambda^+(E)$. Combining these results, we have $\lambda^+(E) = \lambda(E \cap A)$ as desired. A similar argument shows that $\lambda^-(E) = -\lambda(E \cap B)$. ∎

356 BASIC ANALYSIS IV: MEASURE THEORY AND INTEGRATION

Homework

Exercise 16.2.1 *Repeat the proof for λ^+ in Theorem 16.2.1 using the choice $1/5^n$ when you use the Supremum Tolerance Lemma.*

Exercise 16.2.2 *In Lemma 16.2.2, the argument that $\lambda^-(E) = -\lambda(E \cap B)$ was left out. Provide the details now.*

16.3 The Variation of a Charge

A charge λ has associated with it a concept that is very similar to that of the variation of a function. We now define the variation of a charge.

Definition 16.3.1 The Variation of a Charge

> *Let $(X, \mathcal{S})$ be a measure space and λ be a charge on $\mathcal{S}$. For a measurable set E, a mesh in E is a finite collection of disjoint measurable sets inside E, $\{E_1, \ldots, E_n\}$ for some positive integer n. Define the mapping V_λ by*
>
> $$V_\lambda(E) = \sup\left\{\sum_i |\lambda(E_i)| \mid \{E_i\} \text{ is a mesh in } E\right\}$$
>
> *where we interpret the sum as being over the finite number of sets in the given mesh. We say $V_\lambda(E)$ is the total variation of λ on E and V_λ is the total variation of λ.*

Theorem 16.3.1 The Variation of a Charge is a Measure

> *Let $(X, \mathcal{S})$ be a measure space and λ be a finite charge on $\mathcal{S}$. Then V_λ is a measure on $\mathcal{S}$.*

Proof 16.3.1
Given a measurable set E, the Jordan Decomposition of λ implies that for a mesh $\{E_1, \ldots, E_n\}$ in E, $|\lambda(E_i)| \leq \lambda^+(E_i) + \lambda^-(E_i)$. Hence, since λ^+ and λ^- are measures and countably additive, we have

$$\sum_i |\lambda(E_i)| \leq \sum_i \lambda^+(E_i) + \sum_i \lambda^-(E_i)$$
$$\leq \lambda^+(E) + \lambda^-(E) < \infty$$

since λ^+ and λ^- are both finite. We conclude V_λ is a finite mapping.

Since the only mesh in $\emptyset$ is $\emptyset$ itself, we see $V_\lambda(\emptyset) = 0$. It remains to show countable additivity. Let (E_n) be a countable disjoint family in $\mathcal{S}$ and let E be their union. Let $\{A_1, \ldots, A_p\}$ be a mesh in E. Then each A_i is inside E and they are pairwise disjoint. Let $A_{i\,n} = A_i \cap E_n$. Note A_i is the union of the sets $A_{i\,n}$. Then it is easy to see $\{A_{1n}, \ldots, A_{pn}\}$ is a mesh in E_n. For convenience, call this mesh M_n. Then

$$\sum_{i=1}^p |\lambda(A_i)| = \sum_{i=1}^p \sum_n |\lambda(A_{i\,n})| = \sum_n \left(\sum_{i=1}^p |\lambda(A_{i\,n})|\right)$$

The term in parenthesis is the sum over the mesh M_n of E_n. By definition, this is bounded above by $V_\lambda(E_n)$. Thus, we must have

$$\sum_{i=1}^{p} |\lambda(A_i)| \leq \sum_{n} V_\lambda(A_i)$$

To get the other inequality, we apply the Supremum Tolerance Lemma to the definition of $V_\lambda(E_n)$ to find meshes

$$M_n^\epsilon = \{A_{1\,n}^\epsilon, \ldots, A_{p_n\,n}^\epsilon\}$$

where p_n is a positive integer, so that

$$V_\lambda(E_n) < \sum_{i=1}^{p_n} |\lambda(A_{i\,n}^\epsilon)| + \epsilon/2^n$$

It follows that the union of a finite number of these meshes is a mesh of E. For each positive integer N, let

$$\mathcal{M}_N = \bigcup_{i=1}^{N} M_i^\epsilon$$

denote this mesh. Then,

$$\sum_{n=1}^{N} V_\lambda(E_n) < \sum_{n=1}^{N} \left(\sum_{i=1}^{p_n} |\lambda(A_{i\,n}^\epsilon)| + \epsilon/2^n \right)$$

The first double sum corresponds to summing over a mesh of E and so by definition, we have

$$\sum_{n=1}^{N} V_\lambda(E_n) < V_\lambda(E) + \sum_{n=1}^{N} \epsilon/2^n \leq V_\lambda(E) + \sum_{n=1}^{\infty} \epsilon/2^n = V_\lambda(E) + \epsilon$$

Since N is arbitrary, we see the sequence of partial sums on the left hand side converges to a finite limit. Thus,

$$\sum_{n=1}^{\infty} V_\lambda(E_n) \leq V_\lambda(E) + \epsilon$$

Since ϵ is arbitrary, the other desired inequality follows. ∎

Homework

Exercise 16.3.1 *Let μ be Lebesgue measure on $[a, b]$. Prove $V_\mu(E) = \mu(E)$ for all measurable E.*

Exercise 16.3.2 *Let $(X, \mathcal{S})$ be a measure space and λ be a finite measure on $\mathcal{S}$. Then $V_\lambda = \lambda$.*

Exercise 16.3.3 *Prove that a partition P of $[a, b]$ determines many different meshes of $[a, b]$.*

Theorem 16.3.2 $V_\lambda = \lambda^+ + \lambda^-$

BASIC ANALYSIS IV: MEASURE THEORY AND INTEGRATION

Let $(X, \mathcal{S})$ be a measure space and λ be a finite charge on $\mathcal{S}$. Then $V_\lambda = \lambda^+ + \lambda^-$.

Proof 16.3.2

Choose a measurable set E and let $\epsilon > 0$ be chosen. Then, by the Supremum Tolerance Lemma, there is a mesh $M^\epsilon = \{A_1^\epsilon, \ldots, A_p^\epsilon\}$ so that

$$ V_\lambda(E) - \epsilon \; < \; \sum_i |\lambda(A_i^\epsilon)| \leq V_\lambda(E) $$

Let F be the set of indices i in the mesh above where $\lambda(A_i^\epsilon) \geq 0$ and G be the other indices where $\lambda(A_i^\epsilon) < 0$. Let $\mathcal{F}$ be the union over the indices in F and $\mathcal{G}$ be the union over the indices in G. Note we have

$$ V_\lambda(E) - \epsilon \; < \; \sum_i |\lambda(A_i^\epsilon)| = \sum_F |\lambda(A_i^\epsilon)| + \sum_G |\lambda(A_i^\epsilon)| $$

Now in F,

$$ |\lambda(A_i^\epsilon)| \; = \; \lambda^+(A_i^\epsilon) - \lambda^-(A_i^\epsilon) \leq \lambda^+(A_i^\epsilon) $$

and in G,

$$ |\lambda(A_i^\epsilon)| \; = \; \lambda^-(A_i^\epsilon) - \lambda^+(A_i^\epsilon) \leq \lambda^-(A_i^\epsilon) $$

Thus, we can say

$$ \begin{aligned} V_\lambda(E) - \epsilon \; &\leq \; \sum_F \lambda^+(A_i^\epsilon) + \sum_G \lambda^-(A_i^\epsilon) = \lambda^+(\mathcal{F}) + \lambda^-(\mathcal{G}) \\ &\leq \; \lambda^+(E) + \lambda^-(E) \end{aligned} $$

Thus, for all $\epsilon > 0$, we have

$$ V_\lambda(E) \; \leq \; \lambda^+(E) + \lambda^-(E) + \epsilon $$

This implies

$$ V_\lambda(E) \; \leq \; \lambda^+(E) + \lambda^-(E) $$

To prove the reverse, note if $A \subseteq E$ for $E \in \mathcal{S}$, then A itself is a mesh (a pretty simple one, of course) and so $|\lambda(A)| \leq V_\lambda(A)$. Further, $\lambda(E) = \lambda(A) + \lambda(E \setminus A)$. Thus, we have

$$ \begin{aligned} 2\,\lambda(A) \; &\leq \; \lambda(A) + |\lambda(A)| \leq \lambda(E) - \lambda(E \setminus A) + |\lambda(A)| \\ &\leq \; \lambda(E) + |\lambda(E \setminus A)| + |\lambda(A)| \end{aligned} $$

But the collection $\{A, E \setminus A\}$ is a mesh for E and so

$$ 2\,\lambda(A) \; \leq \; \lambda(E) + V_\lambda(E) $$

Next, using the definition of λ^+, we find

$$ 2\,\lambda^+(E) \; \leq \; \lambda(E) + V_\lambda(E) $$

DECOMPOSING MEASURES 359

Finally, using the Jordan Decomposition of λ, we obtain

$$2\,\lambda^+(E) \;\leq\; \lambda^+(E) - \lambda^-(E) + V_\lambda(E)$$

This immediately leads to $\lambda^+(E) + \lambda^-(E) \leq V_\lambda(E)$. ∎

Homework

Exercise 16.3.4 *Let $(X, \mathcal{S})$ be a measure space and assume λ_1 and λ_2 are both finite charges on $\mathcal{S}$. Prove $\lambda_1 + \lambda_2$ is also a finite charge.*

Exercise 16.3.5 *Let $(X, \mathcal{S})$ be a measure space and assume λ_1 and λ_2 are both finite charges on $\mathcal{S}$. Prove $V_{\lambda_1 + \lambda_2} \leq V_{\lambda_1} + V_{\lambda_2}$.*

Exercise 16.3.6 *Let $(X, \mathcal{S})$ be a measure space and assume λ is a finite charge on $\mathcal{S}$. Then $V_\lambda = 0$ implies $\lambda = 0$.*

Exercise 16.3.7 *Let $(X, \mathcal{S})$ be a measure space and assume λ is a finite charge on $\mathcal{S}$. Let c be any real number. Prove $c\lambda$ is also a finite charge and $V_{c\lambda} = |c|V_\lambda$.*

Exercise 16.3.8 *Let $(X, \mathcal{S})$ be a measure space. Let $B = \{\lambda | \lambda$ is a finite charge on $\mathcal{S}\}$. Prove B is a vector space over $\Re$.*

Exercise 16.3.9 *Let $(X, \mathcal{S})$ be a measure space. Let $B = \{\lambda | \lambda$ is a finite charge on $\mathcal{S}\}$. Prove $V_\lambda(X)$ is a norm on B.*

16.4 Absolute Continuity of Charges

Now we are ready to look at absolute continuity in the context of charges.

Definition 16.4.1 Absolute Continuity of Charges

Let $(X, \mathcal{S}, \mu)$ be a measurable space and let λ be a charge on $\mathcal{S}$. Then λ is said to be absolutely continuous with respect to μ if whenever E is a measurable set with $\mu(E) = 0$, then $\lambda(E) = 0$ also. We write this as $\lambda \ll \mu$. The set of all charges that are absolutely continuous with respect to μ is denoted by $AC[\mu]$.

There is an intimate relationship between the absolute continuity of V_λ, λ, λ^+ and λ^-; essentially, one implies all the others.

Theorem 16.4.1 Equivalent Absolute Continuity Conditions for Charges

Let $(X, \mathcal{S}, \mu)$ be a measurable space. Then for the statements
(1): λ^+ and λ^- are in $AC[\mu]$,
(2): V_λ is in $AC[\mu]$, and
(3): λ is in $AC[\mu]$, we have (1) $\Leftrightarrow$ (2) $\Leftrightarrow$ (3).

Proof 16.4.1

(1) $\to$ (2): if $\mu(E) = 0$, then $\lambda^+(E)$ and $\lambda^-(E)$ are also zero by assumption. Applying the Jordan Decomposition of λ, we see $\lambda(E) = 0$ too. Hence, λ is in $AC[\mu]$.
(2) $\to$ (3): if $\mu(E) = 0$, then $V_\lambda(E) = 0$. But, by Theorem 16.3.2, we have both $\lambda^+(E)$ and $\lambda^-(E)$ are zero. Then, applying the Jordan Decomposition again, we have $\lambda(E) = 0$. This tells us λ is absolutely continuous with respect to μ.

360   BASIC ANALYSIS IV: MEASURE THEORY AND INTEGRATION

(3) → (1): Let (A, B) be a Hahn Decomposition of X due to λ. If $\mu(E) = 0$, then $\lambda(E) = 0$ by assumption. Thus, $\lambda(E \cap A) = \lambda(E \cap B) = 0$ as well. By Lemma 16.2.2, we then have that $\lambda^+(E) = \lambda^-(E) = 0$ showing that (1) holds. ∎

There is another characterization of absolute continuity that is useful.

Lemma 16.4.2 $\epsilon - \delta$ Version of Absolute Continuity of a Charge

---

Let λ be a charge on $\mathcal{S}$. Then

$$\lambda \ll \mu \iff \forall \epsilon > 0, \exists \delta > 0 \ni |\lambda(E)| < \epsilon \text{ for measurable } E \text{ with } \mu(E) < \delta$$

Proof 16.4.2

($\Rightarrow$): If λ is absolutely continuous with respect to μ, then note by Theorem 16.4.1 V_λ is also in $AC[\mu]$. We will prove this result by contradiction. Assume the desired implication does not hold for V_λ. Then, there is a positive ϵ so that for all n, there is a measurable set E_n with $\mu(E_n) < 1/2^n$ and $V_\lambda(E_n) \geq \epsilon$.
Let $G_n = \bigcup_{k=n}^{\infty} E_k$ and $G = \bigcap_n G_n$. Then,

$$\mu(G) \leq \mu(G_n) \leq \sum_{k=n}^{\infty} E_k < \sum_{k=n}^{\infty} 1/2^k = 1/2^{n-1}$$

Since this holds for all n, this implies $\mu(G) = 0$. Since V_λ is in $AC[\mu]$, we then have $V_\lambda(G) = 0$. But

$$V_\lambda(G) = \lim_n V_\lambda(G_n) \geq \epsilon$$

This contradiction implies that our assumption that the right-hand side did not hold must be false. Hence, the condition holds for V_λ. It is easy to see that since $V_\lambda = \lambda^+ + \lambda^-$, the condition holds for them also. This then implies the condition holds for $\lambda = \lambda^+ - \lambda^-$.
($\Leftarrow$): We assume the condition on the right-hand side holds. Now let (A, B) be a Hahn Decomposition for X with respect to λ. In particular, if $\mu(E) = 0$, then $\mu(E \cap A) = 0$ also. The condition then implies $\lambda(E \cap A) < \epsilon$. However, the choice of ϵ is arbitrary which then implies $|\lambda(E \cap A)| = 0$. But the absolute values are unnecessary as λ is non-negative on A. We conclude $\lambda^+(E) = \lambda(E \cap A) = 0$. A similar argument then shows $\lambda^-(E) = -\lambda(E \cap B) = 0$. This tells us $\lambda(E) = 0$ by the Jordan Decomposition. ∎

Lemma 16.4.3 The Absolute Continuity of the Integral

Let $(X, \mathcal{S}, \mu)$ be a measure space and f be a summable function. Define the map λ by $\lambda(E) = \int_E f \, d\mu$ for all measurable E. Then, λ is a charge with

$$\lambda^+(E) = \int_E f^+ \, d\mu, \ \lambda^-(E) = -\int_E f^- \, d\mu$$

Moreover, if $P_f = \{x \mid f(x) \geq 0\}$ and $N_f = P_f^C$, then (P_f, N_f) is a Hahn Decomposition for X with respect to λ. Finally, since $\lambda \ll \mu$, we know for all positive ϵ, there is a positive δ, so that if E is a measurable set with $\mu(E) < \delta$, then $\left| \int_E f \, d\mu \right| < \epsilon$.

DECOMPOSING MEASURES

Proof 16.4.3

It is easy to see that $\nu_1 = \int_E f^+ \, d\mu$ and $\nu_2 = \int_E f^- \, d\mu$ define measures and that $\lambda = \nu_1 - \nu_2$. Hence, λ is a charge which is absolutely continuous with respect to μ. It is also easy to see that (P_f, N_f) is a Hahn Decomposition for λ. Now if B is measurable and contained in the measurable set E, we have

$$\lambda(B) \;=\; \int_{B \cap P_f} f^+ \, d\mu \;-\; \int_{B \cap N_f} f^+ \, d\mu$$

$$\leq \; \int_{B \cap P_f} f^+ \, d\mu$$

$$\leq \; \int_{E \cap P_f} f^+ \, d\mu$$

Next, note that $\int_{E \cap P_f} f^+ \, d\mu = \int_E f^+ \, d\mu$ because the portion of E that lies in N_f does not contribute to the value of the integral. Thus, for any $B \subseteq E$, we have

$$\lambda(B) \;\leq\; \int_E f^+ \, d\mu \;=\; \nu_1(E)$$

The definition of λ^+ then implies two things: first, the inequality above tells us $\lambda^+(E) \leq \nu_1(E)$ and second, since $E \cap P_f$ is a subset of E, we know $\lambda(E \cap P_f) \leq \lambda^+(E)$. However, $\lambda(E \cap P_f) = \nu_1(E)$ and hence, $\nu_1(E) \leq \lambda^+(E)$ also. Combining, we have $\lambda^+(E) = \nu_1(E)$.

A similar argument shows that $\lambda^-(E) = \nu_2(E)$.

The last statement of the proposition follows immediately from Lemma 16.4.2. ∎

Homework

Exercise 16.4.1 *Let μ be Lebesgue measure on $[-3, 2]$. Let the new charge λ be defined by $\lambda(E) = \int_E f \, d\mu$ for $f(x) = 4x^3$.*

- *Find f^+ and f^-.*
- *Find P_f and N_f.*
- *Find λ^+ and λ^-.*
- *Compute $\lambda(E)$ for $E = [-3, x] \subset [-3, 2]$ for $x > 0$ and $x < 0$.*

Exercise 16.4.2 *Let μ be Lebesgue measure on $[-3\pi, 4\pi]$. Let the new charge λ be defined by $\lambda(E) = \int_E f \, d\mu$ for $f(x) = \sin(2x)$.*

- *Find f^+ and f^-.*
- *Find P_f and N_f.*
- *Find λ^+ and λ^-.*
- *Compute $\lambda(E)$ for $E = [c, d] \subset [-3\pi, 4\pi]$.*

Exercise 16.4.3 *Let $g(x) = x^4$ on $[-3, 2]$.*

- *Let $V_g(x) = V(g : -3, x)$ for $x \in [-3, 2]$. Prove*

$$V_g(x) \;=\; \begin{cases} 81 - x^4, & -3 \leq x \leq 0 \\ 81 + x^4, & 0 < x \leq 2 \end{cases}$$

362 BASIC ANALYSIS IV: MEASURE THEORY AND INTEGRATION

- *Thus, $g_1 = V_g$ and $g_2 = V_g - g$ is the standard decomposition of g into a difference of monotone functions; i.e. $g = g_1 - g_2$.*

- *Define, as usual, the charge λ by $\lambda(E) = \mu_1(E) - \mu_2(E)$ where $\mu_1(E) = \int_E dg_1$ and $\mu_2(E) = \int_E dg_2$ are Lebesgue - Stieltjes measures.*

- *So for $f(x) = 1$, consider the charge $\nu(E) = \int_E 1 dg_1 - \inf_E 1 dg_2$. Find P_f and N_f.*

- *Find ν^+ and ν^-.*

- *Compute $\nu([-3, x])$ for both $x > 0$ and $x < 0$. Compare this to the results from the first exercise. Note the first exercise could have been written $\lambda(E) = \int_E g' d\mu$. Interesting connection, eh?*

Exercise 16.4.4 *Let μ be Lebesgue measure on $[-5, 6]$. Let the new charge λ be defined by $\lambda(E) = \int_E f d\mu$ for $f(x) = 6x^5$.*

- *Find f^+ and f^-.*

- *Find P_f and N_f.*

- *Find λ^+ and λ^-.*

- *Compute $\lambda(E)$ for $E = [-3, x] \subset [-3, 2]$ for $x > 0$ and $x < 0$.*

Exercise 16.4.5 *Let $g(x) = x^6$ on $[-5, 6]$.*

- *Let $V_g(x) = V(g : -3, x)$ for $x \in [-6, 5]$. Find V_g.*

- *Thus, $g_1 = V_g$ and $g_2 = V_g - g$ is the standard decomposition of g into a difference of monotone functions; i.e. $g = g_1 - g_2$.*

- *Define, as usual, the charge λ by $\lambda(E) = \mu_1(E) - \mu_2(E)$ where $\mu_1(E) = \int_E dg_1$ and $\mu_2(E) = \int_E dg_2$ are Lebesgue - Stieltjes measures.*

- *So for $f(x) = 1$, consider the charge $\nu(E) = \int_E 1 dg_1 - \inf_E 1 dg_2$. Find P_f and N_f.*

- *Find ν^+ and ν^-.*

- *Compute $\nu([-3, x])$ for both $x > 0$ and $x < 0$. Compare this to the results from the previous exercise. Note the previous exercise could have been written $\lambda(E) = \int_E g' d\mu$. Interesting connection, eh?*

16.5 The Radon - Nikodym Theorem

From our work above, culminating in Lemma 16.4.3, we know that integrals of summable functions define charges which are absolutely continuous with respect to the measure we are using for the integration. The converse of this is that if a measure is absolutely continuous, we can find a summable function so that the measure can be found by integration. That is if $\lambda \ll \mu$, there exists f summable so that $\lambda(E) = \int f \, d\mu$. This result is called the Radon - Nikodym theorem and as you might expect, its proof requires some complicated technicalities to be addressed. Hence, we begin with a lemma.

Lemma 16.5.1 Radon - Nikodym Technical Lemma

Let $(X, \mathcal{S}, \mu)$ be a measurable space with $\mu(X)$ finite. Let λ be a measure which is finite with $\lambda(X) > 0$ and $\lambda \ll \mu$. Then there is a positive ϵ and a measurable set A with $\mu(A) > 0$ so that

$$\epsilon \mu(E \cap A) \leq \lambda(E \cap A), \forall E \in \mathcal{S}$$

DECOMPOSING MEASURES

Proof 16.5.1

Pick a fixed $\epsilon > 0$ and assume the set A exists. Let $\nu = \lambda - \epsilon\mu$. Then, ν is a finite charge also. Note, our assumption tells us that

$$\nu(B) \;=\; \lambda(B) - \epsilon\mu(B) \;\geq\; 0$$

for all measurable subsets B of A. Hence, by the definition of ν^-, we must have that $-\nu^-(A) \geq 0$ or $\nu^-(A) \leq 0$. But ν^- is always non-negative. Combining, we have $\nu^-(A) = 0$. This gives us some clues as to how we can find the desired A. Note if (A, B) is a Hahn Decomposition for ν, then we have this desired inequality, $\nu^-(A) = 0$. So, we need to find a positive value of ϵ^* so that when (A, B) is a Hahn Decomposition of

$$\nu^*(A) \;=\; \lambda(A) - \epsilon^*\mu(A)$$

we find $\nu^*(A) > 0$.

To do this, for $\epsilon = 1/n$, let (A_n, B_n) be a Hahn Decomposition for $\nu_n = \lambda - (1/n)\mu$. Let $G = \cup_n A_n$ and $H = \cap_n B_n$. We also know $A_n \cup B_n = X$ and $A_n \cap B_n = \emptyset$ for all n. Further,

$$H^C \;=\; \left(\bigcap_n B_n\right)^C \;=\; \bigcup_n B_n^C \;=\; \bigcup_n A_n \;=\; G$$

We conclude $X = G \cup H$; it is easy to see $G \cap H = \emptyset$. Now, $H \subseteq B_n$ for all n, so $\nu_n(H) = -\nu_n^-(H) \leq 0$ as B_n is a negative set. Hence, we can say

$$\lambda(H) - (1/n)\mu(H) \;\leq\; 0$$

which implies $\lambda(H) \leq (1/n)\mu(H)$ for all n. Since λ is a measure, we then have

$$0 \;\leq\; \lambda(H) \;\leq\; \mu(H)/n$$

which implies by the arbitrariness of n that $\lambda(H) = 0$. Hence,

$$\lambda(X) \;=\; \lambda(G) + \lambda(H) \;=\; \lambda(G)$$

Thus, $\lambda(G) > 0$ as $\lambda(X) > 0$. Since $\lambda \ll \mu$, it then follows that $\mu(G) > 0$ also. Since $G = \cup_n A_n$, it must be true that there is at least one n with $\mu(A_n) > 0$. Call this index N. Then, $\nu_N(E \cap A_N) \geq 0$ as A_N is a positive set for ν_N. This implies

$$\lambda(E \cap A_N) - \frac{\mu(E \cap A_N)}{N} \;\geq\; 0$$

which is the result we seek using $A = A_N$ and $\epsilon = 1/N$. ∎

Theorem 16.5.2 The Radon - Nikodym Theorem: Finite Charge Case

*Let $(X, \mathcal{S}, \mu)$ be a measurable space with μ σ-finite. Let λ be a charge with $\lambda \ll \mu$. Then, there is a summable function f so that $\lambda(E) = \int_E f\, d\mu$ for all measurable E. Moreover, if g is another summable function which satisfies this equality, then $f = g$ μ a.e. The summable function f is called the **Radon - Nikodym derivative** of λ with respect to μ and is often denoted by the usual derivative symbol: $f = \frac{d\lambda}{d\mu}$. Hence, this equality is often written $\lambda(E) = \int_E \frac{d\lambda}{d\mu}\, d\mu$*

BASIC ANALYSIS IV: MEASURE THEORY AND INTEGRATION

Proof 16.5.2
We will do this in three steps.
Step 1: *We assume $\mu(X)$ is finite and λ is a finite measure.*
Step 2: *We assume μ is σ - finite and λ is a finite measure.*
Step 3: *We assume μ is σ - finite and λ is a finite charge.*

As is usual, the proof of **Step 1** *is the hardest.*
Proof Step 1: *Let*

$$\mathcal{F} = \{ f : X \to \Re \mid f \geq 0, \ f \ summable \ and \ \int_E f \, d\mu \leq \lambda(E), \ \forall \, E \in \mathcal{S} \}$$

Note since $f = 0_X$ is in $\mathcal{F}$, $\mathcal{F}$ is nonempty. From the definition of $\mathcal{F}$, we see $\int_X f \, d\mu \leq \lambda(X) < \infty$ for all f in $\mathcal{F}$. Hence,

$$c = \sup_{f \in \mathcal{F}} \int_X f \, d\mu < \infty$$

We will find a particular $f \in \mathcal{F}$ so that $c = \int_X f \, d\mu$. Let $(f_n) \subseteq \mathcal{F}$ be a minimizing sequence: i.e. $\int_X f_n \, d\mu \to c$. We will assume without loss of generality that each f_n is finite everywhere as the set of points where all are infinite is a set of measure zero. Now, there are details that should be addressed in that statement, but we have gone through those sort of manipulations many times before. As an exercise, you should go through them again on scratch paper for yourself. With that said, we will define a new sequence of finite functions (g_n) by

$$\begin{aligned} g_n &= f_1 \vee f_2 \vee \ldots \vee f_n \\ &= \max \{ f_1, \ldots, f_n \} \end{aligned}$$

This is a pointwise operation and it is clear that (g_n) is an increasing sequence of non-negative functions. Since f_1 and f_2 are summable, let A be the set of points where $f_1 > f_2$. Then,

$$\begin{aligned} \int_X f_1 \vee f_2 \, d\mu &= \int_A f_1 \, d\mu + \int_{A^C} f_2 \, d\mu \\ &\leq \int_X f_1 \, d\mu + \int_X f_2 \, d\mu \end{aligned}$$

This tells us $f_1 \vee f_2$ is summable also. A simple induction argument then tells us g_n is summable for all n.

Is $g_n \in \mathcal{F}$? Let E be measurable. Define the measurable sets (E_n) by

$$\begin{aligned} E_1 &= \{ x \mid g_n(x) = f_1(x) \} \cap E, \\ E_2 &= \{ x \mid g_n(x) = f_2(x) \} \cap (E \setminus E_1), \\ &\vdots \\ E_n &= \{ x \mid g_n(x) = f_n(x) \} \cap (E \setminus \cup_{i=1}^{n-1} E_i) \end{aligned}$$

Then, it is clear $E = \cup_i E_i$, each E_i is disjoint from the others and $g_n(x) = f_i(x)$ on E_i. Thus, since each f_i is in $\mathcal{F}$, we have

$$\int_E g_n \, d\mu = \sum_{i=1}^n \int_{E_i} f_i \, d\mu \leq \sum_{i=1}^n \lambda(E_i) = \lambda(\cup_{i=1}^n E_i) = \lambda(E)$$

We conclude each g_n is in $\mathcal{F}$ for all n. Next, if $g = \sup\ g_n$, then $g_n \uparrow g$ and

$$\int_E g_n\, d\mu \leq \lambda(E)\ \leq\ \lambda(X)$$

for all n. Now apply the Monotone Convergence Theorem to see g is summable and

$$\int_E g_n\, d\mu \to \int_E g\, d\mu\ \leq\ \lambda(E)$$

Let's define f by

$$f(x)\ =\ \begin{cases} g(x) & g(x) < \infty, \\ 0 & g(x) = \infty. \end{cases}$$

Since g is summable, the set of points where it takes on the value ∞ is a set of measure 0. Thus, $f = g$ μ a.e. and f is measurable. It is easy to see f is in $\mathcal{F}$.

Moreover, since $f_n \leq g_n$, we have

$$
\begin{aligned}
c\ &=\ \lim_n \int_X f_n\, d\mu \\
&\leq\ \lim_n \int_X g_n\, d\mu \leq c
\end{aligned}
$$

because $g_n \in \mathcal{F}$. Thus,

$$c\ =\ \lim_n \int_X g_n\, d\mu\ =\ \int_X g\, d\mu$$

This immediately tells us that $\int_X f\, d\mu = c$ with $f \in \mathcal{F}$.

Next, define $m : \mathcal{S} \to \Re$ by

$$m(E)\ =\ \lambda(E) - \int_E f\, d\mu$$

for all measurable E. It is straightforward to show m is the difference of two measures and hence is a finite charge. Also, since f is in $\mathcal{F}$, we see m is non-negative and thus is a measure. In addition, since $\lambda \ll \mu$ and the measure defined by $\int_E f\, d\mu$ is also absolutely continuous with respect to μ, we have that $m \ll \mu$ too. Now if $m(X) = 0$, this would imply, since $m(E) \leq m(X)$, that

$$0\ \leq\ \lambda(E) - \int_E f\, d\mu\ \leq m(X)\ =\ 0$$

But this says $\lambda(E) = \int_E f\, d\mu$ for all measurable E which is the result we seek.

Hence, it suffices to show $m(X) = 0$. We will do this by contradiction. Assume $m(X) > 0$. Now apply Lemma 16.5.1 to conclude there is a positive ϵ and measurable set A so that $\mu(A) > 0$ and

$$\epsilon\, \mu(E \cap A)\ \leq\ m(E \cap A), \tag{$*$}$$

for all measurable E. Define a new function h using Equation $*$ by $h = f + \epsilon\, I_A$. Then for a given measurable E, we have

$$\int_E h\, d\mu \;=\; \int_E f\, d\mu + \epsilon\, \mu(E \cap A)$$
$$\leq\; \int_E f\, d\mu + m(E \cap A)$$

by Equation $*$. Now replace m by its definition to find

$$\int_E h\, d\mu \;\leq\; \int_E f\, d\mu + \lambda(E \cap A) - \int_{E \cap A} f\, d\mu$$
$$=\; \int_{E \cap A^C} f\, d\mu + \lambda(E \cap A)$$

Finally, use the fact that f is in $\mathcal{F}$ to conclude

$$\int_E h\, d\mu \;\leq\; \lambda(E \cap A^C) + \lambda(E \cap A) \;=\; \lambda(E)$$

This shows that h is in $\mathcal{F}$. However,

$$\int_X h\, d\mu \;=\; \int_X f\, d\mu + \epsilon\, \mu(A) \;>\; c!$$

which is our contradiction. This completes the proof of **Step 1**.

Proof Step 2*: Now μ is σ finite. This means there is a countable sequence of disjoint measurable sets (X_n) with $\mu(X_n)$ finite for each n and we can write $X = \cup_n X_n$. Let $\mathcal{S}_n$ be the σ - algebra of subsets of X_n given by $\mathcal{S} \cap X_n$. By* **Step 1***, there are summable non-negative functions f_n so that*

$$\lambda(F) \;=\; \int_F f_n\, d\mu$$

for each F in $\mathcal{S}_n$. Now define f by $f(x) = f_n(x)$ when $x \in X_n$. This is a well-defined function and it is easy to see f is measurable. We also know by Theorem 9.8.5 that μ_f defined by $\mu_f(E) = \int_E f\, d\mu$ is a measure. Now if E is measurable, then $E = \cup_n E \cap X_n$, $E = \cup_n E \cap X_n$ and

$$\mu_f(E) \;=\; \int_E f\, d\mu = \int_{\cup_n E \cap X_n} f\, d\mu = \lim_n \int_{\cup_{i=1}^n E \cap X_n} f\, d\mu$$

Then, for any n,

$$\int_{\cup_{i=1}^n E \cap X_i} f\, d\mu \;=\; \sum_{i=1}^n \int_{E \cap X_i} f\, d\mu = \sum_{i=1}^n \int_{E \cap X_i} f_n\, d\mu$$
$$=\; \sum_{i=1}^n \lambda(E \cap X_i) \;=\; \lambda(\cup_{i=1}^n E \cap X_i) \;\leq\; \lambda(E)$$

which is a finite number. Hence, the series of non-negative terms $\sum_n \int_{E \cap X_n} f\, d\mu$ converges to a finite number and

$$\int_E f\, d\mu \;=\; \sum_n \int_{E \cap X_n} f_n\, d\mu \;=\; \lambda(\cup_n E \cap X_n) \;=\; \lambda(E)$$

DECOMPOSING MEASURES

This establishes the result for **Step 2**.

Proof Step 3: *Here, we have μ is σ - finite and λ is a finite charge. By the Jordan Decomposition of λ, we can write*

$$\lambda(E) \;=\; \lambda^+(E) \;-\; \lambda^-(E)$$

for all measurable E. Now apply **Step 2** *to find non-negative summable functions f^+ and f^- so that*

$$\lambda^+(E) \;=\; \int_E f^+ \, d\mu$$
$$\lambda^-(E) \;=\; \int_E f^- \, d\mu$$

Let $f = f^+ - f^-$ and we are done with the proof of **Step 3**.

Finally, it is clear from the proof above, that the Radon - Nikodym derivative of λ with respect to μ, is unique up to redefinition on a set of μ measure 0. ∎

Homework

Exercise 16.5.1 *In the proof of Step 1, we find a sequence of functions (f_n), where $(f_n) \subseteq \mathcal{F}$ is a minimizing sequence: i.e. $\int_X f_n \, d\mu \to c$. We will assume without loss of generality that each f_n is finite everywhere as the set of points where all are infinite is a set of measure zero. There are details that should be addressed in that statement and in this exercise, you should go through them again for yourself.*

Exercise 16.5.2 *In Step 1, we also define a sequence of finite functions (g_n) by*

$$\begin{aligned} g_n \;&=\; f_1 \vee f_2 \vee \ldots \vee f_n \\ &=\; \max\{f_1, \ldots, f_n\} \end{aligned}$$

Provide the details that tell us g_n is summable for all n.

Exercise 16.5.3 *Prove the Radon - Nikodym derivative of λ with respect to μ is unique up to redefinition on a set of μ measure 0.*

We can also prove the Radon- Nikodym theorem for the case that λ is a sigma-finite charge. Of course, this includes the case where λ is a charge with values in $\overline{\mathfrak{R}}$. Recall by Comment 16.1.1 that in this case, the charge λ can only take on ∞ or $-\infty$ values as it must be additive.

Theorem 16.5.3 The Radon - Nikodym Theorem: Sigma-Finite Charge Case

Let $(X, \mathcal{S}, \mu)$ be a measurable space with μ σ - finite. Let λ be a sigma-finite measure with $\lambda \ll \mu$. Then, there is a function f so that

$$\lambda(E) = \int_E f \, d\mu$$

for all measurable E. Moreover, if g is another function which satisfies this equality, then $f = g$ μ a.e. The function f is called the Radon - Nikodym derivative of λ with respect to μ and is often denoted by the usual derivative symbol: $f = \frac{d\lambda}{d\mu}$. Hence, this equality is often written

$$\lambda(E) = \int_E \frac{d\lambda}{d\mu} \, d\mu$$

Proof 16.5.3

We can still do this in three steps.
Step 1*: We assume μ and λ are finite measures.*
Step 2*: We assume μ and λ are sigma-finite measures.*
Step 3*: We assume μ is σ - finite and λ is a sigma-finite charge.*

The proof of **Step 1** *is identical to the one presented in Theorem 16.5.2. The proof of* **Step 2** *is also quite similar. The only difference is that we choose a disjoint countable collection of sets X_n for which both μ and λ are finite.*

Proof Step 2*: Since μ and λ are σ finite, there is a countable sequence of disjoint measurable sets (X_n) with $\mu(X_n)$ and $\lambda(X_n)$ finite for each n and $X = \cup_n X_n$. Let $\mathcal{S}_n$ be the σ - algebra of subsets of X_n given by $\mathcal{S} \cap X_n$. By* **Step 1***, there are summable non-negative functions f_n so that*

$$\lambda(F) = \int_F f_n \, d\mu$$

for each F in $\mathcal{S}_n$. Now define f by $f(x) = f_n(x)$ when $x \in X_n$. This is a well-defined function and it is easy to see f is measurable. If E is measurable, then $E = \cup_n E \cap X_n$, $E = \cup_n E \cap X_n$ and

$$\int_E f \, d\mu = \int_{\cup_n E \cap X_n} f \, d\mu$$

Then, for any n,

$$\int_{\cup_{i=1}^n E \cap X_i} f \, d\mu = \sum_{i=1}^n \int_{E \cap X_i} f \, d\mu = \sum_{i=1}^n \int_{E \cap X_i} f_n \, d\mu$$

$$= \sum_{i=1}^n \lambda(E \cap X_i) = \lambda(\cup_{i=1}^n E \cap X_i) \leq \lambda(E)$$

Now, since λ is sigma-finite, it is possible for this value to be ∞. However, whether it is a finite number or infinite in value, the partial sums defined on the left-hand side are clearly bounded above by this number. Hence, the series of non-negative terms $\sum_n \int_{E \cap X_n} f \, d\mu$ converges and

$$\int_E f \, d\mu = \sum_n \int_{\cup_n E \cap X_n} f_n \, d\mu = \lambda(E \cap X_n) = \lambda(E)$$

DECOMPOSING MEASURES

We can no longer say the function f is summable, of course. This establishes the result for **Step 2**.

Proof Step 3: *Here, we have μ is σ - finite and λ is a sigma-finite charge. From Theorem 16.1.2, we may assume λ has a Jordan Decomposition $\lambda = \lambda^+ - \lambda^-$ with λ^+ a finite measure. Thus, we can write*

$$\lambda(E) \;=\; \lambda^+(E) \,-\, \lambda^-(E)$$

for all measurable E. Now apply **Step 2** *to find non-negative functions f^+ and f^- so that*

$$\lambda^+(E) \;=\; \int_E f^+ \, d\mu,$$
$$\lambda^-(E) \;=\; \int_E f^- \, d\mu$$

Since $\lambda+$ is assumed finite, it is clear f^+ is actually summable and so the addition defined above is well-defined in all cases. We let $f = f^+ - f^-$ and we are done with the proof of **Step 3**. $\blacksquare$

Homework

Exercise 16.5.4 *Let μ be Lebesgue measure on $[-1,2]$ and define $\lambda(E) = \int_E g d\mu$ for $g(x) = x$. Prove λ is a measure and find $\frac{d\lambda}{d\mu}$.*

Exercise 16.5.5 *Let μ be Lebesgue measure on $[1,5]$ and define $\lambda(E) = \int_E g d\mu$ for $g(x) = x^4$. Prove λ is a measure and find $\frac{d\lambda}{d\mu}$.*

Exercise 16.5.6 *Let μ be Lebesgue measure on $[-3,2]$. Let the new charge λ be defined by $\lambda(E) = \int_E f d\mu$ for $f(x) = 4x^3$. Find $\frac{d\lambda^+}{d\mu}$ and $\frac{d\lambda^-}{d\mu}$.*

Exercise 16.5.7 *Let μ be Lebesgue measure on $[-3\pi, 4\pi]$. Let the new charge λ be defined by $\lambda(E) = \int_E f d\mu$ for $f(x) = \sin(2x)$.*

- *Find f^+ and f^-.*

- *Find λ^+ and λ^-.*

- *Find $\frac{d\lambda^+}{d\mu}$ and $\frac{d\lambda^-}{d\mu}$.*

Exercise 16.5.8 *Let $g(x) = x^4$ on $[-3,2]$.*

- *Let $V_g(x) = V(g : -3, x)$ for $x \in [-3,2]$.*

- *Thus, $g_1 = V_g$ and $g_2 = V_g - g$ is the standard decomposition of g into a difference of monotone functions; i.e. $g = g_1 - g_2$.*

- *Define, as usual, the charge λ by $\lambda(E) = \mu_1(E) - \mu_2(E)$ where $\mu_1(E) = \int_E dg_1$ and $\mu_2(E) = \int_E dg_2$ are Lebesgue - Stieltjes measures.*

- *So for $f(x) = 1$, consider the charge $\nu(E) = \int_E 1 dg_1 - \inf_E 1 dg_2$. Find ν^+ and ν^-.*

- *Find $\frac{d\nu^+}{d\mu}$ and $\frac{d\nu^-}{d\mu}$.*

Exercise 16.5.9 *Let μ be Lebesgue measure on $[-5,6]$. Let the new charge λ be defined by $\lambda(E) = \int_E f d\mu$ for $f(x) = 6x^5$. Find $\frac{d\lambda^+}{d\mu}$ and $\frac{d\lambda^-}{d\mu}$.*

Exercise 16.5.10 *Let $g(x) = x^6$ on $[-5,6]$.*

370 BASIC ANALYSIS IV: MEASURE THEORY AND INTEGRATION

- Let $V_g(x) = V(g : -3, x)$ for $x \in [-6, 5]$.

- Thus, $g_1 = V_g$ and $g_2 = V_g - g$ is the standard decomposition of g into a difference of monotone functions; i.e. $g = g_1 - g_2$.

- Define, as usual, the charge λ by $\lambda(E) = \mu_1(E) - \mu_2(E)$ where $\mu_1(E) = \int_E dg_1$ and $\mu_2(E) = \int_E dg_2$ are Lebesgue - Stieltjes measures.

- So for $f(x) = 1$, consider the charge $\nu(E) = \int_E 1 dg_1 - \inf_E 1 dg_2$. Find ν^+ and ν^-.

- Find $\frac{d\nu^+}{d\mu}$ and $\frac{d\nu^-}{d\mu}$.

16.6 The Lebesgue Decomposition of a Measure

Definition 16.6.1 Singular Measures

> Let $(X, \mathcal{S}, \mu)$ be a measure space and let λ be a charge on $\mathcal{S}$. Assume there is a decomposition of X into disjoint measurable subsets U and V ($X = U \cup V$ and $U \cap V = \emptyset$) so that $\mu(U) = 0$ and $\lambda(E \cap V) = 0$ for all measurable subsets E of V. In this case, we say λ is perpendicular to μ and write $\lambda \perp \mu$.

Comment 16.6.1 *If $\lambda \perp \mu$, let (U, V) be a decomposition of X associated with the singular measure λ. We then know that $\mu(U) = 0$ and $\lambda(E \cap V) = 0$ for all measurable E. Note, if E is measurable, then*

$$E = \left(E \cap U \right) \cup \left(E \cap V \right)$$

Thus,

$$\lambda(E) = \lambda\left(E \cap U \right) + \lambda\left(E \cap V \right) = \lambda\left(E \cap U \right)$$

Further,

$$\mu(E) = \mu\left(E \cap U \right) + \mu\left(E \cap V \right) = \mu\left(E \cap V \right)$$

Comment 16.6.2 *If $\lambda \perp \mu$ with $\lambda \neq 0$, then there is a measurable set E so that $\lambda(E \cap U) \neq 0$. But for this same set $\mu(E \cap U) = 0$ as $E \cap A$ is a subset of U. Thus, $\lambda \not\ll \mu$.*

Comment 16.6.3 *If $\lambda \perp \mu$ and $\lambda \ll \mu$, then for any measurable set E, we have $\lambda(E) = \lambda(E \cap U)$. But, since $\mu(E \cap U) = 0$, we must have $\lambda(E \cap U) = 0$ because $\lambda \ll \mu$. Thus, $\lambda = 0$.*

Comment 16.6.4 *It is easy to prove that $\lambda \perp \mu$ implies $V_\lambda \perp \mu$, $\lambda^+ \perp \mu$ and $\lambda^- \perp \mu$. Also, if $\lambda^+ \perp \mu$ and $\lambda^- \perp \mu$, this implies $\lambda \perp \mu$.*

Homework

Exercise 16.6.1 *Prove that $\lambda \perp \mu$ implies $V_\lambda \perp \mu$, $\lambda^+ \perp \mu$ and $\lambda^- \perp \mu$.*

Exercise 16.6.2 *Prove if $\lambda^+ \perp \mu$ and $\lambda^- \perp \mu$, this implies $\lambda \perp \mu$.*

DECOMPOSING MEASURES

Theorem 16.6.1 Lebesgue Decomposition Theorem

Let $(X, \mathcal{S}, \mu)$ be a σ - finite measure space. Let λ be a finite charge on $\mathcal{S}$. Then, there are two unique finite measures, $\lambda_{ac} \ll \mu$ and $\lambda_p \perp \mu$ such that $\lambda = \lambda_{ac} + \lambda_p$.

Proof 16.6.1

We will prove this result in four steps.
Step 1: *λ and μ are finite measures.*
Step 2: *μ is a σ - finite measure and λ is a finite measure.*
Step 3: *μ is a σ - finite measure and λ is a finite charge.*
Step 4: *The decomposition is unique.*
Proof Step 1: *As is usual, this is the most difficult step. We can see, in this case, that $\lambda + \mu$ is a measure. Note that $(\lambda + \mu)(E) = 0$ implies that $\lambda(E)$ is 0 too. Hence, $\lambda \ll (\lambda + \mu)$. By the Radon - Nikodym Theorem, there is then a non-negative $\lambda + \mu$ summable f so that for any measurable E,*

$$\lambda(E) = \int_E f\, d\,(\lambda + \mu)$$

Hence, f is μ and λ summable as well and

$$\lambda(E) = \int_E f\, d\,\lambda + \int_E f\, d\,\mu$$

Let

$$\begin{aligned} A_1 &= \{x \mid f(x) = 1\}, \\ A_2 &= \{x \mid f(x) > 1\}, \text{ and} \\ B &= \{x \mid f(x) < 1\} \end{aligned}$$

Also, for each n, let

$$E_n = \{x \mid f(x) \geq 1 + 1/n\}$$

Then, we see immediately $A_2 = \cup_n E_n$ and $X = A \cup B$. Now, we also have

$$\lambda(E_n) = \int_{E_n} f\, d\,(\lambda + \mu) \geq (1 + 1/n)\left(\lambda(E_n) + \mu(E_n)\right)$$

This implies $\lambda(E_n) \geq (1 + 1/n)\, \lambda(E_n)$ which tells us $\lambda(E_n) \leq 0$. But since λ is a measure, this forces $\lambda(E_n) = 0$. From the same inequality, we also have $\lambda(E_n) \geq \lambda(E_n) + \mu(E_n)$. which forces $\mu(E_n) = 0$ too.
Next, note the sequence of sets (E_n) increases to A_2 and so

$$\lim_n \mu(E_n) = \mu(A_2), \quad \lim_n \lambda(E_n) = \lambda(A_2)$$

Since $\mu(E_n) = \lambda(E_n) = 0$ for all n, we conclude $\mu(A_2) = \lambda(A_2) = 0$.

Also,

$$\lambda(A_1) = \int_{A_1} f\, d\,(\lambda + \mu) = \int_{A_1} 1\, d\,(\lambda + \mu) = \mu(A_1) + \lambda(A_1)$$

which implies $\mu(A_1) = 0$. Let $A = A_1 \cup A_2$. Then, the above remarks imply $\mu(A) = 0$. We

372 BASIC ANALYSIS IV: MEASURE THEORY AND INTEGRATION

now suspect that A and B will give us the decomposition of X, which will allow us to construct the measures $\lambda_{ac} \ll \mu$ and $\lambda_p \perp \mu$. Define λ_{ac} and λ_p by

$$\lambda_{ac} \;=\; \lambda(E \cap B), \quad \lambda_p = \lambda(E \cap A)$$

Then,

$$\lambda(E) \;=\; \lambda(E \cap A) + \lambda(E \cap B) = \lambda_{ac}(E) + \lambda_p(E)$$

showing us the we have found a decomposition of λ into two measures.
Is $\lambda_{ac} \ll \mu$? Let $\mu(E) = 0$. Then $\mu(E \cap B) = 0$ as well. Now, we know

$$\lambda(E \cap B) \;=\; \int_{E \cap B} f \, d\,(\lambda + \mu)$$
$$= \int_{E \cap B} f \, d\lambda + \int_{E \cap B} f \, d\mu$$

However, the second integral must be zero since $\mu(E \cap B) = 0$. Thus, we have

$$\lambda(E \cap B) \;=\; \int_{E \cap B} f \, d\lambda$$

We also have $\lambda(E \cap B) = \int_{E \cap B} 1 \, d\lambda$ and so

$$\int_{E \cap B} 1 \, d\lambda \;=\; \int_{E \cap B} f \, d\lambda$$

Thus,

$$\int_{E \cap B} \left(1 - f \right) d\lambda \;=\; 0$$

But on $E \cap B$, $1 - f > 0$; hence, we must have $\lambda(E \cap B) = 0$. This means $\lambda_{ac}(E \cap B) = 0$ implying $\lambda_{ac} \ll \mu$.

Is $\lambda_p \perp \mu$? Note, for any measurable E, we have

$$\lambda_p(E \cap B) \;=\; \lambda\Big((E \cap B) \cap A \Big) = \lambda(\emptyset) = 0$$

Thus, $\lambda_p \perp \mu$. In fact, we have shown

$$\lambda(E) \;=\; \int_{E \cap B} f \, d\lambda + \lambda_p(E)$$

Proof Step 2*:*
Note that once we find a decomposition $X = A \cup B$ with A and B measurable and disjoint satisfying $\mu(A) = 0$ and $\lambda(E \cap B) = 0$ if $\mu(E) = 0$, then we can use the technique in the proof of **Step 1***. We let $\lambda_{ac}(E) = \lambda(E \cap B)$ and $\lambda_p(E) = \lambda(E \cap A)$. This furnishes the decomposition we seek. Hence, we must find a suitable A and B.*

The measure μ is now σ - finite. Hence, there is a sequence of disjoint measurable sets X_n with $\mu(X_n) < \infty$ and $X = \cup_n X_n$. Let $\mathcal{S}_n$ denote the σ - algebra of subsets $\mathcal{S} \cap X_n$. By **Step 1***, there is a decomposition $X_n = A_n \cup B_n$ of disjoint and measurable sets so that $\mu(A_n) = 0$ and $\lambda(E \cap B_n) = 0$*

DECOMPOSING MEASURES
373

if $\mu(E) = 0$. Since the sets X_n are mutually disjoint, we know the sequences (A_n) and (B_n) are disjoint also. Let $A = \cup_n A_n$ and $B = A^C$ and note $A^C = \cap_n B_n$. Then, since μ is a measure, we have

$$\mu(\cup_{i=1}^n A_i) \;=\; \sum_{i=1}^n \mu(A_i) \;=\; 0$$

for all n. Hence,

$$\mu(A) \;=\; \lim_n \mu(\cup_{i=1}^n A_i) \;=\; 0$$

Next, if $\mu(E) = 0$, then $\mu(E \cap B_n) = 0$ for all n by the properties of the decomposition (A_n, B_n) of X_n. Since

$$E \cap B \;=\; \cap_n \left(E \cap B_n \right)$$

and $\lambda(E \cap B_1)$ is finite, we have

$$\lambda(E \cap B) \;=\; \lim_n \lambda(E \cap B_n)$$

However, each $\lambda(E \cap B_n)$ is zero because $\mu(E) = 0$ by assumption. Thus, we conclude $\lambda(E \cap B) = 0$. We then have the A and B we need to construct the decomposition.

Proof Step 3: *The mapping λ is now a finite charge. Let $\lambda = \lambda^+ - \lambda^-$ be the Jordan Decomposition of the charge λ. Applying* **Step 2**, *we see there are pairs of measurable sets (A_1, B_1) and (A_2, B_2) so that*

$$X \;=\; A_1 \cup B_1, \; A_1 \cap B_1 = \emptyset, \; \mu(A_1) = 0, \mu(E) = 0 \Rightarrow \lambda^+(E \cap B_1) = 0$$

and

$$X \;=\; A_2 \cup B_2, \; A_2 \cap B_2 = \emptyset, \; \mu(A_2) = 0, \mu(E) = 0 \Rightarrow \lambda^-(E \cap B_2) = 0$$

Let $A = A_1 \cup A_2$ and $B = B_1 \cap B_2$. Note $B^C = A$. It is clear then that $\mu(A) = 0$. Finally, if $\mu(E) = 0$, then $\lambda^+(E \cap B_1) = 0$ and $\lambda^-(E \cap B_2) = 0$. This tells us

$$\begin{aligned}
\lambda(E \cap B) \;&=\; \lambda^+(E \cap B) \;-\; \lambda^-(E \cap B) \\
&=\; \lambda^+(E \cap B_1 \cap B_2) \;-\; \lambda^-(E \cap B_1 \cap B_2) \\
&=\; \lambda^+\left((E \cap B_1) \cap B_2 \right) \;-\; \lambda^-\left((E \cap B_2) \cap B_1 \right)
\end{aligned}$$

Both of the terms on the right-hand side are then zero because we are computing measures of subsets of a set of measure 0. We conclude $\lambda(E \cap B) = 0$. The decomposition is then

$$\lambda_{ac}(E) \;=\; \lambda(E \cap B) \;=\; \left(\lambda^+ - \lambda^- \right)(E \cap B),$$

$$\lambda_p(E) \;=\; \lambda(E \cap A) \;=\; \left(\lambda^+ - \lambda^- \right)(E \cap A)$$

Proof Step 4: *To see this decomposition is unique, assume $\lambda = \lambda_1 + \lambda_2$ and $\lambda_{ac} + \lambda_p$ are two Lebesgue decompositions of λ. Then, $\lambda_{ac} - \lambda_1 = \lambda_2 - \lambda_p$. But since λ_1 and λ_{ac} are both absolutely continuous with respect to μ, it follows that $\lambda_{ac} - \lambda_1 \ll \mu$ also. Further, since both λ_2 and λ_p are*

BASIC ANALYSIS IV: MEASURE THEORY AND INTEGRATION

singular with respect to μ, we see $\lambda_2 - \lambda_p \perp \mu$. However, $\lambda_{ac} - \lambda_1 = \lambda_2 - \lambda_p$ by assumption and so $\lambda_{ac} - \lambda_1 \ll \mu$ and $\lambda_{ac} - \lambda_1 \perp \mu$. By Comment 16.6.3, this tells us $\lambda_{ac} = \lambda_1$. This then implies $\lambda_2 = \lambda_p$.

■

Homework

Exercise 16.6.3 *Let $(X, \mathcal{S})$ be a measurable space and λ is a charge on $\mathcal{S}$. Prove if P_1 and P_2 are positive sets for λ, then $P_1 \cup P_2$ is also a positive set for λ.*

Exercise 16.6.4 *Let $g_1(x) = 2x$, $g_2(x) = I_{[0,\infty)}(x)$, $g_3(x) = x\, I_{[0,\infty)}(x)$ and $g_4(x) = arctan(x)$. All of these functions generate Borel - Stieltjes measures on $\Re$.*

(i): Determine which are absolutely continuous with respect to Borel measure. Then, if absolutely continuous with respect to Borel measure, find their Radon - Nikodym derivative.

(ii): Which of these measures are singular with respect to Borel measure?

Exercise 16.6.5 *Let λ and μ be σ - finite measures on $\mathcal{S}$, a σ - algebra of subsets of a set X. Assume λ is absolutely continuous with respect to μ. If $g \in M^+(X, \mathcal{S})$, prove that*

$$\int g\, d\lambda \;=\; \int g\, f\, d\mu$$

where $f = d\lambda/d\mu$ is the Radon - Nikodym derivative of λ with respect to μ.

Exercise 16.6.6 *Let λ, ν and μ be σ - finite measures on $\mathcal{S}$, a σ - algebra of subsets of a set X. Use the previous exercise to show that if $\nu \ll \lambda$ and $\lambda \ll \mu$, then*

$$\frac{d\nu}{d\mu} \;=\; \frac{d\nu}{d\lambda} \frac{d\lambda}{d\mu}, \; \mu\, a.e.$$

Further, if λ_1 and λ_2 are absolutely continuous with respect to μ, then

$$d(\lambda_1 + \lambda_2)/d\mu \;=\; d\lambda_1/d\mu + d\lambda_2/d\mu\, \mu\, a.e.$$

Chapter 17

Connections to Riemann Integration

We need to connect Riemann Integration ideas to Lebesgue Integration.

Theorem 17.0.1 Every Riemann Integrable Function is Lebesgue Integrable and the Two Integrals Coincide

> *Let $f[a, b] \to \Re$ be a Riemann integrable function. If we let $\inf f d\mu = \int_{[a,b]} f d\mu$ represent the Lebesgue integral of f with μ denoting Lebesgue measure on the real line and $\int_a^b f(x)dx$ be the Riemann integral of f on $[a, b]$, then*
>
> $$\int_{[a,b]} f d\mu = \int_a^b f(x)dx$$

Proof 17.0.1

For each positive integer n, let π_n denote the uniform partition of $[a, b]$ which divides the interval into pieces of length $\frac{1}{2^n}$. Hence, $\pi_n = \{a + i\frac{b-a}{2^n} : 0 \le i \le 2^n\}$. This partition defines intervals as follows: $E_i = [x_i, x_{i+1})$ for $1 \le i < 2^n$ and $E_{2^n} = [x_{2^n-1}, x_{2^n}]$. Let m_i and M_i be defined as usual in Chapter 4. Define the simple functions ϕ_n and ψ_n as follows:

$$\phi_n(x) = \sum_{i=1}^{2^n-1} m_i I_{E_i}, \quad \psi_n(x) = \sum_{i=1}^{2^n} M_i I_{E_i}$$

Then $\phi_n \le \phi_{n+1} \le f$ and $\psi_n \ge \psi_{n+1} \ge f$ for all n. Further, these monotonic limits define measurable functions g and h so that $\phi_n \uparrow g$ and $\psi_n \downarrow h$ pointwise on $[a, b]$. Finally, it follows that $g \le f \le h$ on $[a, b]$.

Next, note that

$$\int \phi_n \, d\mu = \sum_{i=1}^{2^n} m_i \mu(E_i) = \sum_{i=1}^{2^n} m_i(x_{i+1} - x_i) = L(f, \pi_n)$$

$$\int \psi_n \, d\mu = \sum_{i=1}^{2^n} M_i \mu(E_i) = \sum_{i=1}^{2^n} M_i(x_{i+1} - x_i) = U(f, \pi_n)$$

Since f is Riemann integrable by assumption, we know $L(f, \pi_n) \to \int_a^b f(x)dx$ and $U(f, \pi_n) \to \int_a^b f(x)dx$. Hence, $\lim_n \int \phi_n d\mu \le \int_a^b f(x)dx$. It follows from Levi's Theorem 9.12.1 that g is

375

376

summable and $\int \phi_n\, d\mu \uparrow \int g d\mu$. We can apply Levi's Theorem again to $-\psi_n$ to conclude h is summable and $\int \psi_n\, d\mu \downarrow \int h d\mu$.

However, we also know that

$$\int (h - g)\, d\mu \;=\; \lim_n \int (psi_n - \phi_n)\, d\mu = \lim_n \left(U(f, \pi_n) - L(f, \pi_n) \right) \;=\; 0$$

We conclude $h = g$ a.e. Since $g \leq f \leq h$, this tells us $f = g = h$ a.e. Since Lebesgue measure is complete and g and h are measurable, we now know f is measurable and summable with

$$\int f\, d\mu \;=\; \lim_n \int \phi_n\, d\mu = \lim_n L(f, \pi_n) = \int_a^b f(x) dx$$

$\blacksquare$

Homework

Exercise 17.0.1 *Let Ψ be the Cantor Singular function. We know Ψ is constant on the middle thirds of the component intervals used in its definition.*

- *Prove Ψ is Riemann integrable on $[0, 1]$.*

- *But it is also Lebesgue integrable and the value of the integrals is the same.*

- *Show $\int_0^1 \Psi(x) ds = \int_{C^C} \Psi d\mu$.*

- *Since $\lambda(E) = \int_E |\Psi d\mu$ defines a measure prove $\int_{C^C} \Psi d\mu = \int_{\cap_n P_n} \Psi d\mu$ and $\int_{\cap_n P_n} \Psi d\mu = \lim_n \int_{P_n} \Psi d\mu$.*

- *Prove $\int_{P_n} \Psi d\mu = (1/2) \sum_{k=1}^n (2/3)^k$.*

- *Finally, prove $\int_{C^C} \Psi d\mu = 1$.*

- *Thus, prove $\int_0^1 \Psi(x) dx = 1 = \int \Psi d\mu$.*

Exercise 17.0.2 *Let $g(x) = (1/2)(x + \Psi(x))$ function and let μ be Lebesgue measure.*

- *Prove g is strictly increasing on $[0, 1]$ and so is Riemann integrable.*

- *Prove $\int_0^1 g(x) dx = \int g d\mu = 3/4$.*

Exercise 17.0.3 *Let f be defined by*

$$f(x) \;=\; \begin{cases} 2x, & 0 \leq x < 1 \\ 3, & 1 = x \\ 5x^2 + 1, & 1 < x \leq 2 \end{cases}$$

Let μ be Lebesgue measure on $[0, 2]$. Compute $\int f d\mu$.

Exercise 17.0.4 *Let f be defined by*

$$f(x) \;=\; \begin{cases} 2, & x = -1 \\ x + 4, & -1 < x < 1 \\ 6, & 1 = x \\ x^2 + 7, & 1 < x < 2 \\ 12, & x = 2 \\ 2x^3, & 2 < x \leq 3 \end{cases}$$

Let μ be Lebesgue measure on $[-1, 3]$. Compute $\int f d\mu$.

Exercise 17.0.5 *Let f_n, for $n \geq 2$, be defined by*

$$f_n(x) = \begin{cases} 1, & x = 1 \\ 1/(k+1), & 1/(k+1) \leq x < 1/k, \ 1 \leq k < n \\ 0, & 0 \leq x < 1/n \end{cases}$$

Let μ be Lebesgue measure on $[0, 1]$. Compute $\int f_n d\mu$.

Exercise 17.0.6 *Let f be the pointwise limit of the sequence of functions (f_n) for $n \geq 2$ defined by*

$$f_n(x) = \begin{cases} 1, & x = 1 \\ 1/(k+1), & 1/(k+1) \leq x < 1/k, \ 1 \leq k < n \\ 0, & 0 \leq x < 1/n \end{cases}$$

Let μ be Lebesgue measure on $[0, 1]$. Compute $\int f d\mu$.

Exercise 17.0.7 *Assume h is continuous on $[0, 2L]$ and f is in $L_2([0, 2L])$. Define*

$$u_n(t) = \sin\left(\left(N + \frac{1}{2}\right)\frac{\pi}{L}t\right)$$

and

$$a_n = \int_0^{2L} h\, u_n d\mu$$

Prove

$$\sum_{i=1}^{\infty} a_i^2 \leq \int_0^{2L} f^2 d\mu < \infty$$

and use that to prove $a_n \to 0$.

We can prove a connection between Riemann - Stieltjes integrals and Lebesgue - Stieltjes integrals also.

Theorem 17.0.2 Riemann - Stieltjes Integrable Functions are Lebesgue - Stieltjes Integrable and the Two Integrals Coincide

Let $f : [a, b] \to \Re$ and $g : [a, b] \to \Re$ be bounded functions. Assume g is monotone increasing and continuous from the right on $[a, b]$. Assume further that f is Riemann - Stieltjes integrable with respect to the integrator g on $[a, b]$. Let μ_g denote the Lebesgue - Stieltjes measure induced by g on $[a, b]$. If we let $\int_{[}a, b] f d\mu_g$ represent the Lebesgue - Stieltjes integral of f and $\int_a^b f(x) dg$ be the Riemann - Stieltjes integral of f with respect to g on $[a, b]$, then

$$\int_{[a,b]} f\, d\mu_g = \int_a^b f(x) dg$$

378

Proof 17.0.2
For each positive integer n, let π_n denote the uniform partition of $[a, b]$ which divides the interval into pieces of length $\frac{1}{2^n}$. Hence, $\pi_n = \{a + i\frac{b-a}{2^n} : 0 \leq i \leq 2^n\}$. This partition defines intervals as follows: $E_i = [x_i, x_{i+1})$ for $1 \leq i < 2^n$ and $E_{2^n} = [x_{2^n-1}, x_{2^n}]$. Let m_i and M_i be defined as usual in Chapter 4. Define the simple functions ϕ_n and ψ_n as follows:

$$\phi_n(x) = \sum_{i=1}^{2^n} m_i I_{E_i}, \quad \psi_n(x) = \sum_{i=1}^{2^n} M_i I_{E_i}$$

Then $\phi_n \leq \phi_{n+1} \leq f$ and $\psi_n \geq \psi_{n+1} \geq f$ for all n Further, these monotonic limits define measurable functions u and v so that $\phi_n \uparrow u$ and $\psi_n \downarrow v$ pointwise on $[a, b]$. Finally, it follows that $u \leq f \leq v$ on $[a, b]$.

Next, note that

$$\int \phi_n \, d\mu_g = \sum_{i=1}^{2^n} m_i \mu_g(E_i) = \sum_{i=1}^{2^n} m_i(g(x_{i+1}) - g(x_i)) = L(f, g, \pi_n)$$

$$\int \phi_n \, d\mu_g = \sum_{i=1}^{2^n} M_i \mu_g(E_i) = \sum_{i=1}^{2^n} M_i(g(x_{i+1}) - g(x_i)) = U(f, g, \pi_n)$$

Since f is Riemann - Stieltjes integrable with respect to g by assumption, we know $L(f, g, \pi_n) \to \int_a^b f(x)dg$ and $U(f, g, \pi_n) \to \int_a^b f(x)dg$. Hence, $\lim_n \int \phi_n d\mu_g \leq \int_a^b f(x)dg$. It follows from Levi's Theorem 9.12.1 that u is summable and $\int \phi_n \, d\mu_g \uparrow \int u \, d\mu_g$. We can apply Levi's Theorem again to $-\psi_n$ to conclude v is summable and $\int \psi_n d\mu \downarrow \int v d\mu_g$.

However, we also know that

$$\int (u - v) \, d\mu_g = \lim_n \int (\psi_n - \phi_n) \, d\mu_g = \lim_n \left(U(f, g, \pi_n) - L(f, g, \pi_n) \right) = 0$$

We conclude $u = v$ a.e. Since $u \leq f \leq v$, this tells us $u = f = v$ a.e. Since Lebesgue measure is complete and u and v are measurable, we now know f is measurable and summable with

$$\int f \, d\mu_g = \lim_n \int \phi_n \, d\mu_g = \lim_n L(f, g, \pi_n) = \int_a^b f(x)dg$$

$\blacksquare$

Homework

Exercise 17.0.8 *It is easy to see this theorem extends to Riemann - Stieltjes integrators g that are of bounded variation. If g is of bounded variation and continuous from the right, we can write $g = u - v$ where both u and v are increasing and right continuous. The function u determines a Lebesgue - Stieltjes measure μ_u; the function v, the Lebesgue - Stieltjes measure μ_v and finally g defines the charge $\mu_g = \mu_u - \mu_v$. Prove $\int f \, d\mu_g = \int_a^b f dg$.*

Chapter 18

Fubini Type Results

The discussion of multiple integration in (Peterson (10) 2020) explains in detail the details for the particular case of Riemann Integration in $\Re^n$. We have built Lebesgue measure on $\Re^k$ so the question that remains is how to connect the Lebesgue measure μ_2 on $\Re^2$ with the product of Lebesgue measure μ_1 on $\Re$ with itself. In other words, we want to make precise what something like $d\mu_2 = d\mu_x \, d\mu_y$ where μ_x is Lebesgue measure on the x axis and μ_y is Lebesgue measure on the y axis. For completeness, let's look back at the multidimensional Riemann Integration setting we explained in (Peterson (10) 2020) without the benefit of measure theoretic concepts. We made do with the idea of subsets having content zero instead. However, we now know a set of content zero is also Lebesgue measure zero.

18.1 The Riemann Setting

In this classical setting, we evaluate double integrals for the area of a region D by rewriting them like this

$$\int_D f \;=\; \int_a^b \int_{f(x)}^{g(x)} F(x,y) dy \, dx$$

where $\int_D f$ is the two dimensional integral we have been developing. Expressing this integral in terms of successive single variable integrals leads to the **iterated** integrals we use to evaluate them in our calculus classes. Thus, using the notation $\int dx$, $\int dy$ etc., to indicate these one dimensional integrals along one axis only, we want to know when

$$\int_D F \;=\; \int \left(\int F(x,y) \, dy \right) dx = \int \int F(x,y) \, dy \, dx$$

$$\int_D F \;=\; \int \left(\int F(x,y) \, dx \right) dy = \int \int F(x,y) \, dx \, dy$$

$$\int_\Omega G \;=\; \int \left(\int \left\{ \int G(x,y,z) \, dy \right\} dx \right) dz = \int \int \int G(x,y,z) \, dy \, dx \, dz$$

$$\int_\Omega G \;=\; \int \left(\int \left\{ \int G(x,y,z) \, dy \right\} dz \right) dx = \int \int \int G(x,y,z) \, dy \, dz \, dx$$

$$\int_\Omega G \;=\; \int \left(\int \left\{ \int G(x,y,z) \, dx \right\} dy \right) dz = \int \int \int G(x,y,z) \, dx \, dy \, dz$$

$$\int_{\Omega} G = \int \left(\int \left\{ \int G(x,y,z)\, dx \right\} dz \right) dy = \int \int \int G(x,y,z)\, dx\, dz\, zdy$$

$$\int_{\Omega} G = \int \left(\int \left\{ \int G(x,y,z)\, dz \right\} dx \right) dy = \int \int \int G(x,y,z)\, dz\, dx\, dy$$

$$\int_{\Omega} G = \int \left(\int \left\{ \int G(x,y,z)\, dz \right\} dy \right) dx = \int \int \int G(x,y,z)\, dz\, dy\, dx$$

where D is a bounded open subset of $\Re^2$ and Ω is a bounded open subset of $\Re^3$. Note in the Ω integration there are many different ways to organize the integration. If we were doing an integration over $\Re^n$ there would in general be a lot of choices! The conditions under which we can do this give rise to what are called Fubini type theorems. If general, you think of the bounded open set V as being in $\Re^{n+m} = \Re^n \times \Re^m$, label the volume elements as dV_n and dV_m and want to know when

$$\int_V F = \int \left(\int F(x_1, \ldots, x_n, x_{n+1}, \ldots, x_{n+m})\, dV_n \right) dV_m$$

$$= \int \left(\int F\, dV_n \right) dV_m = \int \int F\, dV_n\, dV_m$$

18.1.1 Fubini on a Rectangle

Let's specialize to the bounded set $D = [a,b] \times [c,d] \subset \Re^2$. Assume $f : D \to \Re$ is continuous. There are three integrals here:

- The integral over the two dimensional set D, $\int_D f\, dV_2$ where the subscript 2 reminds us that this is the two dimensional situation.

- The integral of functions like $g : [a,b] \to \Re$. Assume g is continuous. Then the one dimensional integral would be represented as $\int_a^b g(x)dx$ in the integration treatment (Peterson (8) 2020) but here we will use $\int_{[a,b]} g\, dV_1$.

- The integral of functions like $h : [c,d] \to \Re$. Assume h is continuous. Then the one dimensional integral $\int_c^d h(y)dy$ is called $\int_{[c,d]} h\, dV_2$. Note although the choice of integration variable y in $\int_c^d g(y)dy$ is completely arbitrary, it makes a lot of sense to use this notation instead of $\int_c^d g(x)dx$ because we want to cleanly separate what we are doing on each axis from the other axis. So as usual, we have to exercise discipline to set up the variable names and notation to help us understand clearly what we are trying to do.

If you need to, go back and refresh your understanding of the higher dimensional Riemann integral as discussed in (Peterson (10) 2020). We will use those ideas and notations freely in what follows.

Specialize to the bounded set $D = [a,b] \times [c,d] \subset \Re^2$. Any partition P of D divides the bounding box R we choose to enclose D into rectangles. If $P = P_1 \times P_2$ where $P_1 = \{a = x_0, x_1, \ldots, x_{n-1}, x_n = b\}$ and $P_2 = \{c = y_0, y_1, \ldots, y_{m-1}, y_m = d\}$, the rectangles have the form

$$S_{ij} = [x_i, x_{i+1}] \times [y_j, y_{j+1}] = \Delta x_i \times \Delta y_j$$

where $\Delta x_i = x_{i+1} - x_i$ and $\Delta y_j = y_{j+1} - y_j$. Thus for an evaluation set σ with points $(z_i, w_j) \in S_{ij}$, the Riemann sum is

$$S(f, \sigma, P) = \sum_{(i,j)\in P} f(z_i, w_j)\, \Delta x_i\, \Delta y_j = \sum_{(i,j)\in P} f(z_i, w_j)\, \Delta y_j\, \Delta x_i$$

FUBINI TYPE RESULTS

Note the order of the terms in the area calculation for each rectangle do not matter. Also, we could call $\Delta x_i = \Delta V_{1i}$ (for the first axis) and $\Delta y_j = \Delta V_{2j}$ (for the second axis) and rewrite as

$$S(f, \boldsymbol{\sigma}, P) \quad = \quad \sum_{(i,j)\in P} f(z_i, w_j)\, \Delta V_{1i}\, \Delta V_{2j} = \sum_{(i,j)\in P} f(z_i, w_j)\, \Delta V_{2j}\, \Delta V_{1i}$$

We can also revert to writing $\sum_{(i,j)\in P}$ as a double sum to get

$$S(f, \boldsymbol{\sigma}, P) \quad = \quad \sum_{i=1}^{n} \left(\sum_{j=1}^{m} f(z_i, w_j)\, \Delta V_{2j} \right) \Delta V_{1i} = \sum_{j=1}^{m} \left(\sum_{i=1}^{n} f(z_i, w_j)\, \Delta V_{1i} \right) \Delta V_{2j}$$

Then the idea is this

- **y first then x**

  For $\sum_{i=1}^{n} \left(\sum_{j=1}^{m} f(z_i, w_j)\, \Delta V_{2j} \right) \Delta V_{1i}$, fix the variables on axis one and suppose

$$\lim_{\|P_2\|\to 0} \sum_{j=1}^{m} f(z_i, w_j)\, \Delta V_{2j} \quad = \quad \int_{z_i \times [c,d]} f(z_i, y) dy$$

To avoid using specific sizes like m for P_2 here, we would usually just say

$$\lim_{\|P_2\|\to 0} \sum_{j\in P_2} f(z_i, w_j)\, \Delta V_{2j} \quad = \quad \int_{z_i \times [c,d]} f(z_i, y) dy$$

Now for this to happen we need the function $f(\cdot, y)$ to be integrable for each choice of x for slot one. An easy way to guarantee this is to assume f is continuous in $(x, y) \in D$ which will force $f(\cdot, y)$ to be a continuous function of y for each x. Let's look at this integral more carefully. We have defined a new function $G(x)$ by

$$G(x) \quad = \quad \int_{[c,d]} f(x, y) dy = \int_c^d f(x, y) dy$$

If $G(x)$ is integrable on $[a, b]$, we can say

$$\lim_{\|P_1\|\to 0} \sum_{i=1}^{n} \left(\lim_{\|P_2\|\to 0} \sum_{j=1}^{m} f(z_i, w_j)\, \Delta V_{2j} \right) \Delta V_{1i} \quad = \quad \lim_{\|P_1\|\to 0} \sum_{i=1}^{n} \left(\int_{z_i \times [c,d]} f(z_i, y) dy \right) \Delta V_{1i}$$

$$= \quad \lim_{\|P_1\|\to 0} \sum_{i=1}^{n} G(z_i) \Delta V_{1i} = \int_a^b G(x) dx$$

We can resay this again. Let ΔV_{ij} be the area $\Delta x_i\, \Delta y_j$. Then

$$\int_D f\, d\boldsymbol{V}_2 \quad = \quad \lim_{\|P1\times P_2\|\to 0} \sum_{(i,j)\in P_1\times P_2} f(z_i, w_j)\Delta V_{ij}$$

$$= \quad \lim_{\|P_1\|\to 0} \sum_{i\in P_1} \left(\lim_{\|P_2\|\to 0} \sum_{j\in P_2} f(z_i, w_j)\, \Delta V_{2j} \right) \Delta V_{1i}$$

$$= \quad \lim_{\|P_1\|\to 0} \sum_{i\in P_1} \left(\int_c^d f(z_i, y)\, dy \right) \Delta V_{1i}$$

BASIC ANALYSIS IV: MEASURE THEORY AND INTEGRATION

$$= \int_a^b \left(\int_c^d f(x,y) \, dy \right) dx = \int_a^b \int_c^d f(x,y) \, dy \, dx$$

- **x first then y**

For $\sum_{j=1}^m \left(\sum_{i=1}^n f(z_i, w_j) \, \Delta V_{1i} \right) \Delta V_{2j}$, fix the variables on axis two and suppose

$$\lim_{\|P_1\| \to 0} \sum_{i=1}^n f(z_i, w_j) \, \Delta V_{1i} = \int_{[a,b] \times w_j} f(x, w_j) dx$$

Of course, this is the same as

$$\lim_{\|P_1\| \to 0} \sum_{i \in P_1} f(z_i, w_j) \, \Delta V_{1i} = \int_{[a,b] \times w_j} f(x, w_j) dx$$

Now for this to happen we need the function $f(x, \cdot)$ to be integrable for each choice of y for slot two. If we assume f is continuous in $(x, y) \in D$ this will force $f(x, \cdot)$ to be a continuous function of x for each y. Let's look at this integral more carefully. We have defined a new function $H(y)$ by

$$H(y) = \int_{[a,b]} f(x,y) dx = \int_a^b f(x,y) dx$$

If $H(y)$ is integrable on $[c,d]$, we can say

$$\lim_{\|P_2\| \to 0} \sum_{j=1}^m \left(\lim_{\|P_1\| \to 0} \sum_{i=1}^n f(z_i, w_j) \, \Delta V_{1i} \right) \Delta V_{2j} = \lim_{\|P_2\| \to 0} \sum_{j=1}^m \left(\int_{[a,b] \times w_j} f(x, w_j) dx \right) \Delta V_{2j}$$

$$= \lim_{\|P_2\| \to 0} \sum_{j=1}^m H(w_j) \Delta V_{2j} = \int_c^d H(y) dy$$

or

$$\int_D f \, d\boldsymbol{V}_2 = \lim_{\|P_1 \times P_2\| \to 0} \sum_{(i,j) \in P_1 \times P_2} f(z_i, w_j) \Delta V_{ij}$$

$$= \lim_{\|P_2\| \to 0} \sum_{j \in P_2} \left(\lim_{\|P_1\| \to 0} \sum_{i \in P_1} f(z_i, w_j) \, \Delta V_{1i} \right) \Delta V_{2j}$$

$$= \lim_{\|P_2\| \to 0} \sum_{j \in P_2} \left(\int_a^b f(x, w_j) \, dx \right) \Delta V_{2j}$$

$$= \int_c^d \left(\int_a^b f(x,y) \, dx \right) dy = \int_c^d \int_a^b f(x,y) \, dx \, dy$$

We have laid out an approximate chain of reasoning to prove our first Fubini result and we can follow this to prove the result. Pay close attention to how we might have to modify this proof for a more general measure theoretic proof.

Theorem 18.1.1 Fubini's Theorem on a Rectangle: 2D

FUBINI TYPE RESULTS

Let $f : [a, b] \times [c, d] \to \Re$ be continuous. Then

$$\int_D f \, d\mathbf{V}_2 = \int_a^b \left(\int_c^d f(x, y) \, dy \right) dx = \int_a^b \int_c^d f(x, y) \, dy \, dx$$

$$\int_D f \, d\mathbf{V}_2 = \int_c^d \left(\int_a^b f(x, y) \, dx \right) dy = \int_c^d \int_a^b f(x, y) \, dx \, dy$$

Proof 18.1.1

For $x \in [a, b]$, let G be defined by $G(x) = \int_c^d f(x, y) dy$. We need to show G is continuous for each x. Note since f is continuous on $[a, b] \times [c, d]$, given $\epsilon > 0$, there is a $\delta > 0$ so that

$$\sqrt{(x' - x)^2 + (y' - y)^2} < \delta \implies |f(x, y) - f(x', y')| < \epsilon$$

In particular, for fixed $y' = y$, there is a $\delta > 0$ so that

$$|x' - x| < \delta \implies |f(x', y) - f(x, y)| < \epsilon/(d - c)$$

Thus, if $|x' - x| < \delta$,

$$
\begin{aligned}
|G(x') - G(x)| &= \left| \int_c^d (f(x', y) - f(x, y)) dy \right| \leq \int_c^d |f(x', y) - f(x, y)| dy \\
&\leq (d - c) \, \epsilon/(d - c) = \epsilon
\end{aligned}
$$

and so we know G is continuous in x for each y. A similar argument shows the function H defined at each $y \in [c, d]$ by $H(y) = \int_a^b f(x, y) dy$ is also continuous in y for each x.

Since f is continuous on $[a, b] \times [c, d]$, and the boundary of D has measure zero, we see the set of discontinuities of f on D is a set of measure zero and so f is integrable on D. Let $P_n = P_{1n} \times P_{2n}$ be a sequence of partitions with $\|P_n\| \to 0$. Then, we know for any evaluation set σ of P, that

$$\int_D f = \lim_{\|P_n\| \to 0} \sum_{i \in P_{1n}} \sum_{j \in P_{2n}} f(z_i, w_j) \, \Delta x_i \, \Delta y_j$$

and following the reasoning we sketched out earlier, we can choose to organize this as **x first then y** or **y first then x**. We will do the case of **x first then y** only and leave the details of the other case to you. Fix $\epsilon > 0$. Then there is N, so that $n > N$ implies

$$\left| \sum_{j \in P_{2n}} \left(\sum_{i \in P_{1n}} f(z_i, w_j) \, \Delta x_i \right) \Delta y_j - \int_D f \right| < \epsilon/2$$

Since $f(x, w_j)$ is continuous in x for each w_j, we know $\int_a^b f(x, w_j) dx$ exists and so for any sequence of partitions Q_n of $[a, b]$ with $\|Q_n\| \to 0$, using as the evaluation set the points z_i, we have

$$\sum_{i \in Q_n} f(z_i, w_j) \, \Delta x_i \quad \to \quad \int_a^b f(x, w_j) dx$$

Since $\|P_n\| \to 0$, so does the sequence P_{1n}, there is $N_1 < N$ so that $n > N_1$ implies

$$\left| \sum_{i \in P_{1n}} f(z_i, w_j)\, \Delta x_i - \int_a^b f(x, w_j)dx \right| = \left| \sum_{i \in P_{1n}} f(z_i, w_j)\, \Delta x_i - H(w_j) \right| \;<\; \epsilon/(4(d-c))$$

Similarly, since $\|P_{2n}\| \to 0$ and H is integrable, there is $N_2 > N_1 > N$ so that

$$\left| \sum_{j \in P_{2n}} H(w_j)\, \Delta y_j - \int_c^d H(y)dy \right| < \epsilon/4$$

where the points w_j form the evaluation set we use in each P_{2n}. Thus,

$$\left| \sum_{j \in P_{2n}} \left(\sum_{i \in P_{1n}} f(z_i, w_j)\, \Delta x_i \right) \Delta y_j - \int_c^d H(y)dy \right|$$

$$= \left| \sum_{j \in P_{2n}} \left(\sum_{i \in P_{1n}} f(z_i, w_j)\, \Delta x_i - H(w_j) + H(w_j) \right) \Delta y_j - \int_c^d H(y)dy \right|$$

$$\leq \left| \sum_{j \in P_{2n}} \left(\sum_{i \in P_{1n}} f(z_i, w_j)\, \Delta x_i - H(w_j) \right) \Delta y_j \right| + \left| \sum_{j \in P_{2n}} H(w_j)\Delta y_j - \int_c^d H(y)dy \right|$$

$$\leq \sum_{j \in P_{2n}} \left| \sum_{i \in P_{1n}} f(z_i, w_j)\, \Delta x_i - H(w_j) \right| \Delta y_j + \left| \sum_{j \in P_{2n}} H(w_j)\Delta y_j - \int_c^d H(y)dy \right|$$

$$< \sum_{j \in P_{2n}} \epsilon/(4(d-c))\Delta y_j + \epsilon/4 = \epsilon/2$$

We can now complete the argument. For $n > N_2$,

$$\left| \int_c^d H(y)dy - \int_d f \right| \;\leq\; \left| \sum_{j \in P_{2n}} \left(\sum_{i \in P_{1n}} f(z_i, w_j)\, \Delta x_i \right) \Delta y_j - \int_D f \right|$$

$$+ \left| \sum_{j \in P_{2n}} \left(\sum_{i \in P_{1n}} f(z_i, w_j)\, \Delta x_i \right) \Delta y_j - \int_c^d H(y)dy \right|$$

$$< \epsilon/2 + \epsilon/2 = \epsilon$$

Thus

$$\int_D f = \int_c^d H(y)dy \;=\; \int_c^d \left(\int_a^b f(x,y)\, dx \right) dy = \int_c^d \int_a^b f(x,y)\, dx\, dy$$

*A very similar argument handles the case **y first then x** giving*

$$\int_D f = \int_a^b G(x)dx \;=\; \int_a^b \left(\int_c^d f(x,y)\, dy \right) dx = \int_a^b \int_c^d f(x,y)\, dy\, dx$$

As you can see, the argument to prove Theorem 18.1.1 is somewhat straightforward but you have to be careful to organize the inequality estimates. A similar approach can prove the more general theorem.

Theorem 18.1.2 Fubini's Theorem for a Rectangle: n dimensional

> Let $f : R_1 \times R_2 \to \Re$ be continuous where R_1 is a rectangle in $\Re^n$ and $R_2 \in \Re^m$. Then
> $$\int_D f dV_{n+m} = \int_{R_1} \left(\int_{R_2} f(x,y)\, dV_m \right) dV_n = \int_{R_1} \int_{R_2} f(x,y)\, dV_m\, dV_n$$
> $$\int_D f dV_{n+m} = \int_{R_2} \left(\int_{R_1} f(x,y)\, dV_n \right) dV_m = \int_{R_2} \int_{R_1} f(x,y)\, dV_n\, dV_m$$
> where x are the variables $x_1, \ldots, x_n$ from $\Re^n$ and y are the variables $x_{n+1}, \ldots, x_{n+m}$.

Proof 18.1.2
For example, $f : [a_1, b_1] \times [a_2, b_2] \times [a_3, b_3] \times [a_4, b_4] \to \Re$ would use $R_1 = [a_1, b_1] \times [a_2, b_2]$ and $R_2 = [a_3, b_3] \times [a_4, b_4]$. The proof here is quite similar. We prove $G(x)$ and $H(y)$ are continuous since f is continuous. Then, since f is integrable, given any sequence of partitions P_n with $\|P_n\| \to 0$, there is N so that $n > N$ implies

$$\left| \sum_{j \in P_{2n}} \left(\sum_{i \in P_{1n}} f(z_i, w_j) \Delta x_i \right) \Delta y_j - \int_D f \right| < \epsilon/2$$

where $P_n = P_{1n} \times P_{2n}$ like usual, except P_{1n} is a partition of R_1 and P_{2n} is a partition of R_2. The summations like $\sum_{j \in P_{2n}}$ are still interpreted in the same way. The rest of the argument is straightforward. ∎

Now let's specialize to some useful two dimensional situations. Consider the situation shown in Figure 18.1.

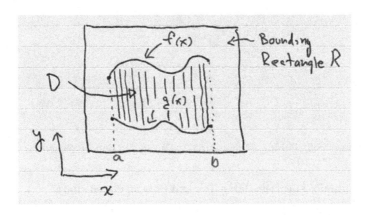

Figure 18.1: Fubini's Theorem in 2D: a top and a bottom curve.

Theorem 18.1.3 Fubini's Theorem: 2D: a Top and Bottom curve

> Let f and g be continuous real-valued functions on $[a, b]$ and $F : D \to \Re$ be continuous on D where $D = \{(x, y) : a \le x \le b,\ g(x) \le y \le f(x)\}$. Then
> $$\int_D F dV_2 = \int_a^b \int_{g(x)}^{f(x)} dy\, dx$$

Proof 18.1.3

Let's apply Theorem 18.1.2 on the rectangle $[a,b] \times [g_m, f_M]$ where g_m is the minimum value of g on $[a,b]$ which we know exists because g is continuous on a compact domain. Similarly f_M is the maximum value of f on $[a,b]$ which also exists. Then the D showing in Figure 18.1 is contained in $[a,b] \times [g_m, f_M]$. We see $\hat{F}$ is continuous on this rectangle except possibly on the curves given by f and g. We also know the set $S_f = \{(x, f(x)) | a \leq x \leq b\}$ and the set $S_g = \{(x, g(x)) | a \leq x \leq b\}$ both have measure zero in $\Re^2$. Thus $\hat{F}$ has a discontinuity set of measure zero and so F is integrable on $[a,b] \times [g_m, f_M]$ which matches the integral of F on D.

Next, since F is continuous on the interior of D, for fixed x in $[a,b]$, $G(y) = F(x,y)$ is continuous on $[g(x), f(x)]$. Of course, this function need not be continuous at the endpoints but that is a set of measure zero in $\Re^1$ anyway. Hence, $\hat{G}$ is integrable on $[g_m, f_M]$ and hence G is integrable on $[g(x), f(x)]$. By Theorem 18.1.2, we conclude

$$\int_{[a,b] \times [g_m, f_M]} F d\mathbf{V}_2 \;=\; \int_D F d\mathbf{V}_2 = \int_a^b \int_{g_m}^{f_M} G(y)\, dy\, dx$$

$$= \int_a^b \int_{g(x)}^{f(x)} F(x,y)\, dy\, dx$$

∎

Next, look at a typical closed curve situation.

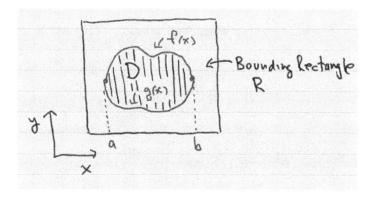

Figure 18.2: Fubini's Theorem in 2D: a top and bottom closed curve.

Theorem 18.1.4 Fubini's Theorem: 2D: a Top and Bottom closed curve

Let f and g be continuous real-valued functions on $[a,b]$ which form the bottom and top halves of a closed curve and $F: D \to \Re$ be continuous on D where $D = \{(x,y) : a \leq x \leq b,\ g(x) \leq y \leq f(x)\}$. Then

$$\int_D F d\mathbf{V}_2 \;=\; \int_a^b \int_{g(x)}^{f(x)} dy\, dx$$

Proof 18.1.4

From Figure 18.2, we see the closed curve defines a box $[a,b] \times [g_m, f_M]$ where g_m is the minimum value of g and f_M is the maximum value of f on $[a,b]$. These exist because f and g are continuous on the compact domain $[a,b]$. The rest of the argument is exactly the same as the proof of Theorem

FUBINI TYPE RESULTS

18.1.3. We have

$$\int_{[a,b]\times[g_m,f_M]} F \, d\boldsymbol{V}_2 \;=\; \int_D F \, d\boldsymbol{V}_2 = \int_a^b \int_{g_m}^{f_M} G(y) \, dy \, dx$$

$$=\; \int_a^b \int_{g(x)}^{f(x)} F(x,y) \, dy \, dx$$

∎

It is easy to see we could prove similar theorems for a region $D = \{(x,y)|c \le y \le d,\ g(y) \le x \le f(y)\}$ for functions f and g continuous on $[c,d]$ and F continuous on D. We would find

$$\int_{[g_m,f_M]\times[c,d]} F \, d\boldsymbol{V}_2 \;=\; \int_D F \, d\boldsymbol{V}_2 = \int_c^d \int_{g(y)}^{f(y)} F(x,y) \, dx \, dy$$

Call the top - bottom region a TB region and the left -right region a LR region. Then you should be able to see how you can break a lot of regions into finite combinations of TB and LR regions and apply Fubini's Theorem to each piece to complete the integration.

Comment 18.1.1 *We can work out similar theorems in Re^3, $\Re^4$ and so forth. Naturally, this process gets complicated as the dimension increases. Still, the principles of the arguments are the same. The boundary surfaces are all of measure zero as they are a strict subspace of $\Re^n$. You should try a few to see how the arguments are constructed. This is how you learn how things work: by doing.*

Homework

Exercise 18.1.1 *Prove Fubini's Theorem for rectangles in $\Re^3$. It is enough to prove it for one of the six cases.*

Exercise 18.1.2 *Since we can approximate these integrals using Riemann sums, can you estimate how computationally expensive this gets using uniform partitions? Would it be easy to compute a Riemann approximation to a 10 dimensional integral over a rectangle?*

Exercise 18.1.3 *This problem shows you how we can use these ideas to do something interesting!*

1. *Compute the two dimensional $I_R = \int_{D_R} e^{-x^2-y^2}$ where $D_R = B(R,(0,0))$ for all R by using the polar coordinate transformation. Can you compute $\lim_{R\to\infty}$ for this result?*

2. *Let $J_R = \int_{S_R} e^{-x^2-y^2} dxdy$. Prove $\lim_{R\to\infty} J_R = \lim_{R\to\infty} I_R$.*

3. *What is the value of $\int_{-\infty}^{\infty} e^{-x^2} dx$?*

Exercise 18.1.4 *Draw a bounded 2D region in $\Re^2$ which is constructed from two TB's and one LR and convince yourself how Fubini's Theorem is applied to compute the integral.*

Exercise 18.1.5 *Draw an arbitrary bounded region in $\Re^2$ and decompose it into appropriate TB and LR regions on which Fubini's Theorem can be applied and convince yourself how the total integral is computed.*

18.2 The Lebesgue Setting

Let's do this in general first. We want a product measure that behaves like we expect for our $\Re^2$ experience. If we let μ_{2D} be the usual two dimensional Lebesgue measure, we already know how it

388   BASIC ANALYSIS IV: MEASURE THEORY AND INTEGRATION

is constructed and we know

$$\mu_{2D}([a,b] \times [c,d]) = (b-a)(d-c), \quad \text{or } \mu_{2D}([a,b] \times [c,d]) = \mu_{1D}([a,b])\,\mu_{1D}([c,d])$$

where μ_{1D} is the usual one dimensional Lebesgue measure. Another way of saying this is in terms of outer measures. We would have

$$\mu_{2D}^*([a,b] \times [c,d]) = (b-a)(d-c), \quad \text{or } \mu_{2D}^*([a,b] \times [c,d]) = \mu_{1D}^*([a,b])\,\mu_{1D}^*([c,d])$$

Now if we replace the μ_{1D}^* that occurs twice, with the Lebesgue outer measure μ^* for the x axis and the Lebesgue outer measure ν^* for the y axis of $\Re^2$, we have

$$\mu_{2D}^*([a,b] \times [c,d]) = \mu^*([a,b])\,\nu^*([c,d])$$

But we can be more general: we want this defining equation to hold for members of a useful covering family for the outer measures. So if we insist this is true for all μ^* and ν^* measurable sets, we want

$$\mu_{2D}^*(A \times B) = \mu^*(A)\,\nu^*(B), \ \forall\, A \in \mathcal{M}_\mu, \ \forall\, B \in \mathcal{M}_\nu$$

where $\mathcal{M}_\mu$ and $\mathcal{M}_\nu$ are both the usual one dimensional Lebesgue measurable sets. But there is no need to think of doing this on the x and y axis of $\Re^2$. Instead we can do this in the product space $X \times Y$ and demand that

$$\mu_{XY}^*(A \times B) = \mu^*(A)\,\nu^*(B), \ \forall\, A \in \mathcal{M}_\mu, \ \forall\, B \in \mathcal{M}_\nu$$

where $\mathcal{M}_\mu$ is the set of μ^* measurable set in X and $\mathcal{M}_\nu$ is the set of μ^* measurable set in Y, respectively. This suggests we define the product outer measure by

$$\mu_{XY}^*(S \times T) = \inf\left\{ \sum_{i=1}^\infty \mu^*(A_i)\,\nu^*(B_i), \ \forall\, S \times T \subset \cup_{i=1}^\infty A_i \times B_i, \ A_i \in \mathcal{M}_\mu, \ B_i \in \mathcal{M}_\nu \right\}$$

for any $S \times T \subset X \times Y$.

The premeasure here is $\tau = \mu^* \times \nu^*$ and the covering family is $\mathcal{M}_\mu \times \mathcal{M}_\nu$ and so μ_{XY}^* is automatically an outer measure on $X \times Y$ and so satisfies the usual properties such as countable subadditivity. Let's state this formally as a definition.

Definition 18.2.1 Product Outer Measure

Let μ^* be an outer measure on a set X and ν^* be an outer measure on the set Y. These outer measures induce the measures μ and ν respectively using the standard construction approach. The product measure $\mu^* \times \nu^*$ is defined on each subset $S \subset X \times Y$ by

$$(\mu^* \times \nu^*)(S) =$$

$$\inf\left\{ \sum_{i=1}^\infty \mu^*(A_i)\,\nu^*(B_i) \mid S \subset \cup_{i=1}^\infty A_i \times B_i, \ A_i \subset X, \ B_i \subset Y \right\}$$

where each A_i and B_i are μ^* and ν^* measurable for each i; i.e. each A_i and B_i are μ and ν measurable sets, respectively.

As mentioned earlier, our previous theorems tell us $\mu^* \times \nu^*$ is an outer measure as it is built from a premeasure and a covering family in the standard way. Now let's recall some notation. We know μ^* is generated by a premeasure τ_μ and a covering family $\mathcal{F}_\mu$ and we don't know anything about them.

FUBINI TYPE RESULTS

We also know ν^* is generated by a premeasure τ_ν and a covering family $\mathcal{F}_\nu$ which we also know nothing about. Hence, we don't know the members of their respective covering families are in their corresponding σ algebras of μ^* and ν^* measurable sets, $\mathcal{M}_\mu$ and $\mathcal{M}_\nu$, respectively. However, we do know the Caratheodory condition is satisfied and so we can say

$$\mu^*(S) \geq \mu^*(S \cap A) + \mu^*(S \cap A'), \forall S \subset X, \quad \nu^*(T) \geq \nu^*(T \cap B) + \nu^*(T \cap B'), \forall T \subset Y$$

where the $'$ indicates set complement relative to the appropriate universe set.

Theorem 18.2.1 $\mu^* \times \nu^*(A \times B) = \mu^*(A)\,\nu^*(B)$ **for A and B μ^* and ν^* Measurable**

> If $A \in \mathcal{M}_\mu$ and $B \in \mathcal{M}_\nu$, then $\mu^* \times \nu^*(A \times B) = \mu^*(A)\nu^*(B)$.

Proof 18.2.1
Since $A \times B \subset A \times B$ for $A \in \mathcal{M}_\mu$ and $B \in \times \mathcal{M}_\nu$, we have immediately that $\mu^* \times \nu^*(A \times B) \leq \mu^*(A)\,\nu^*(B)$. To prove the other inequality, if $A \times B \subset \cup_{i=1}^\infty A_i \times B_i$ is a cover, then

$$
\begin{aligned}
A \times B \;\subset\; & (A \times B) \cap (\cup_{i=1}^\infty A_i \times B_i) \cup (A \times B) \cap (\cup_{i=1}^\infty A_i \times B_i)' \\
= \;& \cup_{i=1}^\infty (A_i \cap A) \times (B_i \cap B) \cup \cup_{i=1}^\infty \cap \cap_{i=1}^\infty (A_i \times B_i)' \\
= \;& \cup_{i=1}^\infty (A_i \cap A) \times (B_i \cap B) \cup \cup_{i=1}^\infty (A_i \cap A) \times (B_i' \cap B) \cup \\
& \cup_{i=1}^\infty (A_i' \cap A) \times (B_i \cap B) \cup \cup_{i=1}^\infty (A_i' \cap A) \times (B_i' \cap B)
\end{aligned}
$$

The last three pieces cannot contain $A \times B$ and so we must have $A \times B = \cup_{i=1}^\infty (A_i \cap A) \times (B_i \cap B)$. Thus we have by monotonicity of the outer measure

$$\mu^* \times \nu^* \left(\cup_{i=1}^\infty A_i \times B_i \right) \geq \mu^* \times \nu^* \left(\cup_{i=1}^\infty (A_i \cap A) \times (B_i \cap B) = A \times B \right)$$

Thus, for all covers,

$$\sum_{i=1}^\infty \mu^*(A_i)\,\nu^*(B_i) \geq \mu^*(A)\,\nu^*(B)$$

which immediately implies the other inequality. We conclude $\mu^* \times \nu^*(A \times B) = \mu^*(A)\,\nu^*(B)$. ∎

To prove $\mu^* \times \nu^*$ is a regular measure, we can use Lemma 11.5.2. We just need to show the sets of the form A and B μ^* and ν^* satisfy the Caratheodory condition for the product outer measure. If that is true, they will be measurable with respect to the product outer measure. The lemma then tells us the product measure is regular.

Theorem 18.2.2 The Product Outer Measure is a Regular Measure

> The product measure $\mu^* \times \nu^*$ is a regular measure.

Proof 18.2.2
Recall to show the product measure is regular, it is enough to show each $A \times B$ for $A \in \mathcal{M}_\mu$ and $B \in \times \mathcal{M}_\nu$ satisfies the Caratheodory condition. Hence we must show for any S from $X \times Y$ we have

$$(\mu^* \times \nu^*)(S) \geq (\mu^* \times \nu^*)((S) \cap (A \times B)) + (\mu^* \times \nu^*)((S) \cap (A \times B)')$$

Consider any cover of S: i.e. $S \subset \cup_{i=1}^{\infty} U_i \times V_i$ for $U_i \in \mathcal{M}_\mu$ and $V_i \in \times \mathcal{M}_\nu$. Then we can write this as

$$
\begin{aligned}
S \ &= \ (S) \cap (A \times B) \cup (S) \cap (A \times B)' \\
&\subset \ (A \times B) \cap (\cup_{i=1}^{\infty} U_i \times V_i) \ \cup \ (A \times B) \cap (\cup_{i=1}^{\infty} U_i \times V_i)' \\
&= \ \cup_{i=1}^{\infty}(U_i \cap A) \times (V_i \cap B) \ \cup \ \cup_{i=1}^{\infty} \cap \cap_{i=1}^{\infty}(U_i \times V_i)' \\
&= \ \cup_{i=1}^{\infty}(U_i \cap A) \times (V_i \cap B) \ \cup \ \cup_{i=1}^{\infty}(U_i \cap A) \times (V_i' \cap B) \ \cup \\
& \quad \ \cup_{i=1}^{\infty}(U_i' \cap A) \times (V_i \cap B) \ \cup \ \cup_{i=1}^{\infty}(U_i' \cap A) \times (V_i' \cap B)
\end{aligned}
$$

Let $\cup_{i=1}^{\infty} U_i^n \times V_i^n$ be a minimizing sequence for the infimum which defines $(\mu^* \times \nu^*)(S)$. Then

$$
(\mu^* \times \nu^*)(S) \ = \ \lim_{n \to \infty} \sum_{i=1}^{\infty} \mu^*(U_i^n)\, \nu^*(V_i^n)
$$

where we know for each n

$$
\begin{aligned}
S \ &\subset \ (A \times B) \cap (\cup_{i=1}^{\infty} U_i^n \times V_i^n) \ \cup \ (A \times B)' \cap (\cup_{i=1}^{\infty} U_i^n \times V_i^n) \\
&= \ \cup_{i=1}^{\infty}(U_i^n \cap A) \times (V_i^n \cap B) \ \cup \ \cup_{i=1}^{\infty}(U_i^n \cap A') \times (V_i^n \cap B) \ \cup \\
& \quad \ \cup_{i=1}^{\infty}(U_i^n \cap A) \times (V_i^n \cap B') \ \cup \ \cup_{i=1}^{\infty}(U_i^n \cap A') \times (V_i^n \cap B')
\end{aligned}
$$

and

$$
\begin{aligned}
(\mu^* \times \nu^*)(S) = \\
\lim_{n \to \infty} \Bigg\{ \sum_{i=1}^{\infty} \mu^*(U_i^n \cap A)\, \nu^*(V_i^n \cap B) &+ \sum_{i=1}^{\infty} \mu^*(U_i^n \cap A) \times \nu^*(V_i^n \cap B') + \\
\sum_{i=1}^{\infty} \mu^*(U_i^n \cap A')\, \nu^*(V_i^n \cap B) &+ \sum_{i=1}^{\infty} \mu^*(U_i^n \cap A')\, \nu^*(V_i^n \cap B') \Bigg\}
\end{aligned}
$$

We can rewrite this as

$$
\begin{aligned}
(\mu^* \times \nu^*)(S) = \\
\lim_{n \to \infty} \Bigg\{ \sum_{i=1}^{\infty} \Big(\mu^*(U_i^n \cap A)\, \nu^*(V_i^n \cap B) &+ \mu^*(U_i^n \cap A)\, \nu^*(V_i^n \cap B') + \\
\mu^*(U_i^n \cap A')\, \nu^*(V_i^n \cap B) &+ \mu^*(U_i^n \cap A')\, \nu^*(V_i^n \cap B') \Big) \Bigg\} = \\
\lim_{n \to \infty} \Bigg\{ \sum_{i=1}^{\infty} \Big((\mu^*(U_i^n \cap A) &+ \mu^*(U_i^n) \cap A')) \, (\nu^*(V_i^n \cap B) + \nu^*(V_i^n \cap B')) \Big) \Bigg\}
\end{aligned}
$$

But these are all measurable sets and so we can write this as

$$
(\mu^* \times \nu^*)(S) \ = \ \lim_{n \to \infty} \sum_{i=1}^{\infty} \mu^*(U_i^n)\, \nu^*(V_i^n)
$$

Now to finish, let

$$
\Theta \ = \ (\mu^* \times \nu^*)((S) \cap (A \times B)) \ + \ (\mu^* \times \nu^*)((S) \cap (A \times B)')
$$

Then we have

$$\mu^* \times \nu^*(S \cap (A \times B)) \ \leq \ (\mu^* \times \nu^*)\left(\bigcup_{i=1}^{\infty}\left(U_i^n \times V_i^n\right) \cap \left(A \times B\right)\right)$$

$$(\mu^* \times \nu^*)(S \cap (A \times B')) \ \leq \ (\mu^* \times \nu^*)\left(\bigcup_{i=1}^{\infty}\left(U_i^n \times V_i^n\right) \cap \left(A \times B'\right)\right)$$

$$(\mu^* \times \nu^*)(S \cap (A' \times B)) \ \leq \ (\mu^* \times \nu^*)\left(\bigcup_{i=1}^{\infty}\left(U_i^n \times V_i^n\right) \cap \left(A' \times B\right)\right)$$

$$(\mu^* \times \nu^*)(S \cap (A' \times B')) \ \leq \ (\mu^* \times \nu^*)\left(\bigcup_{i=1}^{\infty}\left(U_i^n \times V_i^n\right) \cap \left(A' \times B'\right)\right)$$

By the countable subadditivity of the product outer measure, we then have

$$\Theta = (\mu^* \times \nu^*)((S \times T) \cap (A \times B)) + (\mu^* \times \nu^*)((S \times T) \cap (A \times B)') \leq$$

$$\sum_{i=1}^{\infty}\Big\{\mu^*(U_i^n \cap A)\,\nu^*(V_i^n \cap B) + \mu^*(U_i^n \cap A)\,\nu^*(V_i^n \cap B') +$$

$$\mu^*(U_i^n \cap A')\,\nu^*(V_i^n \cap B) + \mu^*(U_i^n \cap A')\,\nu^*(V_i^n \cap B')\Big\}$$

$$= \sum_{i=1}^{\infty}\Big\{\Big(\mu^*(U_i^n \cap A) + \mu^*(U_i^n \cap A')\Big)\Big(\nu^*(V_i^n \cap B) + \nu^*(V_i^n \cap B')\Big)\Big\}$$

$$= \sum_{i=1}^{\infty}\mu^*(U_i^n)\,\nu^*(V_i^n)$$

Taking the limit as $n \to \infty$, we see $\Theta \leq \mu^* \times \nu^*(S)$. This shows the product measure is regular. ∎

Comment 18.2.1 *Note this result is true whether or not the outer measures μ^* and ν^* are regular.*

We already know since the outer measure is regular that we can find a measurable cover, but let's show you how to construct as it is a nice argument.

Theorem 18.2.3 The Existence of a Measurable Cover for a Set in $X \times Y$

Let S be in $X \times Y$ and let μ^* be an outer measure on X and ν^* an outer measure on Y and $\mu^* \times \nu^*$ the corresponding outer measure. Then there is a $\mu^* \times \nu^*$ measurable set F so that $\mu^* \times \nu^*(S) = \mu^* \times \nu^*(F)$ of the form $F = \cap_{j=1}^{\infty}V_j$ with $V_j = \cup_{i=1}^{\infty}A_i^j \times B_i^j$ with $S \subset V_j$ for all j.

Proof 18.2.3
By the definition of the outer measure, there is a sequence of sets $V_j = \cup_{i=1}^{\infty}A_i^j \times B_i^j$ with

$$S \ \subset \ \cup_{i=1}^{\infty}A_i^j \times B_i^j, \quad A_i^j \in \mathcal{M}_\mu, \ B_i^j \in \mathcal{M}_\nu$$

$$\mu^* \times \nu^*(S) \ \leq \ \sum_{i=1}^{\infty}\mu^*(A_i^j) \times \nu^*(B_i^j) < \mu^* \times \nu^*(S) + \frac{1}{j}$$

Let $W = \cap_{j=1}^{\infty} V_j$. This is a $\mu^* \times \nu^*$ measurable set and since the sequence $\{\cap_{j=1}^{n} V_j\}$ is a decreasing sequence of measurable sets with $S \subset W \subset \cap_{j=1}^{n} V_j$ we know for all n

$$\mu^* \times \nu^*(S) \quad \leq \quad \mu^* \times \nu^*(W) \leq \mu^* \times \nu^* \left(\cap_{j=1}^{n} V_j\right) \leq \mu^* \times \nu^*(S) + \frac{1}{n}$$

Hence, by Lemma 9.2.2,

$$\mu^* \times \nu^*(S) \quad \leq \quad \mu^* \times \nu^*(W) = \lim_{n\to\infty} \mu^* \times \nu^* \left(\cap_{j=1}^{n} V_j\right) \leq \mu^* \times \nu^*(S) + \frac{1}{n} = \mu^* \times \nu^*(S)$$

Hence, W is the measurable cover we seek. $\blacksquare$

Theorem 18.2.4 Representing the Product Measure as an Iterated Integral

$$(\mu^* \times \nu^*)(S) \quad = \quad \inf_{S \subset W} \{\rho_{XY}(W)\}, \quad (\mu^* \times \nu^*)(S) = \inf_{S \subset W} \{\rho_{YX}(W)\}$$

where

$$\rho_{XY}(S) \quad = \quad \int_X \left\{\int_Y I_S(x,y)d\nu\right\} d\mu$$

$$\rho_{YX}(S) \quad = \quad \int_Y \left\{\int_X I_S(x,y)d\mu\right\} d\nu$$

Proof 18.2.4
We define for any set $S \subset X \times Y$ of the form $S = \cup_{i=1}^{\infty} A_i \times B_i$ where A_i is μ^* measurable and B_i is ν^* measurable

$$\rho_{XY}(S) \quad = \quad \int_X \left\{\int_Y I_S(x,y)d\nu\right\} d\mu$$

$$\rho_{YX}(S) \quad = \quad \int_Y \left\{\int_X I_S(x,y)d\mu\right\} d\nu$$

If A is μ^* measurable and B is ν^* measurable, an easy calculation shows

$$\int_X I_{A \times B}(x,y)d\mu \quad = \quad \mu(A)I_B(y)$$

$$\int_Y I_{A \times B}(x,y)d\nu \quad = \quad \nu(B)I_A(x)$$

$$\rho_{XY}(A \times B) \quad = \quad \int_X \left\{\int_Y I_{A \times B}(x,y)d\nu\right\} d\mu$$

$$= \quad \nu(B) \int_X I_A(x)d\mu = \mu(A)\nu(B) = (\mu^* \times \nu^*)(A \times B)$$

$$\rho_{YX}(A \times B) \quad = \quad \int_Y \left\{\int_X I_{A \times B}(x,y)d\mu\right\} d\nu$$

$$= \quad \mu(A) \int_Y I_B(y)d\nu = \mu(A)\nu(B) = (\mu^* \times \nu^*)(A \times B)$$

If $S = \cup_{i=1}^{\infty} A_i \times B_i$ where each A_i and B_i are μ^* and ν^* measurable, we can rewrite this as a *countable* **disjoint** *union of sets with $A_i' \times B_i'$ with A_i' and B_i' μ^* and ν^* measurable respectively. Then we have (for convenience we are just writing $\cup$ instead of $\cup_{i=1}^{\infty}$ here)*

$$
\begin{aligned}
\int_X I_{\cup A_i \times B_i}(x,y)d\mu &= \int_X I_{\cup A_i' \times B_i'}(x,y)d\mu \\
&= \sum_{i=1}^{\infty} \int_X I_{A_i' \times B_i'}(x,y)d\mu = \sum_{i=1}^{\infty} \mu(A_i') I_{B_i'}(y) \\
\int_Y I_{\cup A_i \times B_i}(x,y)d\nu &= \int_Y I_{\cup A_i' \times B_i'}(x,y)d\nu \\
&= \sum_{i=1}^{\infty} \int_X I_{A_i' \times B_i'}(x,y)d\nu = \sum_{i=1}^{\infty} \nu(B_i') I_{A_i'}(x) \\
\rho_{YX}(\cup A_i \times B_i) &= \int_Y \left\{ \int_X I_{\cup A_i \times B_i}(x,y)d\mu \right\} d\nu \\
&= \int_Y \left\{ \sum_{i=1}^{\infty} \mu(A_i') I_{B_i'}(y) \right\} d\nu = \sum_{i=1}^{\infty} \nu(B_i')\mu(A_i') \\
\rho_{XY}(\cup A_i \times B_i) &= \int_X \left\{ \int_Y I_{\cup A_i \times B_i}(x,y)d\nu \right\} d\mu \\
&= \int_X \left\{ \sum_{i=1}^{\infty} \nu(B_i') I_{A_i'}(x) \right\} d\mu = \sum_{i=1}^{\infty} \nu(B_i')\mu(A_i')
\end{aligned}
$$

Claim *If $S \subset X \times Y$, then for any $W = \cup_{i=1}^{\infty} A_i \times B_i$ with A_i μ^* measurable and B_i ν^* measurable*

$$
(\mu^* \times \nu^*)(S) = \inf_{S \subset W}\{\rho_{XY}(W)\}, \quad (\mu^* \times \nu^*)(S) = \inf_{S \subset W}\{\rho_{YX}(W)\}
$$

From our work above, we note

$$
\begin{aligned}
\rho_{XY}(\cup_{i=1} A_i \times B_i) &\leq \sum_{i=1}^{\infty} \rho_{XY}(A_i \times B_i) = \sum_{i=1}^{\infty} \mu(A_i)\nu(B_i) \\
\rho_{YX}(\cup_{i=1} A_i \times B_i) &\leq \sum_{i=1}^{\infty} \rho_{YX}(A_i \times B_i) = \sum_{i=1}^{\infty} \mu(A_i)\nu(B_i)
\end{aligned}
$$

By definition, this says

$$
\begin{aligned}
\inf_{S \subset W} \rho_{XY}(W) &\leq \inf_{S \subset W} \sum_{i=1}^{\infty} \mu(A_i)\nu(B_i) = (\mu^* \times \nu^*)(S) \\
\inf_{S \subset W} \rho_{YX}(W) &\leq \inf_{S \subset W} \sum_{i=1}^{\infty} \mu(A_i)\nu(B_i) = (\mu^* \times \nu^*)(S)
\end{aligned}
$$

We also know for any W, we can rewrite it as a disjoint union and obtain

$$
\begin{aligned}
\rho_{YX}(W) &= \sum_{i=1}^{\infty} \nu(B_i')\mu(A_i') \geq (\mu^* \times \nu^*)(S) \\
\rho_{XY}(W) &= \sum_{i=1}^{\infty} \nu(B_i')\mu(A_i') \geq (\mu^* \times \nu^*)(S)
\end{aligned}
$$

394 BASIC ANALYSIS IV: MEASURE THEORY AND INTEGRATION

This implies

$$\inf_{S \subset W} \rho_{XY}(W) \;\; \geq \;\; (\mu^* \times \nu^*)(S), \quad \inf_{S \subset W} \rho_{YX}(W) \geq (\mu^* \times \nu^*)(S)$$

which proves the claim. ∎

We can then find a measurable cover for the set $S \subset X \times Y$ in the context of the ρ_{XY} and ρ_{YX} setting.

Theorem 18.2.5 The Existence of a Measurable Cover for a Set in $X \times Y$ using Iterated Integrals

> *This is very similar to the argument we used to prove Theorem 18.2.3 and so we leave it for you. The idea is this. For $S \subset X \times Y$, there is a set $F = \cap j = 1^\infty \cup_{A_i^j} \times B_i^j$ with each A_i^j and B_i^j μ^* and ν^* measurable respectively with $S \subset \cup_{A_i^j} \times B_i^j$ for all j and $(\mu^* \times \nu^*)(S) = \rho_{XY}(F)$. Further, $(\mu^* \times \nu^*)(F \setminus S) = 0$. Also, there is a set G of this type with $(\mu^* \times \nu^*)(S) = \rho_{YX}(G)$ and $(\mu^* \times \nu^*)(G \setminus S) = 0$.*

Theorem 18.2.6 A Simple Fubini Theorem

> *Let μ^* be an outer measure on X and ν^* an outer measure on Y. Assume the set S is $\mu^* \times \nu^*$ measurable and σ finite with respect to this measure. Then the set $S_y = \{x \mid (x,y) \in S\}$ is μ^* measurable and $S_x = \{y \mid (x,y) \in S\}$ is ν^* measurable and*
>
> $$\int_{X \times Y} I_S d(\mu^* \times \nu^a st) \;\; = \;\; \int_X \left\{ \int_Y I_S(x,y) d\nu \right\} d\mu = \int_Y \left\{ \int_X I_S(x,y) d\mu \right\} d\nu$$
> $$= \;\; \int_X \nu(S_x) d\mu = \int_Y \nu(S_y) d\nu$$

Proof 18.2.5
Let's do the argument using ρ_{XY}. The argument using ρ_{YX} is similar. We first let $S \subset X \times Y$ be $\mu^* \times \nu^*$ measurable with finite measure. Then there is a set W of the form $W = \cap_j \cup_u A_i^j \times B_i^j$ with each A_i^j and B_i^j μ^* and ν^* measurable respectively with $S \subset R$ and $(\mu^* \times \nu^*)(S) = \rho_{XY}(W)$ and $(\mu^* \times \nu^*)(W \setminus S) = 0$.

Note $S_x = \{y|(x,y) \subset S\} \subset W_x = \{y|(x,y) \subset W\}$. This says

$$\{x\} \times S_x = (\{x\} \times \{\cap_j \cup_i B_i^y\}) \cap S \cup \left(\{x\} \times \left\{ \cap_j \cup_i B_i^j \right\} \right) \cap S^C$$

But $(\{x\} \times \{\cap_j \cup_i B_i^j\}) \cap S^C$ is a subset of $W \setminus S$ which is a set of product measure zero. This set is therefore is a subset of a set of measure zero and hence it is measurable with measure zero also. We also see the set $(\{x\} \times \{\cap_j \cup_i B_i^y\}) \cap S$ is $\mu^* \times \nu^*$ measurable which says immediately that $S_x = \cap_j \cup_i B_i^y \cap S$ is ν^* measurable and $\nu^*(S_x) = \nu^*(W_x)$. We can do a similar argument for S_y.

Then

$$(\mu^* \times \nu^*)(S) \;\; = \;\; \rho_{XY}(W) = \int_X \left\{ \int_Y I_R(x,y) d\nu \right\} d\mu$$
$$= \;\; \int_X \left\{ \int_Y I_S(x,y) d\nu \right\} d\mu + \int_X \left\{ \int_Y I_{W \setminus S}(x,y) d\nu \right\} d\mu$$

FUBINI TYPE RESULTS

But

$$\int_Y I_{W \setminus S}(x,y)d\nu \;\; = \;\; \int_Y I_{\{y|(x,y) \in W \setminus S\}} = 0$$

because $\{y|(x,y) \in W \setminus S\}$ *is a subset of a set of product measure* 0 *and hence it is measurable. Thus we have*

$$\int_{X \times Y} I_S d(\mu^* \times \nu^a st) \;\; = \;\; (\mu^* \times \nu^*)(S)$$
$$= \;\; \int_X \left\{ \int_Y I_S(x,y)d\nu \right\} d\mu$$

The argument for the σ finite case is left for you. ■

Here is a reasonable version of iterated integration in this more general case.

Theorem 18.2.7 The General Fubini Theorem

> *Let μ^* be an outer measure on X and ν^* an outer measure on Y. Assume f is measurable with respect to this measure and $\{(x,y) \in X \times Y | f(x,y) \neq 0\}$ is σ finite. Also assume $\int_{X \times Y} f(x,y)d(\mu^* \times \nu^*)$ exists. It can be finite or infinite.*
>
> *Then the function $g(x) = \int_Y f(x,y)d\nu$ is μ^* measurable on X and the function $h(y) = \int_X f(x,y)d\mu$ is ν^* measurable on Y and*
>
> $$\int_{X \times Y} f(x,y)d(\mu^* \times \nu^*) \;\; = \;\; \int_Y \left\{ \int_X f(x,y)d\mu \right\} d\nu = \int_X \left\{ \int_Y f(x,y)d\nu \right\} d\mu$$

Proof 18.2.6
We will leave the proof of this to you. Think of it as an end of the book examination! The proof is in several steps:

- **Step 1**: *Apply Theorem 18.2.6 for the case $f(x,y) = I_S(x,y)$ first.*

- **Step 2**: *Now extend to simple functions.*

- **Step 3**: *Now approximate f by simple functions and use convergence theorems.*

- **Step 4**: *For the most general f, recall $f = f^+ - f^-$ which has been handled in **Step 3**.*

■

Homework

Exercise 18.2.1 *If $W = \cup_{i=1}^\infty A_i \times B_i$ where each A_i and B_i are μ^* and ν^* measurable respectively, prove we can rewrite this as a countable disjoint union of sets $\cup_{i=1}^\infty A_i' \times B_i'$ with A_i' and B_i' μ^* and ν^* measurable respectively.*

Exercise 18.2.2 *Let $S \subset X \times Y$. Prove there is a set $F = \cap j = 1^\infty \cup_{A_i^j} \times B_i^j$ with each A_i^j and B_i^j μ^* and ν^* measurable respectively with $S \subset \cup_{A_i^j} \times B_i^j$ for all j and $(\mu^* \times \nu^*)(S) = \rho_{XY}(F)$. Further, $(\mu^* \times \nu^*)(F \setminus S) = 0$. Also, there is a set G of this type with $(\mu^* \times \nu^*)(S) = \rho_{YX}(G)$ and $(\mu^* \times \nu^*)(G \setminus S) = 0$. This is the proof of Theorem 18.2.5.*

396 BASIC ANALYSIS IV: MEASURE THEORY AND INTEGRATION

Exercise 18.2.3 *Give the argument for the σ finite case in the proof of Theorem 18.2.6.*

Exercise 18.2.4 *Prove* **Step 1** *of Theorem 18.2.7: Apply Theorem 18.2.6 for the case $f(x,y) = I_S(x,y)$*

Exercise 18.2.5 *Prove* **Step 2** *of Theorem 18.2.7: extend to simple functions.*

Exercise 18.2.6 *Prove* **Step 3** *of Theorem 18.2.7: approximate f by simple functions and use convergence theorems.*

Exercise 18.2.7 *Prove* **Step 4** *of Theorem 18.2.7: use $f = f^+ - f^-$ and apply* **Step 3***.*

Exercise 18.2.8 *Show how to use Theorem 18.2.7 for the case of $f(x,y)$ is continuous on $[0,1] \times [0,1]$ using the product measure $d(\mu \times \mu)$ where μ is Lebesgue measure. Compare this to what we do in the Riemann case.*

Exercise 18.2.9 *Let $S = \{(x,y) \in \Re^2 : x^2 + y^2 \leq 1\}$ and assume $f(x,y)$ is continuous on S. Show how to use Theorem 18.2.7 for $F(x,y) = f(x,y)$ if $(x,y) \in S$ and $f(x,y) = \infty$ off S using the product measure $d(\mu \times \mu)$ where μ is Lebesgue measure. Compare this to what we do in the Riemann case.*

Chapter 19

Differentiation

We will discuss the kinds of properties a function f needs to have so that we have a Fundamental Theorem of Calculus type result: $f(b) - f(a) = \int_a^b f' d\mu$ in our setting of measures on the real line.

19.1 Absolutely Continuous Functions

Definition 19.1.1 Absolute Continuity of Functions

> Let $[a, b]$ be a finite interval in $\Re$. $f : [a, b] \to \Re$ is absolutely continuous if for each $\epsilon > 0$, there is a $\delta > 0$ such that if $\{[a_n, b_n]\}$ is any finite or countable collection of non-overlapping closed intervals in $[a, b]$ with $\sum_k (b_k - a_k) < \delta$ then $\sum_k |f(b_k) - f(a_k)| < \epsilon$.

Theorem 19.1.1 Properties of Absolutely Continuous Functions

> Let f be absolutely continuous on the finite interval $[a, b]$. Then
>
> 1. f is continuous on $[a, b]$,
>
> 2. f is of bounded variation on $[a, b]$,
>
> 3. If E is a subset of $\Re$ with Lebesgue Measure 0, then $f(E)$ is Lebesgue measurable also with measure 0; i.e. $\mu(E) = 0 \Rightarrow \mu(f(E)) = 0$.

Proof 19.1.1
Condition (1) is clear.

To prove (2), choose a $\delta > 0$ so that if $\{[a_k, b_k]\}$ is any collection of non-overlapping closed intervals in $[a, b]$ with $\sum_k (b_k - a_k) < \delta$, then $\sum_k |f(b_k) - f(a_k)| < 1$. If $[c, d]$ is any interval in $[a, b]$ with $d - c < \delta$, then for any non-overlapping sequence of intervals $([\alpha_n, \beta_n])$ inside $[c, d]$ whose summed length is less than δ, we must have $\sum |f(\alpha_n) - f(\beta_n)| < 1$. It then follows that $V(f, c, d) \leq 1$.

Now choose the integer N so that $N > \frac{b-a}{\delta}$. Partition $[a, b]$ into N intervals I_1 to I_N. The variation of f on each I_k is less than 1; hence, $V(f, a, b) \leq N < \infty$. This proves condition (2).

To prove (3), let $\epsilon > 0$ be given. Choose $\delta > 0$ so that if $\{[c_k, d_k]\}$ is any collection of non-overlapping closed intervals in $[a, b]$ with $\sum_k (d_k - c_k) < \delta$, then $\sum_k |f(d_k) - f(c_k)| < \epsilon$. By the

398   BASIC ANALYSIS IV: MEASURE THEORY AND INTEGRATION

infimum tolerance lemma, there is an open set G so that $0 = \mu(E) \leq \mu(G) + \delta$. Since G is open, we also know we can write G as a finite or countable union of disjoint open intervals (a_k, b_k); i.e. $G = \cup_k(a_k, b_k)$. Then

$$f(E) \;\subseteq\; f(G) \subseteq f\left(\cup_k[a_k, b_k]\right) \subseteq \bigcup_k [f(u_k), f(v_k)]$$

where u_k and v_k are the points in $[a_k, b_k]$ where f achieves its minimum and maximum, respectively. Hence, $\mu^*(f(E)) \leq \sum_k \mu^*([f(u_k), f(v_k)])$ or $\mu^*(f(E)) \leq \sum_k([f(v_k) - f(u_k)])$. However, the points u_k and v_k determine a closed subinterval of $[a_k, b_k]$ with

$$\sum_k |v_k - u_k| \;\leq\; \sum_k |b_k - a_k| < \delta$$

Since the intervals (a_k, b_k) are disjoint, the intervals formed by u_k and v_k are non-overlapping. Thus, we conclude $\sum_k(f(v_k) - f(u_k)) < \epsilon$. This tells us $\mu^*(f(E)) < \epsilon$. But since ϵ was arbitrary, we see $\mu^*(f(E)) = 0$; thus, $f(E)$ is measurable and has measure 0.   ∎

Homework

Exercise 19.1.1 *Prove the function $f(x) = x\sin(1/x)$ is not absolutely continuous on $[0, 1]$. To do this, choose the sequence of points (x_n) and (y_n) by $\sin(1/x_n) = 1$ and $\sin(1/y_n) = -1$. Then show the $\sum_{k=1}^n |x_n - y_n|$ acts like $\sum 1/k^2$, which converges and the sum $\sum_{k=1}^n |f(x_n) - f(y_n)|$ acts like $\sum 1/k$, which diverges. This will show the function is not absolutely continuous.*

Now let's revisit some Cantor set things.

Exercise 19.1.2 *The mapping $g : [0, 1] \to [0, 1]$ by $g(x) = (\Psi(x) + x)/2$ where Ψ is the Cantor singular function built from the middle thirds Cantor set C is quite nice. We know g is strictly increasing and continuous from $[0, 1]$ onto $[0, 1]$. In an exercise in Chapter 13 we showed $g(\sum_{j=1}^\infty x_j\, r_j) = \sum_{j=1}^\infty x_j\, r_j'$ where $r_j' = (1/2^j + r_j)/2$ where $\sum_{j=1}^\infty x_j\, r_j$ is a representation of a point in C and $\sum_{j=1}^\infty x_j\, r_j'$ is a representation of a point in another Cantor set C'. We proved the Lebesgue measure of $C' = g(C) = 1/2$ even though C has measure zero. Prove this shows the g is not absolutely continuous.*

Exercise 19.1.3 *We showed that $g(x) = \sin(1/x)$ is not of bounded variation in a comment in Chapter 3. Prove this shows g is not absolutely continuous.*

Exercise 19.1.4 *Prove if f is Lipschitz with constant $L > 0$ on $[a, b]$, then f is absolutely continuous.*

Exercise 19.1.5 *Prove if f has a continuous bounded derivative on $[a, b]$, then f is absolutely continuous.*

19.2   Lebesgue-Stieltjes Measures and Absolutely Continuous Functions

If f is continuous and non-decreasing, there is a nice connection between f and the associated Lebesgue-Stieltjes measure μ_f. First, we explore a relationship between the Lebesgue-Stieltjes outer measure and Lebesgue outer measure.

Theorem 19.2.1 $\mu_f^*(E) = \mu^*(f(E))$

DIFFERENTIATION

> *Let f be continuous and non-decreasing on $\Re$ and let μ_f^* be the associated Lebesgue-Stieltjes outer measure. For all sets E in $\Re$, $\mu_f^*(E) = \mu^*(f(E))$.*

Proof 19.2.1

Let E be in $\Re$. Let G be an open set that contains $f(E)$. We know G can be written as a countable disjoint union of open intervals J_n. Since f is continuous and non-decreasing, we must have $I_n = f^{-1}(J_n)$ is an interval as well. Letting $I_n = (a_n, b_n)$ gives $J_n = (f(a_n), f(b_n))$. Noting $E \subseteq \cup_n I_n$ and $\mu(G) = \sum_n \mu(J_n)$, we find

$$\begin{aligned}
\mu_f^*(E) &\leq \mu_f^*(\cup_n I_n) = \mu_f(\cup_n I_n) \\
&\leq \sum_n \mu_f(I_n) = \sum_n \mu(J_n) \\
&= \mu(G)
\end{aligned}$$

This is true for all such open sets G. We therefore conclude $\mu_f^*(f(E)) \leq \mu^*(f(E))$ by Theorem 12.3.1.

The above immediately tells us that if $\mu_f^*(E) = \infty$, we also have $\mu^*(f(E)) = \infty$ and the result holds. We may thus safely assume that $\mu_f^*(E)$ is finite for the rest of the argument. Let $\epsilon > 0$ be given. Let $((a_n, b_n])$ be a collection of half open intervals whose union covers E with $\sum_n (f(b_n) - f(a_n)) < \mu_f^*(E) + \epsilon$. Let $J_n = f((a_n, b_n])$. Again, since f is continuous and non-decreasing, each interval J_n can be written as $J_n = (f(a_n), f(b_n)]$. Since $f(E) \subseteq \cup_n J_n$, we must have

$$\begin{aligned}
\mu^*(f(E)) &\leq \mu^*(\cup_n J_n) \\
&\leq \sum_n (f(b_n) - f(a_n)) < \mu_f^*(E) + \epsilon
\end{aligned}$$

Since ϵ is arbitrary, we see $\mu^*(f(E)) \leq \mu_f^*(E)$. This second inequality completes the argument. ∎

Homework

Exercise 19.2.1 *Let g be defined by*

$$g(x) = \begin{cases} 2x, & 0 \leq x < 1 \\ 5x^2 + 1, & 1 \leq x \leq 2 \end{cases}$$

Let μ_g^* be the associated Lebesgue-Stieltjes outer measure. For all sets E in $[0,2]$, we know $\mu_g^*(E) = \mu^*(g(E))$.

- *Is g absolutely continuous?*

- *Find $g([0,2])$.*

- *Compute $\mu_g^*([0,2])$.*

Exercise 19.2.2 *Let g be defined by*

$$g(x) = \begin{cases} x + 4, & -1 \geq x < 1 \\ x^2 + 7, & 1 \geq x < 2 \\ 2x^3, & 2 \geq x \leq 3 \end{cases}$$

Let μ_g^* be the associated Lebesgue-Stieltjes outer measure. For all sets E in $[-1, 3]$, we know $\mu_g^*(E) = \mu^*(g(E))$.

- *Is g absolutely continuous?*

- *Find $g([-1, 3])$.*

- *Compute $\mu_g^*([-1, 3])$.*

Exercise 19.2.3 *Let g_n, for $n \geq 2$, be defined by*

$$g_n(x) = \begin{cases} 1, & x = 1 \\ 1/(k+1), & 1/(k+1) \leq x < 1/k, \ 1 \leq k < n \\ 0, & 0 \leq x < 1/n \end{cases}$$

Let $\mu_{g_n}^*$ be the associated Lebesgue-Stieltjes outer measure. For all sets E in $[0,1]$, we know $\mu_{g_n}^*(E) = \mu^*(g_n(E))$.

- *Is g_n absolutely continuous?*

- *Find $g_n([0, 1])$.*

- *Compute $\mu_g^*([0, 1])$.*

Exercise 19.2.4 *Let g be the pointwise limit of the sequence of functions (g_n) for $n \geq 2$ defined by*

$$g_n(x) = \begin{cases} 1, & x = 1 \\ 1/(k+1), & 1/(k+1) \leq x < 1/k, \ 1 \leq k < n \\ 0, & 0 \leq x < 1/n \end{cases}$$

Let μ_g^* be the associated Lebesgue-Stieltjes outer measure. For all sets E in $[0, 1]$, we know $\mu_g^*(E) = \mu^*(g(E))$.

- *Is g absolutely continuous?*

- *Find $g([0, 1])$.*

- *Compute $\mu_g^*([0, 1])$.*

There is a very nice connection between a continuous non-decreasing f and its associated Lebesgue-Stieltjes measure μ_f.

Theorem 19.2.2 f is Absolutely Continuous if and only if $\mu_f << \mu$

A continuous non-decreasing function f is absolutely continuous on $[a, b]$ if and only if its associated Lebesgue-Stieltjes measure μ_f is absolutely continuous with respect to Lebesgue measure μ.

Proof 19.2.2
Let f be continuous and non-decreasing on $[a, b]$. By Theorem 19.2.1, we have $\mu_f^*(E) = \mu^*(f(E))$ for all subsets E in $[a, b]$.

If we assume f is absolutely continuous on $[a, b]$, by Condition (3) of Theorem 19.1.1, when E has Lebesgue measure 0, we know $\mu(f(E)) = 0$ as well. Thus, $\mu^*(f(E)) = 0 = \mu_f^*(E)$. But then E is μ_f measurable with μ_f measure 0. This tells us $\mu_f \ll \mu$.

Conversely, if $\mu_f \ll \mu$, we can apply Lemma 16.4.3. Given $\epsilon > 0$, there is a $\delta > 0$, so that $\mu(E) < \delta$ implies $\mu_f(E) < \epsilon$. Hence, for any E which is a non-overlapping countable union of

DIFFERENTIATION

intervals $[a_n, b_n]$ with $\sum_n (b_n - a_n) < \delta$, we have $\sum_n (f(b_n) - f(a_n)) = \mu_f(E) < \epsilon$. This says f is absolutely continuous on $[a, b]$. ∎

Homework

Exercise 19.2.5 *Let g be defined by*

$$g(x) \;=\; \begin{cases} 2x, & 0 \le x < 1 \\ 5x^2 - 3, & 1 \le x \le 2 \end{cases}$$

- *Prove g absolutely continuous.*

- *Prove the Lebesgue - Stieltjes measure μ_g is absolutely continuous.*

Exercise 19.2.6 *Let g be defined by*

$$g(x) \;=\; \begin{cases} x + 4, & -1 \ge x < 1 \\ x^2 + 4, & 1 \ge x < 2 \\ 2x^3 - 8, & 2 \ge x \le 3 \end{cases}$$

- *Prove g absolutely continuous.*

- *Prove the Lebesgue - Stieltjes measure μ_g is absolutely continuous.*

Exercise 19.2.7 *Let $g(x) = x^3$ on $[-2, 6]$.*

- *Prove g absolutely continuous.*

- *Prove the Lebesgue - Stieltjes measure μ_g is absolutely continuous.*

Exercise 19.2.8 *Let $g(x) = \tanh(x)$ on $[-3, 1]$.*

- *Prove g absolutely continuous.*

- *Prove the Lebesgue - Stieltjes measure μ_g is absolutely continuous.*

19.3 Derivatives of Functions of Bounded Variation

We are going to look carefully at the behavior of monotone functions and their rates of change.

Definition 19.3.1 Derived Numbers

We say the extended real number α is a derived number for a function f at the point x_0 in $\mathrm{dom}(f)$ is there is a sequence of nonzero numbers $x_0 + h_n$ in $\mathrm{dom}(f)$ with $h_n \to 0$ so that

$$\lim_n \frac{f(x_0 + h_n) - f(x_0)}{h_n} \;=\; \alpha$$

We use the notation $Df(x)$ to denote the collection of all derived numbers for a function f at the point x in its domain.

If a function f is strictly increasing on $[a, b]$ we want to be able to look quantitatively how the outer measure *size* of E compares to $f(E)$ for any subset E. To do this, we need a technical tool called a *Vitali Cover*.

402   BASIC ANALYSIS IV: MEASURE THEORY AND INTEGRATION

Definition 19.3.2 Vitali Cover

> *Let $\mathcal{I}$ be a collection of non-degenerate closed intervals in $\Re$. Let E be a subset of $\Re$ and $\mathcal{V}$ be a sub-collection of $\mathcal{I}$. If for all $x \in E$ and $\epsilon > 0$, there is a $V \in \mathcal{V}$ so that $x \in V$ and $\mu(V) < \epsilon$, we say $\mathcal{V}$ is a Vitali Cover for E or a Vitali Covering for E.*

We can prove this important result.

Theorem 19.3.1 Vitali Covering Theorem

> *Let $\mathcal{V}$ be a Vitali Cover of a set E in $\Re$. The there is a countable collection (V_n) in $\mathcal{V}$ that is pairwise disjoint and $\mu(E \setminus \cup_n V_n) = 0$.*

Proof 19.3.1

First assume E is bounded. Let J be any open interval containing E and let $\mathcal{V}_0$ be those intervals in $\mathcal{V}$ that are contained in J. Then $\mathcal{V}_0$ is also a Vitali Cover for E. Let V_1 be chosen from $\mathcal{V}_0$. If $\mu(E \setminus V_1) = 0$, we have proven our conjecture. If not, we continue this process using induction. Choose V_2 this way. Let $F_1 = V_1$ and $G_1 = J \setminus F_1$. Then G_1 is open. Define the new collection $\mathcal{V}_1$ by

$$\mathcal{V}_1 \;\; = \;\; \{V \in \mathcal{V}_0 \, : \, V \subseteq G_1\}$$

Since by assumption $E \setminus V_1$ is not empty and $\mathcal{V}_0$ is a Vitali Cover for E, there must be sets in the family $\mathcal{V}_1$. Let

$$\mathcal{S}_1 \;\; = \;\; \sup \{\mu(V) \, : \, V \in \mathcal{V}_1\}$$

The members of a Vitali Cover are non-degenerate which implies $S_1 > 0$. Further, each member of $\mathcal{V}_0$ is in J which tells us S_1 is finite. Choose V_2 from $\mathcal{V}_1$ so that $\mu(V_2) > \frac{S_1}{2}$. Then $V_2 \subseteq G_1$ implying V_1 and V_2 are disjoint. If $\mu(E \setminus V_1 \cup V_2) = 0$ we are done. Otherwise, we continue.

We can now see how to do the induction. Suppose we have chosen the sets $\{V_1, V_2, \ldots, V_n\}$ so that they are pairwise disjoint and $\mu(E \setminus \cup_{i=1}^n V_i) \neq 0$. Then, let $F_n = \cup_{i=1}^n V_i$ and $G_n = J \setminus F_n$. Then G_n is open. Define the new collection $\mathcal{V}_n$ by

$$\mathcal{V}_n \;\; = \;\; \{V \in \mathcal{V}_0 \, : \, V \subseteq G_n\}$$

Since by assumption $E \setminus V_n$ is not empty and $\mathcal{V}_n$ is also a Vitali Cover for E, there must be sets in the family $\mathcal{V}_n$. Let

$$\mathcal{S}_n \;\; = \;\; \sup \{\mu(V) \, : \, V \in \mathcal{V}_n\}$$

The members of a Vitali Cover are non-degenerate which implies $S_n > 0$. Further, each member of $\mathcal{V}_0$ is in J which tells us S_n is finite. Choose V_{n+1} from $\mathcal{V}_n$ so that

$$\mu(V_{n+1}) > \frac{S_n}{2} \tag{19.1}$$

Then again, $V_{n+1} \subseteq G_n$ implying the new collection $\{V_1, V_2, \ldots, V_n, V_{n+1}\}$ is pairwise disjoint.

If this process terminates after a finite number of steps, we have proven the result. Otherwise, we construct a countable number of sets V_n which form a pairwise disjoint collection from $\mathcal{V}_0$. Let $S = \cup_n V_n$. We will show $\mu(E \setminus S) = 0$ proving the result.

DIFFERENTIATION 403

Each V_n has a midpoint. Let W_n be an interval with the same midpoint as V_n which is 5 times the length. Then $\mu(W_n) = 5\mu(V_n)$. Also, by construction $S = \cup_n V_n \subseteq E \subseteq J$ and so

$$\sum_n \mu(W_n) \;=\; 5 \sum_n \mu(V_n) \;\leq\; 5\,\mu(J) \tag{19.2}$$

Now let $x \in E \setminus S$. Then, $x \in \cap_n(E \cap V_n^C)$ implying $x \in E \cap E_n^C$ for all n. But since $E \subseteq J$, this tells us $x \in J \cap E_n^C$ always. From the definition of G_n, it follows that $x \in G_n$ for all n. We conclude $x \in \cap_n G_n$.

Now fix the positive integer i. Since x is in G_i which is open, there is a positive number r so that $B(x; r) \subseteq G_i$. From the definition of a Vitali Cover of E, it follows there is a non-degenerate closed interval V in $\mathcal{V}_0$ with $\mu(V) < \frac{r}{2}$. Hence, V is contained in G_i and so $V \cap V_i = \emptyset$.

Since $x \in S^C$, we see that V cannot be any of the intervals V_n we have constructed. Now all the intervals V_n are non-degenerate and all live in the bounded interval J. Thus, $\sum_i \mu(V_i) < \mu(J)$ which implies $\mu(V_i) \to 0$. From Equation 19.1, it follows that $S_n \to 0$ as well. Choose a value of N so that $S_N < \mu(V)$.
From the definition of S_N, we then have that V cannot be in G_N and so $V \cap F_N \neq \emptyset$.

Let $M = \inf\{j : V \cap F_j \neq \emptyset\}$. It is clear $M > i$. Hence, for this index M, $V \cap F_{M-1} = \emptyset$ implying $V \subseteq G_{M-1}$. Further, $V \cap F_M \neq \emptyset$ and this tells us $V \cap V_M \neq \emptyset$. From the definition of S_{M-1}, it then follows that $\mu(V) \leq S_{M-1} < 2\mu(V_m)$. Since W_M shares the same center as V_M with 5 times the length, this means $V \subseteq W_M$.

Finally, $M > i$, so $V \subseteq \cup_{j=i}^{\infty} W_j$. Since $x \in V$, this shows $x \in \cup_{j=i}^{\infty} W_j$. We conclude $E \setminus S \subseteq \cup_{j=1}^{\infty} W_j$. From Equation 19.2, we know the series $\sum_n \mu(W_n)$ converges. Given any positive ϵ, there is a P so that $\sum_{j=1}^{\infty} \mu(W_j) < \epsilon$ if $i > P$. Since $\mu(E \setminus S) \leq \sum_{j=1}^{\infty} \mu(W_j)$ for all i, we conclude $\mu(E \setminus S) = 0$.
The proof for the case that E is unbounded is left to you. ∎

Homework

Exercise 19.3.1 *Prove Theorem 19.3.1 for the case that E is unbounded.*

Exercise 19.3.2 *Let f be defined by*

$$f(x) \;=\; \begin{cases} 2x, & 0 \leq x < 1 \\ 3, & 1 = x \\ 5x^2 + 1, & 1 < x \leq 2 \end{cases}$$

Find $Df(x)$ on $[0, 2]$.

Exercise 19.3.3 *Let f be defined by*

$$f(x) \;=\; \begin{cases} 2, & x = -1 \\ x + 4, & -1 < x < 1 \\ 6, & 1 = x \\ x^2 + 7, & 1 < x < 2 \\ 12, & x = 2 \\ 2x^3, & 2 < x \leq 3 \end{cases}$$

Find $Df(x)$ on $[-1, 3]$.

404 BASIC ANALYSIS IV: MEASURE THEORY AND INTEGRATION

Exercise 19.3.4 *Let g be defined by*

$$g(x) = \begin{cases} 2x, & 0 \leq x < 1 \\ 5x^2 + 1, & 1 \leq x \leq 2 \end{cases}$$

Find $Dg(x)$ on $[0, 2]$.

Exercise 19.3.5 *Let g be defined by*

$$g(x) = \begin{cases} x + 4, & -1 \geq x < 1 \\ x^2 + 7, & 1 \geq x < 2 \\ 2x^3, & 2 \geq x \leq 3 \end{cases}$$

Find $Dg(x)$ on $[-1, 3]$.

Exercise 19.3.6 *Let g_n, for $n \geq 2$, be defined by*

$$g_n(x) = \begin{cases} 1, & x = 1 \\ 1/(k+1), & 1/(k+1) \leq x < 1/k, \ 1 \leq k < n \\ 0, & 0 \leq x < 1/n \end{cases}$$

Find $Dg_n(x)$ on $[0, 1]$.

Exercise 19.3.7 *Let g be the pointwise limit of the sequence of functions (g_n) for $n \geq 2$ defined by*

$$g_n(x) = \begin{cases} 1, & x = 1 \\ 1/(k+1), & 1/(k+1) \leq x < 1/k, \ 1 \leq k < n \\ 0, & 0 \leq x < 1/n \end{cases}$$

Find $Dg_n(x)$ on $[0, 1]$.

Exercise 19.3.8 *Let $g(x) = x^4$ on $[-5, 3]$. Find $Dg(x)$ on $[-5, 3]$.*

19.4 Measure Estimates

Lemma 19.4.1 $\mu^*(f(E)) \leq p \, \mu^*(E)$

> *Let f be strictly increasing on the interval $[a, b]$ and let $E \subseteq [a, b]$. If at each point $x \in E$, there exists at least one derived number α in $Df(x)$ satisfying $\alpha < p$, then $\mu^*(f(E)) \leq p \, \mu^*(E)$.*

Proof 19.4.1
Let $\epsilon > 0$ be chosen and let G be a bounded open set containing E so that

$$\mu(G) < \mu^*(E) + \epsilon \tag{19.3}$$

For any x_0 in E, there is a sequence (h_n), all $h_n \neq 0$, with $h_n \to 0$ so that the interval $[x_0, x_0 + h_n]$ (if $h_n > 0$) or $[x_0 + h_n, x - 0]$ (if $h_n < 0$) are in G and

$$\frac{f(x_0 + h_n) - f(x_0)}{h_n} < p \tag{19.4}$$

To keep our notation simple, we will simply use the notation $[x_0, x_0 + h_n]$ whether h_n is positive or negative. Let $I_n(x_0) = [x_0, x_0 + h_n]$ and $J_n(x_0) = [f(x_0), f(x_0 + h_n)]$. Since f is strictly increasing, $f(I_n(x_0)) \subseteq J_n(x_0)$ and $J_n(x_0)$ is a non-degenerate closed interval. We also know

DIFFERENTIATION   405

$\mu(I_n(x_0)) = |h_n|$ and $mu(J_n(x_0)) = |f(x_0 + h_n) - f(x_0)|$. It follows from Equation 19.4 that

$$\mu(J_n(x_0)) \;<\; p\,\mu(I_n(x_0)) \;=\; p\,|h_n| \tag{19.5}$$

Since $h_n \to 0$, we then see $\lim_n \mu(J_n(x_0)) = 0$. Let $\mathcal{V}$ be the collection of intervals

$$\mathcal{V} \;=\; \{J_n(x_0) \,:\, x_0 \in E, \, n \in Z^+\}$$

It is easy to see $\mathcal{V}$ is a Vitali Cover for $f(E)$. Thus, by Theorem 19.3.1, there is a countable disjoint family $\{J_{n_i}(x_i) : i \in Z^+\}$ so that

$$\mu\bigg(f(E) \setminus \cup_i J_{n_i}(x_i) \bigg) \;=\; 0 \tag{19.6}$$

From Equation 19.6, we then find

$$\mu^*(f(E)) \;\leq\; \sum_i \mu(J_{n_i}(x_i)) \;<\; p \sum_i \mu(I_{n_i}(x_i)) \tag{19.7}$$

But f is strictly increasing and so the intervals $I_{n_i}(x_i)$ must also be pairwise disjoint. Using Equation 19.3, we infer

$$\sum_i \mu(I_{n_i}(x_i)) \;=\; \mu\bigg(\cup_i I_{n_i}(x_i) \bigg) \;\leq\; \mu(G) \;<\; \mu^*(E) + \epsilon \tag{19.8}$$

Combining, we have $\mu^*(f(E)) < p\mu^*(E) + \epsilon$. Since ϵ is arbitrary, this proves the result.   ∎

Lemma 19.4.2 $\mu^*(f(E)) \geq q\,\mu^*(E)$

> Let f be strictly increasing on the interval $[a,b]$ and let $E \subseteq [a,b]$. If at each point $x \in E$, there exists at least one derived number α in $Df(x)$ satisfying $\alpha > q$, then $\mu^*(f(E)) \geq q\,\mu^*(E)$.

Proof 19.4.2
The proof of this is left to you.   ∎

Homework

Exercise 19.4.1 *In the proof of Lemma 19.4.1, we let $\mathcal{V}$ be the collection of intervals $\{J_n(x_0) : x_0 \in E, \, n \in Z^+\}$. Prove $\mathcal{V}$ is a Vitali Cover for $f(E)$.*

Exercise 19.4.2 *Prove Lemma 19.4.2.*

We are now ready to prove a very nice result: a function of bounded variation is differentiable μ a.e.

Theorem 19.4.3 Functions of Bounded Variation are Differentiable a.e

> Let f be a function of bounded variation on the bounded interval $[a,b]$. Then f has a finite derivative μ a.e.

Proof 19.4.3
Since a function of bounded variation can be decomposed into the difference of two monotone functions, it is enough to prove this result for a non-decreasing function. Further, if f is non-decreasing,

406 BASIC ANALYSIS IV: MEASURE THEORY AND INTEGRATION

then $g(x) = x + f(x)$ is strictly increasing. If we prove the result for g, we will know the result is true for f as well. Hence, we may assume, without loss of generality, that f is strictly increasing in our proof.

Let E_∞ be the set of points in $[a, b]$ where $Df(x)$ contains the value ∞. Then $f(E_\infty) \subseteq [f(a), f(b)]$ since f is strictly increasing. Now for any positive integer q, since ∞ is a derived number at x in E_∞, by Lemma 19.4.2, it follows that $\mu^*(f(E_\infty)) \geq q\mu^*(E_\infty)$. We also know $\mu^*(f(E_\infty)) \leq \mu^*([f(a), f(b)]) = f(b) - f(a)$. Thus,

$$q\,\mu^*(E_\infty) \quad \leq \quad f(b) - f(a), \text{ for all positive integers } q$$

Hence, $\mu^*(E_\infty) = 0$.

Now choose real numbers u and v so that $0 \leq u < v$ and define the set E_{uv} by

$$E_{uv} \quad = \quad \{x \,:\, \exists\, \alpha, \beta \in Df(x) \ni \alpha < u < v < \beta\}$$

Applying Lemma 19.4.1 and Lemma 19.4.2, we have

$$v\mu^*(E_{uv}) \quad \leq \quad \mu^*(f(E_{uv}) \leq u\mu^*(E_{uv})$$

Since $v > u$, we then have $(v - u)\mu^*(E_{uv}) \leq 0$. This implies $\mu^*(E_{uv}) = 0$.

Now, if f is not differentiable at a point x, f either has ∞ as a derived number or it has at least two different finite derived numbers there. In the second case, let the two derived numbers be α_x and β_x. Then there are rational numbers r_1 and r_2 so that $\alpha_x < r_1 < r_2 < \beta_x$. This implies $x \in E_{r_1 r_2}$. Hence, if N is the set of points where f fails to be differentiable in $[a, b]$, we have

$$N \quad \subseteq \quad E_\infty \cup \{E_{pq} \,:\, p, q \text{ rational}\}$$

But the outer measure of all these component sets is 0. Therefore, N has outer measure 0 also which tells us N is measurable and has measure 0. ∎

In the homework, you will construct a function which is not absolutely continuous since it has a derivative nowhere. This is a project in the second semester of (Peterson (8) 2020).

Homework

We are going to build a function which is continuous on $\Re$ but differentiable nowhere. This is a messy construction, but well worth your effort. It is also a nice review of series, uniform convergence in a classical analysis setting.

DIFFERENTIATION   407

Exercise 19.4.3 *Define the function ϕ, for integers n, by*

$$\phi(x) \;=\; \begin{cases} \vdots, & \vdots \\ |x+4n|, & -4n-2 \le x \le -4n+2 \\ \vdots, & \vdots \\ |x+8|, & -10 \le x \le -6 \\ |x+4|, & -6 \le x \le -2 \\ |x|, & -2 \le x \le 2 \\ |x-4|, 2 \le x \le 6 \\ |x-8|, 6 \le x \le 10 \\ \vdots, & \vdots \\ |x-4n|, & 4n-2 \le x \le 4n+2 \\ \vdots, & \vdots \end{cases}$$

Sketch this function nicely so you know what it looks like. This is a sawtooth curve with the peaks occurring at $\ldots, -6, -2, 2, 6, \ldots$ of value 2 and the minima of 0 occurring at the other integers. Prove $\phi(x+4p) = \phi(x)$ for any integer p.

Exercise 19.4.4 *Prove that $|\phi(t) - \phi(s)| = |s-t|$ whenever s and t are in an open interval that does not contain an even integer.*

Exercise 19.4.5 *Define $f_n(x) = (1/4^n)\phi(4^n x)$ and $f(x) = \sum_{n=1}^{\infty} f_n(x)$. Prove f is continuous using the appropriate Weierstrass Test.*

Exercise 19.4.6 *Define a sequence (ϵ_k) as follows: we look at the open intervals whose endpoints are $4^k a$ and $4^k a + \epsilon_k$ for $\epsilon_k = \pm 1$. At least one of these intervals **does not** contain an even integer. We choose ϵ_k to be -1 if the interval $(4^k a - 1, 4^k a)$ **does not** contain an even integer and $\epsilon_k = 1$ if $(4^k a, 4^k a + 1)$ **does not** contain an even integer. If both have an even integer we choose either value. For example consider the intervals $(4^k a - 1, 4^k a)$ and $(4^k a, 4^k a + 1)$ for $a = 0.3$:*

$$4^1(0.3) = 1.2, \text{ intervals } (1.2 - 1, 1.2) = (0.2, 1.2) \text{ and } (1.2, 2.2)$$

*The first interval does **NOT** contain an even integer, so $\epsilon_1 = -1$.*

$$4^2(0.3) = 4.8, \text{ intervals } (4.8 - 1, 4.8) = (3.8, 4.8) \text{ and } (4.8, 5.8)$$

*The second interval does **NOT** contain an even integer, so $\epsilon_2 = +1$.*

$$4^3(0.3) = 19.2$$
$$\text{intervals } (19.2 - 1, 19.2) = (18.2, 19.2) \text{ and } (19.2, 20.2)$$

*The first interval does **NOT** contain an even integer so $\epsilon_3 = -1$. For our $a = 0.3$, $4^k a$ is never an integer, but for other values of a that is possible. If $4^k a$ is an integer Q, then these two intervals would be of the form $[Q-1, Q]$ and $[Q, Q+1]$ and **NEITHER** contain an even integer, and so we can choose either choice $\epsilon_k = \pm 1$.*

You are to do the argument of the example above in detail to find the first five terms ϵ_k for $a = 9.3$ and $a = -1.6$.

Exercise 19.4.7 *Now for $a = 9.3$ and $a = -1.6$, do the following: prove for $1 \le n \le k$, there is no even integer between $4^n(a)$ and $4^n(a) + 4^{n-k}\epsilon_k$.*

Hint 19.4.1 *Do by contradiction. Let the even integer be $2p$ and examine the two cases of $\epsilon_k = \pm 1$. The contradiction comes from how we picked ϵ_k in (1.3).*

For example, if $e_k = 1$, then $[4^n(0.3), 4^n(0.3) + 1]$ **DOES NOT** *contain an even integer. So if there was an integer $2p$ so that $4^n(0.3) < 2p < 4^n(0.3) + 4^{n-k}(1)$, then multiply by 4^{k-n} throughout to get*

$$4^k(0.3) < (2p)4^k < 4^k(0.3) + 1$$

which says an even integer is in this interval which contradicts the way ϵ_k was chosen. The case $\epsilon_k = -1$ is handled similarly.

Exercise 19.4.8 *Now prove for $1 \leq n \leq k$,*

$$|f_n(a + 4^{-k}\epsilon_k) - f_n(a)| \ = \ 4^{-n}\left|\phi(4^n(a) + 4^{n-k}\epsilon_k) - \phi(4^n(a))\right| = 4^{-k}$$

Exercise 19.4.9 *Next use the periodicity of ϕ to prove for $1 \leq k < n$ $f_n(a + 4^{-k}\epsilon_k) = f_n(a)$.*

Exercise 19.4.10 *For the sequence going to zero $h_k = 4^{-k}\epsilon_k$ prove*

$$\frac{f(a + h_k) - f(a)}{h_k} \ = \ \sum_{n=1}^{k} \frac{f_n(a + h_k) - f_n(a)}{h_k}$$

For example, for these indices, we know $|f_n(0.3 + h_k) - f_n(0.3)| = 4^{-k}$ Thus,

$$\frac{f(0.3 + h_k) - f(0.3)}{h_k} \ = \ \sum_{n=1}^{k} \epsilon_k$$

Exercise 19.4.11 *Use the previous exercise to prove that as $k \to \infty$, the $\lim_{k \to \infty} \frac{f(a+h_k)-f(a)}{h_k}$ does not exist. Explain why this tells us $f'(a)$ cannot exist.*

Exercise 19.4.12 *Since we can do this for any value of a, explain why this proves f is not differentiable anywhere and hence f is not absolutely continuous.*

19.5 Extending the Fundamental Theorem of Calculus

We are now getting closer to our *Fundamental Theorem of Calculus* extension. We can now prove a very weak form of the Recapture Theorem in Riemann integration.

Theorem 19.5.1 The Weak Monotone Recapture Theorem or Monotone Functions Have Summable Derivatives

> *Let f be non-decreasing on $[a, b]$. Then f' is measurable and $\int_a^b f' d\mu \leq f(b) - f(a)$. Note this tells us f' is summable.*

Proof 19.5.1
First, extend f to the interval $[a, b+1]$ by setting $f(x) = f(b)$ on $[b, b+1]$. For convenience, we will call this extended f by the same name. Then, let the functions f_n be defined by

$$f_n(x) \ = \ \frac{f(x + \frac{1}{n}) - f(x)}{\frac{1}{n}} = n\left(f(x + \frac{1}{n}) - f(x)\right)$$

DIFFERENTIATION 409

Then, at each point where f' exists, $f_n \to f'$. Since each f_n is measurable as f is monotone, we have the pointwise limit f' is measurable a.e. Now apply Fatou's Lemma to see

$$\int_a^b \liminf_n f_n \, d\mu \;=\; \int_a^b f' \, d\mu \;\leq\; \liminf_n \int_a^b f_n \, d\mu$$

Looking at the definition of the limit inferior, we see $\liminf_n \leq \sup_n$ always. Hence, for all n, we have

$$\int_a^b f' \, d\mu \;\leq\; \int_a^b f_n \, d\mu$$

Since f is monotone, each f_n is Riemann integrable and so the Lebesgue integrals here are Riemann integrals. We will use substitution to finish our argument. Note

$$\int_a^b f_n(x) \, dx \;=\; n \int_a^b \left(f(x + \frac{1}{n}) - f(x) \right) dx$$

$$=\; n \, int_{a+\frac{1}{n}}^{b+\frac{1}{n}} f(x) \, dx \;-\; n \, int_a^b f(x) \, dx$$

$$=\; n \int_b^{b+\frac{1}{n}} f(x) \, dx \;-\; n \int_a^{a+\frac{1}{n}} f(x) \, dx$$

$$=\; f(b) \;-\; n \int_a^{a+\frac{1}{n}} f(x) \, dx$$

because on $[b, b+\frac{1}{n}]$, $f(x) = f(b)$. Finally, on $[a, a+\frac{1}{n}]$, $f(x) \geq f(a)$. Hence, the last integral is bounded below by $n \, f(a) \frac{1}{n} = f(a)$. Hence, $\sup_n \int_a^b f_n \, d\mu \leq f(b) - f(a)$ always. We conclude $\int_a^b f' \, d\mu \leq f(b) - f(a)$ too. ∎

Homework

Exercise 19.5.1 *Consider the function $\Psi(x)$ where Ψ is the Cantor Singular function built from the middle thirds Cantor set.*

- *Look back at Chapter 13 to refresh your mind about the details of the construction of Ψ. We show $\Psi(C) = [0, 1]$. Explain why this says Ψ is not absolutely continuous and the Lebesgue - Stieltjes measure μ_Ψ is not absolutely continuous with respect to μ.*

- *Prove $\Psi'(x) = 0$ a.e.*

- *Show $\int_0^1 \Psi' d\mu < \Psi(1) - \Psi(0)$.*

Exercise 19.5.2 *The function $g(x) = (1/2)(x + \Psi(x))$ where Ψ is the Cantor Singular function built from the middle thirds Cantor set is not absolutely continuous.*

- *Prove $g'(x) = 1/2$ a.e.*

- *Show $\int_0^1 g' d\mu < g(1) - g(0)$.*

If f is absolutely continuous, we can say more. First, we show an absolutely continuous function can be rewritten as the difference of two non-decreasing absolutely continuous functions. We let AC denote the phrase *absolutely continuous* for convenience.

Lemma 19.5.2 AC Functions are the Difference of AC Non-Decreasing Functions

> Let f be absolutely continuous on $[a, b]$. Then there are absolutely continuous non-decreasing functions u and v so that $f = u - v$. In fact, $u = v_f$ and $v = v_f - f$ are the usual choices.

Proof 19.5.2

For any $a \leq \alpha < \beta \leq b$, from Theorem 3.4.1 we know the total variation function is additive on intervals. Recall the total variation function of f is defined by

$$V_f(x) = \begin{cases} 0, & x = a \\ V(f; a, x), & a < x \leq b \end{cases}$$

Thus, it follows that

$$V_f(\beta) - V_f(\alpha) \quad = \quad V(f, \alpha, \beta)$$

Let $\epsilon > 0$ be chosen. Then there is a $\delta > 0$ so that if $((a_n, b_n))$ is a collection of non-overlapping intervals with $\sum_n (b_n - a_n) < \delta$, then $\sum_n |f(b_n) - f(a_n)| < \frac{\epsilon}{2}$. Let π_n be any partition of $[a_n, b_n]$. Then its component intervals form a non-overlapping collection of intervals whose summed length is smaller than δ. Call this collection I_n and let its intervals be labeled $[x_n^i, y_n^i]$ for convenience. In fact, the component intervals in each I_n can be glued together via a union to form a collection whose summed length is also less than δ. Call this larger collection I. Then, the absolute continuity condition for f says

$$\sum_{I_n} |f(y_n^i) - f(x_n^i)| \quad < \quad \frac{\epsilon}{2}$$

However, the choice of partitions π_n is arbitrary. Hence, it follows that

$$\sum_n V(f, a_n, b_n) \quad < \quad \epsilon$$

But, we know the total variation is additive. Hence, we can rewrite the inequality using $V(f, a_n, b_n) = V_f(b_n) - V_f(a_n)$ to get

$$\sum_n V_f(b_n) - V_f(a_n) \quad < \quad \epsilon$$

Thus, v_f is absolutely continuous on $[a, b]$. Moreover, $v = v_f - f$ is also absolutely continuous as sums and differences of absolutely continuous functions are also absolutely continuous. It is then clear that $u = v_f$ and $v = v_f - f$ is a suitable decomposition. ∎

We can now prove a reasonable recapture theorem.

Theorem 19.5.3 The Absolutely Continuous Recapture Theorem

> Let f be absolutely continuous on $[a, b]$. Then
>
> $$\int_a^b f' \, d\mu \quad = \quad f(b) - f(a)$$

Proof 19.5.3

We have proven most of the requisite pieces for this result. From Lemma 19.5.2, we know f can

DIFFERENTIATION

be written as the difference of two absolutely continuous functions. From Theorem 19.4.3, we then know f is differentiable a.e. From Theorem 19.5.1, we know f' is summable. It is now clear, we can assume without loss of generality that our absolutely continuous function f is non-decreasing. To finish, look at the proof of Theorem 19.5.1. Extend f to $[b, b+1]$ as before and consider the same sequence of functions (f_n).

$$f_n(x) \;\; = \;\; \frac{f(x + \frac{1}{n}) - f(x)}{\frac{1}{n}} \;\; = \;\; n\left(f(x + \frac{1}{n}) - f(x) \right)$$

Since f is absolutely continuous, it is also continuous and so this time, we know each f_n is continuous. From the proof of Theorem 19.5.1, we know

$$\int_a^b f_n(x)\, dx \;\; = \;\; f(b) - n \int_a^{a+\frac{1}{n}} f(x)\, dx$$

The integral term above converges to $f(a)$ using the standard Fundamental Theorem of Calculus for continuous integrands. Thus, we know

$$\int_a^b f_n(x)\, dx \;\; \to \;\; f(b) - f(a)$$

Let $\epsilon > 0$ be chosen. Our extension of f to $[b, b+1]$ is still absolutely continuous and so there is a $\delta_1 > 0$ so that given any collection of non-overlapping intervals $([a_n, b_n])$ from $[a, b+1]$ whose summed length is less than δ_1, we have $\sum_n |f(b_n) - f(a_n)| < \frac{\epsilon}{3}$.

Further, since f' is summable, by the absolute continuity of the integral, there is a $\delta_2 > 0$ so that $\int_F |f'|\, d\mu < \frac{\epsilon}{3}$ whenever $\mu(F) < \delta_2$. We can choose $\delta_2 < \delta_1$.

Let E be the set of points in $[a, b]$ where f is differentiable. Then E^C is measure zero and we know $f_n \to f'$ pointwise a.e. From Egoroff's Theorem, Theorem 15.2.1, we then know $f_n \to f'$ almost uniformly. Apply this theorem for tolerance δ_2. Then, there is a measurable set G with measure less than δ_2 so that $f_n \to f'$ uniformly on G^C. Thus, we also know $\int_G f'\, d\mu < \frac{\epsilon}{3}$. From the definition of uniform convergence of G^C, we know there is a positive integer N so that

$$|f'(x) - f_n(x)| \;\; < \;\; \frac{\epsilon}{3(b-a)}, \; \forall\, n > N$$

It is clear that G^C does not contain points of E^C! It then follows that for all $n > N$,

$$\int_{G^C} |f' - f_n|\, d\mu \;\; < \;\; \mu)G^C) \frac{\epsilon}{3(b-a)} < \frac{\epsilon}{3}$$

Now let's combine our pieces. We have for $n > N$,

$$\left| \int_a^b (f_n - f') : d\mu \right| \;\; = \;\; \left| \int_G (f_n - f') : d\mu + \int_{G^C} (f_n - f') : d\mu \right|$$

$$\leq \;\; \int_{G^C} |f_n - f'| : d\mu + \int_G |f_n| : d\mu + \int_G |f'| : d\mu$$

$$< \;\; 2\frac{\epsilon}{3} + + \int_G f_n\, d\mu$$

where we can drop the absolute values of f_n because f is non-decreasing by assumption and so each f_n is non-negative.

412 BASIC ANALYSIS IV: MEASURE THEORY AND INTEGRATION

It remains to show $\int_G f_n \, d\mu < \frac{\epsilon}{3}$ also. Since $\mu(G) < \delta_2 < \delta_1$, we can find an open subset V of (a, b) so that $G \subseteq V$ and $\mu(V) < \delta_1$. Express V as a countable union of disjoint open intervals $((c_i, d_i))$. Pick any x from $[0, 1]$. Then the collection $([c_i + x, d_i + x])$ is a non-overlapping collection of intervals whose summed length is less than δ_1. Hence,

$$\sum_i |f(d_i + x) - f(c_i + x)| < \frac{\epsilon}{3}$$

From the proof of Theorem 19.5.1, we find

$$\int_{c_i}^{d_i} f_n(x) \, dx = n \int_{d_i}^{d_i + \frac{1}{n}} f(x) \, dx - n \int_{c_i}^{c_i + \frac{1}{n}} f(x) \, dx$$

$$= n \int_0^{\frac{1}{n}} (f(d_i + x) - f(c_i + x)) \, dx$$

Summing over i, we have

$$\sum_i \int_{c_i}^{d_i} f_n(x) \, dx = n \int_0^{\frac{1}{n}} \left(\sum_i (f(d_i + x) - f(c_i + x)) \right) dx$$

But the inner summation adds up to less than $\frac{\epsilon}{3}$. We conclude

$$\sum_i \int_{c_i}^{d_i} f_n(x) \, dx < n \int_0^{\frac{1}{n}} \frac{\epsilon}{3} \, dx = \frac{\epsilon}{3}$$

Thus,

$$\int_G f_n \, d\mu \leq \int_V f_n \, d\mu = \sum_i \int_{c_i}^{d_i} f_n(x) \, dx < \frac{\epsilon}{3}$$

This is the last piece to complete the proof of this result. ∎

Homework

Exercise 19.5.3 *Let f be defined by*

$$f(x) = \begin{cases} 2x, & 0 \leq x < 1 \\ 3, & 1 = x \\ 5x^2 + 1, & 1 < x \leq 2 \end{cases}$$

- *Is f absolutely continuous?*

- *Find f' when it exists.*

- *Compute $\int_{[0,2]} f' d\mu$.*

- *Compute $f(2) - f(0)$. Do they match?*

DIFFERENTIATION 413

Exercise 19.5.4 *Let f be defined by*

$$f(x) \;=\; \begin{cases} 2, & x = -1 \\ x + 4, & -1 < x < 1 \\ 6, & 1 = x \\ x^2 + 7, & 1 < x < 2 \\ 12, & x = 2 \\ 2x^3, & 2 < x \le 3 \end{cases}$$

- *Is f absolutely continuous?*

- *Find f' when it exists.*

- *Compute $\int_{[-1,2]} f' d\mu$.*

- *Compute $f(2) - f(-1)$. Do they match?*

Exercise 19.5.5 *Let g be defined by*

$$g(x) \;=\; \begin{cases} 2x, & 0 \le x < 1 \\ 5x^2 - 3, & 1 \le x \le 2 \end{cases}$$

- *Is g absolutely continuous?*

- *Find g' when it exists.*

- *Compute $\int_{[0,2]} g' d\mu$.*

- *Compute $g(2) - g(0)$. Do they match?*

Exercise 19.5.6 *Let g be defined by*

$$g(x) \;=\; \begin{cases} x + 4, & -1 \ge x < 1 \\ x^2 + 4, & 1 \ge x < 2 \\ x^3, & 2 \ge x \le 3 \end{cases}$$

- *Is g absolutely continuous?*

- *Find g' when it exists.*

- *Compute $\int_{[-1,3]} g' d\mu$.*

- *Compute $g(3) - g(-1)$. Do they match?*

Exercise 19.5.7 *Let g_n, for $n \ge 2$, be defined by*

$$g_n(x) \;=\; \begin{cases} 1, & x = 1 \\ 1/(k+1), & 1/(k+1) \le x < 1/k, \; 1 \le k < n \\ 0, & 0 \le x < 1/n \end{cases}$$

- *Is g_n absolutely continuous?*

- *Find g_n' when it exists.*

- *Compute $\int_{[0,1]} g_n' d\mu$.*

- *Compute $g_n(1) - g_n(0)$. Do they match?*

414   *BASIC ANALYSIS IV: MEASURE THEORY AND INTEGRATION*

Exercise 19.5.8 *Let g be the pointwise limit of the sequence of functions (g_n) for $n \geq 2$ defined by*

$$g_n(x) \quad = \quad \begin{cases} 1, & x = 1 \\ 1/(k+1), & 1/(k+1) \leq x < 1/k, \ 1 \leq k < n \\ 0, & 0 \leq x < 1/n \end{cases}$$

- *Is g absolutely continuous?*

- *Find g' when it exists.*

- *Compute $\int_{[0,1]} f' d\mu$.*

- *Compute $g(1) - g(0)$. Do they match?*

19.6   Charges Induced by Absolutely Continuous Functions

We can now establish the linkage between the Lebesgue - Stieltjes charges induced by an absolutely continuous function f on $[a, b]$ and the derivative f'.

Theorem 19.6.1 Characterizing Lebesgue - Stieltjes Measures Constructed from Absolutely Continuous Functions

> *Let f be absolutely continuous on $[a, b]$. Let ν_f be the finite charge induced by f in the Lebesgue - Stieltjes construction process. Then, for all measurable E, we have*
>
> $$\nu_f(E) \quad = \quad \int_E f' \, d\mu$$

Proof 19.6.1
Since f is absolutely continuous, $\nu_f \ll \mu$. From the Radon-Nikodym theorem, there is therefore a summable g so that

$$\nu_f(E) \quad = \quad \int_E g \, d\mu$$

for all measurable E. In particular, for any $x \in (a, b)$,

$$\nu_f((a,x]) \quad = \quad f(x) - f(a) = \int_{(a,x]} g \, d\mu$$

However, we also know

$$f(x) - f(a) \quad = \quad \int_{(a,x]} f' \, d\mu$$

Hence, $\int_a^x (g - f') \, d\mu = 0$. In fact, $\int_I (g - f') \, d\mu = 0$ for any interval in $[a, b]$. Let $h = g - f'$ which is summable. Now let's assume there is a measurable set E in $[a, b]$ with $\int_E h \, d\mu > 0$ and let $\epsilon = \int_E h \, d\mu$. From the continuity of the integral, for this ϵ, there is a $\delta > 0$ so that $\int_G h \, d\mu < \epsilon$ for any measurable set G with $\mu(G) < \delta$.

Choose an open set U in $\Re$ which contains E with $\mu(U) < \mu(E) + \delta$. Write the open set U as a countable union of open intervals $I_n = (a_n, b_n)$. Let $V_n = I_n \cap [a, b]$. Then, $\int_{V_n} h \, d\mu = 0$ as V_n

DIFFERENTIATION

is an interval. From this we conclude, if $V = \cup_n V_n$, that $\int_V h\,d\mu = 0$ as well. By construction, $E \subseteq V \subseteq U$ which implies $V \setminus E \subseteq U$. Thus

$$\mu(E) + \mu(E^C \cap V) + \mu(V^C \cap U) \ = \ \mu(U)$$
$$< \ \mu(E) + \delta$$

It follows that $\mu(V \setminus E) < \delta$ and hence, $\int_{V \setminus E} h\,d\mu < \epsilon$. However,

$$0 \ = \ \int_V h\,d\mu \ = \ \int_E h\,d\mu + \int_{V \setminus E} h\,d\mu$$

It follows immediately that

$$\epsilon \ = \ \left| \int_E h\,d\mu \right| \ = \ \left| \int_{V \setminus E} h\,d\mu \right| < \epsilon$$

This is a contradiction. We conclude $\int_E h\,d\mu = 0$ for all measurable E. In particular, this is true for $\{x \mid h(x) > 0\}$ implying $h^+ = 0$ a.e. and for $\{x \mid h(x) \leq 0\}$ implying $h^- = 0$ a.e. Thus, $h = 0$ a.e. We therefore have shown that $f' = g$ a.e. and we can conclude

$$\nu_f(E) \ = \ \int_E f'\,d\mu$$

∎

Homework

Exercise 19.6.1 *Let f be defined by*

$$f(x) \ = \ \begin{cases} 2x, & 0 \leq x < 1 \\ 3x^2 - 1, & 1 < x \leq 2 \end{cases}$$

- *Is f absolutely continuous?*

- *Find f' when it exists.*

- *Let μ_f be the Lebesgue - Stieltjes measure induced by f. Compute $\mu_f([0,2])$, $\mu_f([0.5,1.5])$ and $\mu_f([0.3,1.9])$.*

Exercise 19.6.2 *Let f be defined by*

$$f(x) \ = \ \begin{cases} 2x + 4, & -1 \leq x < 1 \\ x^2 + 5, & 1 \leq x < 2 \\ 2x^3 - 7, & 2 \leq x \leq 3 \end{cases}$$

- *Is f absolutely continuous?*

- *Find f' when it exists.*

- *Let μ_f be the Lebesgue - Stieltjes measure induced by f. Compute $\mu_f([-1,3])$, $\mu_f([-0.5,1.5])$ and $\mu_f([0.3,2.9])$.*

Exercise 19.6.3 *Let g be defined by*

$$g(x) \ = \ \begin{cases} 2x, & 0 \leq x < 1 \\ 5x^2 - 3, & 1 \leq x \leq 2 \end{cases}$$

BASIC ANALYSIS IV: MEASURE THEORY AND INTEGRATION

- *Is g absolutely continuous?*

- *Find g' when it exists.*

- *Let μ_g be the Lebesgue - Stieltjes measure induced by g. Compute $\mu_g([0,2])$, $\mu_g([0.5,1.5])$ and $\mu_g([0.3,1.9])$.*

Exercise 19.6.4 *Let $g(x) = x^3$ on $[-10,10]$. Let μ_g be the Lebesgue - Stieltjes measure induced by g. Compute $\mu_g([-10,10])$, $\mu_g([-5.5,6.5])$ and $\mu_g([0.3,8.9])$.*

Exercise 19.6.5 *Let $g(x) = \tanh(x)$ on $[-3,5]$. Let μ_g be the Lebesgue - Stieltjes measure induced by g. Compute $\mu_f[a,b]$ for any $[,b]$ in $[-3,5]$.*

Part VII

Summing It All Up

Chapter 20

Summing It All Up

We have now come to the end of this set of notes, which means you should have gone through four volumes of notes on analysis. There is, of course, much more we could have discussed. We have not covered all of the things we wanted to, but we view that as a plus: there is more to look forward to! In particular, in (Peterson (6) 2020) we will do more.

In Figure 20.1, you can see how all this material fits together. Since you have learned the first four volumes, the arrangement of this figure should make it clear to you both how much you now know and how much you can still learn! Since what we have discussed in the first four volumes is still essentially what is a *primer* for the start of learning even more analysis and mathematics at this level, in Figure 20.1 we have explicitly referred to our texts using that label. The first volume is the one in the figure we call **A Primer on Analysis**, the second is **Primer Two: Escaping The Real Line**, the third is **Primer Three: Basic Abstract Spaces**, the fourth (this text) is **Primer Four: Measure Theory** and the fifth is **Primer Five: Functional Analysis**. Keep in mind these new labels as they show up in Figure 20.3 also. The typical way a graduate student takes this sequence is shown in Figure 20.2. There are lots of problems with how a mathematically interested student learns this material. An undergraduate major here would be required to take a two semester sequence in core real analysis, however, given the competition for graduate students in a mathematical sciences graduate program, it is often true that incoming master's degree students are admitted with deficiencies. One often has people come into the program with just one semester of core real analysis. Such students should take the second semester when they come into the program so that they can be exposed to Riemann integration theory and sequences of functions among other important things, but not all want to as it means taking a deficiency course. However, from Figure 20.2, you can see that jumping straight to **Primer Three** on Linear Analysis without proper training will almost always lead to inadequate understanding. Such students will have a core set of mathematics poorly developed which is not what is wanted. Hence, one needs to do as much as one can to train all undergraduate students in two full semesters of core real analysis. Even if students take the full two semesters of core real analysis at the graduate level, the undergraduate curriculum today does not in general offer courses in the following areas:

- Basic Ordinary Differential Equation (ODE) Theory: existence of solutions, continuous dependence of the solution on the data and the structure of the solutions to linear ODEs. Since most students do not take **Primer Two**, they have not been exposed to differential calculus in $\Re^n$, the inverse and implicit function theorem, and so forth. Many universities have replaced this type of course with a course in Dynamical Systems which cannot assume a rigorous background in the theory of ODE and so the necessary background must be discussed in a brief fashion. Again a text such as **Primer Two** designed for both self-study and as a regular class is an important tool for the mathematical science major to have available to them.

419

Figure 20.1: The general structure of the five core basic analysis courses.

- Basic Theoretical Numerical Analysis: the discussion of the theory behind the numerical solution of ODEs, linear algebra factorizations such as QR and optimization algorithms. This is the course where one used to discuss operator norms for matrices as part of trying to understand our code implementations.

- Basic Manifold Theory: the discussion of the extension of $\Re^n$ topology and analysis to the more general case of sets of objects locally like $\Re^n$. This material is pertinent to many courses in physics and to numerical Partial Differential Equations (PDE) as manifolds with a boundary are essential. Note this type of course needs **Primer Two** material, more topology and a more general theory of integration on chains.

All of the above areas of study are much enhanced by knowledge of the material in this fourth primer. We had carefully developed the space $\mathbb{L}_2([a,b])$ so that we could talk intelligently about Hilbert spaces of functions without needing full training in measure theory in (Peterson (9) 2020), but your understanding of Hilbert space structure and deeper connections in all areas are greatly enhanced

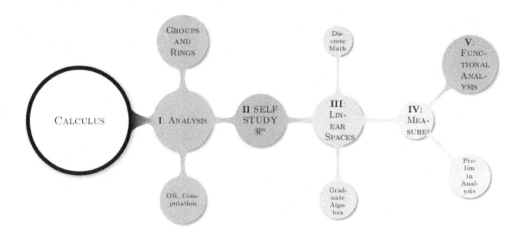

Figure 20.2: The analysis education pathway.

now that you have the training in Primer Four, (Peterson (7) 2020). For example, we can now extend the idea of differential equations to non-classical situations where the solution is not differentiable in the traditional sense with the introduction of absolutely continuous functions which are differential a.e. we can study equations of the form $x' = f(t, x)$ a.e. You are also now ready to explore ideas from topological vector spaces and algebraic topology as they can be used to model signals into a complex biological or engineering system and both the mathematical sciences and physical sciences students have not been exposed to those ideas. Again, it is clear one must inculcate into the students the ability to read and think critically on their own, *outside of class and instructors* so they can become the scientists of the future. The way these proposed analysis courses fit into the grand scheme of things is shown in Figure 20.3. Now that you have read through the first four primers, you can see much better the many pathways you can take from this point on.

So here's to the future!

As always, remember to master the material in any courses you take at your university. Make sure you supplement as needed from books such as ours and learn to think carefully and deeply. You are preparing yourself for a life of research and the better you learn this material the easier it will be for you to read journal papers and think about solving problems that have not been solved before. Study all the proofs carefully as they tell you how to approach new problems. In the proof of each theorem, ask yourself what role each assumption plays in the proof so that you understand why the proof fails if you don't make that assumption. And remember, a new conjecture you have later in your own work always follows this pattern: it is true and you just haven't found the proof yet or it is false and there is a counterexample. This is a tricky thing to navigate and the best advice we can give you is to really, really learn the background material, such as is in the books we have written and know the examples well. The final piece of advice is to learn to redefine your problem: when you are stuck, a lot of times your failure leads to a better formulation of the problem you want to solve. It is the equivalent of walking around a wall instead of just butting into it over and over!

So enjoy your journey and get started!

Figure 20.3: Analysis connections.

Part VIII

References

References

[1] A. Bruckner and J. Bruckner and B. Thomson. *Real Analysis*. Prentice - Hall, 1997.

[2] S. Douglas. *Introduction to Mathematical Analysis*. Addison-Wesley Publishing Company, 1996.

[3] John W. Eaton, David Bateman, Søren Hauberg, and Rik Wehbring. *GNU Octave version 5.2.0 manual: a high-level interactive language for numerical computations*, 2020. URL `https://www.gnu.org/software/octave/doc/v5.2.0/`.

[4] Free Software Foundation. *GNU General Public License Version 3*, 2020. URL `http://www.gnu.org/licenses/gpl.html`.

[5] MATLAB. *Version Various (R2010a) - (R2019b)*, 2018 - 2020. URL `https://www.mathworks.com/products/matlab.html`.

[6] J. Peterson. *Basic Analysis V: Functional Analysis and Topology*. CRC Press, Boca Raton, Florida 33487, 2020.

[7] J. Peterson. *Basic Analysis IV: Measure Theory and Integration*. CRC Press, Boca Raton, Florida 33487, 2020.

[8] J. Peterson. *Basic Analysis I: Functions of a Real Variable*. CRC Press, Boca Raton, Florida 33487, 2020.

[9] J. Peterson. *Basic Analysis III: Mappings on Infinite Dimensional Spaces*. CRC Press, Boca Raton, Florida 33487, 2020.

[10] J. Peterson. *Basic Analysis II: A Modern Calculus in Many Variables*. CRC Press, Boca Raton, Florida 33487, 2020.

[11] H. Sagan. *Advanced Calculus of Real Valued Functions of a Real Variable and Vector - Valued Functions of a Vector Variable*. Houghton Mifflin Company, 1974.

[12] G. Simmons. *Introduction to Topology and Modern Analysis*. McGraw-Hill Book Company, 1963.

[13] K. Stromberg. *Introduction to Classical Real Analysis*. Wadsworth International Group and Prindle, Weber and Schmidt, 1981.

[14] A. Taylor. *General Theory of Functions and Integration*. Dover Publications, Inc., 1985.

Part IX

Detailed Index

Index

Bounded variation
  Always possess right- and left-hand limits, 51
  Are bounded, 44
  Bounded differentiable implies bounded variation, 45
  Closed as a set under addition, 45
  Closed as a set under products, 46
  Continuous and bounded variation if and Only if the variation function V_f and $V_f - f$ are continuous and increasing, 54
  Continuous if and only if the variation function is continuous, 52
  Definition, 44
  Have countable discontinuity sets, 52
  Inverse, 46
  Monotone functions, 32
    Associated u and v properties, 34
    Continuous part, 38
    Discontinuity set and associated u and v, 34
    Estimates, 32
    Have countable number of discontinuities, 33
    Saltus function and the continuous part: example, 40
    Saltus function associated with a monotone function, 36
    Saltus function: properties, 37
  Monotone functions are of bounded variation, 44
  Total variation, 45
  Total variation is additive on intervals, 48
  Variation function, 49

Cantor set
  Cantor function
    $\Psi_a : [0,1] \to [0,1]$ is a non-decreasing continuous map of $[0,1]$ onto $[0,1]$ and is constant on each of $I_{n,k}$ in $[0,1]-C_a$, 306
    For $x \in S, \phi((x_n)) = \sum_{j=1}^{\infty} x_j (1/2^j)$ and $\Psi_a = \phi \circ f_a^{-1}$, 305
  Consequences
    Any Lebesgue measurable set with positive Lebesgue measure must contain a subset which is not in the Lebesgue sigma-algebra, 307

Borel sigma-algebra is strictly contained in the Lebesgue sigma-algebra, 307
  Two functions can be equal a.e. with respect to Borel measure but only one is Borel measurable, 307
Generalized Cantor set, 303
Generating sequence, 301
Generating sequence induces associated $J_{n,k}$ and $I_{n,k}$ intervals, 301
Lebesgue singular function
  $g : [0,1] \to [0,1]$ by $g(x) = (\Psi(x) + x)/2$, 306
  If C_a is a Cantor set, then $g(C_a)$ is also a Cantor set, 306
Measure of a Cantor set, 306
Monotone g_a built from the Cantor function g is strictly increasing and continuous from $[0,1]$ onto $[0,1]$, 306
Outer measure of a Cantor set is the same as its Lebesgue measure, 306
Representation of the Cantor set C_a
  S is the set of all sequences whose values are either 0 or 1, 304
  C_a contains no open intervals, 305
  f_a maps into C_a and is $1-1$ and onto, 304
  Constructing f mapping S to C_a by $f(x) = \sum_{n=1}^{\infty} x_n r_n$, 304
  Entries of C_a are a_n and associated with it is the sequence $r_n - a_{n-1} - a_n$, 304
  For x_i either 0 or 1, define the intervals $J(x_1, ..., x_n) = [\sum_{j=1}^{n} x_j r_j, a_n + \sum_{j=1}^{n} x_j r_j]$, 304
Change of variables in integration
  Proving Fubini in 2 dimensions: a top and bottom halves of a closed curve bound the region, 386
  Proving Fubini on a rectangle in n dimensions, 384
Connections to Riemann Integration
  Every Riemann Integrable function is Lebesgue Integrable and the two integrals coincide, 375
  Riemann - Stieltjes Integrable Functions are Lebesgue - Stieltjes Integrable and the two integrals coincide, 377

430   *INDEX*

Constructing measures
    Algebra of sets, 252
    Constructing via premeasures, 264
        Additive set functions, 266
        Condition for an outer measure to be regular, 266
        Conditions for OMI-F measures, 268
        Conditions for OMI-FE measures, 269
        Definition of OMI-F σ-algebra, 266
        Definition of OMI-FE σ-algebra, 266
        Need additional properties for the premeasure τ and covering family $\mathscr{T}$ to have, 266
        Pseudo-measures, 268
        Regular outer measures, 266
    Example one, 269
    Example two, 270
    Metric outer measures
        Called MOM, 258
        Construct σ algebra $\mathcal{M}$ satisfying Caratheodory condition for MOM μ^* to give a measure on $\mathcal{M}$, 258
        Open sets in a metric space are μ^* measurable if and only if μ^* is a metric outer measure, 262
        Open sets in a metric space are OMI measurable, 259
    Outer measures, 251
        μ^* Measurable sets form an algebra, 253
        μ^* measurable sets, 252
        Caratheodory condition, 252
        Measure induced by an outer measure, 257
        Measure induced by an outer measure is complete, 258
        Properties of μ^* measurable sets, 254
        Set in $\mathcal{M}$ is called **OMI** measurable, where **OMI** stands for outer measure induced, 258
    Premeasures and covering families, 264
Convergence modes
    Almost uniform convergence, 326
        Implies converges a.e so limit function is measurable, 326
    Almost uniform convergence implies convergence in measure, 335
    Cauchy In measure implies completeness, 331
    Continuity of the integral, 341
    Convergence in measure, 326
    Convergence pointwise and pointwise a.e., 325
    Convergence uniformly, 325
    Egoroff's Theorem, 336
    Extracting subsequences
        Cauchy In measure implies a convergent subsequence, 328
        Cauchy sequences In Measure, 328

    p-Norm convergence implies convergence in measure, 334
    p-Summable Cauchy sequence condition I, 340
    p-Summable Cauchy sequence condition II, 342
    p-Summable functions have p-norm arbitrarily small off a set, 339
    p-Summable inequality, 340
    Pointwise a.e. convergence plus Domination implies p-Norm convergence, 338
    Summary, 334
        Convergence relationships on finite measure space four, 345
        Convergence relationships on finite measure space one, 344
        Convergence relationships on finite measure space three, 344
        Convergence relationships on finite measure space two, 344
        Convergence relationships on general measurable space, 345
        Convergence relationships with p-domination five, 346
        Convergence relationships with p-domination four, 346
        Convergence relationships with p-domination one, 345
        Convergence relationships with p-domination six, 346
        Convergence relationships with p-domination three, 346
        Convergence relationships with p-domination two, 345
        Convergent subsequences exist, 346
    Vitali Convergence Theorem, 342

Decomposition of measures
    $\epsilon-\delta$ version of absolute continuity of a charge, 360
    Absolute continuity of charges, 359
    Absolute continuity of the integral, 360
    Equivalent absolute continuity conditions for charges, 359
    Hahn decomposition associated with a charge, 354
    Hahn decomposition characterization of a charge, 355
    Jordan decomposition of a charge, 353
    Jordan decomposition of a finite charge, 349
    Lebesgue decomposition of a measure, 370
    Lebesgue decomposition Theorem, 370
    Positive and negative parts of a charge, 349
    Positive and negative sets for a charge, 349
    Radon - Nikodym technical lemma, 362

INDEX

Radon - Nikodym Theorem: finite charge case, 363

Radon - Nikodym Theorem: sigma-finite charge case, 367

Variation of a charge, 356

Variation of a charge in terms of the plus and minus parts, 357

Variation of a charge is a measure, 356

What happens if the charge is extended real-valued?, 352

Definition

$RS([g, a, b])$, 112

Absolute continuity of a measure, 200

Absolute continuity of charges, 359

Abstract Darboux Integrability, 215

Additive set function, 266

Algebra of Sets, 252

Almost uniform convergence, 326

Caratheodory condition, 252

Cauchy sequence in norm, 232

Cauchy sequences in measure, 328

Charges, 174

Common refinement of two partitions, 30

Complete measure, 191

Complete NLS, 233

Conjugate index pairs, 228

Content of open interval, 273

Continuous almost everywhere, 101

Continuous part of a monotone function, 38

Convergence in measure, 326

Convergence pointwise and pointwise a.e., 325

Convergence uniformly, 326

Darboux integrability, 62

Darboux lower and upper integrals, 62

Darboux upper and lower sums, 58

Derived Numbers, 401

Discontinuity set of a monotone function, 34

Equivalent conditions for the measurability of a function, 156

Equivalent conditions for the measurability of an extended real-valued function, 161

Essentially bounded functions, 240

Extended real number system, 153

Functions

Absolute Continuity, 397

Functions of bounded variation, 44

Generalized Cantor set, 303

Inner product space, 247

Integral of a non-negative measurable function, 185

Integral of a simple function, 182

Lebesgue outer measure, 274

Limit inferior and superior of sequences of sets, 178

Measurability of a function, 155

Measurability of extended real-valued functions, 161

Measures, 174

Borel measure, 290

Metric on a set, 223

Metric outer measure, 258

Monotone functions, 32

Saltus function associated with a monotone function, 36

Norm convergence, 224

Norm on a vector space, 223

Outer measure, 251

Partition, 30

Positive and negative parts of a charge, 349

Positive and negative sets for a charge, 349

Premeasures and covering families, 264

Product Outer Measure, 388

Propositions holding almost everywhere, 181

Pseudo-measure, 268

Refinement of a partition, 30

Regular outer measures, 266

Rewriting Lebesgue outer measure using edge length restricted covers: the M_δ form of μ^*, 283

Riemann - Stieltjes Criterion for integrability, 123

Riemann - Stieltjes Darboux Integral, 122

Riemann - Stieltjes Sum for two bounded functions, 111

Riemann Integrability of a bounded f, 55

Riemann integrability of a bounded function, 17

Riemann sum, 16, 55

Riemann's criterion for integrability, 62

Set of extended real-valued measurable functions, 161

Sets of content zero, 101

Sigma-Algebra, 149

sigma-algebra generated by collection A, 151

Simple functions, 182

Singular measures, 370

Space of essentially bounded functions, 241

Space of p-summable functions, 227

Step function, 115

Summable functions, 202

Upper and lower Riemann - Stieltjes Darboux sums, 121

Upper and lower Riemann - Stieltjes Integrals, 122

Variation of a charge, 356

Vitali Cover, 402

Differentiation results

$\mu^*(f(E)) \geq q\,\mu^*(E)$, 405

$\mu^*(f(E)) \leq p\,\mu^*(E)$, 404

f is absolutely continuous if and only if $\mu_f <<\mu$, 400

Absolute continuity of functions, 397

Absolutely Continuous Recapture Theorem, 410

AC functions are the difference of AC non-decreasing functions, 409

Characterizing Lebesgue - Stieltjes measures constructed from absolutely continuous functions, 414

Derived numbers, 401

Function which is continuous but differentiable nowhere: so not absolutely continuous, 406

Functions of bounded variation are differentiable a.e, 405

Lebesgue-Stieltjes measures and absolutely continuous functions, 398

Properties of absolutely continuous functions, 397

Vitali Cover, 401

Vitali Covering Theorem, 402

Weak Monotone Recapture Theorem or monotone functions have summable derivatives, 408

Fubini Theorems

Measure space setting

Existence of a measurable cover for a set in $X \times Y$, 391

Existence of a measurable cover for the iterated integral version of product measure, 394

General Fubini Theorem, 395

Product outer measure, 388

Product outer measure is a regular measure, 389

Representing the product measure as a double integral, 392

Simple Fubini Theorem, 394

Riemann setting

$\Re^n$, 380

Specialize to the bounded set $D = [a, b] \times [c, d] \subset \Re^2$, 380

Discussion of how to handle the iterated integrals in the $D = [a, b] \times [c, d]$ setting, 380

Fubini's Theorem for a rectangle: n dimensional, 384

Fubini's Theorem: 2D: a top and bottom curve, 385

Proving Fubini in 2 dimensions: a top and bottom halves of a closed curve bound the region, 386

Proving Fubini on a rectangle in 2 dimensions, 382

Further Riemann - Stieltjes Integration results

Existence

Integrand bounded variation and integrator continuous implies Riemann - Stieltjes Integral exists, 137

Integrand continuous and integrator continuously differentiable implies Riemann - Stieltjes Integrable, 137

Integrand continuous and integrator of bounded variation implies Riemann - Stieltjes Integral exists, 136

Integrand Riemann Integrable and integrator continuously differentiable implies Riemann - Stieltjes Integrable, 138

Riemann - Stieltjes Fundamental Theorem of Calculus, 133

Worked out examples of Riemann - Stieltjes computations: five, 142

Worked out examples of Riemann - Stieltjes computations: four, 141

Worked out examples of Riemann - Stieltjes computations: one, 139

Worked out examples of Riemann - Stieltjes computations: six, 143

Worked out examples of Riemann - Stieltjes computations: three, 140

Worked out examples of Riemann - Stieltjes computations: two, 140

Further Riemann Integration Results

If f is non-negative and Riemann Integrable with zero value, then f is zero a.e., 106

Weierstrass Approximation Theorem, 96

Further Riemann Integration results

f continuous and g Riemann Integrable implies $f \circ g$ is Riemann Integrable, 99

Continuous almost everywhere, 101

Example: the composition of Riemann Integrable functions is not Riemann Integrable, 100

If f is continuous and zero a.e., the Riemann integral of f is zero, 106

Riemann - Lebesgue Lemma, 102

Riemann integrable f and g and $d_1(f, g) = 0$ implies f is g a.e., 107

Riemann integrable f and g equal a.e. have the same integral, 107

Riemann Integral limit interchange theorem, 93

Sets of content zero, 101

Examples, 101

Lebesgue - Stieltjes measures

$\mathcal{A}$ is the algebra generated by finite unions of sets from $\mathcal{U}$, 311

μ_g^* is called Lebesgue - Stieltjes outer measure, 315

μ_g is called Lebesgue - Stieltjes Measure, 315

INDEX

σ-algebra $\mathcal{M}_g$, generated by τ_g, contains $\mathscr{A}$ and the induced measure μ_g is regular, 313

g non-decreasing function continuous from the right, 311

Approximating sets with a Lebesgue - Stieltjes Outer measure, 317

Characterizations of Lebesgue - Stieltjes measures, 318

Consequences of τ_g being a pseudo-measure on $\mathcal{A}$, 315

Define τ_g on $\mathcal{U}$, the collection of $(a, b]$, $(-\infty, b]$ and (a, ∞) sets, 311

Extend τ_g to be additive on $\mathcal{A}$, 311

Extension τ_g is a pseudo-measure, 313

Extension of τ_g to be additive on $\mathcal{A}$ is well-defined, 311

First thoughts

Another measure uses premeasure $\tau_g((a, b]) = g(b) - g(a)$, for g monotone and increasing function on $[a, b]$, 309

If λ is a measure, then $h(x) = \lambda([a, x])$ is right continuous, 309

If λ is a measure, then $h(x) = \lambda([a, x])$ need not be left continuous, 309

Lebesgue uses premeasure $\tau((a, b]) = b - a$, 309

Lebesgue - Stieltjes measure of a singleton need not be 0, 316

Premeasure τ_g generates the outer measure μ_g^*, 315

Restriction of μ_g^* to μ_g^* measurable sets gives a measure μ_g, 315

Use g to define $\tau - g$, 311

Lebesgue measure on $\Re^k$, 273

Approximation of a summable function with a continuous function, 293

Borel measure

Finite measure Borel sets can be approximated by open sets, 291

Finite measure Borel sets can be approximated ny closed sets, 290

Borel measures, 290

Continuous approximation of a characteristic function, 292

Continuous approximation of a simple function, 293

Existence of a non-Lebesgue measurable set, 297

It is translation invariant, 295

Lebesgue outer measure, 273

$M_\delta = \mu^*$, 284

Rewriting Lebesgue outer measure using edge length restricted covers: the M_δ form of μ^*, 283

Sums over finite Lebesgue covers of the closure of an interval dominate its content: motivation for the proof in $\Re^2$, 275

Sums over finite Lebesgue covers of the closure of an interval dominate its content: the actual proof in $\Re^k$, 280

Approximate finite Lebesgue covers of $\overline{I}$, 283

Borel σ-algebra of subsets is contained in $\mathcal{M}$, 286

Content of an open interval, 273

Definition, 274

Induces a measure but covering family is not an algebra, 274

Lebesgue measure is regular, 287

Lebesgue outer measure is a metric outer measure: LOM is a MOM, 285

Measurability and approximation conditions, 288

Outer measure of an interval is the same as the content, 283

Outer measure of the closure of an interval is the same as the content, 282

Sums over finite Lebesgue covers of the closure of an interval dominate its content, 274

Metric space of finite measurable sets, 297

Metric space of finite measurable sets is complete, 298

Non-measurable set Lemma 1, 296

Non-measurable set Lemma 2, 296

Space $\mathcal{L}_1([a, b], \mathcal{M}, \mu)$ is separable, 294

Lemma

$M_\delta = \mu^*$, 284

$\epsilon - \delta$ version of absolute continuity of a charge, 360

$\mu^*(f(E)) \geq q\,\mu^*(E)$, 405

$\mu^*(f(E)) \leq p\,\mu^*(E)$, 404

τ_g is a pseudo-measure on $\mathcal{A}$, 313

f zero on (a, b) implies zero Riemann Integral, 88

Absolute continuity of the integral, 360

AC functions are the difference of AC non-decreasing functions, 409

Approximate finite Lebesgue covers of $\overline{I}$, 283

Characterizing limit inferior and limit superior of sequences of sets, 178

Condition for outer measure to be regular, 266

Continuity of the integral, 341

Continuous functions of finite measurable functions are measurable, 168

Continuous functions of measurable functions are measurable, 170

De Morgan's Laws, 149

Disjoint decompositions of unions, 179

INDEX

Essentially bounded functions bounded above by their essential bound a.e., 242

Essentially bounded functions that are equivalent have the same essential bound, 242

Extended real-valued measurability in terms of the finite part of the function, 162

Extending τ_g to additive is well-defined, 311

Finite jump step functions as integrators, 120

Function f zero a.e. if and only if its integral is zero, 200

Function measurable if and only if positive and negative parts measurable, 160

Fundamental infimum and supremum equalities, 66

Hahn decomposition characterization of a charge, 355

Infimum Tolerance Lemma, 31

Limit inferior and limit superior of monotone sequences of sets, 178

Measure of monotonic sequences of sets, 176

Monotonicity, 175

Monotonicity of the abstract integral for non-negative functions, 189

Non-measurable set Lemma 1, 296

Non-measurable set Lemma 2, 296

One jump step functions as integrators: one, 115

One jump step functions as integrators: three, 119

One jump step functions as integrators: two, 117

Outer measure of interval equals content of interval, 283

Outer measure of the closure of interval equals content of interval, 282

p-Summable Cauchy sequence condition I, 340

p-Summable Cauchy sequence condition II, 342

p-Summable functions have p-Norm arbitrarily small off a set, 339

p-Summable inequality, 340

Pointwise infimums, supremums, limit inferiors and limit superiors are measurable, 165

Products of measurable functions are measurable, 166

Properties of extended real-valued measurable functions, 163

Properties of measurable functions, 158

Properties of simple function integrations, 186

Radon - Nikodym technical lemma, 362

Real number conjugate indices inequality, 228

Riemann Integrable f and g on $[a, b]$ match on (a, b) implies Riemann Integrals match, 89

Sums over finite Lebesgue covers of $\overline{I}$ dominate content of I, 274

Supremum Tolerance Lemma, 31

Upper and lower Darboux integral is additive on intervals, 73

Upper and lower Riemann - Stieltjes Darboux Integral is additive on intervals, 126

Measurable functions and spaces

$M(X, \mathcal{S})$ is closed under multiplication, 166

Approximation of non-negative measurable functions by monotone sequences, 168

Borel sigma algebra, 152

Characterizing open sets of $\Re$, 154

Continuous functions of finite measurable functions are measurable, 167

Continuous functions of measurable functions are measurable, 170

DeMorgan's Laws, 149

Extended Borel sigma-algebra, 153

If S is a sigma-algebra and $A \in S$, S is called an S measurable set, 149

Intersection of sigma-algebras, 151

Limit inferior and limit superior of measurable extended real-valued functions, 165

Measurable extended real-valued functions, 161

Equivalent conditions for the measurability of an extended real-valued function, 161

Measurable functions, 155

Examples, 156

Examples: indicator or characteristic functions, 157

Examples: monotone functions, 157

Function is measurable if and only if its positive and negative parts are measurable, 160

Three equivalent ways of proving a function is measurable, 156

Pointwise infimums and supremums of measurable extended real-valued functions, 165

Pointwise limits of measurable extended real-valued functions are measurable, 165

Properties of measurable extended real-valued functions, 163

Properties of measurable functions, 158

Set of measurable extended real-valued functions

Extended real-valued measurability in terms of the finite part of the function, 162

Set of measurable extended real-valued functions $M(X, \mathcal{S})$, 161

INDEX

Sigma paces are sets plus a sigma-algebra, 149

Sigma-algebras, 149

Sigma-algebras: examples, 150

Sigma-algebras: the smallest sigma-algebra that contains a collection of subsets is the sigma-algebra generated by the collection, 151

Measure

    Borel, 315

    Lebesgue - Stieltjes, 315

    Measurable cover, 266

Measure and integration

    L_1 equivalence relation $f \sim g$ if $f = g$ μ a.e., 225

    L_1 semi-norm, 224

    $L_1(X, \mathcal{S}, \mu)$ is a linear space, 210

    $L_p(X, \mathcal{S}, \mu)$ is the set of functions that are p-summable, 227

    $L_p(X, \mathcal{S}, \mu)$

        L_p semi-norm, 231

        Sequences that Converge in p-norm possess subsequences converging pointwise a.e., 237

        Cauchy - Bunyakovskiĭ - Schwartz Inequality, 229

        Cauchy sequences in norm, 232

        Complete normed linear spaces, 233

        Conjugate index pairs p and q, 228

        Equivalence classes under $f \sim g$ if $f = g$ μ a.e, 232

        Hölder's Inequality $1 < p < \infty$, 228

        It is a complete normed linear space, 233

        It is a normed linear space, 232

        It is a vector space, 231

        Minkowski's Inequality: $1 \leq p < \infty$, 229

        Real number conjugate indices inequality, 228

        Sequences that converge in $\mathcal{L}_p$ are Cauchy, 233

    $\mathcal{L}_1(X, \mathcal{S}, \mu)$ is a normed linear space, 225

    $\mathcal{L}_1(X, \mathcal{S}, \mu)$ is the set of equivalence classes of $L_1(X, \mathcal{S}, \mu)$, 225

    Absolute continuity of a measure, 200

    Abstract integration is additive, 196

    Alternative abstract integration schemes, 214

        $DAI(f(x) \equiv c)$ is $c\mu(X)$, 217

        $DAI(f) = 0$ if $\mu(X) = 0$, 217

        $DAI(f, m_0, M_0)$ is independent of the choice of m_0 and M_0: so call it $DAI(f)$, 217

        Abstract Darboux Integrability, 215

        Abstract Darboux Integral absolute inequality, 220

        Abstract Darboux Integral is additive, 219

        Abstract Darboux Integral is monotone, 218

        Abstract Darboux Integral Is scalable, 220

        Abstract Darboux Integral lower and upper bounds, 217

Abstract Darboux Integral measures, 217

Abstract Darboux Integral measures are absolutely continuous, 220

Comparing upper and lower sums for any two partitions, 216

Comparing upper and lower sums with refinements, 216

Lower abstract Darboux Integral equals the Upper abstract Darboux Integral, 216

Lower abstract Darboux Integral is dominated by the upper abstract Darboux Integral, 216

Measurable sets induced by a partition, 214

Upper and lower abstract Darboux Integrals are finite, 215

Upper and lower Darboux sums, 215

Characterizing the limit inferior and limit superior of sequences of sets, 178

Charges, 174

Complete measures, 191

    Equality a.e. can imply measurability even if the measure is not complete, 192

    Equality a.e. implies measurability if the measure is complete, 191

Constructing measures from non-negative measurable functions, 199

Convergence Theorems

    Abstract integration is additive, 196

    Extended Monotone Convergence Theorem, 195

    Fatou's Lemma, 198

    Monotone Convergence Theorem, 194

Counting measures, 238

    ℓ_p spaces, 239

    Cauchy - Bunyakovskiĭ - Schwartz Inequality: ℓ_2 sequence spaces, 240

    Hölder's Inequality: ℓ_p sequence spaces: $1 < p < \infty$, 239

    Minkowski's Inequality: ℓ_p sequence spaces: $1 \leq p < \infty$, 240

Disjoint decompositions of unions, 179

Equality a.e. problems, 190

Essentially bounded functions, 240

    L_∞ semi-norm, 243

    $\mathcal{L}_\infty(X, \mathcal{S}, \mu)$ is a complete normed linear space, 245

    $\mathcal{L}_\infty(X, \mathcal{S}, \mu)$ is a normed linear space, 244

    $\mathcal{L}_\infty(X, \mathcal{S}, \mu)$: the space of essentially bounded functions, 241

    Essentially bounded functions bounded above by their essential bound a.e., 242

    Alternate characterization of essentially bounded functions, 241

    Essentially bounded functions that are equivalent have the same essential bound, 242

INDEX

Hölder's Inequality: $p = 1$, 246

Extended Monotone Convergence Theorem Two, 201

Extending integration to extended real-valued functions
  Summable function equal a.e. to another function with measure complete implies the other function is also summable, 204
  Summable function equal a.e. to another measurable function implies the other function is also summable, 204
  Summable functions: $L_1(X, \mathcal{S}, \mu)$, 202
  Summable implies finite a.e., 203

Function f zero a.e. if and only if its integral is zero, 200

Hilbert space $\mathcal{L}_2(X, \mathcal{S}, \mu)$
  Complete in the norm induced by the inner product, 248
  Definition of an inner product, 247
  Inner product, 247
  It is a complete normed linear space, 247

Integral of a non-negative measurable function, 184

Integral of a simple function, 182

Integrals of summable functions create charges, 207

Lebesgue's Dominated Convergence Theorem, 211

Levi's Theorem, 205

Limit inferior and limit superior of monotone sequences of sets, 178

Limit inferior and limit superior of sequences of sets, 178

Measures, 174
  σ-finite, 174
  Examples, 174
  Motivations, 173

Metric on a set, 223

Monotonicity of the abstract integral for non-negative functions, 189

Non-measurable Lebesgue sets: no proof yet, 190

Norm convergence, 224

Norms on a vector space, 223

Properties of $L_1(X, \mathcal{S}, \mu)$, 209

Properties of measures
  Measure of monotonic sequences of sets, 176
  Monotonicity, 175

Properties of simple function integrations, 186

Propositions holding almost everywhere, 181

Simple functions, 182

There are Lebesgue measurable sets that are not Borel sets, 190

Two functions equal a.e. with respect to Borel measure but only one is Lebesgue measurable, 191

Monotone functions
  Continuous at x from left if and only if $u(x) = 0$, 34
  Continuous at x From right if and only if $v(x) = 0$, 34
  Continuous at x if and only if $u(x) = v(x) = 0$, 34
  Left-hand jump at x, $u(x)$, 34
  Right-hand jump at x, $v(x)$, 34
  Saltus function
    Properties, 37
  Total jump at x, $u(x) + v(x)$, 34

Partitions
  Gauge or norm of a partition, 31

Proposition
  Chocolate makes one happier, 11
  Refinements and Common Refinements, 30

Riemann Integration
  As a linear functional, 29
  Common refinements, 30
  Gauge or norm of a partition, 31
  Infimum Tolerance Lemma, 31
  Partitions of $[a, b]$, 30
  Refinements of partitions, 30
  Supremum Tolerance Lemma, 31

Riemann Integration overview
  Basic setup of a Riemann integral, 15
  Cauchy Fundamental Theorem of Calculus, 22
  Definition of Riemann integrability, 17
  Existence of the Riemann Integral, 17
  Fundamental Theorem of Calculus, 19
  Idea of an antiderivative, 15
  Indefinite integral of f is also the antiderivative, 23
  Integration using inverse images of functions values, 18
  Riemann integral of functions which have points of discontinuity
    Jump discontinuities, 25
    Removable discontinuities, 24
  Riemann sums, 16
  Symbol for the Antiderivative of f is $\int f$, 23
  Symbol for the definite integral of f on $[a, b]$ is $\int_a^b f(t)\, dt$, 23

Theorem
  $DAI(f(x) \equiv c)$ is $c\mu(X)$, 217
  $DAI(f) = 0$ if $\mu(X) = 0$, 217
  $DAI(f, m_0, M_0)$ is independent of the choice of m_0 and M_0, 217
  $L(f, m_0, M_0) \geq U(f, m_0, M_0)$, 216

INDEX

$L(f, m_0, M_0) \leq U(f, m_0, M_0)$, 216

$L(f, \pi_1) \leq U(f, \pi_2)$, 61, 216

$L(f, g, \pi_1) \leq U(f, g, \pi_2)$, 122

L_1 semi-norm, 224

L_∞ semi-norm, 243

L_p is a vector space, 231

L_p semi-norm, 231

$RI([a, b])$ is a vector space and $RI(f; a, b)$ is a linear mapping, 56

V_f and $V_f - f$ are monotone for a function f of bounded variation, 50

π' refines π implies $L(f, \pi) \leq L(f, \pi')$ and $U(f, \pi) \geq U(f, \pi')$, 216

$\pi \preceq \pi'$ implies $L(f, \pi) \leq L(f, \pi')$ and $U(f, \pi) \geq U(f, \pi')$, 59

$\pi \preceq \pi'$ implies $L(f, g, \pi) \leq L(f, g, \pi')$ and $U(f, g, \pi) \geq U(f, g, \pi')$, 122

$\mathcal{L}_1$ is separable, 294

$\mu^* \times \nu^*(A \times B) = \mu^*(A) \, \nu^*(B)$ for A and B μ^* and ν^* measurable, 389

$\mu_f^*(E) = \mu^*(f(E))$, 398

$f \in BV([a, b]) \cap C([a, b])$ if and only if V_f and $V_f - f$ are continuous and increasing, 54

f Riemann - Stieltjes Integrable with respect to g of bounded variation implies integrable with respect to V_g and $V_g - g$, 127

f bounded and continuous at all but finitely many points implies f is Riemann Integrable, 91

f bounded and continuous at all but one point implies f is Riemann Integrable, 90

f bounded variation and g continuous implies Riemann - Stieltjes Integral exists, 137

f continuous and g Riemann Integrable implies $f \circ g$ is Riemann Integrable, 99

f continuous and g bounded variation implies Riemann - Stieltjes Integral exists, 136

f is absolutely continuous if and only if $\mu_f << \mu$, 400

μ^* measurable sets form an algebra, 253

μ^* measurable sets properties, 254

Absolutely Continuous Recapture Theorem, 410

Abstract Darboux Integral absolute inequality, 220

Abstract Darboux Integral is additive, 219

Abstract Darboux Integral is monotone, 218

Abstract Darboux Integral is scalable, 220

Abstract Darboux Integral measures, 217

Abstract Darboux Integral measures are absolutely continuous, 220

Abstract Darboux Integral zero implies $f = 0$ a.e., 220

Abstract Integral Darboux lower and upper bounds, 217

Abstract integration is additive, 196

Almost uniform convergence implies convergence in measure, 335

Almost uniform convergence implies pointwise convergence and so limit function is measurable, 326

Alternate characterization of essentially bounded functions, 241

Antiderivative of f, 81

Approximation of a summable function with a continuous function, 293

Approximation of non-negative measurable functions by monotone sequences, 168

Approximation of the Riemann Integral, 83

Average value for Riemann Integrals, 81

Bounded differentiable implies bounded variation, 45

Bounded Variation implies Riemann Integrable, 72

Cauchy - Schwartz Inequality, 229

Cauchy - Schwartz Inequality: ℓ_2 sequence spaces, 240

Cauchy in measure implies a convergent subsequence, 328

Cauchy in measure implies completeness, 331

Cauchy's Fundamental Theorem, 82

Characterizing Lebesgue - Stieltjes measures constructed from absolutely continuous functions, 414

Conditions for OMI-F measures, 268

Conditions for OMI-FE measures, 269

Constant functions are Riemann Integrable, 72

Constructing measures from non-negative measurable functions, 199

Constructing outer measures via premeasures, 264

Continuous approximation of a ample function, 293

Continuous approximation of a characteristic function, 292

Continuous implies Riemann Integrable, 71

Convergence relationships on finite measure space four, 345

Convergence relationships on finite measure space one, 344

Convergence relationships on finite measure space three, 344

Convergence relationships on finite measure space two, 344

Convergence relationships on general measurable space, 345

Convergence relationships with p-domination five, 346

Convergence relationships with p-domination four, 346

INDEX

Convergence relationships with p-domination one, 345

Convergence relationships with p-domination six, 346

Convergence relationships with p-domination three, 346

Convergence relationships with p-domination two, 345

Convergent subsequences exist, 346

Egoroff's Theorem, 336

Equality a.e. can imply measurability even if the measure is not complete, 192

Equality a.e. implies measurability if the measure is complete, 192

Equivalent absolute continuity conditions for charges, 359

Every Riemann Integrable function is Lebesgue Integrable and the two integrals coincide, 375

Existence of a measurable cover for a set in $X \times Y$, 391

Existence of a measurable cover for a set in $X \times Y$ using iterated integrals, 394

Extended Monotone Convergence Theorem, 195

Extended Monotone Convergence Theorem Two, 201

Fatou's Lemma, 198

Fubini's Theorem for a rectangle: n dimensional, 384

Fubini's Theorem on a rectangle: 2D, 382

Fubini's Theorem: 2D: a top and bottom closed curve, 386

Fubini's Theorem: 2D: a top and bottom curve, 385

Function of bounded variation continuous if and only if V_f is continuous, 52

Function of bounded variation is the difference of two increasing functions, 50

Functions
  Properties of absolutely continuous functions, 397

Functions of bounded variation always possess right- and left-hand limits, 51

Functions of bounded variation are bounded, 44

Functions of bounded variation are closed under addition, 45

Functions of bounded variation are differentiable a.e, 405

Functions of bounded variation have countable discontinuity sets, 52

Fundamental abstract integration inequalities, 209

Fundamental Riemann - Stieltjes Integral estimates, 125

Fundamental Riemann Integral estimates, 57

Fundamental Theorem of Calculus, 77

General Fubini Theorem, 395

Hölder's Inequality: $1 < p < \infty$, 228

Hölder's Inequality: $p = 1$, 246

Hölder's Inequality: sequence spaces: $1 < p < \infty$, 239

Hahn decomposition associated with a charge, 354

If f is continuous and zero a.e., the Riemann Integral of f is zero, 106

If f is non-negative and Riemann Integrable with zero value, then f is zero a.e., 106

Inner product on the space of square summable equivalence classes, 247

Integrals of summable functions create charges, 207

Integrand continuous and integrator continuously differentiable implies Riemann - Stieltjes Integrable, 137

Integrand Riemann Integrable and integrator continuously differentiable implies Riemann - Stieltjes Integrable, 138

Integration by parts, 85

Inverses of functions of bounded variation, 47

Jordan decomposition of a charge, 353

Jordan decomposition of a finite charge, 349

Lebesgue - Stieltjes measures
  Characterizations of Lebesgue - Stieltjes measures, 318

Lebesgue Decomposition Theorem, 371

Lebesgue measure is regular, 287

Lebesgue measure is translation invariant, 295

Lebesgue outer measure is a metric outer measure, 285

Lebesgue's Criterion for the Riemann Integrability of bounded functions, 101

Lebesgue's Dominated Convergence Theorem, 211

Leibnitz's rule, 86

Levi's Theorem, 205

Linearity of the Riemann - Stieltjes Integral, 112

Mean Value Theorem for Riemann Integrals, 80

Measurability and approximation conditions, 288

Measure induced by an outer measure, 257

Measure induced by an outer measure is complete, 258

Measures and metric Spaces
  Approximating sets with a Lebesgue - Stieltjes outer measure, 317

Measures and metric spaces
  Finite measure Borel sets can be approximated by closed sets, 290

INDEX

Finite measure Borel sets can be approximated by open sets, 291

Metric space of all finite measure sets is complete, 298

Metric space of finite measurable sets, 298

Minkowski's Inequality, 229

Minkowski's Inequality: sequence spaces: $1 \leq p < \infty$, 240

Monotone Convergence Theorem, 194

Monotone function has a countable number of discontinuities, 33

Monotone functions
Partition sum estimate, 32

Monotone functions are of bounded variation, 44

Monotone implies Riemann Integrable, 72

Non-Lebesgue measurable set, 297

Open Set Characterization Lemma, 154

Open sets in a metric space μ^* measurable if and only if μ^* is a metric outer measure, 262

Open sets in a metric space are OMI measurable, 259

p-Norm convergence implies convergence in measure, 334

Pointwise a.e. convergence plus domination implies p-norm convergence, 338

Pointwise Limits of Measurable Functions Are Measurable, 165

Product outer measure is a regular measure, 389

Products and reciprocals of functions Riemann - Stieltjes Integrable with respect to g of bounded variation are also integrable, 130

Products of functions of bounded variation are of bounded variation, 46

Properties of f_c, 38

Properties of the Riemann - Stieltjes Integral, 124

Properties of the Riemann Integral, 67

Radon - Nikodym Theorem: finite charge case, 363

Radon - Nikodym Theorem: sigma-finite charge case, 367

Recapture Theorem, 82

Representing the Cantor Set, 304

Representing the product measure as an iterated integral, 392

Riemann - Lebesgue Lemma, 102

Riemann - Stieltjes Fundamental Theorem of Calculus, 133

Riemann - Stieltjes Integrable functions are Lebesgue - Stieltjes Integrable and the two integrals coincide, 377

Riemann - Stieltjes Integral Equivalence Theorem, 123

Riemann - Stieltjes Integral exists on subintervals, 126

Riemann - Stieltjes Integral for two bounded functions, 111

Riemann - Stieltjes Integral is additive on subintervals, 127, 130

Riemann - Stieltjes Integral is order preserving, 126

Riemann - Stieltjes Integration by parts, 113

Riemann Integrable f and g equal a.e. have the same integral, 107

Riemann Integrable f and g satisfy $d_1(f, g) = 0$ implies f is g a.e., 107

Riemann Integral equivalence theorem, 63

Riemann Integral exists on subintervals, 75

Riemann Integral is additive on subintervals, 75

Riemann Integral is order preserving, 58

Riemann Integral Limit Interchange Theorem, 93

Sequences of equivalence classes in L_p that converge are Cauchy, 233

Sequences that converge in p-norm possess subsequences converging pointwise a.e., 237

Set of equivalence classes of L_1 is a normed linear space, 225

Set of equivalence classes of L_∞ is a normed linear space, 244

Set of equivalence classes of L_p is a normed linear space, 232

Simple Fubini Theorem, 394

Space of equivalence classes of L_∞ is a Banach space, 245

Space of equivalence classes of L_p is a Banach space, 233

Space of square summable equivalence classes is a Hilbert space, 248

Substitution in Riemann integration, 85

Summable Function equal a.e. to another function with measure complete implies the other function is also summable, 204

Summable Function equal a.e. to another measurable function implies the other function is also summable, 204

Summable functions form a linear space, 210

Summable implies finite a.e., 203

Total variation is additive on intervals, 48

Total variation of a function of bounded variation, 45

Two Riemann Integrable functions match at all but finitely many points implies integrals match, 89

Upper and lower abstract Darboux Integrals are finite, 215

Upper and lower Darboux integrals are finite, 61

Upper and lower Riemann - Stieltjes Darboux Integrals are finite, 122

Variation function of a function f of bounded variation, 49

Variation of a charge in terms of the plus and minus parts, 357

Variation of a charge is a measure, 356

Vitali Convergence Theorem, 342

Vitali Covering Theorem, 402

Weak Monotone Recapture Theorem or monotone functions have summable derivatives, 408

Weierstrass Approximation Theorem, 96

Theory of Riemann - Stieltjes Integration

Bounded variation integrators, 127

Products and reciprocals of functions Riemann - Stieltjes Integrable with respect to g of bounded variation are also integrable, 130

Existence

Comparing upper and lower Riemann - Stieltjes Darboux sums on any two partitions, 122

Comparing upper and lower Riemann - Stieltjes Darboux sums on refinements, 122

Finite jump step functions as integrators, 120

One jump step functions as integrators: one, 115

One jump step functions as integrators: three, 119

One jump step functions as integrators: two, 117

Riemann - Stieltjes criterion for integrability, 123

Riemann - Stieltjes Darboux Integral, 122

Riemann - Stieltjes Integral Equivalence Theorem, 123

Step functions, 115

Upper and lower Riemann - Stieltjes Darboux Integrals, 122

Upper and Lower Riemann - Stieltjes Darboux sums, 121

Fundamental Riemann - Stieltjes Integral estimates, 125

Integrand and the integrator, 112

Linearity of the Riemann - Stieltjes Integral, 112

Properties of the Riemann - Stieltjes Integral, 124

Riemann - Stieltjes Integral exists on subintervals, 126

Riemann - Stieltjes Integral for two bounded functions, 111

Riemann - Stieltjes Integral is additive on subintervals, 127, 130

Riemann - Stieltjes Integral is order preserving, 126

Riemann - Stieltjes Integration by parts, 113

Riemann - Stieltjes sum, 111

Upper and lower Riemann - Stieltjes Darboux Integral is additive on intervals, 126

Theory of Riemann Integration

$RI([a, b])$ is a vector space and $RI(f; a, b)$ is a linear mapping, 56

Antiderivative of f, 81

Approximation of the Riemann Integral, 83

Average Value for Riemann Integrals, 81

Bounded Variation implics Riemann Integrable, 72

Cauchy's Fundamental Theorem, 82

Constant functions are Riemann Integrable, 72

Continuous implies Riemann Integrable, 71

Existence

Comparison of Darboux lower and upper sums for refinements, 59

Comparison of upper and lower sums on any two partitions, 61

Darboux integrability of a bounded function, 62

Darboux Upper and lower sums, 58

Riemann Integral Equivalence Theorem, 63

Riemann's Criterion, 62

Upper and lower Darboux integrals, 61

Fundamental Riemann Integral estimates, 57

Fundamental Theorem of Calculus, 77

Idea of an antiderivative is intellectually distinct from the Riemann integral of a bounded function f, 81

Integration by Parts, 85

Leibnitz's Rule, 86

Mean Value Theorem for Riemann Integrals, 80

Monotone implies Riemann Integrable, 72

Properties

Fundamental infimum and supremum equalities, 66

Properties of the Riemann Integral, 67

Riemann Integral exists on subintervals, 75

Upper and lower Darboux Integral is additive on intervals, 73

Recapture Theorem, 82

Riemann Integrability of a bounded function, 55

Riemann Integral is additive on subintervals, 75

Riemann Integral is order preserving, 58

Riemann sum, 55

Substitution in Riemann Integration, 85

INDEX

When Do Two Functions Have the Same Integral?, 88

    f bounded and continuous at all but finitely many points implies f is Riemann Integrable, 91

    f bounded and continuous at all but one point implies f is Riemann Integrable, 90

    Riemann Integrable f and g on $[a, b]$ match on (a, b) implies Riemann Integrals match, 89

    Two Riemann Integrable functions match at all but finitely many points implies integrals match, 89

When Do Two Functions have The Same Integral?

    f zero on (a, b) implies zero Riemann Integral, 88

Part X

Appendix: Undergraduate Analysis Background Check

Presented below are study guides and course description material from the first volume (Peterson (8) 2020). This is material you should know before you take a course in measure theory. This is also material many students are a bit deficient in, however, if you have studied carefully the material in (Peterson (8) 2020), and (Peterson (9) 2020) you will be well prepared. Although we do not teach the material of (Peterson (10) 2020) anymore, the discussions there on things like the inverse and implicit function theorems and Riemann integration in multiple dimensions are well worth the effort of self-study. If you have not taken courses explaining that kind of material or read the first and third volumes, you will need to do that before you can really appreciate the abstract nature of the present material.

Appendix A

Undergraduate Analysis Part One

This course, based on (Peterson (8) 2020), is about learning how to think carefully about mathematics. This is the first one where we really insist you begin to master one of the big tools of our trade: abstraction. Now, of course this is not easy as most of you have not really needed to learn how to do this in your previous courses. However, the time has come, and in this book, we work very hard to help you learn how to think well and deep about mathematical issues. You will learn how to do proofs of many propositions and learn a lot of facts about how functions of a real variable x behave. A lot of what we learn can be generalized (and for good useful reasons as we will see) to more abstract settings and we'll set pointers to the future about that as much as we can. But most of all, we will consistently challenge you to **think** about the *how* and *why* of all that we do. It is a lot of fun, of course! Here is a list of what is typically done in this class. You should understand all these things. In the first part of

the text which is Semester One of undergraduate analysis, we go over the following blocks of material.

Part One These are our beginning remarks you are reading now which are in Chapter 1.

Part Two: Understanding Smoothness
Here we are concerned with developing continuity and the idea of differentiation for both calculus on the real line and calculus in $\Re^2$.

- We think learning abstraction is a hard thing, so we deliberately start slowly. You have probably seen treatments of the real line and mathematical induction already, but in Chapter 2 we start at that basic level. We go over induction carefully and explain how to organize your proofs. We also work out the triangle inequality for numbers and introduce the important concept of the infimum and supremum of sets of numbers.

- In Chapter 3, we introduce sequences of real numbers and lay out a lot of notational machinery we will use later.

- In Chapter 4, we prove the important Bolzano - Weierstrass Theorem for bounded sequences and bounded sets and introduce sequential compactness. We also discuss carefully the limit inferior and superior of sequences in two equivalent ways.

- In Chapter 5, we go over more set theory and introduce topological compactness and finally prove the full equivalence that sequential and topological compactness on the real line are equivalent to the set being closed and bounded.

- In Chapter 6, we define limits of functions and explore limit inferior and limit superiors carefully.

- In Chapter 7, we talk about continuity.

- In Chapter 8, we go over the consequences of continuity of a function on an interval such the existence of global extreme values when the domain is compact.

- In Chapter 9, we step back and talk about other ways to guarantee the maximum and minimum of a function without requiring continuity and compactness of the domain. We introduce convex functions and subdifferentials.

- In Chapter 10, we introduce the idea of a derivative and talk about this level of smoothness in general.

- The many properties of differentiation are discussed in Chapter 11. We also introduce the trigonometric functions.

- In Chapter 12, the consequences of differentiation are explored such as Taylor polynomials and remainders.

- In Chapter 13, we use our knowledge of sequences and limits to develop exponential and logarithmic functions. We develop e^x first and then use it to define $\ln(x)$ as the inverse function of e^x.

- In Chapter 14, we go over standard extremal theory for functions of one variable.

- In Chapter 15, we introduce calculus ideas in $\Re^2$. This includes a nice development of simple surface graphing tools to let us see tangent planes and so forth visually. We develop tools to explore tangent plane approximation error

and discuss carefully the theorems that let us know when mixed partials match and when the local behavior of the first order partials guarantees the differentiability of the function.

- In Chapter 16, we develop standard extremal theory of functions to two variables. We also take a deeper look at symmetric matrices and their connection to extremal behavior.

A partial list of these topics is given below. All should look familiar!

Basics:

1. Mathematical Induction
2. Bounded sets.
3. Infimum and supremum of a bounded set.
4. **Theorem:** If S is bounded above and $M = \sup S$, then for any $y < M$, there is an $x \in S$ such that $y < x \leq M$. (Similar type theorem for infimums).
5. **Theorem:** Every nonempty set bounded above has a supremum. Every nonempty set bounded below has an infimum.
6. **Theorem:** Triangle Inequality and Reverse Triangle Inequality.

Functions, Sequences and Limits:

1. Functions $f : S \to \Re$, where S is a subset of $\Re$.
2. Let $f : S \to \Re$; $\lim_{x \to x_0} f(x)$.
3. Let $f : S \to \Re$.
   (a) S is unbounded above: $\lim_{x \to \infty} f(x)$, convergence or divergence.
   (b) S is unbounded below: $\lim_{x \to -\infty} f(x)$, convergence or divergence.
4. Sequences
   (a) $f : N_0 \to \Re$; N_0 is the set of integers starting at n_0.

UNDERGRADUATE ANALYSIS PART ONE

(b) Let $f(n) = a_n$. Then the sequence can be written as $\{a_n\}_{n_0}^\infty \equiv \{a_n\}$, where it is usually understood where the sequence begins.

  (a) $\lim_{n\to\infty} a_n = A$, convergence or divergence.

  (b) **Squeezing Lemma**.

  (c) **Theorem:** Uniqueness of Limits.

5. Operations with Sequences; $lim_{n\to\infty} \equiv \lim a_n$.

  (a) **Theorem:** If sequence converges to A, so does any subsequence.

  (b) **Theorem:** If sequence converges, sequence forms bounded set.

  (c) **Theorem:** $\lim(a_n + b_n)$.

  (d) **Theorem:** $\lim(a_n - b_n)$.

  (e) **Theorem:** $\lim(a_n b_n)$.

  (f) **Theorem:** $\lim(\frac{a_n}{b_n})$.

  (g) **Theorem:** $\lim|a_n| = |lim(a_n)|$.

  (h) **Theorem:** If $a_n \geq b_n$, then $\lim a_n \geq \lim b_n$.

6. Limits of Functions

  (a) Limits from the right and the left.

  (b) Operations with Limits of Functions: (All of the items under operations with sequences translated to the function setting).

7. Monotone Sequences

  (a) Increasing and Decreasing Sequences.

  (b) Non-increasing and non-decreasing Sequences.

  (c) **Completeness Axiom** Non-decreasing sequences bounded above converge and $\lim a_n = \sup a_n$.
    Theorem: (Similar statement for non-increasing sequences).

8. Monotone Functions

  (a) Increasing and decreasing functions.

  (b) Non-decreasing and non-increasing functions.

Continuity and Limits Revisited:

1. Continuity of $f : S \to \Re$ at $a \in S$.

2. Discontinuity of $f : S \to \Re$ at $a \in S$.

3. Uniform continuity of $f : S \to \Re$ on S.

4. Operations with continuous functions.

  (a) Same items as in operations with functions.

  (b) **Theorem:** Intermediate Value Theorem.

  (c) **Theorem:** Inverse Functions.

5. Cluster Points and Accumulation Points.

  (a) Cluster Point of a sequence.

  (b) Cluster point of a function.

  (c) **Theorem:** Bolzano - Weierstrass Theorem. Every bounded infinite set has at least one accumulation point and every bounded sequence has at least one cluster point.

  (d) **Theorem:** Cauchy Criterion Sequences.

  (e) **Theorem:** Cauchy Criterion Functions.

6. Limit Inferior

7. Limit Superior

8. **Theorem:** Let $[a, b]$ be finite. If f is continuous on $[a, b]$, then f is bounded on $[a, b]$.

9. **Theorem:** Let $[a, b]$ be finite. Then there is point $x_0 \in [a, b]$ with $f(x_0) \geq f(x) \forall x \in [a, b]$ (Continuous function achieves maximum on finite interval)

10. **Theorem:** Let $[a, b]$ be finite. Then f continuous on $[a, b]$ implies f is uniformly continuous on $[a, b]$.

Derivatives: Here are some derivative ideas.

1. Derivative of a function

2. **Theorem:** f has derivative at a implies f is continuous at a.

3. **Theorem:** Chain Rule.

4. Secant Lines, Tangent Lines.

5. **Theorem:** If f attains an extremum at a, then if $f'(a)$ exists, $f'(a) = 0$.

6. **Theorem:** Rolle's Theorem.

7. **Theorem:** Mean Value Theorem.

8. **Theorem:** If $f'(x) = g'(x)$, then $f(x) g(x) + c$, where c is a constant.

9. **Theorem:** f non-decreasing implies $f'(a) \geq 0$ at every point a where f is differentiable.

10. **Theorem:** $f'(x) \geq 0$ on interval implies f is non-decreasing on the interval. If $f'(x) > 0$, then f is increasing on the interval.

11. **Theorem:** Cauchy Mean value Theorem.

12. **Theorem:** L'Hopital's Rule.

13. **Theorem:** Taylor's Theorem with Remainder.

14. Maximum and Minimum Values of f.

    (a) Critical Points of f.

    (b) **Theorem:** First Derivative Test.

    (c) **Theorem:** Second Derivative Test.

A-1 Sample Exams

Then, to check yourself, you should be able to do any of these sample exams.

A-1.1 Exam 1

Instructions:
This is a closed book and closed notes test. You will need to give us all the details of your arguments as that is the only way we can decide if partial credit is warranted.

Part 1: Definitions (24 Points)

1. (3 Points) Give a precise mathematical definition of the meaning of a sequence of real numbers.

2. (3 Points) Give a precise mathematical definition of the phrase "the sequence $\{a_n\}$ converges to a".

3. (9 Points) Let S be a nonempty set of real numbers. Give a precise mathematical definition of the following phrases:

   (a) U is an upper bound of S.

   (b) U is the least upper bound of S.

   (c) u is a maximal element of S.

4. (3 Points) State the Principle of Mathematical Induction.

5. (3 Points) State the Triangle Inequality for Real Numbers.

6. (3 Points) Give a precise mathematical definition of the phrase "the limit of the function f(x) as x tends to infinity is A".

Part 2: Short Answer (36 Points)
You must determine whether or not these statements are true. If the statement is true, you MUST give us the reason why it is true; if the statement is false, you must give us a counterexample.

1. (4 Points) Is it true that all sequences which converge are bounded?

2. (4 Points) Is it true that if a sequence is bounded it must converge?

3. (4 Points) If $| a_n | \to | a |$, is it true that $a_n \to a$?

4. (4 Points) If $a_n \to a$, is it true that $| a_n | \to | a |$?

Show all your work on the short calculational exercises below. You may use a calculator if you wish.

UNDERGRADUATE ANALYSIS PART ONE

1. (5 Points) Find the inf and sup of the set
$$S = \{x \in (0, \frac{\pi}{2}) \mid \tan(x) < e^{-x}\}$$
You can indicate these points graphically.

2. (5 Points) Give an example of a function whose domain is the real numbers which has a finite limit as x tends to infinity.

3. (5 Points) Give an example of a sequence of real numbers which does not converge.

4. (5 Points) Give an example of a nonempty set of real numbers for which the inf and sup are the same.

Part 3: Proofs (40 Points)

Provide careful proofs of the following propositions. You will be graded on the mathematical correctness of your arguments as well as your use of language, syntax and organization in the proof.

1. (20 Points)
Proposition: For all $n \geq 1$,

$$\frac{1}{1 \cdot 2} + \ldots + \frac{1}{n \cdot (n+1)}$$
$$= \frac{n}{n+1}$$

2. (20 Points)
Proposition: If the sequence $\{a_n\}$ is defined by
$$a_n = \frac{-4n^2 - 2n + 3}{n^2 - 3n}, \quad n \geq 2$$
then the sequence converges.

A-1.2 Exam 2

Instructions:
This is a closed book and closed notes test. You will need to give us all the details of your arguments as that is the only way we can decide if partial credit is warranted.

Part 1: Definitions (36 Points)

Give precise mathematical definitions of the following mathematical concepts using the $\epsilon - \delta$ formalism:

1. (3 Points) "The sequence $\{a_n\}$ converges to a".

2. (4 Points) "The function $f : D \subseteq R \to R$ has a limit A at the point $a \in D$".

3. (4 Points) "The function $f : D \subseteq R \to R$ is **continuous** at the point $a \in D$".

4. (4 Points)"The function $f : D \subseteq R \to R$ is **uniformly continuous** in D".

Give precise mathematical definitions of the following mathematical concepts:

1. (3 Points) "The sequence $\{a_n\}$ is a **monotone** sequence".

2. (3 Points) "The function $f : [0, 1] \to R$ is **monotone increasing**".

State precisely the following mathematical theorems and/or axioms:

1. (5 Points) The **Completeness Axiom**.

2. (5 Points) The **Intermediate Value Theorem**.

3. (5 Points) The **Bolzano - Weierstrass Theorem**.

Part 2: Short Answer (30 Points)

You must answer the following questions. If the answer is **YES**, you MUST give us the reason why; if the answer is **NO**, you must also give a reason.

1. (3 Points) If the sequence $\{a_n\}$ is increasing and bounded above by M, does the sequence converge?

2. (3 Points) If the sequence $\{a_n\}$ is increasing and bounded above by M, does the sequence converge to M?

3. (3 Points) If the sequence $\{a_n\}$ is bounded above by M and below by L, does the sequence converge?

BASIC ANALYSIS IV: MEASURE THEORY AND INTEGRATION

4. (3 Points) If the sequence $\{a_n\}$ bounded above by M and below by L, does the sequence have at least one convergent subsequence?

5. (4 Points) Consider $f : [0, 1] \to R$. Is it possible for f^2 to be continuous on $[0, 1]$ and f to be discontinuous at each point in $[0, 1]$?

6. (4 Points) Consider $f : (0, 1) \to R$. Is it possible for f to be continuous only at $x = .5$ and discontinuous everywhere else in $(0, 1)$?

7. (4 Points) Consider $f : (0, 1) \to R$. Is it possible for f to be continuous on the irrational numbers in $(0, 1)$ and discontinuous on the rational numbers in $(0, 1)$?

Show all your work on the short calculational exercise below. You may use a calculator if you wish.

1. (6 Points) Prove that there is a solution to the equation

$$x + 10 sin(x) = 0$$

on the interval $[1.0, 4.5]$.

Part 3: Proofs (34 Points)

Provide careful proofs of the following propositions. You will be graded on the mathematical correctness of your arguments as well as your use of language, syntax and organization in the proof.

1. (17 Points)
   Proposition: If the function $f : D \subseteq R \to R$ is **continuous** at $a \in D$, then there exists $\delta > 0$ and $M > 0$ such that $|f(x)| < M$ if $|x - a| < \delta$.

2. (17 Points)
   Proposition: If the sequence $\{a_n\}$ is defined by

$$a_1 = 1,$$
$$a_{n+1} = \frac{2a_n + 3}{4}, \quad n > 1$$

then the sequence converges.

A-1.3   Exam 3

Instructions:
This is a closed book and closed notes test. You will need to give us all the details of your arguments as that is the only way we can decide if partial credit is warranted.

Part 1: Definitions (28 Points)
Give precise mathematical definitions of the following mathematical concepts using the appropriate mathematical formalism:

1. (4 Points) "The **Cauchy Criterion** for a sequence $\{a_n\}$".

2. (4 Points) "A is a **cluster point** of the sequence $\{a_n\}$".

3. (4 Points) "The **limit superior** of the sequence $\{a_n\}$".

4. (4 Points) "The function $f : D \subseteq R \to R$, where D is the domain of the function f, has a derivative at $x_0 \in D$".

State precisely the following mathematical theorems:

1. (4 Points) **The Mean Value Theorem.**

2. (4 Points) The **Bolzano - Weierstrass Theorem.**

3. (4 Points) The **Rolle's Theorem.**

Part 2: Short Answer (38 Points)
You must answer the following questions. If the answer is **YES** or **NO**, you MUST give us the reason why; (e.g., the complete statement of a relevant theorem, a counterexample, etc.)

1. (4 Points) Is it possible for a function to be continuous at a point but not differentiable there?

2. (4 Points) Is it possible for a function to be differentiable at a point but not continuous there?

3. (4 Points) Is $f(x) = \sqrt{(x)}$ **uniformly continuous** on $[0, 1]$?

4. (4 Points) Is is possible for a function that is continuous on $[2, 5]$ to be unbounded?

UNDERGRADUATE ANALYSIS PART ONE

5. (4 points) Does every continuous function on $[0, 1]$ achieve a maximum value?

Show all your work on the short calculational exercises below. You may use a calculator if you wish.

1. (6 Points) If

$$f(x) = x^2, \ x \leq 1$$
$$= 2x - 1, \ x > 1$$

find $f'(x)$ at all points where the derivative exists.

2. (6 points) Find all cluster points of the sequence $\{a_n\}$, where for $n \geq 1$,

$$a_n = (-1)^n sin(\frac{n\pi}{4}) + \frac{1}{3^n}$$

and verify that each is indeed a cluster point.

3. (6 points) Use the Mean Value Theorem to show that if $x < y$, then

$$e^x(y - x) < e^y - e^x$$
$$< < e^y(y - x)$$

Part 3: Proofs (34 Points)

Provide careful proofs of the following propositions. You will be graded on the mathematical correctness of your arguments as well as your use of language, syntax and organization in the proof.

1. (17 Points)
Proposition: Let $f : (a, b) \to R$ and assume that $f'(x) < 0$ on (a, b). Then, f is **strictly decreasing** on (a, b).

2. (17 Points)
Proposition: If the sequence $\{a_n\}$ satisfies

$$|a_{n+1} - a_n| < \frac{1}{3^n}, \ n > 0,$$

then the sequence converges.

A-1.4   Final

Instructions:
This is a closed book and closed notes test. You will need to give us all the details of your arguments as that is the only way we can decide if partial credit is warranted.

Part 1: Definitions (66 Points)
Give precise mathematical definitions of the following mathematical concepts using the appropriate mathematical formalism:

1. (6 points) $f : [a, b] \to R$ is **continuous** at $a \in (a, b)$.

2. (6 points) $f : [a, b] \to R$ is **uniformly continuous** in $[a, b]$.

3. (6 points) The **Cauchy Criterion** for convergence of a sequence $\{a_n\}$.

4. (6 points) A is a **cluster point** of the sequence $\{a_n\}$.

5. (6 points) μ is the **least upper bound** of the nonempty and bounded set S.

6. (6 points) The lim inf $\{a_n\}$, when $\{a_n\}$ is a bounded sequence.

State precisely the following mathematical theorems or axioms:

1. (6 Points) The **Bolzano - Weierstrass Theorem**.

2. (6 Points) The **Completeness Property** of the real numbers.

3. (6 Points) **Rolle's Theorem**.

4. (6 points) **The First Derivative Test**.

5. (6 points) The **Intermediate Value Theorem**.

Part 2: Short Answer (66 Points)
Answer the following questions. If the answer is **YES** or **NO**, you MUST give us the reason why; (e.g., the **complete** statement of a relevant theorem, a counterexample, etc.)

1. (4 points) Is it possible for a continuous function on $[1, \infty)$ to be unbounded?

BASIC ANALYSIS IV: MEASURE THEORY AND INTEGRATION

2. (4 points) Is it necessarily true that the limit inferior of a sequence is a cluster point of the sequence?

3. (4 points) Is it possible for a function to be continuous and differentiable at only one point?

4. (4 points) If a sequence is bounded, does it have to converge?

5. (4 points) If the first derivative of a function is zero at a point, does the function have to have a maximum or minimum at that point?

6. (4 points) If the second derivative of a function exists at a point, is the first derivative of the function continuous at that point?

7. (10 points)
Use the Intermediate Value Theorem to prove that $x^5 + 2x^3 + x - 3$ has at least one real root. Use Rolle's Theorem to show that the root is unique.

8. (12 points)
Let the sequence $\{a_n\}$ be recursively defined by the formula

$$a_1 \;=\; 1,$$
$$a_{n+1} \;=\; \left(1 - \frac{1}{(n+1)^2}\right) a_n$$

Show that $\{a_n\}$ converges.

9. (10 points)
Find $lim_{x \to 0+} \frac{tan(x)-x}{x^3}$.

10. (10 points)
Show that

$$\left| sin(x) - \left(x - \frac{x^3}{6} + \frac{x^5}{120}\right)\right|$$
$$< \frac{1}{5040}, \text{ for } |x| \le 1$$

Part 3: Proofs (68 Points)

Provide careful proofs of the following propositions. You will be graded on the mathematical correctness of your arguments as well as your use of language, syntax and organization in the proof.

1. (17 points)
If $\{a_n\} \to a$, $\{b_n\} \to b$ and

$\{c_n\} \to c$ use an $\epsilon - \delta$ argument to prove that $\{a_n + b_n + c_n\} \to a + b + c$.

2. (17 points)
If $f : R \to R$ is defined by $f(x) = 3x^2 - 2x + 1$, use an $\epsilon - \delta$ argument to prove that f is continuous at $x = 2$.

3. (17 points)
Prove that if $f : [a, b] \to R$ has a bounded derivative $f'(x)$ on (a, b), then f is uniformly continuous on (a,b).

4. (17 points)
Prove for all positive integers

$$\frac{d}{dx}(x^n) \;=\; n\, x^{n-1}$$

Appendix B

Undergraduate Analysis Part Two

The intent of this course is to continue to teach you the basic concepts of analysis. We assume that you have all had a careful and complete grounding in the necessary skills to *read*, *write* and *understand* the process of proving a mathematical proposition. Here is a list of what is typically done in this class. You should understand all these ideas.

In the second part of (Peterson (8) 2020), we go over the following blocks of material.

Part Three: Integration, Function Sequences

In this part, we discuss integration theory as well as sequence and series of functions. This includes going over Fourier series and applying these ideas to some ordinary and partial differential equations.

- In Chapter 17, we introduce uniform continuity and some of its consequences.

- In Chapter 18, we begin discussing the important idea of the completeness of a space by introducing the idea of Cauchy sequences.

- In Chapter 19, we can use the idea of Cauchy sequences to better understand the behavior of the special sequences of partial sums of

a given sequence that give rise to the idea of a series. We discuss some of the standard tests for convergence and define the classic ℓ^p spaces.

- In Chapter 20, we explore series more by adding the ratio and root tests to our repertoire. We also make the study of the ℓ^p spaces more concrete by proving the Hölder's and Minkowski's Inequalities. This allows us to talk out norms and normed spaces a bit.

- In Chapter 21, we define the Riemann Integral and explore some basic properties.

- In Chapter 22, we add a discussion of Darboux Integration and prove the equivalence of it with Riemann Integration and what is called the Riemann Criterion. We also prove some more properties of integrable functions.

- In Chapter 23, we prove the most important theorem yet, which is the Fundamental Theorem of Calculus, which connects the idea of differentiation to the Riemann Integral. This leads to the Cauchy Fundamental Theorem of Calculus which is the one that allows us to evaluate many Riemann Integrals easily. We also use the Fundamental Theorem of Calculus to define $\ln(x)$ and then develop its inverse e^x. This is backward from the order we did this development in the first part. We also start to discuss when two functions have the same integral value.

- In Chapter 24, we begin our discussion of pointwise and uniform convergence of sequences of functions and interchange theorems. We also discuss the Weierstrass Approximation Theorem.

- In Chapter 25, we discuss the ideas of the convergence of sequences of functions in the context of the sequence of partial sums we get in constructing series of functions. In particular, we go over power series in detail and look at the interchange theorems for power series.

- In Chapter 26, we have enough tools to study when two functions have the same Riemann Integral and what kinds of functions are Riemann Integrable carefully.

- In Chapter 27, we discuss Fourier Series. This is important because we cannot handle the convergence issues here with our usual ratio tests and so forth. So new ideas are needed such as sequences of orthonormal functions in a vector space. Hence, more abstraction but for a good cause! We also introduce a fair bit of code to help with the calculations.

- In Chapter 28, we apply Fourier Series ideas to both ordinary and partial differential equations. We also show how to apply power series to ordinary differential equations when the coefficients are simple polynomials.

A partial list of these topics is given below. All should look familiar!

Integration:

$f : [a, b] \to \Re$ is bounded function on the finite interval $[a, b]$.

1. Partition P of $[a, b]$.

2. Refinements of Partition P' of P.

3. Norm of Partition P, $\|P\|$.

4. Upper sum $\overline{S}_P(f) \equiv U(P, f)$.

5. Lower sum $\underline{S}_P(f) \equiv L(P, f)$.

6. Fundamental Inequalities
$L(P, f) \leq U(P, f)$.
$L(P, f) \leq L(P', f)$.
$U(P', f) \leq U(P, f)$.
$L(Q, f) \leq U(P, f)$ for any two partitions P and Q of $[a, b]$.

7. Upper Integral
$\overline{\int_a^b} f(x)dx = \inf \{U(P, f) : P \text{ is a partition of}[a, b]\}$.

8. Lower integral
$\underline{\int_a^b} f(x)dx = \sup \{L(P, f) : P \text{ is a partition of}[a, b]\}$.

9. $\underline{\int_a^b} f(x)dx \leq \overline{\int_a^b} f(x)dx$.

10. $\inf_{x \in [a,b]}(f(x))(b - a) \leq \underline{\int_a^b} f(x)dx \leq \overline{\int_a^b} f(x)dx \leq \sup_{x \in [a,b]}(f(x))(b - a)$.

11. Algebra of upper and lower integrals

    (a) **Theorem:** $\overline{\int_a^b}(f(x)+c)dx = \overline{\int_a^b} f(x)dx + c(b - a)$.
    (Similar for lower integrals)

    (b) **Theorem:** $\overline{\int_a^b} f(x)dx = \overline{\int_a^c} f(x)dx + \overline{\int_c^b} f(x)dx$, for any $c \in [a, b]$.
    (Similar for lower integrals)

12. **Theorem:** (Weak Fundamental Theorem)
$\frac{d}{dx}(\overline{\int_a^x} f(t)dt)_{|c} = \frac{d}{dx}(\underline{\int_a^x} f(t)dt)_{|c} = f(c)$,
at any point $c \in [a, b]$ where f is continuous.

13. **Theorem:** f is Riemann Integrable if $\overline{\int_a^b} f(x)dx = \underline{\int_a^b} f(x)dx$ This common value is denoted $\int_a^b f(x)dx$.

14. What functions are Riemann Integrable?

    (a) **Theorem:** f monotone $\Rightarrow f$ Riemann Integrable.

    (b) **Theorem:** f continuous $\Rightarrow f$ Riemann Integrable.

15. **Theorem:** Necessary and Sufficient Condition for integral to exist

    (a) **Theorem:** f Riemann Integrable $\Leftrightarrow \forall \epsilon > 0 \exists$ partition P such that $U(P, f) - L(P, f) < \epsilon$.

16. Upper and Lower integral estimates

    Theorem: f bounded on $[a, b] \Rightarrow \forall \epsilon > 0 \exists \delta(\epsilon)$ so that
    $U(P, f) < \overline{\int_a^b} f(x)dx + \epsilon$

and $L(P, f) > \int_a^b f(x)dx - \epsilon$ for any partition P with $\|P\| < \delta(\epsilon)$.

17. Riemann Sum
$S(P, f, \xi)$

18. **Theorem:** $J = \lim_{\|P\| \to 0} S(P, f, \xi)$

19. **Theorem:** f Riemann Integrable $\Leftrightarrow J = \int_a^b f(x)dx$.

20. Properties of Riemann Integrable

 (a) **Theorem:** f Riemann Integrable $\Rightarrow F(x) = \int_a^x f(t)dt$ is continuous on $[a, b]$.

 (b) **Theorem:** Subset Property.

 (c) **Theorem:** Intermediate Point Property.

 (d) **Theorem:** $f(x) \geq g(x) \Rightarrow \int_a^b f(x)dx \geq \int_a^b g(x)dx$.

 (e) **Theorem:** Integral of linear combinations of functions is linear combination of integral of functions.

 (f) **Theorem:** f integrable $\Rightarrow f^+$ and f^- integrable.

 (g) **Theorem:** f integrable $\Rightarrow | f |$ integrable

 (h) **Theorem:** Integral Estimate.
 $| \int_a^b f(x)dx | \leq \int_a^b | f(x) | dx$

 (i) **Theorem:** f, g integrable $\Rightarrow fg$ integrable.

21. Fundamental Theorem of Calculus

 (a) Primitive (antiderivative, indefinite integral) of f

 (b) **Theorem:** Fundamental Theorem of Calculus
 f continuous and F primitive of $f \Rightarrow \int_a^b f(x)dx = F(b) - F(a)$

 (c) When does integrating f' recapture f?
 Theorem: f differentiable and f' integrable $\Rightarrow \int_a^b f'(x)dx = f(b) - f(a)$.

 (d) **Theorem:** Integration by Substitution

 (e) **Theorem:** $\frac{d}{dx}(\int_{u(x)}^{v(x)} f(t)dt) = v'(x)f(v(x)) - u'(x)f(u(x))$.

22. **Theorem:** Integration by Parts

Infinite Series:

1. Definition of a series:
 $\{a_n\}_{n_0}^\infty$ is a sequence.

 (a) Partial Sums: $S_p = \sum_{n_0}^{n_0+p-1} a_n$
 Sum of the first p terms of the sequence.

 (b) **Theorem:** If $\lim_{p \to \infty} S_p$ exists, series converges.
 Let $\sum_{n_0}^\infty a_n \equiv \lim_{p \to \infty} S_p$, where $\sum_{n_0}^\infty a_n$ denotes the sum of the series.

 (c) **Theorem:** If $\lim_{p \to \infty} S_p$ does not exist, series diverges.

 (d) Absolute convergence.

 (e) Conditional convergence.

2. Summing a series

 (a) Telescoping terms.

 (b) Partial sums.

3. Tests for Convergence and/or divergence

 (a) **Theorem:** n^{th} term test.

 (b) **Theorem:** Cauchy Criterion specialized to series.

 (c) **Theorem:** Absolute convergence implies original series converges.

4. **Theorem:** Tests for Convergence and/or divergence: series of non-negative terms:

 (a) **Theorem:** p series.

 (b) **Theorem:** geometric series.

 (c) **Theorem:** comparison test.

 (d) **Theorem:** limit comparison test.

 (e) **Theorem:** root test.

 (f) **Theorem:** ratio test.

5. Tests for Series with Variable signs

 (a) **Theorem:** Alternating Series Test (AST).

(b) Apply any of tests for series of non-negative terms to series of absolute values–checks for absolute convergence. If this fails, only option is AST.

Sequence and Series of Functions:
$\{f_n\} \equiv \{f_n\}_{n_0}^{\infty}$ is a sequence of functions defined on $[a, b]$.

1. Pointwise convergence of f_n to a function f on $[a, b]$.

2. Uniform convergence of $\{f_n\}$ on $[a, b]$

3. **Theorem:** $\{f_n\}$ converges uniformly to f on $[a, b] \Leftrightarrow M_n = \sup_{x \in [a,b]} \mid f_n(x) - f(x) \mid \to 0$ as $n \to \infty$.

   (a) Find f via pointwise limit first.

   (b) Once pointwise limit f known, compute M_n limit. If limit nonzero, convergence not uniform.

4. **Theorem:** Cauchy Criterion for uniform convergence

5. **Theorem:** Weierstrass K-Test
   This is for series of functions $\sum_{n_0}^{\infty} u_n(x)$.
   Let $K_n = \sup_{x \in [a,b]} \mid u_n(x) \mid$.
   Then $\sum_{n_0}^{\infty} K_n$ converges $\Rightarrow$ series of functions converges uniformly on $[a, b]$.

6. Consequences of Uniform Convergence

   (a) **Theorem:** If a sequence of continuous functions converges uniformly on $[a, b]$, then limit function is continuous.

   (b) **Theorem:** If series of continuous functions converges uniformly on $[a, b]$, then its sum is continuous function.

   (c) **Theorem:** Interchange of integration and limit
   $f_n \to$ uniformly f on $[a, b]$ where each f_n is integrable. Then f is also integrable and

$\lim \int_a^b f_n = \int_a^b \lim f_n$.
Also **Theorem:** (each f_n is continuous), **Theorem:** (each f_n is monotone), **Theorem:** (applied to series of functions).

(d) When does $f_n' \to f'$?
Theorem: A very technical statement but a very nice theorem because it pulls together a lot of results.

Taylor Series:

1. Power Series
   (a) Radius of convergence.
   (b) Interval of convergence.
   (c) What happens at endpoints of interval of convergence?

2. Power series converges absolutely inside interval of convergence and diverges outside interval of convergence. Must check behavior at endpoints explicitly.

3. Finding Radius of Convergence
   (a) **Theorem:** Ratio Test:
   $q = \lim \mid \frac{a_{n+1}}{a_n} \mid$.
   $R = 0$ if $q = \infty$.
   $R = \infty$ if $q = 0$.
   $R = \frac{1}{q}$ if q finite.
   (b) Root test

4. Properties of Power Series
   (a) **Theorem:** Power series converges uniformly on any closed subinterval contained in the interval of convergence.

   (b) **Theorem:** Let interval of convergence be $(-R, R)$. Then power series converges at R implies power series converges uniformly on $[0, R]$. (Similar result for convergence at other endpoint).

   (c) **Theorem:** If power series has interval of convergence $(-R, R)$, so does derived series.

   (d) **Theorem:** If $R > 0$, derivative of a power series is given by the derived series

UNDERGRADUATE ANALYSIS PART TWO 457

and the interval of convergence is same for both series. (same type result for integral of power series).

(e) **Theorem:** If $R > 0$, power series has derivatives of all orders and interval of convergence is always the same.

(f) **Theorem:** $f(x) = \sum a_n x^n$ and radius of convergence $R > 0$. Then $a_n = \frac{f^{(n)}(0)}{k!}$.

(g) **Theorem:** Power series expansions are unique.

5. Operating on Power Series

(a) **Theorem:** Integrating series.

(b) **Theorem:** Differentiating series.

(c) **Theorem:** Summing Series.

6. Taylor Series

(a) **Theorem:** Taylor's theorem with remainder revisited.

(b) **Theorem:** Taylor series expansion of arbitrary function.

7. Finding Taylor Series

(a) Finding pattern in derivatives.

(b) Using geometric series tricks.

(c) Using recursion trick for ratios of functions.

8. Using Taylor Series

(a) Integrating a function.

(b) Evaluating a limit.

B-1 Sample Exams

Then, to check yourself, you should be able to do any of these sample exams.

B-1.1 Exam 1

Instructions:
This is a closed book and closed notes test. You will need to give us all the details of your arguments as that is the only way we can decide if partial credit is warranted.

Part 1: Definitions (15 Points)

1. (15 Points) **Carefully** define and discuss the integral of a bounded function $f : [a, b] \to R$, where $[a, b]$ is a closed and finite interval. You will need to explain in detail all relevant symbols and terms.

Part 2: Short Answer (51 Points)
You must determine whether or not these statements are true. If the statement is true, you MUST give us the reason why it is true; if the statement is false, you must give us a counterexample.

1. (7 Points) Let $[a, b]$ be a closed and finite interval and assume $f, g :$ $[a, b] \to R$ are not **integrable** on $[a, b]$. Is it necessarily true that $f + g$ is not integrable on $[a, b]$?

Show all your work on the short calculational exercises below. You may use a calculator if you wish.

1. (17 Points) Consider $f(x) = x^2$ on the interval $[0, 1]$. Let P_n be the uniformly spaced partition of $[0, 1]$ consisting of n equal subintervals. Find $\overline{S}_{P_n} - \underline{S}_{P_n}$ and show that

$$\overline{S}_{P_n} - \underline{S}_{P_n} \to 0 \text{ as } n \to \infty.$$

Does this prove that $f(x) = x^2$ is integrable on $[0, 1]$? Explain your answer carefully.

2. Determine whether or not the following functions are integrable on $[0, 1]$ giving full reasons for your answers.

(a) (7 points)

$$f(x) = x^2 - 3x + 7$$

(b) (7 points)

$$f(x) = \frac{\pi}{2}, \ 0 \le x < .25$$
$$= 1.0, \ x = .25$$
$$= .6, \ .25 < x < .75$$
$$= .17, \ x = .75$$
$$= .038, \ .75 < x \le 1.0$$

BASIC ANALYSIS IV: MEASURE THEORY AND INTEGRATION

(c) (7 points)

$$f(x) = 2, \ x \in Q$$
$$= -9, \ x \in I$$

where Q and I indicate the rational and irrational numbers respectively.

(d) Differentiate the function f : $[1, 10] \to R$ defined by:

    i. (3 points)

$$f(x) = \int_1^x (1 + t^2)^{-2} \, dt$$

    ii. (3 points)

$$f(x) = \int_1^{x^2} (2 + t^3)^{-4} \, dt$$

Part 3: Proofs (34 Points)

Provide careful proofs of the following propositions. You will be graded on the mathematical correctness of your arguments as well as your use of language, syntax and organization in the proof.

1. (17 Points)
   Proposition: Let $[a, b]$ be a finite interval and let $f : [a, b] \to R$ be continuous and non-negative on $[a, b]$, i.e. $f(x) \geq 0$, $a \leq x \leq b$. Then if $\exists \ x_0$, $a < x_0 < b \ni f(x_0) > 0$, this implies that $\int_a^b f(x) \, dx > 0$.

2. (17 Points)
   Carefully state and prove a theorem for calculating

$$\frac{d}{dx} \left(\int_a^{u(x)} f(t) \, dt \right)$$

B-1.2   Exam 2

Instructions:
This is a closed book and closed notes test. You will need to give us all the details of your arguments as that is the only way we can decide if partial credit is warranted.

Part 1: Definitions (24 Points)

Carefully and precisely define the following mathematical phrases:

1. (6 Points)

$$\sum_{n=0}^{\infty} a_n = S.$$

In the phrases below, I is a finite interval, $\{f_n\}$ is a sequence of real-valued functions defined on I and f is a real-valued function defined on I.

1. (6 Points)

$$f_n(x) \ \to \ f(x) \text{ pointwise on } I.$$

2. (6 Points)

$$f_n(x) \ \to \ f(x) \text{ uniformly on } I.$$

3. (6 Points)

$$\sum_{n=0}^{\infty} f_n(x)$$

converges uniformly on I to the function $f(x)$.

Part 2: Short Answer (61 Points)

You must determine whether or not these statements are true. If the statement is true, you MUST give us the reason why it is true; if the statement is false, you must give us a counterexample.

1. (5 Points) Let I be a finite interval and $\{f_n\}$ a sequence of real-valued functions defined on I. If $f_n \to f$ pointwise on I, is it necessarily true that f is continuous on I?

2. (5 Points) If $a_n \to 0$ is it necessarily true that $\sum_{n=0}^{\infty} a_n$ converges?

For each of the following series, tell whether or not it is **absolutely convergent, conditionally convergent** or **divergent**, giving full reasons for your answers.

UNDERGRADUATE ANALYSIS PART TWO

1. (6 Points)

$$\sum_{n=0}^{\infty} \frac{2^n}{3\left(5^{n+1}\right)}$$

2. (6 Points)

$$\sum_{n=3}^{\infty} \frac{n^3 + 6n - 1}{n^4 - 17}$$

3. (6 Points)

$$\sum_{n=2}^{\infty} \frac{n!}{(2n-2)!}$$

4. (6 Points)

$$\sum_{n=1}^{\infty} \frac{1}{(2n)^n}$$

5. (6 Points)

$$\sum_{n=0}^{\infty} \left(-\frac{1}{2}\right)^n$$

6. (6 Points)

$$1 - \frac{1}{2} - \frac{1}{4} + \frac{1}{8}$$
$$- \frac{1}{16} - \frac{1}{32} + \frac{1}{64} - \ldots$$

(plus, then 2 minuses)

Show all your work on the short calculational exercises below. You may use a calculator if you wish.

1. (15 Points) Consider for all $n \geq 1$,

$$f_n(x) = \frac{n(x^2)}{1 + n(x^2)}$$

Discuss the **pointwise** and **uniform** convergence of f_n on the intervals:

(a) $[-1, 1]$

(b) $[1, 4]$

Part 3: Proofs (15 Points)

Provide careful proofs of the following propositions. You will be graded on the mathematical correctness of your arguments as well as your use of language, syntax and organization in the proof.

1. (15 Points)
If $\sum a_n$ and $\sum b_n$ both converge absolutely, then $\sum (a_n + b_n)$ converges absolutely.

B-1.3   Exam 3

Instructions:
This is a closed book and closed notes test. You will need to give us all the details of your arguments as that is the only way we can decide if partial credit is warranted.

Part 1: Convergence of Series (36 Points)

Find all the values of x at which the following series converge giving **full** reasons for your answers.

1. (12 Points)

$$\sum_{n=1}^{\infty} \frac{x^{2n}}{n\,3^n}$$

2. (12 Points)

$$\sum_{n=1}^{\infty} \frac{n!\,x^n}{n^n}$$

3. (12 Points)

$$\sum_{n=1}^{\infty} a_n\,x^n, \quad a_n = 1, \text{ if n is even}$$

$$= \frac{1}{n}, \text{ if n is odd}$$

Part 2: Taylor Series (48 Points)

Show all your work on the short calculational exercises below.

1. (16 Points) Let $f(x) = \frac{e^x - 1}{x}$, for $x \neq 0$.

(a) Find the power series that represents f for $x \neq 0$.

(b) Find the power series that represents f' for $x \neq 0$.

2. (16 points) Sum the series

$$\sum_{n=1}^{\infty} \frac{x^{n+1}}{n+1}$$

3. (16 Points) Let $f(x) = \int_0^x t\, sin(t)dt$. Find the Taylor series centered at $x = 0$ for f.

Part 3: Proofs (16 Points)

Provide careful proofs of the following proposition. You will be graded on the mathematical correctness of your arguments as well as your use of language, syntax and organization in the proof.

(16 Points)
Let $\{f_n\}$ and $\{g_n\}$ be two sequences of real-valued functions defined on the finite interval I. If $f_n \to f$ uniformly on I and $g_n \to g$ uniformly on I, then $(f_n + g_n) \to (f + g)$ uniformly on I.

B-1.4 Final

Instructions:
This is a closed book and closed notes test. You will need to give us all the details of your arguments as that is the only way we can decide if partial credit is warranted.

Part 1: Definitions (60 Points)

Give precise mathematical definitions of the following mathematical concepts using the appropriate mathematical formalism:

1. Convergence Issues
   Let $[a, b]$ be a finite interval of the real line with $\{f_n\}$ denoting a sequence of real-valued functions, $f_n : [a, b] \to \Re$ and f indicating another real-valued function, $f : [a, b] \to \Re$. Define the following concepts:

   (a) (3 points) $f_n \to f$ pointwise for $x \in [a, b]$.

   (b) (3 points) $f_n \to f$ uniformly on $[a, b]$.

2. Elementary Topology in $\Re^n$: Define the following terms:

   (a) (3 points) S is an open set in $\Re^n$.

   (b) (3 points) S is an closed set in $\Re^n$.

   (c) (3 points) $\mid x \mid$ for $x \in \Re^n$.

   (d) (3 points) The boundary of a set S in $\Re^n$.

3. Integration Theory:
   $f : [a, b] \to \Re$ is a bounded function on the finite interval $[a, b]$. Define the following concepts:

   (a) (3 points) P is a partition of $[a, b]$.

   (b) (3 Points) P' is a refinement of the partition P of $[a, b]$.

   (c) (3 Points) The Upper sum $\overline{S}(P, f)$ of the bounded function f on $[a, b]$ for partition P.

   (d) (3 Points) The Lower sum $\underline{S}(P, f)$ of the bounded function f on $[a, b]$ for partition P.

   (e) (3 Points) The Upper Integral of f on $[a, b]$.

   (f) (3 Points) The Lower Integral of f on $[a, b]$.

   (g) (3 Points) The integral of f on $[a, b]$.

4. Series:
   Let $\sum_{n_0}^{\infty} a_n$ denote an infinite series: Define the following concepts:

   (a) (3 Points) The nth partial sum of the series.

   (b) (3 Points) The convergence of the series.

5. Theorems
   State precisely the following mathematical theorems or axioms:

   (a) (5 Points) The Cauchy - Schwartz Inequality.

   (b) (5 Points) The Fundamental Theorem of Calculus.

   (c) (5 Points) Taylor's Theorem with Remainder.

Part 2: Short Answer (50 Points)

Answer the following questions. If the answer is **YES** or **NO**, you MUST give us the reason why; (e.g., the **complete** statement of a relevant theorem, a counterexample, etc.)

1. Let $f : [a, b] \to \Re$ for the finite interval $[a, b]$.

UNDERGRADUATE ANALYSIS PART TWO

(a) (5 Points) If f is not continuous on $[a, b]$, does that necessarily imply f is not integrable on $[a, b]$?

(b) (5 Points) If f is continuous on $[a, b]$, does that necessarily imply that f is integrable on $[a, b]$?

(c) (5 Points) If f is bounded on $[a, b]$, does that necessarily imply that f is integrable on $[a, b]$?

(d) (5 Points) If f^2 is integrable on $[a, b]$, does that necessarily imply that f is integrable on $[a, b]$?

2. Let $\sum a_n$ denote an infinite series.

(a) (5 Points) If $\lim_{n \to \infty} a_n = 0$, does that necessarily imply that the series $\sum a_n$ converges?

(b) (5 Points) If $\lim_{n \to \infty} \frac{a_{n+1}}{a_n} = 1$, does that necessarily imply that the series $\sum a_n$ diverges?

3. Let $f(x) = \sum a_n x^n$ denote a power series and R denote its radius of convergence.

(a) (5 Points) Let $R > 0$. If the interval $[a, b]$ is strictly contained in the interval $(-R, R)$, is it necessarily true that the power series $\sum a_n x^n$ converges uniformly in $[a, b]$?

(b) (5 Points) Is it possible for the radius of convergence of the f' series to be different from R?

4. Let $[a, b]$ be a finite interval and let $f : [a, b] \to \Re$ denote a function with derivatives of all orders. Let R be the radius of convergence of the Taylor series of f.

(a) (5 Points) Is it possible for f to have a Taylor series expansion with $R = 0$?

(b) (5 Points) Suppose f is defined on the interval $[-3, 3]$ and f fails to be continuous at the points -1.7 and 2.2. Is it possible for R to be 2.5?

Part 3: Calculational Exercises (56 Points)

Provide complete solutions with all appropriate detail to the following computational exercises.

1. (14 Points) Let

$$
\begin{aligned}
f_n(x) &= \frac{\sin(nx)}{nx}, \ x \neq 0 \\
&= 1, \ x = 0
\end{aligned}
$$

Determine whether the sequence $\{f_n\}$ converges uniformly on the following intervals:

(a) $[1, 6]$

(b) $[-1, 1]$

2. (14 Points) Determine whether or not the function

$$
f(x, y) = \frac{x^3 \tan(x \, (y^4))}{x^4 + y^4}
$$

is continuous at the point $(0, 0)$.

3. (14 Points) Let

$$
f(x) = \int_0^x \frac{1}{1 + t^8} dt
$$

(a) Find $f'(x)$

(b) Express $f(.5)$ without using integrals.

4. (14 Points) Find the Taylor series expansion of $f(x) = \tan^{-1}(x)$ about base point 0 and determine its radius of convergence.

Part 4: Proofs (34 Points)

Provide careful proofs of the following propositions. You will be graded on the mathematical correctness of your arguments as well as your use of language, syntax and organization in the proof.

1. (17 points)

Let $[a, b]$ be a finite interval of the real line with $\{f_n\}$ denoting a sequence of real-valued functions, $f_n : [a, b] \to \Re$. Assume $f_n \to 0$

uniformly on $[a, b]$ and the function $g : [a, b] \to \Re$ is bounded on $[a, b]$. Then $g f_n \to 0$ uniformly on $[a, b]$.

2. (17 points)
Let $f : R \to R$ be continuous on the finite interval $[a, b]$. If $\int_a^x f(x) \, dx = 0$ for all $x \in [a, b]$, then $f(x) = 0$ for all $x \in [a, b]$.

Part XI

Appendix: Linear Analysis Background Check

Appendix C

Linear Analysis

Presented below are study guides and course description material based upon portions of what is covered in (Peterson (9) 2020). The intent of the text is to develop your ability to think abstractly using a standard element of the practicing analyst's toolkit, abstract spaces. In addition, we try hard to continue to help develop your abilities to read and write good mathematics at the professional level. In this text, we go over the following blocks of material.

Part Two: Metric Spaces Here we are concerned with metric spaces and how we construct their completion.

- In Chapter 2, we spend a lot of time proving something most of you have taken for granted for a long time. The existence of $\Re$ as what is know as a complete metric space where Cauchy sequences always converge to an element of $\Re$. You should go through this carefully at this point in your education so you understand it really is a construction process and the thing we call $\Re$ is what we choose to identify with a space of equivalence classes of the right things. We also talk a lot about general metric spaces and the familiar friends we call the sequence spaces. But now we very explicitly explore everything more abstractly.

- Chapter 3 is a long chapter on how we complete metric spaces using

the fact that we have the complete metric space $\Re$ to build on. We put a lot of effort into completing various spaces of integrable functions for our use later. You should know this sort of completion is also done using integration based on measures but that is discussed in (Peterson (7) 2020) and so that point of view is inaccessible for now. But please keep this in mind as the proper way to proceed is to know both approaches.

Part Three: Normed Linear Spaces We discuss the idea of normed linear spaces carefully.

- In Chapter 4, we go back to covering the ideas of vector spaces again. In the first two books (Peterson (8) 2020) and (Peterson (10) 2020) we have gone over these ideas quite a bit and tried to give you extensive examples of how this stuff is used. But here, we want to be more abstract such as proving all vector spaces have a basis by using a non-constructive approach involving Zorn's Lemma. So this is uncharted territory.

- Chapter 5 looks at normed linear spaces again. We have discussed these a fair bit in the first two books, which are mostly modeled on $\Re^n$ with occasional side trips to function spaces. But now we want to introduce norms properly and discuss important ideas such a Schauder bases for various spaces. We also want to revisit compactness again and stress how our vision of what that means changes in infinite dimensional spaces.

- Chapter 6 is about linear operators between normed linear spaces and we no longer focus mostly on finite dimensional spaces. Hence matrix representations are not as useful. We also talk about the extension of the eigenvalue - eigenvector problem to this setting.

Part Four: Inner Product Spaces In this part, we go over many ideas loosely built around the idea of inner product spaces.

- Chapter 7 we introduce the inner product on a vector space and work out the eigenvalue and eigenvector structure of a Hermitian matrix using matrix norm ideas. We show such matrices always have an orthonormal basis of eigenvectors. This is almost the same as the proof we used for symmetric matrices in (Peterson (10) 2020). It is also almost the same proof we use later for the case of a self - adjoint linear operator which we do later. Please pay attention to how these proofs need to be modified to make the transition to the infinite dimensional setting. We also introduce Hilbert spaces and spend a lot of time completing the integrable functions with the least squares norm so we can characterize the Hilbert space of square integrable functions better. We finish with some properties of inner product space.

- Chapter 8 works out the detail of complete orthonormal sequences in a Hilbert space and projections.

- Chapter 9 is a first basic discussion of dual spaces and the double dual of a space. We find the duals of a few sequence spaces too. We also introduce the idea of weak convergence.

- In Chapter 10 we discuss the Hahn - Banach theorem and its extensions. We use Zorn's Lemma here again.

- Chapter 11 We now have better tools at our disposal and we can talk about reflexive spaces intelligently. We also work out the

dual of the set of continuous functions which is a non-trivial result which requires we introduce Riemann - Stieltjes integration in a fairly quick way. We cover that topic more fully in (Peterson (7) 2020) so if you want, you can look ahead at that discussion. We finish with sesquilinear forms and adjoints.

- Chapter 12 presents some important theorems from linear functional analysis called the Open Mapping and Closed Graph Theorem. To do that properly, we also have to discuss a very abstract idea called first and second category metric spaces.

Part Five: Operators Here, we want to work carefully the theory of self -adjoint linear operators and one example of them: the Stürm - Liouville Operators.

- Chapter 13 works out the details of the Stürm - Liouville Operators. We reintroduce a few ideas from ODEs too to help you remember some of those details. We look carefully at the eigenvalue - eigenfunction behavior here.

- In Chapter 14 we look carefully at the general details of the self-adjoint operators and work out their eigenvalue - eigenfunction properties just like we did for symmetric and Hermitian matrices earlier. The tools we use are a bit different as the spaces are infinite dimensional.

Part Six: Topics in Applied Modeling We want to finish with a non-standard application of these ideas to game theory.

- Chapter 15 discusses bounded charges on rings and fields of subsets.

- Chapter 16 is an introduction to games of transferable utility. In our discussion, we even have to look at a Riemann Integral extension, so great fun!

C-1 Sample Exams

Then, to check yourself, you should be able to do any of these sample exams.

C-1.1 Exam 1

Instructions:
This is a closed book and closed notes test. You will need to give all the details of your arguments as that is the only way partial credit may be warranted.

Part 1: Definitions (30 Points)
Carefully and precisely define the following mathematical concepts:

1. (6 Points) Let X be a set.
   (a) (3 Points) $d : X \times X \to \Re$ is a metric on X.
   (b) (3 Points) $d : X \to \Re$ is a norm on X.

2. (9 Points) Let X be a set.
   (a) (3 Points) X is a vector space over the field of real numbers.
   (b) (3 Points) X is a metric space over the field of real numbers.
   (c) (3 Points) X is a normed space over the field of real numbers.

3. (9 Points) Let X be a set.
   (a) (3 Points) X is a separable metric space over the field of real numbers.
   (b) (3 Points) $\{x_n\}$ is a Cauchy sequence of elements in the normed space X over $\Re$.
   (c) (3 Points) X is a complete metric space over the field of real numbers.

4. (6 Points) Let X be a normed space set and M be a subset of X.
   (a) (3 Points) M is a compact subset of X.
   (b) (3 Points) M is a subspace of X properly contained in X and Y is the normed space X/M with the usual induced norm.

Part 2: Short Answer (44 Points)
There are several types of questions here:

True or False: You must determine whether or not these statements are true. If the statement is true, you MUST give us the reason why it is true; if the statement is false, you must give us a counterexample.

Discussion: A careful discussion and reasoned answer to the given question is required.

1. (3 Points) Let X be a normed space and assume that the set $\{x \in X \mid \|x\| \le 1\}$ is compact. Is it possible for X to be finite dimensional?

2. (3 Points) Let X be a normed space and let Y be a finite dimensional subspace of X that is not closed. Is this possible?

3. (8 Points) Let X be the set of all continuous functions defined on the finite interval $[a, b]$. Describe a norm on X which makes X complete and a norm on X which makes X not complete.

4. (3 Points) Is there a procedure which enables us to complete an arbitrary metric space?

5. (3 Points) Is there a procedure which enables us to complete an arbitrary normed space?

6. (6 Points) Give an example of an infinite dimensional metric space which is not separable.

7. (6 Points) Is it necessarily true that the metric in an arbitrary metric space X comes from a norm on X?

8. (12 Points) Let $X = \Re^7$. Let $L = \{< x_1, x_2, x_3, x_4, x_5, x_6, x_7 > \in \Re^7 \mid x_1 + x_5 + 23.0x_7 = 0\}$.
   (a) (2 Points) Show L is a vector space.
   (b) (2 Points) What is the dimension of L?
   (c) (2 Points) What is the dimension of X/L?

BASIC ANALYSIS IV: MEASURE THEORY AND INTEGRATION

(d) (2 Points) What are the elements of X/L?

(e) (2 Points) Is L closed?

(f) (2 Points) Is L compact?

Part 4: Proofs (26 Points)

Provide careful proofs of the following propositions. You will be graded on the mathematical correctness of your arguments as well as your use of language, syntax and organization in the proof.

1. (13 Points) Let X be the set of all continuous functions defined on the finite closed interval $[a, b]$ with the maximum metric. Let ϕ be a given element of X and define $T X \to X$ by

$$(T(x))(t) = \int_a^t \phi(s)x(s)\, ds$$

If $\{x_n\}$ is a Cauchy sequence in X, prove that there exists an element y in X such that $T(x_n) \to y$.

2. (13 Points) Let X be the set of all continuous functions defined on the finite closed interval $[0, 1]$ with the maximum metric. Let S be the set of functions

$$S = \left\{ e^t,\ e^{2t}, e^{3t} \right\}$$

Prove that S is a linearly independent set in X.

C-1.2 Exam 2

Instructions:
This is a closed book and closed notes test. You will need to give all the details of your arguments as that is the only way partial credit may be warranted.

Part 1: Definitions and Short Answers (34 Points)

Carefully and precisely define the following mathematical concepts and answer the short questions.

1. (6 Points)

(a) Precisely define the meaning of a **metric**.

(b) Precisely define the **metric space** X over the field of real numbers.

(c) Is it required that X be a vector space?

2. (10 Points)

(a) Precisely define the meaning of a **norm**.

(b) Precisely define the **Normed Space** X over the real numbers.

(c) Is it required that X be a vector space?

(d) Define the **metric** on X induced by the norm.

(e) Is it true that all metrics can be derived from a norm?

3. (10 Points)

(a) Precisely define the meaning of an **inner product**.

(b) Precisely define the **inner product space** X over the real numbers.

(c) Define the norm on X induced by the inner product.

(d) Define the metric on X induced by the inner product.

(e) State the **Schwartz Inequality** for this inner product space.

4. (8 Points)

(a) Let X be a vector space. Precisely define the meaning of the algebraic dual space of X, X^*.

(b) Let X be a normed space. Precisely define the meaning of the dual space of X, X'.

(c) Precisely define the norm of an element of X'.

(d) Let X be a normed space. Precisely define the meaning of the dual space of X', X''.

Part 2: Short Answer (30 Points)

A careful discussion and reasoned answer to the given question is required.

1. (15 Points) Characterize the bounded linear functionals on ℓ^1.

2. (15 Points) Characterize the bounded linear functionals on c_0.

Part 4: Proofs (36 Points)
Provide careful proofs of the following propositions. You will be graded on the mathematical correctness of your arguments as well as your use of language, syntax and organization in the proof.

1. (18 Points) Let X be the set of all continuous functions defined on the finite closed interval $[a, b]$ with the maximum metric. Let ϕ be a given element of X that satisfies for all positive integers n

$$\|\phi^n\| \leq \frac{\|\phi\|}{n}$$

Define for all positive integers n $T_n X \to X$ by

$$(T_n(x))(t) = \int_a^t \phi^n(s) x(s)\, ds$$

Prove that the sequence of linear operators $\{T_n\}$ converges to the zero element in the space of all bounded linear operators from X to X.

2. (18 Points) Assume for all positive integers n,

$$x_n = \{\xi_1^n, \xi_2^n, \ldots,\} \in \ell^1$$

and

$$x = \{\xi_1, \xi_2, \ldots,\} \in \ell^1$$

satisfy

$$f(x_n) \to f(x), \ \forall f \in (\ell^1)'$$

Prove that for all positive integers k

$$\xi_k^n \to \xi_k$$

C-1.3 Exam 3

Instructions:
This is a closed book and closed notes test. You will need to give all the details of your arguments as that is the only way partial credit may be warranted.

Part 1: Definitions and Short Answers (35 Points)
Carefully and precisely define the following mathematical concepts and answer the short questions.

1. (4 Points) Define what it means for a inner product space X to be complete.

2. (15 Points) Completion (No Details Here–Just Sketch!!)

   (a) (5 Points) State, without proof, the basic ideas behind completing a metric space.

   (b) (5 Points) State, without proof, the extra things that need to be done to complete a normed space.

   (c) (5 Points) State, without proof, the extra things that need to be done to complete an inner product space.

3. (16 Points)

   (a) (4 Points) Define what it means for the set M to be orthonormal in the inner product space X.

   (b) (4 Points) Define what it means for M to be a total orthonormal set in the inner product space X.

   (c) (4 Points) State Bessel's Inequality.

   (d) (4 Points) Discuss the conditions under which Parseval's Relation holds.

470   *BASIC ANALYSIS IV: MEASURE THEORY AND INTEGRATION*

Part 2: Short Answer (35 Points)

A careful discussion and reasoned answer to the given question is required.

1. (5 Points) Let x be an element of the Hilbert space H. Let M be the orthonormal sequence (e_i) in H. Consider the sum $\sum_{i=0}^{\infty} < x, e_i > e_i$. Is it necessarily true that this sum converges in H?

2. (5 Points) Let x be an element of the Hilbert space H. Let M be the orthonormal sequence in H. Consider the sum $\sum_{i=0}^{\infty} < x, e_i > e_i$. Is it necessarily that if this sum converges, the sum is x?

3. (5 Points) Let Y be a subspace of the Hilbert space H. Is it necessarily true that

$$ H \;=\; Y \oplus Y^{\perp} \; ? $$

4. (5 Points) Is it necessarily true that an orthonormal set M in an inner product space X is linearly independent?

5. (5 Points) If M is an uncountable orthonormal set in an inner product space X, is it possible that $< x, m > \neq 0$ for all m in M?

6. (5 Points) If M is a total subset in the inner product space X, is it necessarily true that $M^{\perp} = 0$?

7. (5 Points) Let M be a nonempty, convex and complete subset of the inner product space X. Let x be an element of X not in M. Is it possible for the minimum distance from x to M to be achieved by two different points of M?

Part 3: Proofs (30 Points)

Provide careful proofs of the following propositions. You will be graded on the mathematical correctness of your arguments as well as your use of language, syntax and organization in the proof.

1. (12 Points) Let M be a total orthonormal set in the Hilbert space

H. Assume x in H satisfies $< x, m > = 0$ for all m in M. Prove that $x = 0$.

2. (18 Points) Let X be the inner product space $C[0, 1]$ with the usual L^2 inner product. Recall that this means

$$ \| \, x \, \|_2 \;=\; \sqrt{ \int_0^1 \, | \, x(s) \, |^2 \; ds } $$

Let f and ϕ be chosen arbitrarily from X and define the operator $T : X \to X$ by and

$$ (T(x))(t) = $$
$$ < x, f > + \int_0^t x(s)\phi(s)ds $$

Prove that T is a linear operator and that

(a) for all t in $[0, 1]$,

$$ | \, (T(x))(t) \, | \leq $$
$$ \| \, x \, \|_2 \, (\| \, f \, \|_2 + \| \, \phi \, \|_2). $$

(b)

$$ \| \, T \, \| \leq \| \, f \, \|_2 + \| \, \phi \, \|_2 $$

C-1.4   Final

Instructions:

This is a closed book and closed notes test. You will need to give all the details of your arguments as that is the only way partial credit may be warranted.

Part 1: Definitions and Short Answers (60 Points)

Carefully and precisely define the following mathematical concepts and answer the short questions.

1. (5 Points) A metric space X.

2. (5 Points) A normed space X.

3. (5 Points) An inner product space X.

LINEAR ANALYSIS

4. (5 Points) A complete inner product space X.

5. (5 Points) A function of bounded variation on the interval $[0, 1]$.

6. (5 Points) The Riemann–Stieltjes Integral of functions in $C[0, 1]$.

7. (5 Points) The algebraic dual of a vector space X, X^*.

8. (5 Points) The continuous dual of normed space X, X'.

9. (5 Points) An orthonormal sequence in an inner product space X.

10. (5 Points) A total orthonormal sequence in a Hilbert space X.

11. (5 Points) A sesquilinear form.

12. (5 Points) The adjoint of an operator in a Hilbert Space H.

Short Answer: (80 Points)

1. (4 Points) Is it necessarily true that a projection operator P defined for the closed subspace X of the Hilbert space H satisfies the equation $P(I - P)x = 0$ for all x in H?

2. (6 Points) For each of the following spaces X, we know what space is isomorphic to its dual space X'. What is that space?

    (a) $l^p, 1 < p < \infty$

    (b) l^1

    (c) c

3. (15 Points) For each of the following spaces X, we know how to characterize a given linear functional in X'. Describe that characterization carefully:

    (a) $f \in (l^p)', 1 < p < \infty$

    (b) $f \in (l^1)'$

    (c) $f \in (c_0)'$

    (d) $f \in (C[0, 1])'$

    (e) $f \in H'$, where H is any Hilbert space.

4. (5 Points) Let H be a Hilbert space and $T : H \to H$ an operator. Is it necessarily true that the adjoint of T always exists?

5. (5 Points) Let H be a Hilbert space and $T : H \to H$ an operator. Is it possible for the distinct operators S_1 and S_2 on H to both satisfy for all x, y in H

$$< Tx, y > \ = \ < x, S_1 y >$$
$$< Tx, y > \ = \ < x, S_2 y > \ ?$$

6. (15 Points:) We seek solutions x from $C[0, 1]$ to the differential equation below where f is an arbitrary element of $C[0, 1]$.

$$x'(t) + 3(x'(t))^2 + 2 \int_0^t x(s)\, ds$$
$$= f(t)$$

Convert this mapping into the form

$$T : dom(T) \subset X \ \to \ X$$

by specifying the information below:

(a) A definition of T.

(b) A definition of the domain of T, $dom(T)$.

Finally, answer the following questions:

(a) Is $dom(T)$ a vector space?

(b) Is T a linear operator?

(c) Is T a linear functional?

7. (15 Points:) Let Π be the partition of the interval $[0, 1]$ consisting of $P + 1$ points given by

$$\Pi = \{t_0 = 0 < t_1 < \ldots < t_P = 1\}$$

Let A denote the $2x2$ matrix

$$A \;=\; \begin{bmatrix} 5 & 2 \\ 2 & 2 \end{bmatrix}$$

For a given pair f and g from $C([0,1])$, let $V(f,f,t_i) \equiv V(f,g,i)$ be the row vector $\{f(t_i), g(t_i)\}$. Consider the problem of finding f and g to minimize:

$$\sum_{i=0}^{P} V(f,g,i)\, A\, V(f,g,i)'$$

Convert this mapping into the form

$$T : dom(T) \subset X \;\rightarrow\; X$$

by specifying the information below:

(a) A definition of T.

(b) A definition of the domain of T, $dom(T)$.

Finally, answer the following questions:

(a) Is $dom(T)$ a vector space?

(b) Is T a linear operator?

(c) Is T a linear functional?

8. (15 Points:) Let P be a photograph which has been discretized into an array P of size 640×480. Each element P_{ij} in the array can take on one of 256 discrete values which represents a gray-scale image of the original photograph. We need to compress this information into a matrix of size 320×240. Starting in the upper left hand corner of P, we can compress the original image by looking at 2×2 sub-matrices of P and applying the compression operation we choose to convert each of these sub-matrices into some scalar. The $1,1$ entry in the compressed digitized image thus comes from applying our compression operator to

the 2×2 sub-matrix coming form the upper left corner

$$P_{11},\ P_{12}$$
$$P_{21},\ P_{22}$$

The $1,2$ entry in the compressed array comes from compressing the 2×2 sub-matrix

$$P_{13},\ P_{14}$$
$$P_{23},\ P_{24}$$

and so on. For example, given the array

$$a,\ b$$
$$c,\ d$$

a primitive compression operation consists of applying the averaging operation given by computing the real number

$$\frac{a+b+c+d}{4.0}$$

and then mapping that real number to the closest integer in the range $\{0,1,\dots,255\}$.

Convert this averaging mapping of a given discretized photograph P in

$$X = \Re_{256}^{640 \times 480}$$

(the subscript 256 reminds us that these are matrices that can only have entries out of the integers $\{0,1,\dots,255\}$) into the form:

$$T : dom(T) \subset X \;\rightarrow\; X$$

by specifying the information below:

(a) A definition of T.

(b) A definition of the domain of T, $dom(T)$.

Finally, answer the following questions:

(a) Is $dom(T)$ a vector space?

(b) Is T a linear operator?

(c) Is T a linear functional?

Part 4: Proofs (60 Points)

Provide careful proofs of the following propositions. You will be graded on the mathematical correctness of your arguments as well as your use of language, syntax and organization in the proof.

1. (15 Points)

Let X be the set of all 640×480 matrices whose entries lie in the set of integers $\{0, 1, \ldots, 255\}$. Define the function $d : X \times X \to \Re$ by assigning to each matrix p and q from X the number $d(p, q)$ defined by

$$d(p, q) = \max\{| \, p_{ij} - q_{ij} \, |\}$$

Prove that d is a metric on X.

2. (15 Points)

Let H be a Hilbert space and $T : H \to H$ be an operator with nontrivial kernel K. Let P be the projection operator onto K and Q be the projection operator onto $K^{\perp}$. Prove that

(a) $H = K \oplus K^{\perp}$

(b) The operator $S = TQ$ is invertible on H

3. (15 Points)

Let (e_n) be a total orthonormal sequence in the Hilbert space H. Let a given nonzero element f have the expansion

$$\begin{aligned} f &= a_0 + a_1 e_1 + a_2 e_2 + \ldots \\ &= \sum_{n=0}^{\infty} a_n e_n \end{aligned}$$

Prove that

$$\begin{aligned} \frac{f}{\| \, f \, \|} &= \sum_{n=0}^{\infty} b_n e_n \\ b_n &= \frac{a_n}{\sqrt{\sum_{n=0}^{\infty} a_n^2}} \end{aligned}$$

4. (15 Points)

Let (f_n) be the sequence of functions in $C[0, 1]$ defined by $f_n(t) = t^n$ for all t in the interval $[0, 1]$.

(a) Prove that f_n converges to the function f defined by $f(t) = 0$ for all t in $[0, 1]$ when $C[0, 1]$ is endowed with the L^2 inner product.

(b) Prove that f_n does not converge to f when we simply use pointwise convergence on the interval $[0, 1]$.

Part XII

Appendix: Preliminary Examination Check

Appendix D

The Preliminary Examination in Analysis

The qualifying or preliminary examination in analysis is designed to see if you know enough about the key concepts of analysis to be well prepared for your research career. If you study and learn these ideas, these examinations are no more than a sanity check: clearly you should be competent in core areas of mathematics before you begin your own work. These exams cover the material in (Peterson (8) 2020), (Peterson (9) 2020) and (Peterson (7) 2020). The material in (Peterson (10) 2020) is not usually taught now and all we can say is you should know this too but that is why these texts are designed for self-study.

D-1 Sample Exams

Then, to check yourself, you should be able to do any of these sample exams.

D-1.1 Exam 1

This examination consists of ten (10) questions worth ten (10) points each. Credit is given for partial answers so make sure that you show all relevant work.

1. (10 Points) For a series of real numbers, prove that absolute convergence implies convergence.

2. (10 Points) Let $(X, || \cdot ||)$ denote a normed linear space and $(x_n)_{n=1}^{\infty}$ be a sequence in X.

    (a) What does it mean to say that the series $\sum_{n=1}^{\infty} x_n$ converges?

    (b) Prove that if absolute convergence of a series implies convergence then X is complete.

3. (10 Points) Define $T_n : \ell^2 \to \ell^2$ by

$$T_n(\xi_j) = \eta_j$$

where

$$\eta_j = \begin{cases} 0 & 1 \leq n \\ \xi_{j-n} & n < j \end{cases}$$

    (a) Verify that each T_n is a bounded linear map from ℓ^2 into ℓ^2.

    (b) Compute $|| T_n ||$.

    (c) Verify that T_n is weakly operator convergent to the 0 operator (i.e. verify for every bounded linear functional f on ℓ^2 and every x in ℓ^2 that $\lim_{n \to \infty} f(T_n x) = 0$).

    (d) Verify that T_n is not strongly operator convergent to the 0 operator (i.e. verify that there is an x in ℓ^2 such that $T_n x$ does not converge to 0).

4. (10 Points) Let $(e_n)_{n=1}^{\infty}$ denote an orthonormal sequence in a Hilbert space $(X, < \cdot >)$.

    (a) If x is in X and S_n is the span of $\{e_1, ..., e_n\}$, explain how to find the point y_n in S_n closest to x; i.e. the projection of x on S_n.

    (b) Prove Bessel's Inequality:

$$\sum_{n=1}^{\infty} |< x, e_n >|^2 \leq || x ||^2$$

    for all x in X.

    (c) Prove the series $\sum_{n=1}^{\infty} < x, e_n > e_n$ converges.

BASIC ANALYSIS IV: MEASURE THEORY AND INTEGRATION

(d) Prove that $\sum_{n=1}^{\infty} <x, e_n> e_n$ is the projection of x on the closure of the span of $\{e_1, e_2, ..., e_n, ...\}$.

5. (10 Points) Define $f : D(f) \subseteq \Re^2 \to \Re$ by $f(t, x) = t^2 x^2$ where

$$D(f) = \{(t, x) \mid |t| \le a, |x-1| \le b^3\}$$

with $a, b > 0$.

(a) Verify that f is continuous on $D(f)$.

(b) Verify that $a^2(b+1)^2$ is an upper bound on the range of f.

(c) Verify that $k = 2a^2(b+1)$ is a Lipschitz constant for f on $D(f)$ with respect to the x variable.

(d) Let

$$0 < \beta <$$
$$\min(a, \frac{1}{2a^2(b+1)}, \frac{b}{a^2(b+1)^2})$$

and

$$r = \beta a^2(b+1)^2$$

and define

$$X = \{x : [-\beta, \beta] \to \Re \mid$$
$$x \in C([-\beta, \beta]), \ |x(t) - 1| \le r\}$$

Prove that (X, d_∞) is a complete metric space where $d_\infty(x, y) = \sup_{|t| \le \beta} |x(t) - y(t)|$.

(e) If $|t| \le \beta$ and x is in X, prove that $(t, x(t))$ is in $D(f)$.

(f) Define T on X by

$$(Tx)(t) = 1 + \int_0^t s^2 x^2(s) ds$$

for $|t| \le \beta$. Prove that $T(X) \subseteq X$.

(g) Prove that T is a contraction mapping on X.

6. (10 Points) Let $X = [-1, 1]$.

(a) Give a precise definition of the Lebesgue integral of f on X.

(b) For the function $f(x) = 1/x \sin(1/x)$ discuss the existence of the Riemann and Lebesgue integral on X.

(c) In many engineering applications an integral of importance is the Cauchy–Principal–Value (C.P.V.). Formulate a definition for the C.P.V. for the Lebesgue integral over X assuming one singular point at $x = 0$. Will $f(x)$ as defined by part 6b be Lebesgue integrable in your C.P.V. sense?

7. (10 Points)

(a) Show that if f is integrable on a set A, then given $\epsilon > 0$, there exists a $\delta > 0$ such that

$$|\int_E f(x) \, d\mu| < \epsilon$$

for every measurable set $E \subset A$ of measure less than δ.

(b) Let $g(x) \ge 0$ be integrable on A and define $\nu : P(A) \to \Re$ via

$$\nu(S) = \int_S g(x) \, d\mu.$$

Show that ν defines a measure on A which is absolutely continuous with respect to μ.

8. (10 Points) Let $\{f_n\}$ define a sequence of measurable functions mapping from X to $\Re$. Define the concept of convergence of $\{f_n\} \to f$ in:

(a) L^p norm.

(b) Almost everywhere.

(c) In measure.

9. (10 Points)

(a) State the Fundamental Theorem of Calculus for the Riemann Integral (F.T.C.R.I.).

(b) Is F.T.C.R.I. valid for Lebesgue integrals or can a more general statement be made?

THE PRELIMINARY EXAMINATION IN ANALYSIS

10. (10 Points) Let f be a monotone function on $[a, b]$. Show that for each $c \in (a, b)$ the limit of $f(x)$ exists as $x \to c^-$ and also $x \to c^+$. Hence prove that a function of bounded variation on $[a, b]$ can have only a countable number of discontinuities. Hint: If f is monotone with bounded variation, the number of points where $\mid f(c^+) - f(c^-) \mid > 1/n$ is finite.

D-1.2   Exam 2

This examination consists of ten (10) questions worth ten (10) points each. Credit is given for partial answers so make sure that you show all relevant work.

1. (10 Points)

   For the two real numbers $a > 0$ and b, consider the function $f_{a,b}$ defined below:

   $$f_{a,b}(x) \; = \; \begin{cases} 0 & x < b \\ e^{\frac{-a}{(x-b)^2}} & x \geq b \end{cases}$$

   - Prove that $f'_{a,b}$ exists at b and that it has value 0.

   - Prove that $f''_{a,b}$ exists at b and that it has value 0 also.

   - What can you conjecture about the existence and value of higher order derivatives of $f_{a,b}$ at b? What technique would seem appropriate to prove your conjecture?

   - Draw a careful graph of $f_{a,b}$. Label all important stuff.

2. (10 Points) In addition to the notation of Problem 1 above, for given real numbers a and b, let g be defined as

   $$g_{a,b}(x) \; = \; \begin{cases} 0 & x \geq b \\ e^{\frac{-a}{(x-b)^2}} & x < b \end{cases}$$

   - Define for all positive a and b, $h_{a,b}(x) = f_{a,-b}(x)\, g_{a,b}(x)$. Draw this function carefully labeling all

important points.   What is the closure of the set on which h is nonzero?

- For each positive integer n, define the function F_n by

$$F_n(x) \; = \; f_{a,n}(x)\, g_{a,n+1}(x)$$

This then defines a sequence of continuous functions $S \; = \; \{F_n\}_{n=1}^{\infty}$.
Let $C([0, \infty))$ denote the set of all functions which are continuous of the domain $[0, \infty)$. Endow this set of functions with the usual L_2 inner product and induced norm.

- Prove S is an orthogonal set of functions in $C([0, \infty))$.
- Prove that for all n,

$$\int_n^{n+1} F_n(t)dt =$$
$$\int_0^1 \left[e^{\frac{-a}{u^2}}\, e^{\frac{-a}{(u-1)^2}} \right] dt$$

- Show how you can redefine each F_n so as to make the set S an orthonormal set in $C([0, \infty))$.
- If it were true that the inner product of a function ϕ with F_n is zero for all n implies that ϕ is the zero function, what could you say about the set S?

3. (10 Points)

   Let f be the function $f(t) = t$ and g be the function $g(t) = 2t$ on the domain $D = [0, \infty)$.

   - Prove that f and g are uniformly continuous on D.

   - Prove that fg is not uniformly continuous on D.

   - Prove that fg is uniformly continuous on any finite interval $[a, b]$ contained in D.

BASIC ANALYSIS IV: MEASURE THEORY AND INTEGRATION

4. (10 Points) Find the real numbers a and b which minimize the value of the integral

$$\int_{-1}^{1} |\, ax^2 + b\sin(x) - x^3 \,|^2 \, dx$$

One way to do this problem is to perform the integration and come up with a function of the two variables a and b and then use a standard $2 - D$ optimization strategy. We are not interested in this approach! There is a better way that uses the abstract principles discussed in linear analysis. Use this approach to solve the problem.

5. (10 Points) Let C^2 denote the set of all functions x which are twice differentiable on the real line $\Re$. Let C^0 denote the set of all functions that are continuous on $\Re$. Consider the differentiable equation $x'' - 2x' + x$ with boundary conditions $x(0) = x(1) = 0$.

- Let D be the set of functions which are in C^2 and satisfy the boundary conditions. Prove that D is a vector space over $\Re$.
- Prove that the operator $T : D \to C$ defined by

$$(T(x)) = x'' - 2x' + x$$

is a linear operator on D.
- Prove that T is an unbounded operator if we endow both D and C with the sup-norm.
- Prove that T is a bounded operator if we endow C^2 with an appropriate norm.

6. (10 Points)

- Prove that the Cantor set in $[0, 1]$ has Lebesgue measure zero.
- Is it possible for a function f : $[0, 1] \to \Re$ which has uncountably many points of discontinuity to be Riemann integrable?

- Give an example of a function f : $[0, 1] \to \Re$ which is continuous on $[0, 1]$, is differentiable a.e. on $[0, 1]$ but $\int_0^1 f'(t)dt \neq f(1) - f(0)$.

7. (10 Points)

- Give an example of a sequence of Lebesgue measurable functions $f_n : \Re \to \Re$ such that $|\, f_n(x) \,| \leq 2$ for all x in $\Re$, converges pointwise to a function f on $\Re$ but $\lim_n \int_\Re f_n(x)dx \neq \int_\Re f(x)dx$.
- State an additional condition on the functions f_n in Part (a) above which implies $\lim_n \int_\Re f_n(x)dx = \int_\Re f(x)dx$.

8. (10 Points) Let f be a non-negative Lebesgue measurable function on $\Re$. If $\int_\Re f d\lambda = 0$, prove that $f = 0$ a.e. on $\Re$.

9. (10 Points)

- Let Ω be a nonempty set, $\{x_n\}_{n=1}^\infty$ be a sequence of distinct points in Ω, and $\{a_n\}_{n=1}^\infty$ be a sequence of non-negative numbers. For $E \subseteq \Omega$, define

$$\gamma(E) = \sum_{x_n \in E} a_n$$

Show that γ is a measure on $\mathcal{P}(\Omega)$.
- Let $(\Omega, \mathcal{A}, \mu)$ be a finite measure space. For sets A and B in $\mathcal{A}$, verify that

$$\mu(A \cup B) = \mu(A) + \mu(B)$$
$$-\mu(A \cap B)$$

Why is the finiteness assumption necessary?

10. (10 Points) Let f be a function in $L^1(\Re)$ and $\{E_n\}_{n=1}^\infty$ be a sequence of Lebesgue measurable sets such that

$$E_{n+1} \subseteq E_n$$

THE PRELIMINARY EXAMINATION IN ANALYSIS 481

for all positive integers n. Prove that

$$\int_{\cap_{n=1}^{\infty} E_n} f d\lambda = lim_{n \to \infty} \int_{E_n} f d\lambda$$

D-1.3   Exam 3

This examination consists of ten (10) questions worth ten (10) points each. Credit is given for partial answers so make sure that you show all relevant work.

1. (10 Points) This is a problem about classical continuity and differentiability of functions. Let Q and Ir denote the rational and irrational numbers respectively. Define $f : \Re \to \Re$ by

   (a)

   $$f(x) = \left\{ \begin{array}{ll} x & x \in Q \\ -x & x \in Ir \end{array} \right.$$

   Prove f is continuous at 0 using an ϵ and δ argument.

   (b) Define $f : \Re \to \Re$ by

   $$f(x) = \left\{ \begin{array}{ll} 1 & x \geq 0 \\ -1 & x < 0 \end{array} \right.$$

   Prove f' does not exist at 0 using an ϵ and δ argument.

2. (10 Points)

   (a) Let f_n be a sequence of functions $f_n : \Re \to \Re$ be defined by

   $$f_n(x) = \frac{1}{2^n} \cos(3^n x)$$

   - Show that this sequence of functions converges pointwise to a function f also defined on $\Re$.
   - Prove that this sequence of functions also converges uniformly to the same function f defined in the first part of this question. Prove that the limit function is continuous on $\Re$.

   (b) Define the sequence of functions F_k on $\Re$ by

   $$F_k(x) = \sum_{n=0}^{k} \frac{1}{2^k} \cos(3^k x)$$

   - Show that this sequence of functions converges pointwise to a function F also defined on $\Re$.
   - Prove that this sequence of functions also converges uniformly to the same function F defined in the first part of this question. Prove that the limit function is continuous on $\Re$.

3. Let

   - X_0 denote the vector space of all functions which are continuous on the interval $[0, 1]$.
   - X_1 denote the vector space of all functions which have continuous first derivatives on the interval $[0, 1]$.
   - X_2 denote the vector space of all functions which have continuous second derivatives on the interval $[0, 1]$.

   Define $T : X_2 \to X_0$ by $T(x) = -x''$.

   - Prove that T is a linear mapping from X_2 to X_0 independent of the choice of norm on these spaces.
   - Find norms to use on X_0 and X_2 so that T is a bounded linear operator.
   - Find norms to use on X_0 and X_2 so that T is an unbounded linear operator.

4. (10 Points) Let (X, d) be a metric space.

   (a) Prove an arbitrary union of open sets in X is also an open set.

   (b) Prove that an arbitrary finite intersection of open sets in X is also an open set.

5. (10 Points)

BASIC ANALYSIS IV: MEASURE THEORY AND INTEGRATION

(a) Let X denote the set of all possible real-valued sequences. Find a metric for X which is not induced by any norm on X.

(b) Let Y be the set of all real-valued sequences which are bounded. Find a countably infinite linearly independent subset of Y.

6. (10 Points)

(a) Explain what it means for a subset of the real numbers to be Lebesgue measurable and how to calculate the Lebesgue measure of a set.

(b) Explain why there exist subsets of R which are not Lebesgue measurable.

7. (10 Points) Let f be a mapping from the finite interval $[a, b]$ into $\Re$ and assume there is a constant K such that $|f(x) - f(y)| \le K|x - y|$ for all x and y in $[a, b]$.

(a) Must f be differentiable on $[a, b]$? Explain.

(b) Must f be Lebesgue measurable on $[a, b]$? Explain.

(c) Must f be absolutely continuous on $[a, b]$? Explain.

8. (10 Points) Let $F : R \times [a, b] \to \Re$. Assume

- for each t in $[a, b]$ that $F(\cdot, t)$ is Lebesgue Measurable,

- there is a Lebesgue integrable function g such that $|F(x, t)| < g(x)$ for all x in $\Re$ and t in $[a, b]$, and

- for each x in $\Re$, $F(x, \cdot)$ is continuous on $[a, b]$.

Prove that the function $f : [a, b] \to \Re$ defined by $f(t) = \int_{\Re} F(x, t) \, dx$, where $\int$ represents the Lebesgue integral, is continuous on $[a, b]$.

9. (10 Points) Suppose that $f : \Re \to \Re$ is a Lebesgue integrable function. For each real number a prove that $\int_{\Re} f(x) \, dx$ equals $\int_{\Re} f(x + a) \, dx$.

10. (10 Points) Let $f : [a, b] \to \Re$ be a given function.

(a) State the first and second fundamental theorems for Lebesgue Integrals.

(b) Give an example where f is Lebesgue integrable but not Riemann integrable.

(c) Give an example where the derivative of f is Lebesgue integrable but the second fundamental theorem fails. Explain why the derivative is integrable. Is the derivative Riemann integrable? Explain.

D-1.4 Exam 4

This examination consists of ten (10) questions worth ten (10) points each. Credit is given for partial answers so make sure that you show all relevant work.

1. (10 Points)

(a) Prove the sequence $\frac{n^2 + 3n + 2}{4n^2 + 5n + 8}$ converges using an $\epsilon - N$ technique.

(b) Prove the function

$$f(x) = \begin{cases} x^2 + 3x + 9, & x > 1 \\ 12, & x = 1 \\ -x + 14, & x < 1 \end{cases}$$

has a limit at $x = 1$ using an $\epsilon - \delta$ technique.

(c) Prove the function $f(x) = 2x^3 + 3x + 10$ has a derivative at $x = 2$ using an $\epsilon - \delta$ technique.

2. (10 Points) Prove that $f(x) = \frac{x}{x^2 + 1}$ is uniformly continuous on $\Re$.

3. (10 Points) Let $f(x) = 3x^2 + 4x + 1$. Prove that

(a) f is absolutely continuous on $[2, 6]$ directly from the definition of absolute continuity.

(b) f is of bounded variation on $[2, 6]$ directly from the definition of bounded variation.

THE PRELIMINARY EXAMINATION IN ANALYSIS

4. (10 Points) Let $f(x) = \int_1^{x^2} \sin(\Psi(t))\,(\Psi(t) + t)\,dt$ on $[0,1]$ where Ψ is the standard Cantor function defined on the "middle third" Cantor set.

   (a) Explain why the integrand is continuous on $[0,1]$.

   (b) Explain why the integrand is strictly monotone increasing on $[0,1]$.

   (c) Explain why f is differentiable on $[0,1]$.

   (d) Compute f' on $[0,1]$ explaining how you are able to do so.

5. (10 Points) Let f be Lebesgue measurable on $[0,1]$.

   (a) Prove that if $\{C_\alpha\}$ is an arbitrary collection of subsets of $\Re$, then $f^{-1}(\cup C_\alpha) = \cup f^{-1}(C_\alpha)$.

   (b) Define ν on the Lebesgue measurable sigma-algebra $\mathcal{S}$ by $\nu(E) = \mu(f^{-1}(E))$ where μ is standard Lebesgue measure. Prove that ν is a measure on $\mathcal{S}$.

6. (10 Points) Let V be an n dimensional vector space of $\Re$ and V^* its dual space. Let E be a basis for V and F be the standard dual basis on V^* associated with E.

   (a) Prove $V^* \times V$ is dimension $2n$ and that $\{F, E\}$ is a basis for $V^* \times V$.

   (b) Let B be the collection of all real-valued bilinear maps on $V^* \times V$. Prove that B is a vector space of dimension n^2 and find a basis for the space B.

7. (10 Points) Let $(X, \mathcal{S}, \mu)$ be a measure space. Let f be a measurable non-negative function. Let Q be the simple functions on X and define $J(f) = \inf\{\int \phi d\mu : \phi \in Q, \phi \geq f\}$. Finally, let $\int f$ denote the usual Lebesgue integral. Prove that $J(f)$ is finite if and only if $\int f$ is finite.

8. (10 Points) Let $C([0,1])$ be the set of all continuous functions and $C^1([0,1])$, the set of all continuously differentiable

functions on $[0,1]$. For both spaces, let the norm be the usual sup norm. Define $L : C^1([0,1]) \to C([0,1])$ by $L(x) = \frac{dx}{dt}$. Further, define $G : C([0,1]) \to C^1([0,1])$ as the function whose value at t is $(G(f))(t) = \int_0^t f(s)ds$.

   (a) Prove L is an unbounded and G is a bounded linear operator.

   (b) Prove $\{L(x) = f,\ x(0) = 0\}$ if and only if $G(f) = x$.

   (c) Prove G maps compact subsets of $C([0,1])$ into compact subsets of $C^1([0,1])$.

9. (10 Points) Let B be a compact subset of $\Re^2$, $x_0 \neq 0$ a fixed vector, and define the function C by $C(t) = dist(t x_0, B)$. Let f be a function that is Lipschitz of degree one on $\Re$. Define the function D by $C(t) = dist(f(t)\,x_0, B)$.

   (a) Prove C is a continuous function.

   (b) Prove C is a measurable function.

   (c) Prove D is measurable function.

10. (10 Points) Let $(X, \mathcal{S}, \mu)$ be a measure space and assume f is a summable function. Define $\nu(E) = \int_E f^+\,d\mu$ for all E in $\mathcal{S}$. Prove directly that ν is an absolutely continuous measure.

D-1.5 Exam 5

This examination consists of ten (10) questions worth ten (10) points each. Credit is given for partial answers so make sure that you show all relevant work.

1. (10 Points) Let $\lim_{n \to \infty} x_n = x$ in $\Re$. Let $\mathcal{A}$ be a family of closed sets with the property that if $A \in \mathcal{A}$ then there is a positive integer N so that $x_k \in A$ if $k \geq N$. Prove that $x \in \cap\{A : A \in \mathcal{A}\}$.

2. (10 Points) Let $f : \Re \to \Re$ be a bounded continuous function. Assume $f(U)$ is an open set for all open subsets U of $\Re$. Let V be any nonempty open set of $\Re$ and let $\alpha = \inf_{v \in V} f(v)$ and $\beta = \sup_{v \in V} f(v)$.

   • Prove α and β exist and are finite.

- Prove if $y \in V$, then $\alpha < f(y) < \beta$.

3. (10 Points) Let $f : A \subseteq \Re \to \Re$. Prove f is uniformly continuous of A iff for all pairs of sequences (x_n) and (y_n) in A that satisfy $\lim_{n\to\infty}(x_n - y_n) = 0$, we have $\lim_{n\to\infty}(f(x_n) - f(y_n)) = 0$.

4. (10 Points) Let $\Omega = \{f \in C([0,1]) | \int_0^1 f(x)dx \in (0,3)\}$. Prove Ω is an open set in the sup-norm topology. This means that given f_0 in Ω, there is an $r > 0$ so that $B(f_0; r) = \{g \in C([0,1]) | \||g - f_0\||_\infty < r\}$ is contained in Ω.

5. (10 Points) Let f and g be real-valued functions defined on $[a, b]$.

   - Discuss whether or not the following statement is true: A function f if Riemann Integrable on $[a, b]$ iff $|f|$ is Riemann Integrable on $[a, b]$.

   - If f and g are Lebesgue Integrable on $[a, b]$, prove that $max\{f, g\}$ is Lebesgue Integrable on $[a, b]$.

6. (10 Points) Let A contained in $[0, 1]$ be Lebesgue measurable. Let $r > 0$ be any positive real number and define the set rA by $rA = \{rx | x \in A\}$. Prove that rA is Lebesgue measurable and if μ denotes Lebesgue measure, $\mu(rA) = r\,\mu(A)$.

7. (10 Points) Let (f_n) be a sequence of real-valued functions defined on the set X which are measurable with respect to a sigma-algebra S of subsets of X with an associated measure λ. Assume $f_n \to f$ pointwise on X and there is an S measurable summable real-valued dominating function F so that $|f_n| \le F$ for all n. Prove that $\nu : S \to \Re$ defined by $\nu(E) = \lim_n \int_E f_n d\lambda$ defines a charge which is absolutely continuous with respect to λ. You are expected to use known properties of the Lebesgue Integral in your argument and to clearly state these results when they are used.

8. (10 Points) Let ℓ^2 denote the Hilbert space of real-valued square summable sequences with the usual inner product and norm.

   - Prove that $f = (\frac{1}{n} \cos(\frac{n\pi}{2}))$ and $g = (\frac{1}{n} \sin(\frac{n\pi}{2}))$ are linear independent in ℓ^2.

   - Let $h = (\frac{1}{n} (-1)^n)$. Let V be the subspace of ℓ^2 spanned by f and g. Find the unique z in V which solves the minimization problem $\inf_{v \in V} \||h - v\||$.

9. (10 Points) Let (X, ρ) and (Y, ξ) be normed linear spaces. A subset B in X is bounded if there is an $R > 0$ so that $\rho(x) < R$ for all x in B. Prove that a linear mapping $T : X \to Y$ is continuous on X iff $T(B)$ is a bounded set in Y whenever B is a bounded set in X.

10. (10 Points) Define $T : \Re^3 \to \Re^3$ by

$$T(x_1, x_2, x_3) = \left(\frac{x_1}{5} + \frac{x_2}{2} + \frac{x_3}{4} - 1, \right.$$
$$\frac{x_1}{3} - \frac{x_2}{3} - \frac{x_3}{4} + 2,$$
$$\left. \frac{x_1}{4} + \frac{2x_2}{15} + \frac{x_3}{4} + 3 \right)$$

   - Prove that T is a contraction with respect to an appropriate norm on $\Re^3$.

   - Prove, using the completeness property of $\Re^3$, that T has a fixed point.

D-1.6 Exam 6

This examination consists of ten (10) questions worth ten (10) points each. Credit is given for partial answers so make sure that you show all relevant work.

1. (10 Points) Let $X = (C([0,1]), \|\cdot\|_\infty)$ and $Y = (C([0,1]), \|\cdot\|_1)$. Define

$$A = \{ \text{ continuous } f \text{ on } [0,1] | \||f\||_\infty < 2\}$$

and

$$B = \{ \text{ continuous } f \text{ on } [0,1] | \||f\||_1 < 2\}$$

THE PRELIMINARY EXAMINATION IN ANALYSIS   485

(i): Determine if A is an open set in Y.

(ii): Determine if B is an open set in X.

2. For a subset A of $\Re$, and real numbers a and b, define the set $a\,A\,+\,b$ to be the set of numbers of the form $a\,x\,+\,b$ for all possible x in A. Let μ^* denote Lebesgue outer measure and μ denote Lebesgue measure. Prove

(i): $\mu^*(a\,A\,+\,b) = |a|\,\mu^*(A)$.

(ii): If A is Lebesgue measurable, so is $a\,A\,+\,b$.

3. (10 Points) Consider the set of x and y in $\Re^2$ which satisfy

$$\begin{bmatrix} a & b \\ c & d \end{bmatrix} \begin{bmatrix} x \\ y \end{bmatrix} \leq \begin{bmatrix} \alpha \\ \beta \end{bmatrix}.$$

for real numbers a, b, c, d, α and β. Prove this is a measurable subset of $\Re^2$.

4. (10 Points) Define the mapping $T\,:\,\ell_2 \to \ell_2$ by

$$T(x) = (0,\,x_1,\,(1/2)x_2,\,(1/3)x_3,\,\ldots)$$

where we let x be denoted by the sequence $(x_1, x_2, \ldots)$.

(i): Prove T is a bounded linear operator on ℓ_2.

(ii): Prove if (u_n) is a bounded sequence in ℓ_2, then the image sequence $(T(n))$ is also bounded.

5. (10 Points) Let f be uniformly continuous on (a, b). Prove f is bounded on (a, b). Then find an example that shows continuous and bounded on (a, b) is not enough to guarantee uniform continuity on (a, b).

6. (10 Points) Construct the Cantor set Δ starting from the interval $[0, 1]$ and by removing the middle $1/3$ of the interval and by continuing this process. Show that Δ contains no (nonempty) open intervals.

7. (10 Points) Let $f\,:\,R \to R$ be continuous. Show that $\{x \mid f(x) > 0\}$ is an open subset of R and that $\{x \mid f(x) = 0\}$ is a closed subset of R.

8. (10 Points) Let $y \in (\ell_\infty, \|\cdot\|_\infty)$ and define $g : (\ell_1, \|\cdot\|_1) \to (\ell_1, \|\cdot\|_1)$ by $g(x) = (x_n\,y_n)_{n=1}^\infty$. Verify that $g(x)$ belongs to l_1 and determine if g is continuous.

9. (10 Points) Define a linear operator $T\,:\,C([0,1]) \to C([0,1])$ by $\Big(T(f)\Big)(t) = \int_0^t s\,f(s)\,ds$. Is T bounded with respect to $\|\cdot\|_\infty$ on $C([0,1])$? If so, determine $\|T\|_{op}$.

10. (10 Points) Let $X = [0,1]$ and let (f_n) be a sequence of functions in $C([0,1])$, the space of all functions continuous on $[0, 1]$. Assume there is a function f defined on $[0, 1]$ so that $f_n \to f$ [ptws]. For any $\epsilon > 0$ and positive integer n, let

$$V_n(\epsilon) = \{\,x \in X\,:\,|f_n(x) - f(x)| \leq \epsilon\}.$$

Also, let

$$U(\epsilon) = \bigcup_n V_n^\circ(\epsilon),$$

where $V_n^\circ(\epsilon)$ denotes the interior of $V_n(\epsilon)$. Prove if f is continuous at a point x, then

$$x \in \bigcap_k U(\frac{1}{k}).$$

D-1.7   Exam 7

This examination consists of ten (10) questions worth ten (10) points each. Credit is given for partial answers so make sure that you show all relevant work.

1. (10 Points) Let the sequence $(a_n)_{n=1}^\infty$ be defined by

$$\begin{aligned} x_1 &= a,\ 1 < a < \infty \\ x_{n+1} &= a\,\frac{1+x_n}{a+x_n} \end{aligned}$$

Then show

(a) $x_n \leq a,\ n \geq 1$

(b) $x_n \geq 1, \ n \geq 1$

(c) $x_{n+1} \leq x_n, \ n \geq 1$

From these facts, conclude $x_n \to \sqrt{a}$.

2. (10 Points) Let (X, d) and (Y, ρ) be metric spaces. Assume $f, g : X \to Y$ are continuous on X. Define $h : X \to \Re$ by

$$h(x) = \rho(f(x), g(x)).$$

Then h is continuous on X.

3. (10 Points) Consider the sequence of functions $f_n : [0, 2\pi] \to \Re$ defined by

$$f_n(x) = \sin(nx).$$

It is known that

(a) $\int_0^{2\pi} \sin^2(nx)dx = \pi, \ n \geq 1$

(b) $\int_0^{2\pi} \sin(nx)\sin(mx)dx = 0, \ n \neq m$

Prove that **no** subsequence of $(f_n)_{n=1}^{\infty}$ can converge a.e. on $[0, 2\pi]$ via the following argument:

(a) Assume $(f_{n_k})_{k=1}^{\infty}$ is a convergent subsequence. Define the sequence $(g_k)_{k=1}^{\infty}$ by

$$g_k(x) = (\, f_{n_k}(x) - f_{n_{k-1}}(x)\,)^2.$$

Show $g_k \to 0$ a.e. on $[0, 2\pi]$ and $\int_0^{2\pi} g_k(x)dx = 2\pi, \ k \geq 1$.

(b) Use the dominated convergent theorem to show our assumption that $(f_{n_k})_{k=1}^{\infty}$ converges must be false.

4. (10 Points) If $a_n \geq 0$, for all $n \geq 1$ and $\sum_{n=1}^{\infty} a_n$ converges, prove that

$$\sum_{n=1}^{\infty} \frac{\sqrt{a_n}}{n}$$

also converges.

5. (10 Points) Discuss the pointwise and uniform convergence of the sequence of functions $(f_n)_{n=1}^{\infty}$ defined by

$$f_n(x) = \frac{n^2 x}{1 + n^3 x^2}, \ x \in \Re$$

on the intervals

(a) $[-1, 1]$

(b) $[1, 2]$

(c) $[a, \infty)$, for $a > 0$.

6. (10 Points) Let $(X, || \cdot ||)$ be a normed linear space. Assume $f : X \to \Re$ is a linear functional which is not identically zero. If f is continuous on X, then the nullspace $N = f^{-1}(0)$ is closed and N is **not** dense in X.

7. (10 Points) Let $(X, < \cdot, \cdot >)$ be a real inner product space. Let $x, y \neq 0$ be chosen in X and let M be the subspace generated by x. Then

$$x \perp y \iff dist(y, M) = ||\, y\, ||.$$

8. (10 Points) Let P_a be the generalized Cantor set for the usual sequence $(a_n)_{n=0}^{\infty}$. Prove the Lebesgue measure of P_a is given by

$$\mu(P_a) = \lim_{n \to \infty} 2^n \, a_n.$$

9. (10 Points) Let (X, S, μ) be a measure space and $f : X \to \Re^*$ be summable. Define $\nu : S \to \Re^*$ by

$$\nu(E) = \int_E f \, d\mu, \ \forall E \in S.$$

Prove that ν is a signed measure on S.

10. (10 Points) Let $X = C([0, 1])$ with the sup norm

$$||f||_0 = max_{x \in [0,1]}|f(x)|$$

and $Y = C^1([0, 1])$ with the norm

$$||f||_1 = ||f||_0 + ||f'||_0$$

(a) Let $T : Y \to X$ by $(T(f))(x) = f'(x)$. Prove that T is a bounded linear operator on Y.

(b) Let $S : dom(S) = Y \subset X \to X$ by $(S(f))(x) = f'(x)$. Prove that S is an unbounded linear operator on $dom(S) \subset X$.

THE PRELIMINARY EXAMINATION IN ANALYSIS

487

D-1.8 Exam 8

This examination consists of ten (10) questions worth ten (10) points each. Credit is given for partial answers so make sure that you show all relevant work.

1. (10 pts.) Let the sequence of functions on $\Re$, (f_k), be defined by

$$f_k(t) = \frac{1}{2^k} \cos(3^k\, t),$$

for positive integers k and $t \in \Re$.

Define $f : \Re \to \Re$ by

$$f(t) = \sum_{k=1}^{\infty} f_k(t).$$

Prove that f is continuous on $\Re$.

2. (10 pts.) Define the mapping T with domain $C([-1,1])$ and range $C([-1,1])$ by

$$(T(f))(t) = \int_{-1}^{t} s^2 f(s)ds.$$

Prove that T is a bounded linear operator on $C([-1,1])$ (endowed with the sup-norm) and prove that the operator norm of T is less than 1.

3. (10 pts.) Give an example of a function which is differentiable and continuous at just one point and provide a proof of its properties.

4. (10 pts.) Find a basis for the subspace determined by the equation $3x + 4y + 2z = 0$ and determine the point in this subspace closest to the point $(2, 2, 2)$.

5. (10 pts) Let (X, S, μ) be a finite measure space. If f is μ-measurable, define the sets

$$E_n = \{x \in X \mid (n-1)$$
$$\le \; | f(x) | < n\}.$$

Prove that the sets E_n are μ-measurable and that f is summable if and only if $\sum_{n=1}^{\infty} n\, \mu(E_n) < \infty$.

6. (10 Points) Consider the series $\sum_{n=1}^{\infty} \frac{3}{4^n}$.

   (a) Prove that this series converges.

   (b) Note that if we let $a_n = \frac{3}{4^n}$, we can rewrite a_n as

$$a_n = (\frac{1}{8} + \frac{7}{8})a_n$$
$$= \frac{3}{8 \times 4^n} + \frac{21}{8 \times 4^n}.$$

   Denote this split by $a_n = b_{2n-1} + b_{2n}$ where $b_{2n-1} = \frac{3}{8 \times 4^n}$ and $b_{2n} = \frac{21}{8 \times 4^n}$. Then the original series is equivalent to $\sum_{n=1}^{\infty} a_n = \sum_{n=1}^{\infty} b_n$. Prove that $\limsup \frac{b_{n+1}}{b_n} = 7$ and that $\liminf \frac{b_{n+1}}{b_n} = \frac{1}{28}$.

   (c) Discuss why the following statement of the ratio test with positive terms is incorrect and see if you can fix it. A proof of your new statement is **NOT** necessary; just comment what change or changes might be needed.

   Ratio Test: Let $\sum a_n$ be a series of positive terms and assume that $\rho = \limsup \frac{a_{n+1}}{a_n}$ exists and is finite. Then,

      i. $\rho < 1$ implies that the series converges.

      ii. $\rho > 1$ implies that the series converges.

      iii. $\rho = 1$ gives us no information about the behavior of the series.

7. (10 Points) Let $C([-1,1])$ denote the set of all real-valued functions that are continuous on $[-1, 1]$ endowed with the usual sup-norm. That is,

$$C([-1,1]) = \{f : [-1,1] \to \Re \mid$$
$$f \text{ is continuous on } [-1,1]\},$$

with norm

$$\| f \|_\infty = \sup\{| f(t) | : -1 \le t \le 1\}.$$

Assume (f_j) and (g_j) are sequences of functions in $C([-1,1])$ which satisfy $\sum \| f_j \|_\infty^2 < \infty$ and $\sum \| g_j \|_\infty^2 < \infty$. Prove this variant of Minkowski's inequality:

$$\left[\sum \| f_j + g_j \|_\infty^2 \right]^{\frac{1}{2}} \leq$$

$$\left[\sum \| f_j \|_\infty^2 \right]^{\frac{1}{2}} + \left[\sum \| g_j \|_\infty^2 \right]^{\frac{1}{2}}$$

8. (10 Points) Let $C([-1,1])$ denote the set of all functions that are continuous on $[-1,1]$ endowed with the usual sup-norm. That is,

$$C([-1,1]) = \{f : [-1,1] \to \Re \mid$$
$$f \text{ is continuous on } [-1,1]\},$$

with norm

$$\| f \|_\infty = \sup\{| f(t) | : -1 \leq t \leq 1\}.$$

Let A be an $n \times n$ matrix function whose entries come from $C([-1,1])$; that is: A_{ij} is a continuous function of $[-1,1]$. Let X represent the set of all such matrix functions; that is:

$$X = \{A : [-1,1] \to M_n \mid$$
$$A_{i,j} \in C([-1,1]), \ 1 \leq i,j \leq n\}.$$

Define the mapping $d : X \times X \to \Re$ by

$$d(A, b) = \left[\sum_{i=1}^n \sum_{j=1}^n \| A_{i,j} - B_{i,j} \|_\infty^2 \right]^{\frac{1}{2}}$$

(a) Prove that (X, d) is a metric space.

(b) Prove that X is complete with respect to d.

9. (10 Points) Let $f : \Re \to \Re$ be a lower semicontinuous function. This means that for each x in $\Re$, if $x_n \to x$, then

$$\lim_{x_n \to x} \inf f(x_n) \geq f(x).$$

Define $f^* : \Re \to \Re$ and $f^{**} : \Re \to \Re$ by

$$f^*(y) = \sup\{xy - f(x) \mid x \in \Re\}$$

$$f^{**}(x) = \sup\{xy - f^*(y) \mid y \in \Re\}$$

The function f^{**} is the double Fenchel transform of the function f and has important applications in optimization theory.

(a) Prove f^* is lower-semicontinuous and Lebesgue measurable.

(b) Prove that f^{**} is lower semicontinuous, Lebesgue measurable and satisfies $f^{**}(x) \leq f(x)$ for all x in $\Re$.

10. (10 pts.) Consider the $m \times m$ matrix

$$A = \begin{matrix} \alpha_1, \ \alpha_2, \ \ldots, \ \alpha_n \\ \sigma_1, 0, \ \ldots, 0 \\ 0, \ \sigma_2, \ \ddots, \ 0 \\ 0, \ 0, \ \ldots, \ \sigma_{n-1} \end{matrix}$$

All the parameters α_i are positive and all the σ_j lie in $(0, 1]$ (that is all σ_j are positive and bounded by one). A is called a **Leslie** matrix. It is known that the characteristic equation for the eigenvalues of A has the form:

$$p(\lambda) =$$
$$\lambda^n - \alpha_1 \lambda^{n-1} - \alpha_2 \sigma_1 \lambda^{n-2} \ldots$$
$$-\alpha_{n-1}\sigma_1 \cdots \sigma_{n-2}\lambda^1$$
$$-\alpha_n \sigma_1 \cdots \sigma_{n-1}.$$

(a) Prove that a Leslie matrix has a unique positive eigenvalue λ^* using the following steps: Define the function f by

$$f(\lambda) = 1 - \frac{p(\lambda)}{\lambda^n}.$$

and then show

i. f is monotone decreasing.

ii. $\lim_{\lambda \to 0} f(\lambda) = +\infty$.

iii. $\lim_{\lambda \to \infty} f(\lambda) = 0$.

iv. From the above facts conclude that there is a unique value λ^* where $f(\lambda^*) = 1$.

v. λ^* is the only positive eigenvalue of p.